Materials and Working Mechanisms of Secondary Batteries

Chuan-zheng Yang · Yuwan Lou · Jian Zhang ·
Xiaohua Xie · Baojia Xia

Materials and Working Mechanisms of Secondary Batteries

Science Press
Beijing

Springer

Chuan-zheng Yang
Shanghai University
Shanghai, China

Yuwan Lou
Shanghai Institute of Microsystem
Shanghai, China

Jian Zhang
Shanghai Institute of Microsystem
Shanghai, China

Xiaohua Xie
Shanghai Institute of Microsystem
Shanghai, China

Baojia Xia
Shanghai Institute of Microsystem
Shanghai, China

ISBN 978-981-19-5954-7 ISBN 978-981-19-5955-4 (eBook)
https://doi.org/10.1007/978-981-19-5955-4

Jointly published with Science Press
The print edition is not for sale in China (Mainland). Customers from China (Mainland) please order the print book from: Science Press.
Co-Publisher ISBN 978-7-03-036458-6

Translation from the Chinese language edition: "Lvse Erci Dianchide Cailiao Biaozheng he Dianji Guocheng Jili" by Chuan-zheng Yang et al., © Science Press 2015. Published by Science Press. All Rights Reserved.

This Springer imprint is published by the registered company Springer Nature Singapore Pte Ltd.
The registered company address is: 152 Beach Road, #21-01/04 Gateway East, Singapore 189721, Singapore

Preface I

Material Characterization and Mechanism of Electrode Process of Green Secondary Battery was published in Chinese by China Science Press in 2015. It introduces in detail the battery (chemical power supply) research group of the New Energy Center of Shanghai Institute of Microsystems and Information Technology, Chinese Academy of Sciences, under the leadership of Prof. Baojia Xia; the crystal structure, fine structure, and microstructure of active materials for hydrogen-nickel and lithium-ion green battery were studied by X-ray and other experimental methods. And its changes in the processes of battery charge-discharge, cycle and storage, and these changes are closely related to the performance of the battery, revealing the chemical physics phenomena in this kind of these secondary batteries. The de-intercalation theory and electrical conductivity mechanism in the charge and discharge process of this kind of secondary battery and the physical mechanism in cycle and storage processes are proposed and expounded. The domestic response is good.

Colleagues in the chemical industry know that the 2019 Nobel Prize in Chemistry is awarded to three scientists who have made great contributions to the research and development of lithium batteries. They are Prof. John. Goodenough of the University of Texas at Austin in the USA; Prof. M. Stanley Whittingham, State University of New York, Binghamton, USA; and Prof. Akira Yoshino, Nagoya University, Japan. This event itself shows that the invention, development, and application of lithium-ion batteries are of great significance in science and technology, social progress, and the improvement of human quality of life in today's world.

The papers based on this series of studies are published in domestic journals in Chinese, but their foreign counterparts do not know about it. In order to communicate with foreign counterparts, throw a brick to attract jade, and tell the story of China's new energy research and development to foreign counterparts.

The whole book was translated into English by C. Z. Yang and X. H. Xie and J. Zhang and renamed *Materials and Working Mechanisms of Secondary Batteries*; delete Chap. 22 《De-intercalation Theory and De-intercalation Stress of Ionic Conductivity of Secondary Battery》; change Chap. 14 to Chap. 20; and modify and supplement some of the contents of the original book.

Recommended by Ms. Liu Feng Juan of Science Press in Beijing and published by Springer-Verlay and Science Press in Beijing. Heartfelt thanks to Springer-Verlay and heartfelt thanks to Ms. Guo Rui of Springer in China and to Gowtham.Chakravarthy and Vidyaa Shri K. in India and Nobuko Hirota and Taeko Sato in Japan. as well as the relevant comrades in Singapore, Heartfelt thanks to Ms. Liu Feng Juan of Science Press in Beijing.

It should be noted that in the translation work, the automatic translation software "Tenxun Mr. Translator-online Translation" is used to translate into English paragraph by paragraph. After completing the translation of each section and chapter, according to the order of the paragraphs to the full text, we carefully revised the improper use of professional terms, inconsistencies, and inappropriate word relations, et al. I would like to thank Tenxun Mr. Translator for the help of online translation.

If this book can be published in English, it will bring convenience to those who do not understand Chinese but can read English and promote exchanges between international colleagues. We earnestly look forward to the valuable comments put forward by our international counterparts and express our deep gratitude.

Shanghai, China
December 16, 2022

Chuan-zheng Yang
Yuwan Lou
Jian Zhang
Xiaohua Xie
Baojia Xia

Preface II

Chemical power supply, referred to as "battery", is a device that converts chemical energy into electric energy directly through chemical reaction. Nickel-hydrogen (MH/Ni) battery and lithium-ion battery are the two most important chemical power sources, and they have no environmental pollution, so they are called green secondary battery, which has been widely used and will become the power battery of new energy and hybrid electric vehicles.

The device that converts light energy and heat energy into electricity directly through physical change is called physical power source, such as semiconductor solar cell, isotope temperature difference cell. Nuclear batteries are also physical power sources.

In the process of directly converting chemical energy into electric energy, chemical power supply must have two necessary conditions:

(1) The process of losing electrons (oxidation) and obtaining electrons (reduction) in chemical reactions must be separated on positive and negative electrodes. Therefore, it is different from the general oxidation-reduction reaction.
(2) When the oxidation reaction and reduction reaction occur at the two electrodes respectively, the electron must do work through the external circuit, so it is different from the microcell effect of electrochemical corrosion.

Several related books have been published in China, the representative of which are:

(1) Tang You-Gen, **Ni-MH Battery**, Chemical Industry Press, 2007.5.
(2) Chen Jun, Tao Zhan-Liang, **Ni-MH secondary battery**, Chemical Industry Press, 2006.4.
(3) Qi Lu, **Lithium-Ion Secondary Battery for Electric vehicles** (2nd Edition), Science Press, 2013.4.
(4) Wu Yu-Ping, Yuan Xiang, Dong Chao, Duan et al., **Lithium-Ion Battery-Application and practice** (2nd Edition), Chemical Industry Press, 2012.1.
(5) Li Guo-Xin, Editor-in-Chief, **Introduction to New Chemical Power supply Technology**, Shanghai: Shanghai Science and Technology Press, 2007.

These books basically introduce the basic principle, structure, characteristics, application, development status, and trend of batteries, as well as the preparation, properties, and electrolytes of related materials, and basically do not deal with the problem of X-ray diffraction. *Materials and Working Mechanisms of Secondary Batteries* is an introduction to the research results of X-ray diffraction in the characterization of electrode active materials and process mechanism of charge-discharge, cycle, and storage for secondary battery. It is the result of close cooperation between experts in material physics, X-ray analysis and electrochemistry, chemical power supply, and it is an important achievement in the intersection of disciplines.

It is very necessary to reveal what physical phenomena exist in the process of activation, charge-discharge, circulation, storage, etc. in this chemical system, and the study working physical mechanism too is very necessary, which not only has theoretical and academic significance, but also has practical significance.

The study of the "mechanism of secondary battery" must start with the changes of positive and negative active materials, diaphragms, and electrolytes in the electrode (chemical) process. In particular, we should focus on the change law of crystal structure, fine structure, and microstructure of positive and negative active materials in the electrode (chemical) process and connect these structural changes with the changes of battery performance. In order to correctly and comprehensively reveal these changes of active materials, the correct and comprehensive characterization of related materials is the first problem to be solved.

X-ray diffraction and scattering are the most powerful and convenient experimental analysis methods to characterize and study the structure and microstructure of positive and negative active materials in secondary batteries. Therefore, this book is entitled *Materials and Working Mechanisms of Secondary Batteries*. In short, this book systematically introduces the methods and applications of characterizing and studying battery active materials by X-ray diffraction and scattering, as well as these methods used to study the physical phenomena and physical mechanism of secondary (MH/Ni and Li-ion) batteries in various electrode processes.

The book is divided into six parts, a total of 21 chapters. The first part consists of two chapters, namely the main test methods and the main methods of diffraction data processing; the second part (Chaps. 3–8) is the preparation technology and X-ray characterization of active materials, including β-Ni(OH)$_2$ and AB$_5$ hydrogen storage alloys, synthesis mechanism of LiMeO$_2$ materials, mixed occupation of Ni/Li atoms in Li(Ni, Co, Mn)O$_2$ materials, superstructure in Li (Ni$_{1/3}$Co$_{1/3}$Mn$_{1/3}$)O$_2$ materials, as well as LiFePO$_4$ and 2H-graphite. The activation and charge-discharge processes are classified in the third part (Chaps. 9–12), including the solid electrolyte interface film on the graphite surface of lithium-ion battery, hydrogen-nickel (MH/Ni) battery, graphite/LiMeO$_2$ battery, and graphite/LiFePO$_4$ battery. The fourth part (Chaps. 13–16) is the cycling process mechanism of three kinds of battery. The mechanism of storage process of battery is included in the fifth part (Chaps. 18–19). The two chapters of the sixth part, in which Chap. 20 introduces the effect of β-Ni(OH)$_2$ additive to improve the performance of MH/Ni battery. Chapter 21 introduces the method and mechanism of improving the performance of secondary battery.

The features and innovations of this book can be summarized as follows:

(1) *Materials and Working Mechanisms of Secondary Batteries* is based on the most widely used secondary (hydrogen nickel and lithium ion) batteries. X-ray diffraction is mainly used as an experimental means to carry out the characterization of electrode active materials and the research results and monographs on the chemical physical behavior and physical mechanism of batteries in the electrode process. It is a systematic and in-depth research for many years. It is the first time at **home and abroad** that he has written an academic work with great innovation in theory or a lot of discoveries in experiments. It is the research result of interdisciplinary and close cooperation between experts of material physics, X-ray analysis and electrochemistry, chemical power supply.

(2) In order to correctly and comprehensively reveal the changes of active materials in each electrode process, the correct and comprehensive characterization of related materials is the first problem to be solved, and it is the basis to study the structural and microstructure changes of active materials in the electrode process. In order to facilitate the study of the changes of positive and negative electrode active materials during the electrode process, the preparation and modification of important electrode active materials, such as β-$Ni(OH)_2$, AB_5 hydrogen storage alloy, $LiCoO_2$, $Li(Ni_xCo_yMn_{1-x-y})O_2$, $LiFePO_4$, and graphite, were introduced.

(3) In order to study the problems in the process of battery electrode by X-ray diffraction and other related methods, the authors and collaborators proposed and established a series of X-ray analyses methods to characterize various active materials, and related papers have been published. And a lot of testing work has been done for related factories and companies (such as Shanghai Shanshan Technology Co., Ltd., Jiangsu Haisida) by using the established testing method. A series of theories and methods of data processing and analysis have been developed and established.

    (i) General theory and least square solution method of separating X-ray diffraction lines due to microcrystal-micro-stress, microcrystalline-stacking fault, micro-strain-stacking fault double broadening effect and microcrystalline-micro-strain-stacking fault triple broadening effect, and the calculation program.

    (ii) The simulation calculation and analysis method of mixed occupation of Ni/Li atoms in 3a and 3b crystallographic sites of materials such as $LiNiO_2$, Li (Ni,Co) O_2, Li (Ni,Mn) O_2, and $Li(Ni,Co,Mn)O_2$ is studied, and a series of linear equations for solving the mixed occupying parameter X from the root square of diffraction intensity ratio measured by experiments are established.

    (iii) A new method for determining the stacking disorder degree of hexagonal graphite affecting the de-intercalation performance of Li atoms, and so on.

(4) Instead of comprehensively carrying out all the materials (positive and negative materials, diaphragm materials, and electrolytes) that make up the battery, X-ray and other methods are used to focus on the crystal structure, fine structure, microstructure, electrode surface morphology, and composition of the positive and negative active materials and diaphragm materials, as well as their changes in the electrode process, and a series of important innovative achievements have been obtained. The physical behaviors of positive and negative active materials and diaphragms in the electrode chemical process were discussed, and the related physical mechanism was clarified by connecting these changes with the related battery performance.

(5) The author found for the first time that the internal stress in the material, including macroscopic stress and microscopic stress, is caused by the de-intercalation of conductive ions in the positive and negative active materials during charge and discharge. Therefore, the study of the electrode process of the battery from the perspective of stress (strain) diffraction analysis has become a new highlight of this book.

(6) Many new results which have never been reported before have been obtained in the study of the electrode process mechanism of hydrogen-nickel battery. What is most worth mentioning is the ion de-intercalation theory and conductive physical mechanism of the charge and discharge process of hydrogen-nickel (AB_5/β-$Ni(OH)_2$) batteries.

The working principle is generally written as the following chemical reaction formula.

	At the positive pole	At the negative pole
Charge	β-$Ni(OH)_2 + OH^- \rightarrow \beta$-$NiOOH + H_2O + e^-$	$M + H_2O + e^- \rightarrow$ MH (hydride) $+ OH^-$
Discharge	β-$NiOOH + H_2O + e^- \rightarrow \beta$-$Ni(OH)_2 + OH-$	MH (hydride) $+ OH^- \rightarrow M + H_2O + e^-$

Total Reaction

$M + \beta$-$Ni(OH)_2 \Leftrightarrow$ MH (hydride) $+ \beta$-NiOOH.

The above description of how MH/Ni batteries work has been written into teaching books and academic monographs.

From the above, it can be seen that the conductive mechanism is:

(1) The β-$Ni(OH)_2$→β-NiOOH phase transition occurs during charging, and the H^+ ion is supplied by this phase transition.
(2) The transition of AB_5→AB_5H_x (hydride) occurs during the charging process.
(3) The above two phase transitions are completely reversible in the process of charge and discharge.

The above conductive mechanism can be called phase transition theory. It can also be seen from the above that under the action of electric field, the driving force of β-$Ni(OH)_2$→β-NiOOH phase transition is oxidation, and the driving force of β-NiOOH→β-$Ni(OH)_2$ is reduction. It should be noted that the redox referred to here is for Ni, the β-$Ni(OH)_2$ of bivalent Ni is oxidized to trivalent Ni by β-NiOOH, discharge, and the β-NiOOH of 3 valent Ni is reduced to β-$Ni(OH)_2$ of 2 valent Ni. Both oxidation and reduction occur at the positive electrode during charging and discharge, respectively.

Through the on-line (in situ) and quasi-dynamic X-ray study of the first charge and discharge of MH/Ni battery, it is found that the phase transition of β-$Ni(OH)_2$→β-NiOOH does not occur during the charging process, and only when fully charged and overcharged, part of β-$Ni(OH)_2$ transits into γ-NiOOH, and the two phases of β-$Ni(OH)_2$ and γ-NiOOH coexist all the time. In the charging process, the hydrogen ion is not supplied by the β-$Ni(OH)_2$→β-NiOOH phase transition, but by the hydrogen atom leaving the lattice position of β-$Ni(OH)_2$; on the negative side, at the beginning, the hydrogen atom is intercalated into the AB_5 lattice to form a solid solution, and the AB_5H_x hydride is precipitated only when the volume of the AB_5 changes by a certain percentage due to the intercalation of the hydrogen atom, or after the hydrogen atom reaches a certain concentration. In short, the physical conduction mechanism of MH/Ni battery is the directional migration of hydrogen ions between electrodes formed by detachment and intercalation in positive and negative active materials.

It can be seen that during the charging process, it is not β-$Ni(OH)_2$ that is oxidized to β-NiOOH, but at the positive electrode side, the hydrogen atom leaving the β-$Ni(OH)_2$ lattice loses an electron (oxidation) to form a H^+ ion, and on the negative side, H^+ ion acquires an electron (reduction) and is embedded in the AB_5 lattice. Whether it is the charging process, or the discharge process, oxidation and reduction occur at both positive and negative poles. Therefore, the following new reactions can be written:

	At the positive pole	At the negative pole
Charge		
	β-$Ni(OH)_2$→$Ni(OH_{1-x})_2$+ $2xH^+$	AB_5+$2xH^+$+e^-→AB_5–2xH (solid solution) $0 < X < 0.50$

Overcharged

$$2\beta\text{-}Ni(OH)_2 \rightarrow \beta\text{-}Ni(OH)_2 + \gamma\text{–}NiOOH + H^+ + e^- \quad AB_5 + H^+ + e^- \rightarrow AB_5H \text{ (hydride)}$$

Discharge

$$\beta\text{-}Ni\,(OH_{1\text{-}x})_2 + 2xH^+ + 2xe^- \rightarrow \beta\text{-}Ni(OH)_2 \quad AB_5\text{–}2xH \rightarrow AB_5 + 2xH^+ + e^-$$

Total Charge-Discharge Reaction

$\beta\text{-}Ni(OH)_2 + AB_5 \leftrightarrow Ni(OH_{1\text{-}x})_2 + AB_5\text{-}2xH$ (solid solution).

Total Overcharge and Discharge Reaction

$\beta\text{-}Ni(OH)_2 + AB_5 \leftrightarrow \beta\text{-}Ni(OH)_2 + \gamma\text{-}NiOOH + AB_5Hx$ (hydride).

This is the de-intercalation-intercalation theory of conductivity in the charging and discharging process of Ni-MH battery.

(7) Many new results which have never been reported before have also been obtained in the study of the charge and discharge process of lithium-ion batteries. What is most worth mentioning is the fine structure change and conductive physical mechanism of positive and negative active materials during the first charge and discharge of lithium-ion (2H-graphite/$LiCoO_2$) battery. At present, the charge and discharge reaction of lithium-ion battery is generally written as

$$LiCoO_2 + 6C \longleftrightarrow Li_{1-x}CoO_2 + Li_xC_6$$

The reaction at the positive pole:

$$LiCoO_2 \longleftrightarrow Li_{1-x}CoO_2 + xLi^+ + xe^-$$

The reaction at the negative pole:

$$xLi^+ + 6C + e^- \longleftrightarrow Li_xC_6$$

The working mechanism of the above lithium-ion battery is as follows:

(i) The conductivity of lithium-ion battery depends entirely on the directional migration of lithium-ion Li^+ between positive and negative electrodes.
(ii) After lithium is embedded in the negative electrode, lithium atoms form Li_xC_6 compounds with carbon atoms by covalent bond.

(iii) The electrode reaction in the charge and discharge process of the battery is completely reversible. It is also believed that during the charging process, the $LiCoO_2$ of R-3m structure becomes monoclinic $Li_{1-x}CoO_2$ and finally becomes hexagonal CoO_2. However, the above working mechanism of lithium-ion battery only explains the origin of battery conductivity, but there is no clear explanation for how lithium ion migrates.

The author's research results show that during the battery charging, lithium is embedded in the graphite layer and preferentially enters the gap between the hexagonal grid surfaces of the carbon atom, resulting in the increase of the lattice parameters a and c of the graphite, the increase of the micro-strain ε and the change of the stacking disorder P, and the Li-C compound begins to form in the negative electrode after the battery is charged to 20%. When the battery is charged, the Li atoms at position (000) in the positive active material $LiCoO_2$ are preferentially separated from the crystal lattice. With the increase of the lithium removal amount of the $LiCoO_2$, the lattice parameter a decreases, c increases, and the micro-strain ε also increases. There is no structural phase transition of $LiCoO_2$ during the whole charging and discharging process.

The conductive mechanism of lithium-ion battery is as follows: when charging, the migration of lithium ion starts from the interface between negative electrode and electrolyte; when discharging, the migration starts from the interface between positive electrode and electrolyte; in the process of charge and discharge, lithium intercalation and de-lithium in positive and negative active materials have a process from the surface of the electrode to the inner layer, and the charge and discharge process of the battery is not completely reversible. Therefore, the reaction of the conductive process is newly written as:

Overall response

$LiCoO_2+C \leftrightarrow Li_{1-x}CoO_2+Li\text{-}C_x(\text{solid solution}) \leftrightarrow Li_{1-0.25}CoO_2+LiC_{24} \leftrightarrow Li_{1-0.5}CoO_2+LiC_{12} \leftrightarrow Li_{1-1}CoO_2+LiC_6$

Two phases (two phases with the same structure but different lattice parameters or different structures) may coexist in the negative electrode.

The reaction in the position electrode

$LiCoO_2 \leftrightarrow Li_{1-x}CoO_2+xLi^++xe^-$ (x increases with the increase of charging depth until $x = 1$)

The reaction in the negative electrode

$xLi^++6C+xe^- \leftrightarrow Li\text{-}xC$ (solid solution) $\leftrightarrow LiC_{24} \leftrightarrow LiC_{12} \leftrightarrow LiC_6$

The fifth and sixth points mentioned above are made clear, that is, through the determination of the charging process and discharge process, the fine structure of positive and negative active materials [including lattice parameters (macro-strain), micro-strain] and fine structure parameters, the law of their variation with charge-discharge depth is studied and analyzed, and it is found in the investigation of its cause. It is caused by the de-intercalation of hydrogen atom (hydrogen-nickel battery) or lithium atom (Li-ion battery) in positive and negative active materials. According to the principle of oxidation and reduction of chemical power source (battery) in the process of charge and discharge, both charging process and discharge process occur on the positive and negative electrodes, respectively, which produces mass condensation to the conductive phase transition theory, because the oxidation and reduction of the phase transition process take place on the positive electrode during charging and discharging, respectively, and the experimental study shows that the oxidation and reduction of the phase transition process take place on the positive electrode during charging and discharging, respectively. During the charging process, there was no phase transition between β -Ni(OH)$_2$ (hydrogen-nickel battery) and $LiCoO_2$ (graphite/$LiCoO_2$ battery). On this basis, it is proposed that during the charging process, the hydrogen atom (Ni-MH battery) detaches from the β-Ni(OH)$_2$ lattice and loses an electron (oxidation) at the positive-electrolyte interface to become an ion, which enters the electrolyte and moves directionally to the negative electrode under the action of an electric field, and reaches the electrolyte-negative interface to obtain electrons (reduced) from the outer circuit into atoms, which are embedded in the negative electrode, while the discharge is contrary to the above process.

Similarly, for lithium-ion batteries, such as graphite/$LiCoO_2$, graphite/Li (Ni, Co, Mn)O_2, and graphite/$LiFePO_4$, lithium atoms are de-intercalated at the positive and negative electrodes, and the main body of oxidation and reduction is lithium atoms. The de-intercalation theory of conductive ions is a new and important contribution to the electrochemistry and chemical physics of chemical power sources in this book.

(8) In addition to the latest results mentioned in (6) and (7), there are many important results about the process mechanism of secondary battery electrode (cycle, storage), which cannot be mentioned in the preface. After reading each chapter, you will find that each chapter has new results and conclusions.

(9) As far as the whole book is concerned, the emphasis is not only on the electrical conductivity mechanism of positive and negative active materials and diaphragms in the process of charge and discharge, the behavior of cycle and storage, the attenuation mechanism of cycle performance and storage performance, but also according to the understanding of physical mechanism. The methods of improving the charge and discharge performance of the battery and improving the cycle performance and storage performance of the battery were studied and explored, and effective results were obtained. Chapters 20 and 21 introduce the research results in this area, which are of great practical significance and engineering application value.

(10) Another important feature of this book is its systematicness and comprehensiveness. Comprehensive and correct X-ray characterization methods are proposed not only for the electrode active materials used in secondary batteries, β-Ni(OH)$_2$, AB_5 (such as $LaNi_5$), $LiCoO_2$, Li ($Ni_xCo_yMn_{1-x-y}$)O_2, $LiFePO_4$, and hexagonal graphite, but also for four typical batteries studied, including MH-Ni, graphite-$LiCoO_2$, graphite-Li($Ni_xCo_yMn_{1-x-y}$)O_2, and graphite-$LiFePO_4$. Not only the charge and discharge process of the four

types of batteries, but also the cycle process and storage process of the four types of batteries were studied systematically.

Therefore, no matter in the research methods, research content, or results of each research content, there is some innovation, reflecting the innovation of the whole book. Therefore, the publication of this book is of great significance both theoretically and academically and has important application value in practice.

This book can be used as a reference for engineers and technicians in the development, production and development of battery active materials, as well as for teachers, senior students, master's, and doctoral students in chemistry and chemical engineering, electrochemistry, fine chemical industry, chemical power supply, and other majors in colleges and universities. It can also be used as a reference for related majors in research institutes. In particular, the researchers and graduate students engaged in the research and development of chemical power supply refer to the work.

It should be noted that this book is really a collective achievement, and the editor is only writing and finalizing it. We would like to thank doctoral and master's degree students of Prof. Xia Bao-Jia of Shanghai Institute of Microsystems and Information Technology, Chinese Academy of Sciences, and Prof. Jie Jing-Ying (now Shanghai Space Power Research Institute). Comrades Liu Hui, Yan Jian, Li Jia, Liu Hao-Han, Li Yu-Xia, and Wang Ba-Quo are worth mentioning, as well as Li Hui and Liu Fang, senior engineers of Shanghai Shanshan Science and Technology Development Company. I would like to take this opportunity to pay tribute and heartfelt thanks to all of you.

Heartfelt thanks to the recommendation for this book of Prof. Li Guo-Xin of Shanghai Space Power Research Institute, Prof. Jiang Chuan-Hai of Shanghai Jiaotong (Transportation) University, and Prof. Wang Ran of Shanghai Aowei Technology Development Co., Ltd.

As a way to attract jade, introduce this book to international colleagues. It is hoped that colleagues can use a variety of effective methods to study various problems in chemical power supply. As the author's level is limited, please criticize and correct the inadequacies in the book.

Shanghai, China
June 2022

Chuan-zheng Yang
Yuwan Lou
Jian Zhang
Xiaohua Xie
Baojia Xia

Contents

About the Authors

Chuan-zheng Yang, male, born in August 1939, Professor, from Xinhuang County, Hunan Province.

Graduated from Shanghai University of Science and Technology in June 1963, majoring in metal physics. From 1963.07 to 1988.09, he was engaged in the research of material physics and the application of X-ray diffraction and electron microscope at Shanghai Institute of Metallurgy, Chinese Academy of Sciences. From October 1988 to May 1993, at the invitation of EXXON Research and Engineering Company and Biosym Technology Co., Ltd., USA, he engaged in the cooperative research of synchrotron radiation and neutron diffracted scattering at Brookhaven National Laboratory (BNL), on long Island, USA. From 1993.06 to 1999.08, he taught in the Physics Department of Shanghai University and retired.

Courses such as "**X-ray analysis of materials**", "**Laser spectroscopy**", "**Introduction to Laser Spectroscopy**" (for foreign students), "**theories and methods of material structure research**", "**Basic application of synchrotron radiation**", **and** "**lectures on the frontiers of applied physics**" have been offered to graduate students successively. He has published more than 60 related papers in various journals and magazines. The X-ray study of crystal structure and defects in material science won the fourth prize of National Natural Science in 1982 (ranked second), and the remote-controlled X-ray topography camera won the third prize of major scientific research achievements in Shanghai in 1984 (ranking first).

He was the first member and secretary-general of X-ray professional committee of Chinese physical society (1982–1998), the first member and secretary-general of Shanghai X-ray professional committee of Shanghai physical society (1982–1992), the director of Shanghai metal society, and the deputy director of materials professional committee. He is currently a senior member of the X-ray and Synchrotron Radiation Committee of Shanghai Physics Society.

Hired by Shanghai Institute of Microsystems and Information Technology (formerly Metallurgy), Chinese Academy of Sciences from March 2003 to October 2011 conducted a lot of research on nano-materials and battery active materials and electrode processes, issued more than 50 papers, and won four invention patents.

From November 2011 to now, he is a consultant professor of residual stress analysis and shot peening laboratory in the School of Materials Science and Engineering, Shanghai Jiaotong University.

Chines Books that have been published are (Total ~5200 pages, ~7.15 million words):

(1) **Phase Diffraction Analysis**, (Metallurgical Industry Press, 1989.9.
(2) **The Basis of X-Ray Diffraction of Crystals**, Nanjing University Press, 1962.10.
(3) **Basic Applied Technology of Synchrotron Radiation X-Ray**, Shanghai Science and Technology Press, 2009.1.
(4) **X-Ray Analysis of Nanomaterials**, Chemical Industry Press, 1st Edition 2010.6; 2nd Edition. 2019.6
(5) **X-Ray Diffraction and Scattering Analysis of Materials**, Graduate textbook, higher Education Press, 2010.3.
(6) **X-Ray Diffraction Technology and Its Application**, East China University of Science and Technology Press, 2010.8.
(7) **Neutron Diffraction Technology and Its Applications**, Science Press, 2013.7.
(8) **Internal Stress Diffraction Analysis**, Science Press, 2013.11.

(9) **Record of Hardships and Happy Walking in Scientific Research Journey**, Shanghai University Press, 2015.4.
(10) **Material Characterization and Electrode Process Mechanism of Green Secondary Battery**, Science Press, 2015.13.
(11) **Shot Peening Strengthening and X-Ray Diffraction Characterization of Materials**, Science Press, 2019.
(12) **Modern Test-Analysis and Research of Materials**, higher Education Press, Postgraduate textbooks, 2022
(13) **Electrodeposition Materials and Their Modern Test-Analysis and Characterization**, Science Press, 2022.

Yuwan Lou, female, doctor, Professor, born in July 1971, Xinxiang City, Henan Province. Technical expert of Technological Innovation Strategic Alliance of Shanghai New Energy Vehicle Testing Industry.

From 1989.09 to 1993.07, she received a bachelor's degree from the Department of Applied Chemistry of Harbin Institute of Technology; a master's degree from the Department of Applied Chemistry of Henan normal University from 1999.06 to 2002.10; and a doctorate from Shanghai Institute of Microsystems and Information Technology, Chinese Academy of Sciences from 2003 to 2007.

In 1993.07 ~ 2002.02, state-owned 75th factory, assistant engineer, engineer; In 2002.03 ~ 2014.08, engineer, associate researcher, Shanghai Institute of Microsystems and Information Technology, Chinese Academy of Sciences; in 2008 ~ 2010.08, Shanghai International Automobile City postdoctoral Innovation practice Base, postdoctoral.

She has been engaged in the research of chemical power supply and related materials since 1993. She has participated in more than 20 projects at or above the provincial and ministerial level and presided over 10 projects. She has published more than 30 papers and 20 patents in domestic and foreign journals and has good experience in the combination of industry, university, and research. She won the third prize for technological invention in Shanghai in 2012 (ranked second).

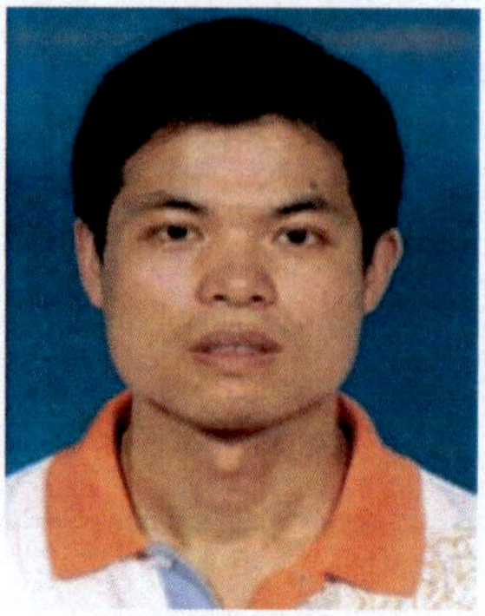

Jian Zhang, male, associate professor, was born in November 1977 in Minhou County, Fujian Province.

From 1997.09 to 2001.07, he received a bachelor's degree from the Department of Applied Chemistry, Harbin Institute of Technology; from 2002.09 to 2004.07, he received a master's degree from the Department of Applied Chemistry, Harbin Institute of Technology; from August 2004 to now, he is a research intern and an assistant researcher of Shanghai Institute of Microsystems and Information Technology, Chinese Academy of Sciences.

Since 2004, he has been engaged in the research of lithium-ion battery and its materials, participated in more than 20 provincial and ministerial projects, presided over 6 projects, published more than 20 papers in domestic and foreign journals, and applied for more than 10 invention patents. It has a good foundation in the synthesis, structure, and performance characterization and design and manufacture of lithium-ion battery materials and has good experience in the combination of industry, university, and research.

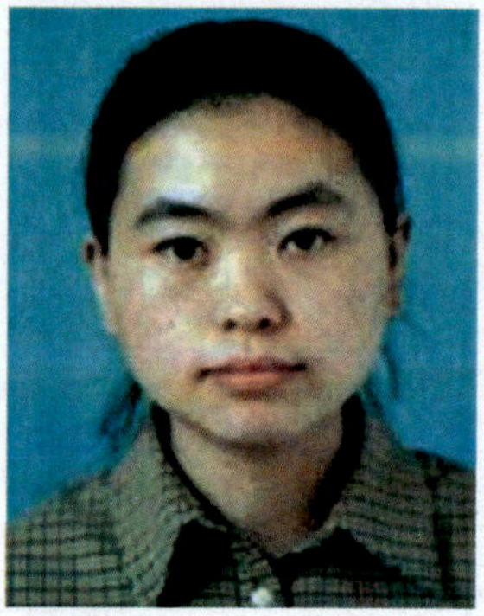

Xiaohua Xie, female, doctor, associate professor, born in July 1979 in Haicheng City, Liaoning Province.

From 1997.09 to 2001.07, he received his bachelor's degree from the Department of Chemical Engineering of Harbin Engineering University; from 2001.09 to 2004.03, he received his master's degree from the School of Chemical Engineering of Harbin Engineering University; from 2004.09 to 2008.06, he received his doctorate from Shanghai Institute of Microsystems and Information Technology, Chinese Academy of Sciences.

From 2008.07 to now, assistant researcher and associate researcher, Shanghai Institute of Microsystems and Information Technology, Chinese Academy of Sciences.

Since 2001, he has been engaged in the research of lithium-ion battery and related materials. He has participated in more than 10 provincial and ministerial projects and presided over 5 projects. More than 30 papers have been published in domestic and foreign journals, and 13 invention patents have been applied for, of which 4 have been authorized.

Baojia Xia, male, professor, doctoral supervisor, head of the research group. Born in September 1961 in Taizhou City, Jiangsu Province.

Member of China Alkaline Battery Standardization Committee and Electrochemistry Committee, Director of Chemical and physical Power Association, Director of solid State Ion Society, Expert of National Science and Technology Award, Visiting Professor of Shanghai Society of Chemistry and Chemical Engineering, and Leader of Battery Group, National Center of Nanotechnology and Applications.

From 1982.09 to 1986.07, he received a bachelor's degree from the Department of Applied Chemistry of Harbin Institute of Technology; from 1986.09 to 1989.07, he received a master's degree from the Department of Applied Chemistry of Harbin Institute of Technology; after graduating from the Department of Applied Chemistry of Harbin Institute of Technology in 1989, he stayed on to teach, focusing on courses such as "Chemical Power supply" and "Progress of Electrochemical Engineering" and engaged in the research of aluminum-air battery pack and Ni-MH battery. From 1995.09.07 to 1999.07, he received a doctorate from the Department of Applied Chemistry, Harbin Institute of Technology; from December 1999 to December 2001.12, he was a postdoctoral fellow of Shanghai Institute of Microsystems and Information Technology, Chinese Academy of Sciences; since he left his post in 2002, he has been the project manager of the knowledge innovation project of the Chinese Academy of Sciences and has successively served as an executive deputy director of the energy room, the director of the personnel education department, the director of the energy department, and the assistant to the director.

He was mainly engaged in the research and industrialization of the application basis and key technologies of power and energy storage batteries and their materials and presided over the completion of nearly 20 national, provincial and ministerial-level projects. He has won the third prize of technological invention in Shanghai (ranked first), the third prize of scientific and technological progress of the Ministry of Aerospace (ranked third), the second prize of Heilongjiang Provincial Education Commission, the excellent teacher of the Chinese Academy of Sciences, and the second prize of scientific and technological progress in Jiading area. He was selected as the outstanding discipline leader of Shanghai, the outstanding contribution individual of Zhangjiang High-tech Zone, and the leading talent of Jiading District. In the past ten years, more than 160 papers have been published, 28 invention patents have been granted, and more than 20 graduate students have been trained.

Chapter 1
Experimental Methods of Material Characterization and Mechanism Research

1.1 Preparation and Performance Test of Battery [1, 2]

The secondary batteries studied in this book include nickel-hydrogen batteries and lithium-ion batteries. However, there are mainly two types of battery structure: 2025 button battery and 18,650 cylindrical batteries.

1.1.1 Preparation of Batteries

1. **Assembly of 2025 button battery**
   The electrode sheet coated on one side is stamped into a disk with a diameter of 1.4 cm. In a glove box in argon atmosphere, separate the electrode from the lithium with a Celgard 2320 diaphragm and place it in a 2025 button battery case. The electrolyte was injected into it, and the composition was 1 mol/L $LiPF_6$ EC:DMC:DEC (volume ratio 1:1:1). Finally, seal the battery on the top.
2. **Preparation of hydrogen/nickel battery**
   The positive electrode is plaster electrode, and nickel foam is used as conductive current collector. The basic production process is as follows: Dry powder (active material, conductive agent, additive, etc.) is mixed with deionized water evenly → adding binder to stir to make paste → rolling paste (using a sizing machine to apply the slurry to the spot-welded nickel foam matrix) → drying in an 80 °C oven → to adjust the weight of the plate → and crush it to a certain filling density.

$$\rho = \frac{G}{a \times b \times c} \tag{1.1}$$

C. Yang et al., *Materials and Working Mechanisms of Secondary Batteries*,
https://doi.org/10.1007/978-981-19-5955-4_1

ρ is the filling density of the plate (g/cm^3); G is the weight of the plate (g); a is the width of the plate (cm); b is the length of the plate (cm), and c is the thickness of the plate. G does not include the weight of the polar ear.

The negative electrode adopts the dry forming process, and the hydrogen storage alloy powder of a certain particle size is directly pressed onto the three-dimensional copper mesh conductive matrix by a counter-rolling machine. The basic manufacturing process is as follows: The hydrogen storage alloy powder of a certain particle size is rolled onto the three-dimensional copper mesh conductive matrix. → The surface of the cutting plate → welding electrode ear → surface is coated with a certain concentration of PTFE → dry in an 80 °C oven.

The electrolytes with different proportions of KOH, NaOH, and LiOH were prepared with deionized water.

The battery assembly process is as follows: positive electrode-diaphragm-negative plate and together winding → shell → rolling groove → liquid injection → welding cap → sealing → cleaning → thermoplastic packaging.

The battery assembly process is as follows: positive electrode-diaphragm-negative plate and winding → shell → rolling groove → liquid injection → welding cap → sealing → cleaning → thermoplastic packaging.

3. **Assembly of 18,650 cylindrical battery**

   $LiMeO_2$ (or $LiFePO_4$) was used as positive active material in this study. 18,650 lithium-ion battery with artificial hexagonal graphite was used as negative electrode. The preparation method is as follows:

   (1) The binder polyvinylidene fluoride (PVDF) is fully dissolved in N-methyl pyrrolidone (NMP) solvent after stirring, and the positive active material and conductive agent superconducting carbon black (Super P) are added according to the mass ratio of 96:4. The pulping was fully stirred and mixed, and it was uniformly coated on the aluminum foil on the GF-TU400-04 gap coater. Then, it was dried in a vacuum incubator at 80 °C for 12 h. Finally, the polar plate is rolled to a specified thickness on a C33 300 twin-roll film mill.

   (2) The styrene-butadiene rubber (SBR) was fully dissolved in deionized water after stirring, and graphite and conductive agent superconducting carbon black (Super *P*) were added according to the mass ratio of 98:2. The pulping was fully stirred and mixed, and it was uniformly coated on the copper foil on the GF-TU400-04 gap coater. Then, it was dried in a vacuum incubator at 40 °C for 12 h. Finally, the polar plate is rolled to a specified thickness on a C33 300 twin-roll film mill.

   (3) The positive and negative plates coated on both sides were cut into poles with widths of 56 mm and 58 mm, respectively. The PC ultrasonic spot-welding machine is used to weld aluminum and nickel ears on the positive and negative electrodes, respectively. On the Jiatuo semi-automatic winding machine, the positive and negative electrodes are separated by Celgard ®2320 diaphragm and wound into a core. Put it into the aluminum 18,650 battery shell, weld the negative electrode lug on the bottom of the shell, roll

the groove on the battery shell on the GC35-W automatic grooving machine, install the gasket, and weld the cap. After vacuum drying at 80 °C for 12 h, the battery was injected into an electrolyte (1 mol/L $LiPF_6$ EC:DMC:DEC (Volume ratio 1:1:1)) in a glove box in argon atmosphere (Mikrouna, Super 1220/750/900). Finally, seal the battery.

1.1.2 Performance Test of MH/Ni Battery [1, 2]

(1) **0.2C capacity Test and calculation of utilization ratio of Nickel hydroxide**
After the formation of the battery, the battery capacity was tested by 0.2C charge–discharge cycle, and the charge–discharge curve was recorded. The charge–discharge cycle system was as follows: In the environment of 20 ± 2 °C, the battery was charged with 0.2C current for 7 h and then discharged with 0.2C current to the termination voltage 1.0 V after static 30 min. After 3 cycles, the capacity of the last time was taken (The discharge capacity C_0 was calculated by the product of discharge time and discharge current.) The weight of nickel hydroxide in the battery is $G_1(g)$. The formula for calculating the utilization rate of nickel hydroxide in the battery is as follows:

$$\text{Utilization rate} = \frac{289G_1}{C_0} \times 100\% \quad (1.2)$$

289 in the above formula is the theoretical specific capacity of nickel hydroxide (mAh/g).

(2) **Rate performance test**
1C (0.6 A) current was charged for 1.2 h, and $-\Delta V = 10$ mV was used as parallel control condition; it was left at rest for 1 h and then discharged to 1.0 V, 0.8 V, and 0.8 V at 1C, 5C, and 10C current, respectively. The discharge capacities C_1, C_5, and C_{10} of the battery were calculated by the product of discharge time and discharge current.

(3) **Cycle life test**
The normal temperature life test was carried out in the environment of 20 °C ± 2 °C, and the high temperature life test was carried out in the environment of 60 °C ± 2 °C.

Cycle life testing system: 1C current charging 1.2 h, with $-\Delta V = 10$ mV as parallel control condition. Then, the 2C current is discharged to 1.0 V, which is counted as a cycle, the open circuit shelving time is not set between the charge and discharge process, and the number of cycles and the capacity of each discharge are recorded. According to the provisions of the international standard GB/T11013-1996, when the discharge capacity attenuates to 60% of the nominal value of $1C_5$ discharge capacity (300 mA/g), the cycle numbers are its cycle life.

(4) **Internal resistance test**
Using Harbin Zimu DK-3000 precision battery internal resistance meter to measure the internal resistance of discharge state at different stages of battery cycle.

(5) **Cyclic voltammetry method**
Cyclic voltammetry (CV), also known as triangular wave linear potential scanning, is an electrochemical method to measure the change of electrode current by controlling the change of electrode potential at a constant speed. According to the potential-current curve, a lot of information about the reaction process of the electrode can be obtained. For example, it can detect the electrochemical activity of substances, measure the redox potential of substances, and investigate the reversibility and reaction mechanism of electrochemical reactions; on the other hand, through the analysis of the shape of the scanning curve, the electrode reaction parameters can be estimated.

For the electrode process with diffusion control, the peak current will appear in the cyclic voltammetry curve. When the potential scan crossed the equilibrium potential of the electrode reaction, with the increase of the overpotential on the electrode, the electrode reaction speed increased and the current increased, but at the same time, it also led to the decrease of the reactant concentration on the electrode surface and the increase of the product concentration. With the continuous scanning of potential, the electrode reaction is controlled by diffusion, the diffusion thickness increases, the flow rate decreases, the reaction current decreases, and the peak value appears. The cathodic reduction peak appears in the negative direction, corresponding to the reduction of the oxidized species on the electrode surface, and the oxidation peak in the positive scanning direction corresponds to the oxidation of the reduced species. By analyzing the data of cyclic voltammetry curve, the information of peak current (i_p), peak potential (φ_p), reaction kinetic parameters, and reaction mechanism can be obtained.

The cyclic voltammetry test was carried out on a comprehensive electrochemical tester made in Switzerland (Autolab 302N). The scanning rate is $0.1\ mV{\cdot}s^{-1}$, starting from the open circuit voltage.

(6) **Electrochemical Impedance Spectroscopy**
Electrochemical impedance spectroscopy (EIS), or AC (Alternating Current) impedance spectroscopy, is an important method to study the electrode process kinetics and interface structure and has been widely used in the study of all kinds of batteries. It causes the corresponding signal by applying a small sine wave AC signal to the tested system, and the information about the dynamics of the electrode process can be obtained from the relationship between the two. In the process of measurement, because the alternating signal with small amplitude basically does not change the state of the measured system, this method can be used to accurately study the relationship between the kinetic parameters of each electrode process and the electrode state.

Electrochemical impedance spectroscopy (EIS) is a frequency domain measurement method, which can obtain impedance spectra with a wide frequency range to study the electrode system, so it can obtain more kinetic information and electrode interface structure information than other electrochemical methods. By fitting the electrochemical impedance spectrum through the equivalent circuit, information such as ohmic resistance, electrode surface film resistance and related capacitance, charge transfer resistance, and electric double-layer capacitance can be obtained in different frequency ranges. Therefore, electrochemical impedance spectroscopy shows unique advantages in the study of electrode/electrolyte interface process and is widely used to study the lithium removal and surface characteristics of carbon materials and transition metal oxide electrodes for lithium-ion batteries.

The electrochemical impedance spectroscopy (Nyquist curve) was drawn on the comprehensive electrochemical tester made in Switzerland. The frequency range of EIS test is 10^5–10^{-2} Hz, and the amplitude of AC signal applied is 5 mV. The impedance data obtained in the experiment are fitted by Nova 1.6 software.

1.2 Diffractometer Method for X-Ray Diffraction Analysis [3, 4]

1.2.1 The Structure of Modern X-Ray Powder Diffractometer

The powder (polycrystalline) X-ray diffractometer is composed of X-ray generator, goniometer, and detection-recording system. The core component of the powder diffractometer is the goniometer. The linear X-ray beam emitted by the linear focal source passes through the vertically divergent Soller aperture S_1, When the horizontal divergent slit DS is injected into the sample C, the diffracted X-ray produced by the sample enters the detector probe through the receiving slit RS, the receiving vertical divergent Soller aperture S_2, and the anti-scattering slit SS, as shown in Fig. 1.1. The distance from the focus of the anode target of the X-ray tube to the center of the diffractometer is equal to the distance from the center of the sample to the receiving slit RS, which is called the diffractometer radius r. Figure 1.1 shows the experimental light path and layout of the X-ray powder diffractometer. The center of the sample is strictly coincident with the axis of the diffractometer, and the θ scan is performed around the axis of the diffractometer, while RS and S_2 are scanned around the axis of the diffractometer together with the counter. During the scanning process, the RS is always tangent to the diffractometer circle with the diffractometer axis as the center and the radius r.

As can be seen from Fig. 1.2, unlike the negative mounted on the focus circle in the focus camera, in the powder diffractometer, the detector does not move along the focus circle, but along the concentric circle of the diffractometer. There are only two

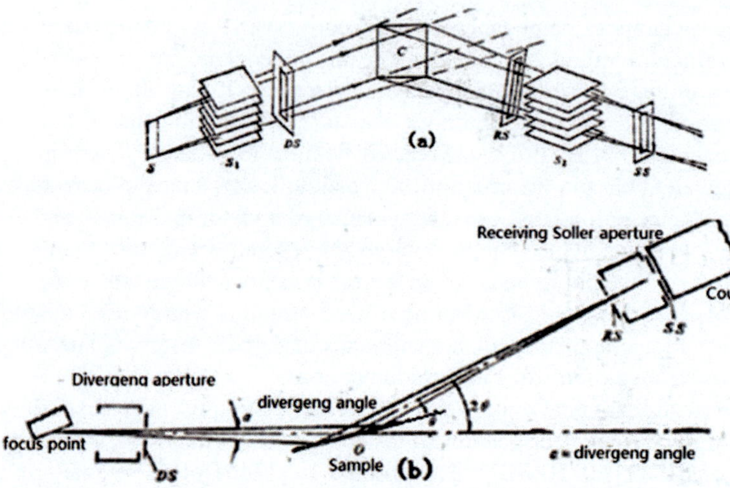

Fig. 1.1 Experimental layout of X-ray powder diffractometer. Optical layout of **a** goniometer, **b** focusing circle plane (plane perpendicular to the axis of diffractometer)

focal points between the diffractometer circle and the focusing circle, one of which is the intersection point F of the target surface, and the other is variable, that is to say, no matter how the diffraction condition is changed, only one diffraction ray can focus on the diffractometer circle under certain conditions. Therefore, the detector moving along the concentric circle of the goniometer circle can only measure the radius R of the focusing circle pair by pair of diffraction lines.

The radius R of the focus circle also changes. The relationship between the radius of the focus circle and the Bragg angle θ can be obtained from Fig. 1.2:

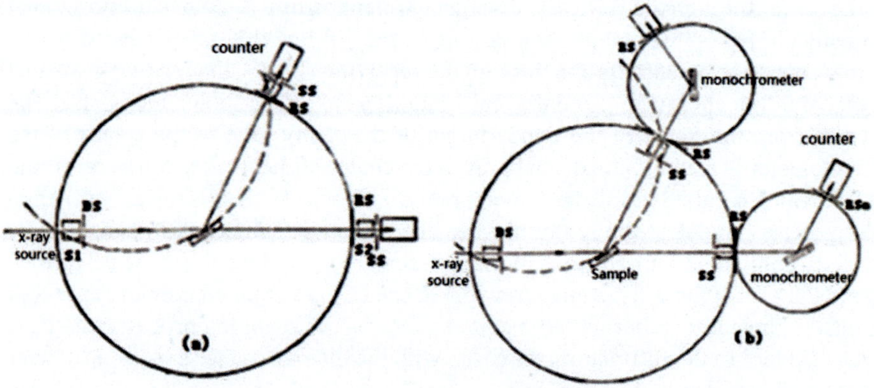

Fig. 1.2 Diffraction geometry of X-ray powder diffractometer (The radius of focus circle decreases with the increase of 2θ.) **a** Crystal-free monochromator **b** crystal monochromator with diffracted beam

Table 1.1 Comparison of horizontal scan diffractometer and vertical scan diffractometer

Powder diffractometer scanning mode	Horizontal scanning	Vertical scanning
Plane of focal	Horizontal plane	Vertical plane
X-ray tube	Fixed target: horizontal installation Foldable target: installed horizontally or vertically. Horizontal installation can only be using two windows	Vertical installation can be using four windows
The utilization rate of space	Low	High
2θ scanning angle range	−100° 165°	−40° 165°
Attachment installation	Easy to refit, and special attachments such as structural analysis, high and low temperature, high pressure, and reactor. Special attachments	So on can be installed in principle, but it is more difficult

$$\frac{r}{2R} = \cos\left(\frac{\pi}{2} - \theta\right) = \sin\theta$$
$$R = \frac{r}{2\sin\theta} \tag{1.3}$$

When $\theta = 0.00°$, the radius of focus circle is infinite, and the radius of focus circle decreases gradually with the increase of θ angle, and when $\theta = 90.00°$, $R = r/2$. Therefore, the goniometer can be regarded as an asymmetric focusing camera in which the radius of the focus circle varies with θ.

The powder diffractometer can be divided into two types: horizontal scanning and vertical scanning, that is, the scanning plane can be horizontal plane or vertical plane, both of which are perpendicular to the axis of the diffractometer.

The comparison between the two is shown in Table 1.1.

1.2.2 The Working Mode of Powder Diffractometer

Modern powder diffractometer has two working modes: wavelength dispersion and energy dispersion. They are introduced as follows:

1. **Wavelength dispersive powder diffraction**
   Wavelength dispersion diffraction is the powder diffraction which usually uses monochromatic (characteristic) X-ray incidence and counter tube (Geiger tube, scintillation tube, proportional counter tube, etc.) as detectors. The diffraction condition must conform to Bragg's law: $2d\mathrm{Sin}\theta = n\lambda$. The following scanning methods are given in Table 1.2.

Table 1.2 Scanning modes and characteristics of wavelength dispersive X-ray powder diffractometer

Scanning mode	Main features	Main application
Reflective $\theta/2\theta$ linkage	The diffraction surface is almost parallel to the sample surface, quasi-focusing geometry	Wide-angle diffraction and wide-angle scattering
Reflective 2θ scanning	Grazing incidence, non-focusing geometry, changing the grazing angle can change the depth of the diffraction, the azimuth of the diffraction surface varies with 2θ, and the orientation of the diffraction surface varies with 2θ	Wide-angle diffraction and scattering of thin film samples
θ scan	Fixed 2θ, only θ scan	One-dimensional polar density measurement
θ–θ scanning	The sample is not moving, and the radiation source and the detector do θ–θ scanning synchronously	The most suitable liquid sample
Transmission 2θ scanning	Fixed at −90°, only 2θ scan	Non-destructive analysis of thicker samples

Modern powder diffractometer can still be divided into two recording modes: continuous scanning and step scanning (also divided into fixed time-counting and fixed count-timing). In fact, there is no difference between the two scanning modes, and continuous scanning is also step, where the step size is called sampling width.

At present, many experimenters do not pay enough attention to the application of transmission geometric diffraction technology. The following is discussed: The geometry of transmission diffraction is shown in Fig. 1.3, where (a), (b), and (c) show the incident light of parallel beam, divergent beam, and convergent beam, respectively.

Careful observation and analysis of Fig. 1.3a and b show that (1) the diffraction lines are obviously broadened, and some of the diffraction lines will overlap, which may actually become halo and/or halo overlap; (2) this phenomenon of broadening and overlap will become more serious with the increase of the thickness of the sample; and (3) the direction of diffraction peak displacement and peak broadening develops to low angle or high angle with the increase of sample thickness, which depends on taking the diffractometer axis as the center, and the thickness increases to the outgoing ray direction or to the incoming ray direction. In order to restrain or eliminate this broadening, the diffraction data which can be compared with the standard data can be obtained by adding a cut-off slit

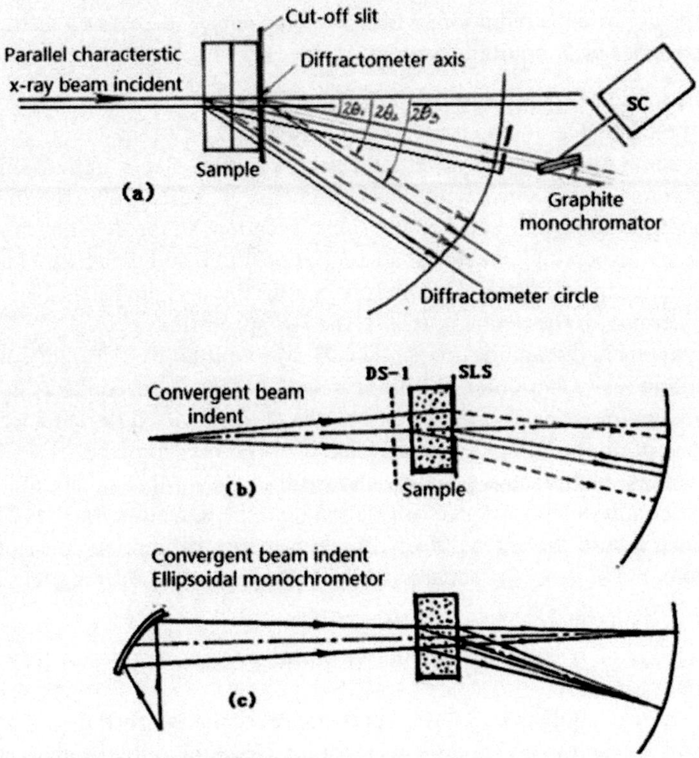

Fig. 1.3 Thick sample transmission Laue symmetrical geometry, **a** parallel beam incident, showing obvious broadening of the diffraction line, **b** divergent beam incident, showing the defocusing effect of the diffracted beam, **c** converging beam incident, showing the focusing effect of the diffracted beam

(SLS) and a divergent slit (DS-1) on the exit surface of the sample. If the convergent beam at RS can be focused by using the ellipsoidal crystal Monochromator, it is ideal [5].

2. **Energy dispersive powder diffraction**
If the white (continuous wavelength) X-ray incidence is used, the crystal planes with different d values are in the same orientation and the direction of the incident rays is the same, then the direction of the diffraction lines of each crystal plane with different d values is the same, so the detector must be fixed at a selected 2θ position. Each diffraction line obeys

$$2d \sin\theta = 12.3985/E \tag{1.4}$$

where E is the energy of the incoming ray, the unit is keV, the unit of d is Å, and $\sin\theta$ is fixed, so that it is in the same azimuth. When X-rays with different energy are diffracted by crystal planes with different d values, the incoming rays should be continuous radiation, which is called energy dispersion diffraction. It has two modes of operation:

(1) The mode of measuring the diffraction intensity of different energy at the same time: The incident ray is the X-ray with different energy incident at the same time, and the different d-valued crystal planes in the same direction choose the X-ray with different energy which satisfies the diffraction condition to be diffracted to the same direction, so the detector must be an energy dispersion detector at a fixed 2θ position, and the energy and intensity of each diffraction line are given after the acceptance and subsequent processing of the detector, that is, the *I-E* spectrum.

(2) X-ray energy scanning mode: The X-rays emitted by the light source are continuously distributed and are scanned by the spectroscopic crystal before incident into the pattern. In other words, X-rays with different energies are incident on the sample one after another. The crystal planes with different d values in the same position choose the energy that meets the diffraction conditions to produce diffraction, and the detectors fixed at $2\theta_S$ position measure the intensity of different energy diffraction lines one after another. The $I \sim 2\theta_m$ pattern is given, where θ_m is the Bragg angle of the spectroscopic crystal.

3. **Comparison of wavelength dispersion and energy dispersion diffraction methods**
Here, we are not going to make a comprehensive comparison between wavelength dispersion diffraction and energy dispersion diffraction, only the characteristics of the two transmission methods are listed in Table 1.3. From the practical point of view of non-destructive detection, energy dispersion diffraction is much superior and convenient in terms of ultra-thick sample effect and diffraction center position; from the point of view of effective penetration thickness and applicability, the two methods have their own scope of application: Wavelength dispersion diffraction can detect the 4–20 mm range of sample thickness (related to target elements); energy dispersion diffraction can detect sample thickness ~ 20 cm order of magnitude. In the case of transmission, energy dispersion diffraction can replace wavelength dispersion diffraction, but wavelength dispersion diffraction cannot replace energy dispersion diffraction.

1.3 X-Ray Photoelectron Spectroscopy [6–9, 11]

Ultraviolet photoelectron spectroscopy (UPS) and X-ray photoelectron spectroscopy (XPS) of atoms excited by conventional ultraviolet or X-ray have been more mature and widely used, especially XPS energy as an important means of chemical analysis, so it is called “electron spectroscopy for chemical analysis” (ESCA). The study of synchrotron radiation photoelectron spectroscopy (SR-PS) began in the early 1970s. At present, there are four working modes of photoelectron spectroscopy:

Table 1.3 Comparison of wavelength dispersion diffraction and energy dispersion diffraction

Compare items		Transmission wavelength dispersion diffraction	Transmission energy dispersion diffraction
1	Incident ray	Monochromatic radiation	Continuous X-rays in a certain energy range
2	Diffraction line	The same characteristic radiation as the incident ray, each diffraction line. The itinerary in the sample is different	Select X-ray diffraction in a certain energy range. The stroke of each diffraction in the sample is the same
3	Diffraction pattern feature	It is similar to the standard pattern of the same radiation	Related to the selected 2θ position
4	Ultra-thick sample effect	The effect of line broadening and overlap is serious, which can be reduced and overcome by cutting slit	There is no line broadening and overlap effect
5	The position of several diffraction centers	It should be at the intersection of the X-ray and the exit plane of the detector, which is difficult to achieve in practice	It can be located at any position where the X-ray intersects the detector, which is not difficult to achieve in practice
6	Reception and recording of diffraction patterns	The counter is scanned for 2θ, or a zero-dimensional or one-dimensional detector is used	Energy scanning, the counter is fixed in the 2θ position or the energy dispersion detector fixed in the 2θ position
7	Effective penetration thickness	Small, related to the radiation wavelength used	Much bigger, 3–5 times
8	Applicability	AgK_α or MoK_α radiation can be used for on-line inspection of drugs and explosives in small packages	Use Au or W target, 60 or higher KV tube pressure Large packages of explosives can be detected online
9	Price	The scanning of counter tube is similar to that of energy scanning, and there is not much difference between using one-dimensional detector and using energy dispersion detector	

(1) **The energy distribution curve (EDC) mode**, that is, under the condition that the incident photon energy hv is fixed, the distribution of photoelectron signal intensity $I(E)$ excited by synchrotron radiation soft X-ray with its energy E can be measured by electron energy spectrometer, and the variation of photoelectron signal intensity E with hv can be investigated. A more complete image is given.

(2) **Constant initial state spectrum (CIS)**, that is, the method of synchronous scanning of incident photon hv and photoelectron energy Ef by using the photon energy of synchrotron radiation and the analyzer energy of photoelectron spectrometer.

(3) **The constant end state (CFS) mode** is that the incident photon energy hv is continuously scanned during the spectrum measurement, but the electron energy E_f accepted by the analyzer is always fixed at a certain value, so the yield spectrum (YS) and partial yield spectrum (PYS) of the constant end state (CFS) spectrum play a role in measuring the density of states and angular momentum of the surface electronic states, and they are the best method to study the initial state electronic structure and partial yield spectrum.

(4) **Photoelectron diffraction**, that is, when receiving photoelectrons in a certain direction, there are not only photoelectrons emitted directly from adsorbed atoms, but also photoelectrons scattered by substrate atoms. The interference between the directly emitted photoelectrons and the scattered photoelectrons will cause the photoelectron intensity to vary with the photon energy or angle. The photoelectron diffraction experiment with synchrotron radiation is to measure the relationship between the photoelectron intensity and the photon energy.

1.3.1 Energy and Intensity of Photoelectron Spectrum

The photoelectron kinetic energy corresponding to the companion peak caused by the fall excitation is:

$$E_e^* = hv - \left(E_f^* - E_i^*\right) \tag{1.5}$$

E_i^* and E_f^* are, respectively, the energy of the initial state and the final state for the wrestling. The energy interval between the companion peak and the main peak is

$$\Delta E = E_e - E_e^* = E_f^* - E_f \tag{1.6}$$

The intensity of the photoelectron peak on the spectrum refers to the area of the photoelectron peak, which corresponds to the intensity of the photon signal without inelastic scattering, that is, the number of photoelectrons emitted. For the photoelectron peak whose energy of element i is E_{ei}, the intensity I_i can be expressed as

$$I_i = I_0 C_i \sigma_i \lambda_{T(Eei)} D_{(Eei)} \tag{1.7}$$

where I_0 is the photon energy of the incident ray, C_i is the concentration of element i, σ_i is the cross section of the photoelectric effect, that is, the photoexcitation probability, the $\lambda_{T(Eei)}$ is average free path of the photoelectron with the energy of E_{ei} in the main material of the sample, and $D_{(Eei)}$ is the detection efficiency of the detector for the photoelectron with the energy of E_{ei}.

1.3.2 Chemical Analysis of X-Ray Photoelectron Spectroscopy

Because the atomic structure of each element in the periodic table is different from that of other elements, it is easy to identify the elements on the surface of the sample by determining the position of one or more photoelectron peaks of an element. Even if the electron binding energies of the adjacent elements of the same energy level in the periodic table are quite different, all the elements in the periodic table except *H* can be identified.

It can be known from Formula (1.7) that if I_0, σ, λ_T, and D_{Eei} know all, the concentration C_i can be calculated according to the measured I_i, but usually do not know, and theoretical calculation is needed, so it is not a simple task to make quantitative analysis.

1.3.3 Valence State Study

The premise that photoelectron emission can occur is that the photon energy of the incident light source must be large enough to excite the electrons at the valence band or core level of the source to the final state where the energy is higher than the vacuum level, that is to say, the energy must be conserved in the process of photoelectron emission, that is

$$E_{f(k)} = E_{i(k)} + hv \tag{1.8}$$

It can be seen that the energy required for excited valence electrons is low, so UV photoelectron spectroscopy, also known as valence band photoelectron spectroscopy, is the study of atomic valence state, as well as the distribution of state density, the dispersion of energy band in wave vector space, the symmetry of wave function, and so on.

The binding energy of electrons in the inner shell of an atom is affected by the distribution of charges inside and outside the nucleus. Any change in the distribution of these charges will shift the position of the photoelectron spectrum. The binding energy shift caused by atoms in different chemical environments is called chemical shift. According to the determination of the chemical shift of the photoelectron peak, the structure and chemical information of the analytical sample can be obtained. The photoelectron produced by excited inner electrons is called core photoelectron spectroscopy. According to the difference of the binding energy of the excitation core, the light source can be vacuum ultraviolet radiation with photon energy of 100 eV, soft X-ray with energy of hundreds to 1–2 kV, or hard X-ray. Hard X-ray is often used for core state excitation of elements in metals and their compounds, so it is called XPS.

The binding energy of electrons at the core level E_b, also known as ionization energy I_k, is equal to the negative value of its self-consistent field (SCF) orbital energy according to Koopmans approximation, that is,

$$E_b = I_k = -E_k^{\mathrm{SCF}} \tag{1.9}$$

This approximation is quite good, because the two main factors that cause the high $I_k(-E_k^{\mathrm{SCF}})$ electron relaxation and electron correlation, to some extent, cancel each other out.

In the energy system, under the simplest approximation that embodies the Koopmans rule and gives a basic description of the charge distribution in the system, the ionization energy difference-chemical shift of the two groups of core photoelectron signals can be expressed as

$$I_A - I_B = (E_B - E_A) + e(V_B - V_A) \tag{1.10}$$

Here, the subscripts A and B are used to distinguish the different valence states of atoms of the same element. The orbital energies E_A and E_B strictly correspond to a hypothetical state of valence-atom, which is obtained by moving the atom adiabatically from the chemical environment to the free space. V_A and V_B are the electrostatic potentials of atoms at their respective positions. It is worth noting that the sign of the "non-local" term $e(V_B{-}V_A)$ is usually opposite to that of the "local" term $(E_B{-}E_A)$, which is an important factor affecting the chemical shift. It can be seen that the difference of the chemical environment around the atom shows that the position of the photoelectron peak is different in the elemental substance and different compounds, and the energy difference between them can be zero to more than ten electron volts. This is the basis for the study of atomic valence state by core photoelectron spectroscopy.

1.3.4 Example of Valence State Study—Study on the Valence State of Anodic Ion in the Synthesis of $Li(Ni_{0.6}Co_{0.2}Mn_{0.2})O_2$

Figure 1.4 shows the photoelectron spectra of the precursor $Mn(OH)_2$, the intermediate product (Lncmm) at 500 °C and the final product at 900 °C during the synthesis of $Li(Ni_{0.6}Co_{0.2}Mn_{0.2})O_2$. From Fig. 1.4a, it can be seen that there is no Li1s peak in the precursor, and the Li1s peak of the intermediate product and the final product is very obvious, the former is relatively wide, and the latter tends to split. From Fig. 1.4b, we can see that the Ni exists in the precursor in the form of +2, and the Ni2p peak of both the intermediate and the final product moves to the low energy direction, while the latter moves more. Generally speaking, the valence of Li varies from +1 to >+ 1 in the intermediate and final product. On the other hand, the price of Ni has changed

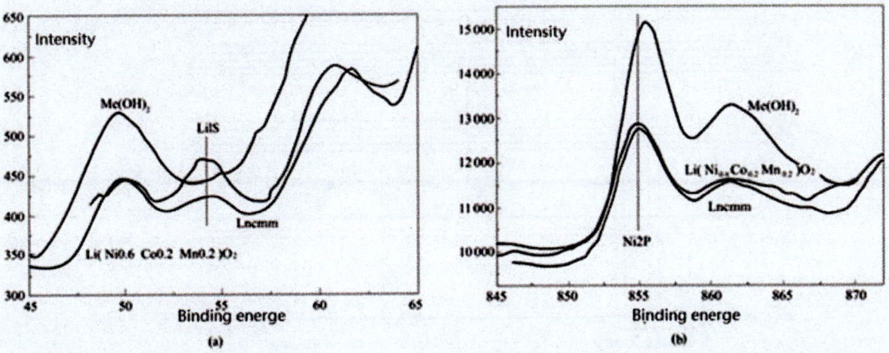

Fig. 1.4 Photoelectron spectroscopy of three main stages in the synthesis of $Li(Ni_{0.6}Co_{0.2}Mn_{0.2})O_2$: precursor [$Me(OH)_2$] and intermediates (Lncmm) and the final product **a** Li1s **b** Ni2p

from +2 to <+ 2. This may be related to mixed occupancy of Li and Ni in distorted $Li(Ni_{0.6}Co_{0.2}Mn_{0.2})O_2$ (intermediate product) and distorted $Li(Ni_{0.6}Co_{0.2}Mn_{0.2})O_2$ (final products). For more information about their mixed occupancy, please refer to Chap. 4.

1.4 X-Ray Emission Spectrum and Its Application [11, 12]

1.4.1 X-Ray Emission Spectrum

X-rays can be excited by high-energy electron beams or other high-energy ions such as X-rays. The X-ray excited sample with *W* target is conventional fluorescence X-ray analysis, and the fluorescence X-ray analysis of sample excited by synchrotron radiation X-ray [12], which began in the 1970s, has the following characteristics compared with conventional fluorescent X-ray:

(1) Selective excitation, the incident ray wavelength for excitation is not limited by the target elements, and the sample is excited by a monochromatic X-ray whose wavelength is less than the absorption limit of the element to be measured and/or less than the absorption limit of each element in the sample. The latter can make a full element analysis of the sample, if the former can provide the relative ratio of the fluorescence intensity of the elements to be measured and other elements, simplify the structure, and reduce the spectral background.
(2) Because of the high brightness of synchrotron radiation and making use of the characteristics of synchrotron radiation, measures are taken to reduce the background, which greatly increases the detection limit. The system detection has exceeded the order of ppm to dozens of ppm, and now, it has developed to the order of ppb, which is much higher than the detection limit of 0.1‰ of electron probe.

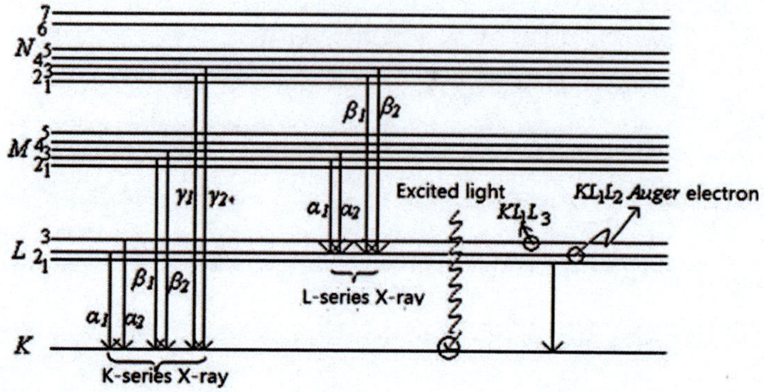

Fig. 1.5 Relaxation of excited atoms: energy level diagrams of X-ray emission and Auger electron emission

(3) Due to the good collimation and small size of synchrotron radiation X-ray, it can realize micro-area fluorescence X-ray analysis, and the spatial resolution has reached the order of μ_m. With the application of the third-generation synchrotron radiation light source, the spatial resolution can reach 50 nm.
(4) Various parameters of the light source of synchrotron radiation can be calculated, which can accurately calculate the intensity and energy spectrum structure of incident light for fluorescence analysis, which is helpful to improve the accuracy and accuracy of non-standard sample analysis.

When the sample is excited by high-energy electron beam or X-ray, the electron in the inner shell of the atom leaves a vacancy due to ionization, and the process in which the atom releases energy by the transition from the outer electron to this energy level emits characteristic X-rays; another relaxation process is that the electron ionization of the *A* shell produces a vacancy, and the electron of the *B* shell transitions to the vacancy of the *A* shell, resulting in the electron emission of the *C* shell, which is called ABC Auger electron emission. Figure 1.5 shows a schematic diagram of the energy level transitions of characteristic X-rays and characteristic Auger electrons emitted by atoms, for example, *K*-layer ionization, *L*-layer electron transition to *K*-layer transition emits K_α X-ray, *M* shell transition to *K*-layer transition emits K_β X-ray, and so on. It can also excite *K*-series Auger electrons, such as KL_1L_2 and KL_1L_3, if it is *L*-shell electron ionization, it will emit *L*-series characteristic X-rays and *L*-series characteristic Auger electrons.

1.4.2 Chemical Analysis of X-Ray Emission Spectrum

The qualitative analysis of elements is simple, as long as the energy (or wavelength) of each peak in the spectrum is calibrated and compared with the data of the characteristic X-ray spectrum of each element. Quantitative analysis is similar to the empirical coefficient method and basic parameter method in conventional fluorescence X-ray, which is called standard sample method and non-standard sample method.

The so-called standard sample method is to use the pure element as the standard sample to measure the strength I_j and I_{jj}^o of the selected characteristic line for the sample and the standard sample, respectively, and replace them with the following formula:

$$C_j = \frac{Q_j I_j(\mu_{s.i} + \mu_{s.f} \sin\psi/\sin\varphi)\rho_s}{1 - \exp\{-[\mu_{s.i} + \mu_{s.f} \sin\psi/\sin\varphi]\rho_s T \sin\psi\}} \tag{1.11}$$

$$Q_j = \frac{1}{I_{jj}^o}(\mu_{s.i}^o + \mu_{s.f}^o \sin\psi/\sin\varphi)/\rho_s^o \tag{1.12}$$

The mass fraction C_j of the elements per unit mass of the sample can be obtained.

In the form, ψ, φ the angle between the incident ray and the surface of the sample and the angle of fluorescent X-ray emitting on the surface of the sample, respectively;

$\rho_s T$ is the thickness of the sample excited by incident radiation and the mass per unit area of the sample with uniform thickness.

$\mu_{s.i}/\rho_s$, $\mu_{s.f}/\rho_s$ the mass absorption coefficients of the sample to incident rays and fluorescent X-rays, respectively.

$\mu_{s.i}^o/\rho_s^o$, $\mu_{s.f}^o/\rho_s^o$ the mass absorption coefficients of pure element standard samples for incident X-rays and fluorescent X-rays, respectively.

It can be seen from the above that the determination of element concentration depends on people's knowledge of incident radiation and fluorescence radiation mass absorption of pure element standard samples and unknown component samples. Sparks has reviewed the determination methods of these coefficients and the mathematical treatment of fluorescence measurement.

The non-standard sample method does not need the standard sample of the intermediate standard component or any empirical coefficient, and the result is obtained through a series of parameter calculation. The spectral distribution of incident radiation is the first parameter necessary for the standard-free method, which is convenient and simple for synchrotron radiation. Secondly, for each sample, there are basic absorption and secondary fluorescence, so it must be considered:

(1) The penetration of primary radiation with wavelength λ in the sample.
(2) The primary excitation of elements in the d_x layer.
(3) The secondary fluorescence of the elements in the d_y layer caused by the secondary excitation of the characteristic radiation produced by the primary excitation in the d_x layer.
(4) The absorption of the new radiation by the matrix. The calculation starts with the X-ray relative intensity measurement of each element to estimate the content of these elements, calculates the intensity of the hypothetical component that should be observed, compares this value with the measured value, and then modifies the hypothetical component. From this, a set of estimated intensity is calculated. This is repeated until it is assumed that the X-ray intensity of the

component is consistent with the measured X-ray intensity to a predetermined degree of accuracy, and then, the analysis results are printed.

For the thin film sample, the monochromatic light is excited and the energy dispersion spectrometer is used to measure it. The content of element I in the test sample C_i is calculated according to the following formula.

$$c_i = \frac{I_i 4\pi R^2}{p_o D_i \sum_k \left[(u_k/\rho)_o w_k f_k\right]_i \rho_s T/\sin\phi} \tag{1.13}$$

$\sum_k \left[(u_k/\rho)_o w_k f_k\right]_i$ item is constant for each element. Expressed by ℓ_i, $4\pi R^2/(p_o D_i \rho_s T/\sin\phi)$ item is only related to the measurement, make it K, then the above formula can be written.

$$c_i = I_i/Kl_i \tag{1.14}$$

When the standard curve or standard addition method is used, the denominator $(K\ell_i)$ will be eliminated, that is, the strength is proportional to the content.

The characteristic X-ray spectrum is almost independent of the physical state of the element or the valence of the atom, which is one of the advantages of X-ray fluorescence analysis. This feature is not applicable to K-series spectral lines of low atomic order elements and L-series or M-series spectral lines of plateau sub-order elements. When there is a certain change in the atomic valence electrons of the above-mentioned elements, that is, the drift and shape distortion of the spectral lines occur. On the contrary, the fine measurement of the drift and shape of spectral lines is an effective method to study the electronic states of atoms.

1.5 X-Ray Absorption Spectrum [10–11]

1.5.1 Absorption Limit

The experimental results show that the mass absorption coefficient μ_m of matter to X-ray has the following relationship with wavelength and atomic number Z of matter.

$$\mu_m \propto \lambda^3 Z^3 \tag{1.15}$$

Figure 1.6 shows the relationship between the absorption coefficient μ_m and the incident X-ray wavelength (energy). Generally speaking, the absorption coefficient decreases sharply with the decrease of the wavelength, but a series of absorption peaks appear. For Pt, $\lambda_K = 0.1582$, $\lambda_{L1} = 0.8940$, $\lambda_{L2} = 0.9348$, $\lambda_{L3} = 1.0731$ Å, these wavelengths (or energy) corresponding to the sudden increase peak are called absorption limits.

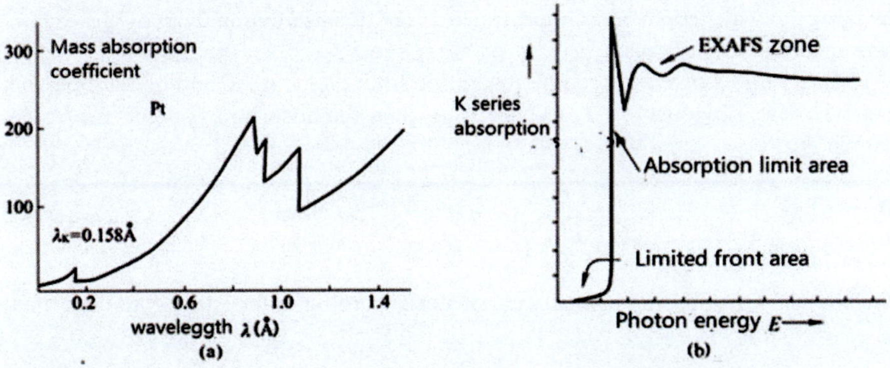

Fig. 1.6 **a** The relationship between μ_m and λ of platinum, **b** amplification near the *K* absorption limit

The absorption limit near the *K*-series is magnified as shown in Fig. 1.6b, which is divided into the front limit region, the absorption limit region, and the extended region, which is also called the extended X-ray absorption fine structure(EXAFS) region. The phenomenon of sudden increase in absorption is explained as follows: When an atom captures an X-ray photon and ionization occurs, the energy of the photon must be equal to or greater than the binding energy of the struck electron. When the incident wavelength is longer (the energy is small) and the energy of the photon is less than the electron binding energy of a certain shell, the electron of the shell cannot be knocked out, but when the energy of the incident photon is exactly equal to or slightly greater than the binding energy of the shell electron, the photon will be absorbed by a large amount of matter, and the absorption coefficient will increase suddenly. Let the electron binding energy of the *K* shell be W_K, then the absorption limit wavelength λ_K of the *K* system is

$$\lambda_k = \frac{hc}{w_k} \tag{1.16}$$

1.5.2 *Chemical Qualitative and Quantitative Analysis by X-Ray Absorption Spectroscopy*

From the previous discussion, it is known that the energy and wavelength of the absorption limit are the characterization of the elements. If the sample is composed of multiple elements, when the total absorption spectrum of the sample is determined, the absorption spectrum of the characteristic absorption limit lines containing each element can be obtained, and the elements in the sample can be judged by calibrating the energy or wavelength of each spectral line. The quantitative analysis by absorption limit method is as follows:

Let the sample be composed of absorption limit element An and non-absorption limit element, obviously = 1, 2,…, $n \neq A$, then the absorption coefficient of the sample is

$$\left(\frac{\mu}{\rho}\right)_m = w_A\left(\frac{\mu}{\rho}\right)_A + w_i\left(\frac{\mu}{\rho}\right)_i \tag{1.17}$$

Then, the transmission intensity attenuates according to the exponential law, that is

$$I = I_0 e^{-\left(\frac{\nu}{\rho}\right)_m \rho_m} = I_0 e^{-\left[\left(\frac{\mu}{\rho}\right)_A w_A + \left(\frac{\mu}{\rho}\right)_i w_i\right]\rho_m t} \tag{1.18}$$

When the above and below of absorption limit of element A are measured, respectively, the mass absorption coefficient of element A is $\left(\frac{\mu}{\rho}\right)A - \text{up}$, $\left(\frac{\mu}{\rho}\right)A - \text{down}$, respectively, but the mass absorption coefficient of non-absorption limit element is unchanged. Therefore, there are

$$I_{\text{above}} = I_0 e^{-\left[\left(\frac{\mu}{\rho}\right)_{A.\text{above}} w_A + \left(\frac{\mu}{\rho}\right)_i w_i\right]\rho_m t} \tag{1.19}$$

$$I_{\text{below}} = I_0 e^{-\left[\left(\frac{\mu}{\rho}\right)_{A\,\text{below}} w_A + \left(\frac{\mu}{\rho}\right)_i w_i\right]\rho_m t} \tag{1.20}$$

Division of two forms

$$\frac{I_{\text{above}}}{I_{\text{below}}} = \mathrm{e}^{W_A\left[\left(\frac{\mu}{\rho}\right)_{A\,\text{bel.}} - \left(\frac{\mu}{\rho}\right)_{A\,\text{ab.}}\right]\rho_m t} \tag{1.21}$$

$$\text{Let } \left(\frac{\mu}{\rho}\right)_{A\,\text{below}} - \left(\frac{\mu}{\rho}\right)_{A.\text{above}} = k_A,\quad \rho_m t = M_m, \tag{1.22}$$

$$W_A = \frac{l_n(I_{\text{ab.}}/I_{\text{bel.}})}{k_A M_A} \tag{1.23}$$

Therefore, the weight percentage W_A of the absorption limit element can be measured, where $\rho_m t$ is the density and thickness of the sample, and M_m is the mass per unit area of the sample. Similarly, W_i can be obtained by measuring above and below the absorption limit of element i.

1.5.3 Near Limit Structure

In the absorption limit region shown in Fig. 1.6b, that is, in the 30–60 eV energy region above the threshold (absorption limit), the absorption spectrum shows strong absorption characteristics and structure, which is called X-ray absorption near-limit structure (XANES—X-ray Absorption Near Edge Structure). It is caused by the multiple scattering of excited photoelectrons in the surrounding environment. By analyzing this spectral structure, we can not only obtain the information of the atomic geometric configuration around the local cluster (Cluster) around the absorbing atom, but also reflect the electronic state structure of the lower position above the Fermi level.

The theoretical methods to explain the characteristics of XANES spectrum are as follows:

(1) Multiple scattering theory.
(2) Independent particle model or single-electron approximation, in which Hartree–Fock approximation has been successfully applied to the calculation of electronic structure for many years; density functional theory puts forward another effective single-electron theory, which can draw more accurate total energy and charge density distribution of ground state.

The above two theories and methods are not introduced here because of their complexity.

1.5.4 Study on Fine Structure and Local Structure of Extended X-Ray Absorption

The extended X-ray absorption fine structure (Extended X-ray Absorption Fine structure-EXAFS) oscillates with the increase of incident X-ray photon energy in the 30–1000 eV range at the high-energy side of the X-ray absorption limit. The amplitude of this oscillation is very small and is generally only a few percent of the absorption coefficient at the absorption limit. This phenomenon was discovered in the 1930s, but it was not until the 1970s, when Cheng et al. [13] made the Fourier transformation of the EXAFS function to connect it with the structure of matter (atomic arrangement), that the EXAFS spectral method became a new and important method for structural analysis, and there have been some books.

According to EXAFS theory, when an electron is bound to the inner layer of an atom by a nucleus, when it absorbs X-ray photons and is excited, the energy of X-ray photons absorbed depends on the initial and final states of the electrons. The initial state is determined by the internal initial state energy level close to the nucleus, which corresponds to the experimental absorption limit. As for the final state, it can be explained by the point of view of fluctuation. The excited photoelectron can be regarded as an outward propagating electron wave, which is backscattered by the neighboring atoms around the absorbing atom, which is an inward propagating wave. The final state is the result of the interference and superposition of external and inward electron waves. The difference of the final state is reflected in the fluctuation of the absorption coefficient with the incident X-ray photon energy E, which is called EXAFS. Because: (1) the frequency of the EXAFS peak is related to the coordination distance of the absorbing atom, (2) the coordination distance near the absorbing atom can be measured from the phase shift, and (3) the amplitude of the EXAFS peak depends on the coordination number of the nearest neighbor, the scattering ability of the nearest neighbor atom and the form and distance of the bonding force. Therefore, the EXAFS spectrum contains rich structural information.

At present, the experimental technology of synchrotron radiation for EXAFS and XANES research, data processing, and interpretation are relatively mature and have been widely used in the study of various materials, such as the local structures of metals and their alloys, silicates, amorphous materials, destroyers, semiconductors, and superconductors.

References

1. Lou Yu, Wan. 2007. XRD study on the microstructure of active material for positive and negative electrodes of metal hydride-nickel battery. Doctoral paper of Shanghai Institute of Microsystems and Information Technology.
2. Li, Jia. 2010. Study on storage performance of lithium-ion battery. Doctoral paper of Shanghai Institute of Microsystems and Information Technology.
3. Yang, C.Z., D.C. Xie, G.Z. Chen, and F.M. Zhong. 1988. Phase diffraction analysis. Metallurgical Industry Press, Beijing, 201–265.
4. Zhang, J.Z., and C.Z. Yang. 1992. The basis of X-ray diffraction of crystals. Nanjing University Press, 1–9.
5. Yang, C.Z., J. Zhang, and G.F. Cheng. 2009. Study on the depth effect of sample filling in Laue transmission geometry of powder X-ray diffractometer. Paper collection of essays on Chinese users of Japanese-Confucianism, 11–14.
6. Cheng, G.F., C.Z. Yang, and J. Zhang. 2009. Study on the influence of filling depth on the test results of polycrystalline powder X-ray diffractometer. *Journal of Analysis and Testing* 28 (3): 342–344.
7. Briggs, D. 1984. Handbook of X-ray and Ultraviolet Photoelectron Spectroscopy Heyden & Son Ltd, 1977, 1984.
8. Ma, Li-Dun, and F.J. Yang (eds.). 2007. Introduction to the application of synchrotron radiation. Shanghai Fudan University Press.
9. Cheng, G.F., C.Z. Yang, and Y.H. Huang. 2019. X-ray analysis of nanomaterials. Chemical Industry Press, Beijing, 2010.6.
10. Gu, B.Y., and K.Q. Lu. 1991. Near-edge structure theory of X-ray absorption. *Progress in Physics* 11 (1): 106–125.
11. Wang, Q.W., and W.H. Liu. 1994. X-ray absorption fine structure and its application. Science Publication Beijing.
12. Jiang, C.H., and C.Z. Yang. 2010. Ray diffraction-scattering analysis of materials. Higher Education Press Beijing.
13. Cheng, G.F., Y.H. Huang, and C.Z. Yang. 2009. Technical basis of synchrotron radiation X-ray application. Shanghai Science and Technology Press.

Chapter 2
X-Ray Diffraction Analysis Methods of Material Characterization and Mechanism Research

2.1 Qualitative Analysis of Phase [1–3]

Phase is a solid substance that has a certain crystal structure and can be characterized by a chemical formula (or has a certain range of composition). For example, it is also iron, which can exist in three phases: α-Fe of body-centered cubic structure, γ-Fe of face-centered cubic structure and high temperature, δ-Fe of body-centered cubic structure. After heat treatment, the structure of polymer materials and other condensed solids often includes crystalline phase, transition phase, metastable phase, and amorphous phase.

2.1.1 The Principle and Method of Phase Qualitative Analysis

Any crystalline substance, whether it is single crystal or polycrystal, has a specific crystal structure type, cell size, the number of atoms, ions or molecules in the cell, and their location, so specific polycrystal X-ray diffraction patterns can be given. In fact, no two different crystalline substances can give exactly the same diffraction pattern, just as it is impossible to find two people with identical fingerprints. On the other hand, the diffraction pattern of the unknown mixture is the sum of the diffraction patterns of each phase in the mixture, and the d value and relative intensity (I/I_1) of the diffraction lines of each phase remain unchanged, which is the basis of phase qualitative analysis (phase identification) by various diffraction methods.

C. Yang et al., *Materials and Working Mechanisms of Secondary Batteries*,
https://doi.org/10.1007/978-981-19-5955-4_2

The basic method of qualitative analysis is to compare the diffraction patterns of unknown phases with those of known substances. This method was first created by Hanawalt and its collaborators. At first, they collected diffraction data of more than 1000 compounds as a basic reference. Later, the American Society for material testing and the Society for X-ray and Electron diffraction published the first set of diffraction data cards in 1942, which was later updated year by year. By 1963, a total of thirteen groups were published, and then, a group was published each year, which was divided into organic and inorganic parts, called ASTM cards. In 1969, the Joint Committee on Powder diffraction Standards (JCPDS), an international organization, was established to edit and publish powder diffraction card sets, referred to as PDF card sets, with the cooperation of relevant national organizations. In 1998, the ICDD card group was collected, edited, and published by the International Center Diffraction Data (ICDD), and the database was established by Window. Figure 2.1 shows the Fe_3O_4 standard powder diffraction card.

PDF# 19-0269：QM=Common（=）；的 Diffractometer；I=（Unknown） PDF Card

Magnetite，sys
Fe_3O_4 ($Fe^{+2}2Fe^{3+}4O^{-2}$)

Radiation=CuK。1 Lambda=1.5406 Filter=
Calibration= 2Theta=19.269～144.848 I/Ic(RIR)=4.9
Ref: Level-1PDF

Cubic，Fd-3m（227） Z=8 mp=
Cell：8.396×8.396×8.396<90.0×90.0×90.0> P.S=
Density(c)=5.173 Density(m)= Mwt= Vol=591.9
Ref: lbid

Strong Lines：2.53/X1.48/4 2.97/3 1.62/3 2.10/2 1.09/1 1.71/1 1.28/1

26 Lines，Wavelength to Compute Theta=1。54056（Cu）I%=（Unknown）

	d	I(f)	hkl	2theta	theta	I/(2d)		d	I(f)	hkl	2theta	theta	I(/2d)
1	4.5820	8.0	111	19.269	9.135	0.1031	14	1.1221	4.0	642	86.702	43.351	0.4456
2	2.9670	30.0	220	30.095	15.047	0.1685	15	1.0930	12.0	731	89.617	44.981	0.4575
3	2.5320	100.0	311	35.422	17.711	0.1975	16	1.0496	6.0	800	94.425	47.213	0.4764
4	2.4243	8.0	222	37.052	19.526	0.2062	17	0.9696	2.0	660	102.224	51.112	0.5053
5	2.0993	20.0	400	43.052	21.526	0.2382	18	0.9695	6.0	751	105.218	52.609	0.5157
6	1.7146	10.0	422	53.391	26.695	0.2916	19	0.9632	4.0	662	106.205	53.102	0.5191
7	1.6158	30.0	511	56.942	28.471	0.3094	20	0.9388	4.0	840	11.269	55.102	0.5191
8	1.4845	40.0	440	62.515	31.257	0.3368	21	0.8952	2.0	664	119.736	59.368	0.5585
9	1.4192	2.0	531	65.743	32.871	0.3523	22	0.8802	6.0	931	122.118	61.059	0.5681
10	1.3277	4.0	620	70.924	35.462	0.3766	23	0.8569	8.0	844	128.032	64.016	0.5835
11	0.2807	10.0	533	73.948	36.974	0.3904	24	0.8233	4.0	1020	138.032	69.325	0.6073
12	1.2659	4.0	622	74.960	37.480	0.3950	25	0.8117	6.0	951	143.235	71.617	0.6160
13	1.2119	2.0	444	78.929	39.464	0.4126	26	0.8080	4.0	1022	144.848	72.424	0.6188

Fig. 2.1 There is a standard powder diffraction card (i.e., PDF-4) of Magnetite (Fe_3O_4) in PC hard disk

2.1.2 Application of Jade Qualitative Phase Analysis System [3, 4]

The so-called semi-automatic or fully automatic retrieval is a qualitative analysis system using PDF database and related programs (such as Jade), which is introduced as follows:

(1) Double-click MDI Jade6.5 with the left button, see Fig. 2.2, enter the Jade6.5 program, and the figure in Fig. 2.3 appears.

Click the left button to find the folder and File Name you want to analyze. For example, click Hung TS on disk D to show Fig. 2.4.

Left-click 1221-0.raw (or click 1221-0.raw + Read) to get the diffraction pattern you want to analyze, see Fig. 2.5.

Right BG left-click Apply and Strip K-alpha2 to remove K-alpha2 components or left-click Apply and Remove to remove both background and K-alpha2 components. Right-click S/M or , the figure in Fig. 2.6 appears.

(2) Left-click General → Reset, return to the unset state, set Subfile to Search and Search/Match Filters, see Fig. 2.6, that is, the periodic table appears, see Fig. 2.7.

Select the elements in the sample and divide them into necessary (O) and possible (La and H) cases; left-click OK. Figure 2.8 appears. Then, left-click OK, and then, Fig. 2.9 appears.

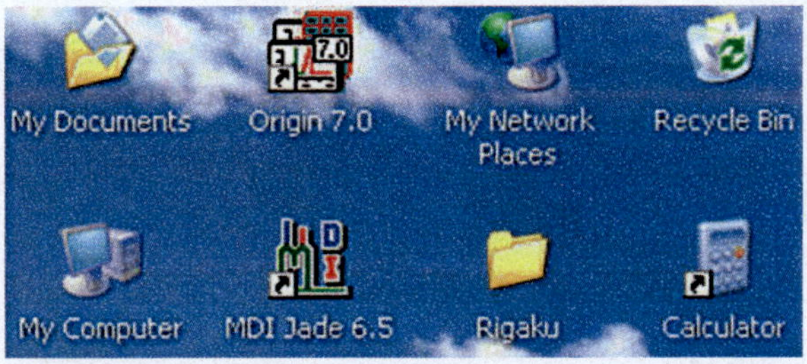

Fig. 2.2 Computer desktop

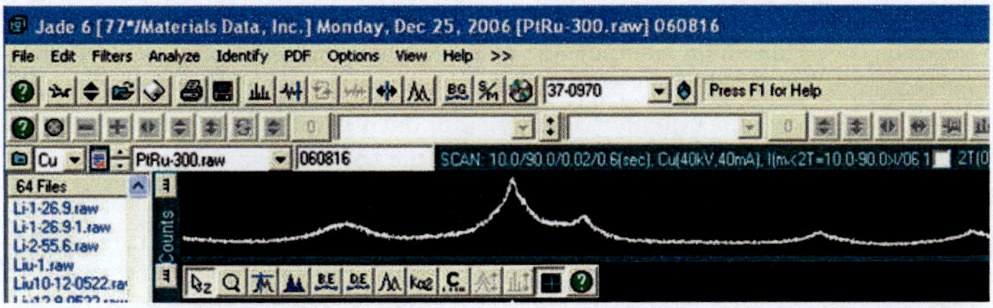

Fig. 2.3 Jade program interface

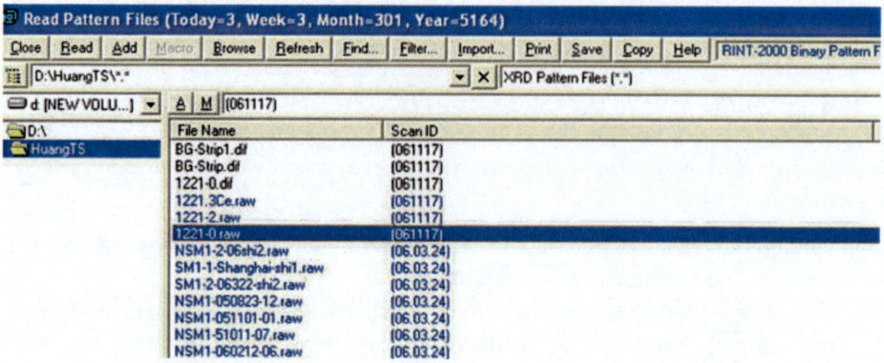

Fig. 2.4 Open a file

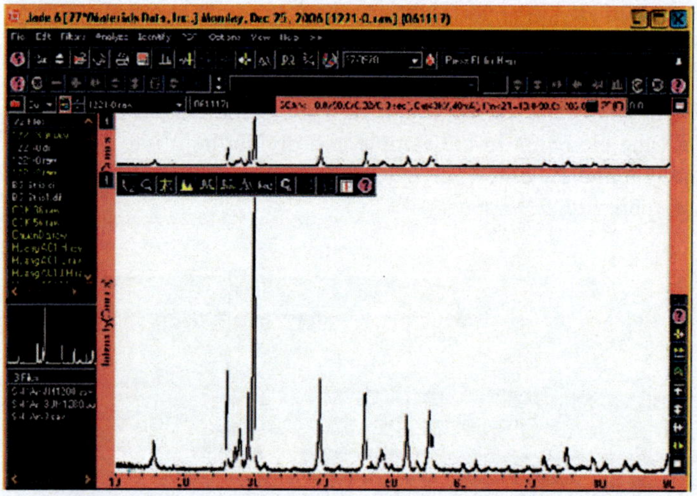

Fig. 2.5 Diffraction pattern

Then, left-click No or Yes, (No means no background, while Yes deducts the background), the English names, chemical formulas, PDF card numbers, space groups, and lattice parameters of many possible phases appear, as shown in Fig. 2.10.

Left-click to select (item by item or skip) the phase that may match with the unknown pattern and click on the left side of the better phase, that is, check, and then right-click  to eliminate the unselected phase, leaving only La_2O_3 and $La(OH)_3$ phases, as shown in Fig. 2.11. In this way, **retrieval/matching** can be carried out.

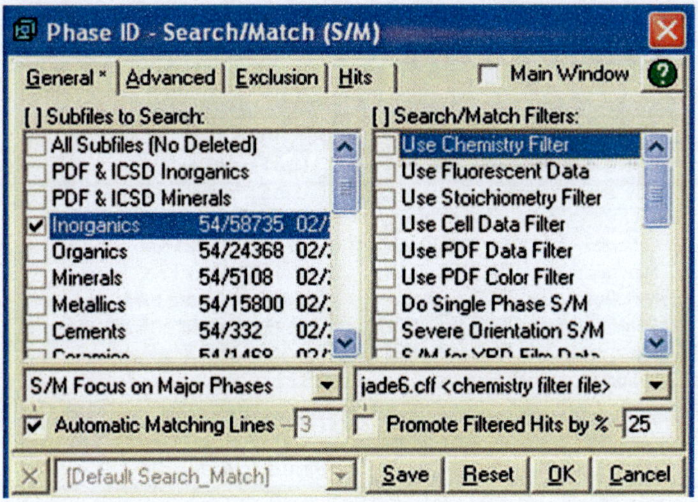

Fig. 2.6 Phase ID dialog box

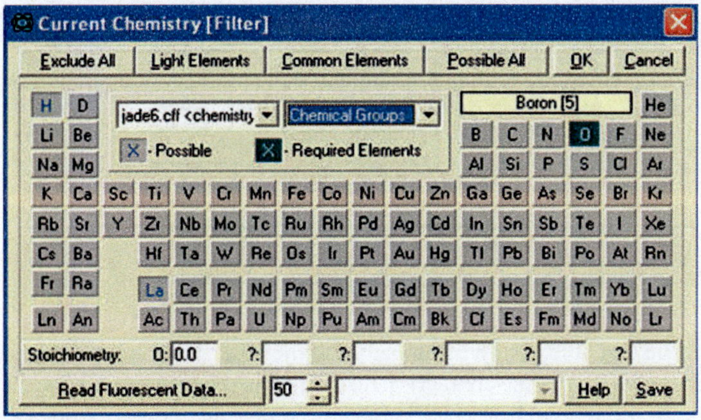

Fig. 2.7 Periodic table of elements

2.2 Quantitative Analysis of Phase [1, 3]

2.2.1 The Principle and Intensity Formula of Phase Quantitative Analysis

As mentioned in Sect. 2.1.1, the polycrystalline X-ray diffraction pattern of the unknown mixture is sum of the diffraction pattern of each phase in the mixture.

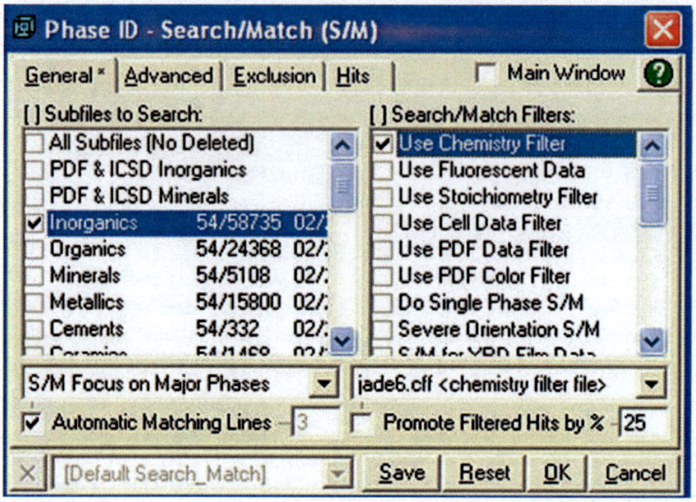

Fig. 2.8 "Phase ID" dialog box

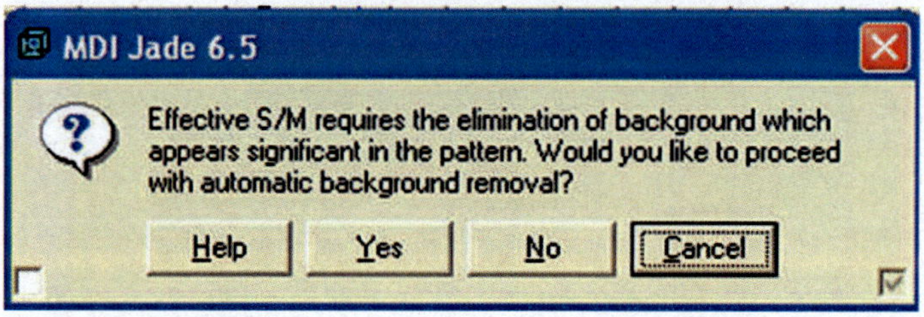

Fig. 2.9 Confirm the interface

The *d* value and the relative intensity of the diffraction lines of each phase remain unchanged, that is, the characteristic diffraction pattern of each phase remains unchanged, but the relative intensity between the phases in the mixture varies with the percentage content of each phase in the mixture. Therefore, we can determine the percentage content of each phase in the mixture by measuring and analyzing the relative intensity of each phase. In order to solve this problem, we must first know the relationship between the intensity and percentage content of each phase. X-ray diffractometer is mainly used for quantitative phase analysis.

The diffraction intensity of polycrystalline samples can only be dealt with by kinematic diffraction theory. Generally, starting from the scattering intensity of a free electron to X-ray, the scattering intensity of a multi-electron atom to X-ray is discussed, and then, the X-ray scattering intensity of a unit cell and a small crystal is studied. Finally, the expression of diffraction integral intensity of polycrystalline sample is derived. Here, we only start with the expression of the integral intensity of a certain diffraction line of a single-phase sample.

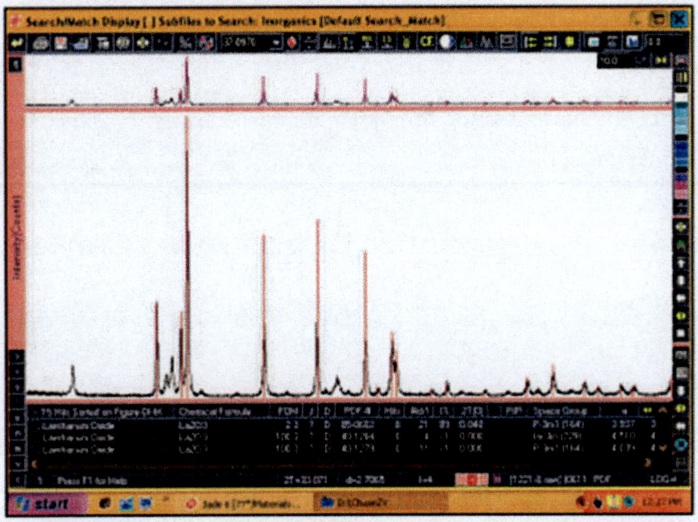

Fig. 2.10 Patterns and parameters of possible phases

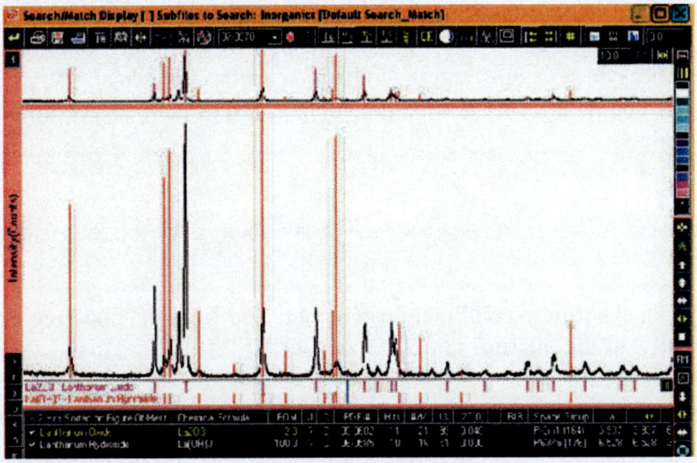

Fig. 2.11 Patterns and parameters of La_2O_3 and $La(OH)_3$ two phase

In the experimental work with X-ray diffractometer, if the sample is multiphase, the integral intensity I_{hkl} of a *hkl* diffraction line of phase *i* is affected by the absorption of the whole mixture, and the diffraction volume V_i of this phase is a part of the total diffraction volume V_{total}. If the linear absorption coefficient of the mixture sample is $\overline{\mu}_l$ and the volume fraction of phase *i* is f_i, then the diffraction intensity I_i of a *hkl* of phase *I* (omitting the subscript *hkl*) is

$$I_i = \frac{RK_i\overline{V}}{2\overline{\mu}_l} \cdot f_i = \frac{RK_i\overline{V}}{2\overline{\mu}_m\overline{\rho}} \cdot f_i \tag{2.1}$$

If the density and mass fraction of phase i are, respectively, ρ_i and x_i, $x_i = \frac{W_i}{W} = \frac{\overline{V} f_i \rho_i}{\overline{V}\overline{\rho}} = \frac{f_i \rho_i}{\overline{\rho}}$, insert (2.1)

$$I_i = \frac{RK_i\overline{V}\overline{\rho}}{2\overline{\mu}_l \rho_i} x_i = \frac{RK_i\overline{V}}{2\overline{\mu}_m \rho_i} x_i \tag{2.2}$$

here

$$R = \left(\frac{I_0}{32\pi r} \cdot \frac{e^4\lambda^3}{m^2c^4}\right) \tag{2.3}$$

$$K_{hkl} = \left(N^2 \cdot P_{hkl} \cdot F_{hkl}^2 \cdot \frac{1 + \cos^2 2\theta_{hkl}}{\sin^2\theta_{hkl} \cdot \cos\theta_{hkl}} \cdot \mathrm{e}^{-2M}\right) \tag{2.4}$$

R and K_{hkl} are called physical-instrumental constants and phase-experimental parameters, respectively. For the density of mixed samples, Eqs. (2.1) and (2.2) are the diffraction intensity formulas directly related to the content of i phase (volume fraction f_i, mass fraction x_i). They are the starting point of quantitative phase analysis.

If the element component P and its content ω_p of the mixed sample are known, the absorption coefficient of the mixed sample can be obtained according to the following formula.

$$\overline{\mu}_l = \overline{\rho\mu}_m = \overline{\rho}\sum_{p=1}^{m} \omega_p \mu_{mp} \tag{2.5}$$

Similarly, the absorption coefficient can be obtained from the phase component i and its content x_i of the mixture sample.

$$\overline{\mu}_l = \overline{\rho\mu}_m = \overline{\rho}\sum_{i=1}^{m} x_i \mu_{mi} \tag{2.6}$$

The mass absorption coefficient μ_{mi} of i phase is obtained. If the chemical composition or chemical formula of this phase is known, the μ_{mi} can be calculated according to Formula (2.6). The volume fraction is:

$$f_i = \frac{V_i}{\overline{V}} = V_i / \left(\sum_{i=1}^{n} V_i\right) \tag{2.7}$$

It is worth noting that at first glance, Eqs. (2.1) and (2.2) indicate that the diffraction intensity is linearly related to the content (f_i or x_i) of the phase. But in practice, this is

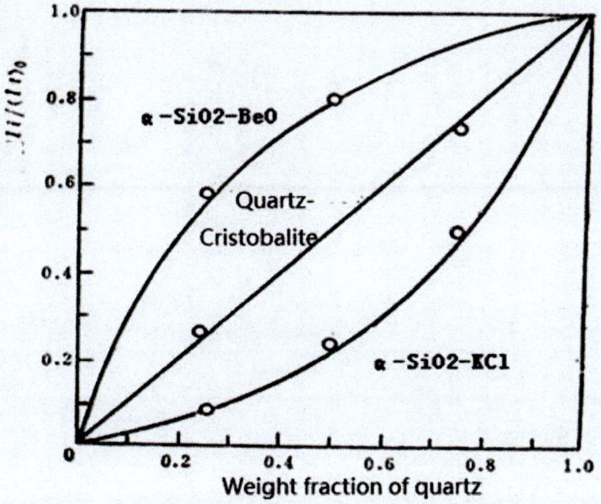

Fig. 2.12 Quantitative analysis curve of quartz

often not necessarily the case (see Fig. 2.12). This is because the diffraction intensity is also related to the total diffraction volume and the absorption coefficient of the sample, while the diffraction volume and absorption coefficient (or) are related to the content of the phase. This can be seen in Fig. 2.12.

The quartz-cristobalite line is a straight line because both of them are isomers of SiO_2, and the diffraction volume and mass absorption coefficient of the mixed samples do not change with their relative contents. The other two are nonlinear because the diffraction volume and absorption coefficient vary with the relative content of the two phases.

2.2.2 Standard Sample Methods for Quantitative Phase Analysis and Their Comparison

In practical work, K_i is difficult to calculate theoretically in many cases, so many X-ray analysts use different experimental techniques and data processing methods, either to avoid complicated calculation or to simplify the calculation, so there are a variety of quantitative phase analysis methods. The methods can be divided into two categories: standard sample method and non-standard sample method.

1. **Comparison of the characteristics for standard sample method**
   Table 2.1 gives the working equations and key points of various standard sampling methods in a comparative way.
2. **Experimental comparison of standard sample method** [5]
   Table 2.2 shows the results of a group of samples determined by different standard sample methods. Comparing the contents and data of these standard sample methods, the following conclusions can be obtained:

Table 2.1 Comparison of standard methods for quantitative phase analysis by X-ray diffraction

Methods		Working equation	Symbol description	Method description
Internal standard method	Alexander method	$\frac{I_i'}{I_s'} = \left(\frac{K_i}{K_s} \cdot \frac{\rho_s}{\rho_i} \cdot \frac{1-x_s}{x_s}\right)x_i$	I_i', I_s' are diffraction intensity of the *i* and *s* phase, respectively, after added standard samples *s*	Add a group of x_i with the same x_s to the pure phase *s* which does not exist in the sample. In the same reference sample, make. The working curve of $I_i'/I_s' \sim x_i$
	Z–Y method	$\sum_{i=1,\neq s}^{n}\left(\frac{I_{sK}}{I_{iK}} \cdot \frac{I_{iJ}}{I_{sJ}} \cdot x_{iK}\right) = 1$ $\sum_{i=1}^{n} x_{iK} = 1$	The subscript lowercase and uppercase letters represent the phase number and sample number, respectively. The strength is after adding the intensity sample	There are *n* samples to be tested, and there are *n* phases in each sample. The pure phase *s*, which does not exist in the sample, is added to each sample with the same x_s
Increment method	Copeland method	$\frac{I_i'}{I_j'} = \frac{K_i}{K_j} \cdot \frac{\rho_j}{\rho_i} \cdot \frac{x_i+x_{is}}{x_i}$	x_{is} is fraction on the original sample as 1	The phase *i* to be tested is added to the sample to be tested, and each phase makes a series of increments. For $\left(I_i'/I_j'\right) \sim x_{is}$ the drawing, the intercept between the phase *I* and the x_{is} axis is x_i

(continued)

Table 2.1 (continued)

Methods		Working equation	Symbol description	Method description
	Bezjak method	$x_i = \frac{x_{is}}{\left[\frac{I_i'}{I_j'} / \frac{I_i}{I_j}\right] - 1}$		Make an incremental sample and measure the original sample and incremental sample, respectively I_i/I_j and $I_i^{'}/I_j^{'}$
	Popovic method	$x_i = \frac{x_{is}\ R_{ji}}{P(1 - R_{ji})}$ $P = 1 - \sum x_{is}$, $R_{ji} = \frac{I_j^{'}}{I_i^{'}} \cdot \frac{I_i}{I_j}$		The original samples and incremental samples were determined by adding $(n-1)$ pure phase together
External standard method	Lerocux method	$x_i = \frac{I_i}{(I_i)_0} \cdot \left(\frac{\overline{\mu}_l}{\mu_{li}}\right)^2 \cdot \frac{\rho_i}{\rho}$	$(I_i)_0$ and I_0 intensity of pure *i* phase and *i* phase in the sample, respectively	Pure phase as external standard
	Karlak–Burnett method	Solve the problem of overlapping lines The working equation is complex		Need to test n pure phases, 1:1:1 …. 1 the mixed sample and the original sample have a total of 2 samples

(continued)

Table 2.1 (continued)

Methods		Working equation	Symbol description	Method description
	Simplified external standard method	$\frac{I_i}{I_m} = \left(\frac{I_i}{I_m}\right)_{1:1} \frac{x_i}{x_m}$ $\sum_{i=1}^{n} x_i = 1$	$\frac{I_i}{I_m}$ is the intensity ratio of i and m phases in the original sample	The external standard sample is 1:1:1:… Made of n pure phases of 1. Test the original sample and external standard sample, respectively
Scavenging method of matrix absorption effect	Scavenging agent (K value) meth od	$x_i = \left(\frac{K_f}{K_i} \cdot \frac{\rho_i}{\rho_f}\right) \cdot \frac{I_i}{I_f} \cdot \frac{x_f}{1 - x_i}$ $= \left(\frac{I_f}{I_i}\right)_{1:1} \frac{I_i}{I_f} \cdot \frac{x_f}{1 - x_i}$	x_f is to add the sample to be tested and eliminate it The mass fraction of the agent; $\left(\frac{I_f}{I_i}\right)_{1:1}$ is intensity ratio of reference sample of f phase and i phase two according to 1:1 ratio	The pure phase which does not exist in the sample to be tested is added to the sample to be tested, and the phase to be tested and the eliminator are prepared into a reference sample at 1:1

(continued)

Table 2.1 (continued)

Methods		Working equation	Symbol description	Method description
	The method of using incremental phase as scavenging agent	$x_f = \frac{x_f^a}{\left[\left(I_f^a/I_i^a\right)/(f_f/I_i)\right]-1}$ $x_i = \frac{1-x_f}{\sum_{i=1,i\neq f}^{n}\left[(K_i/K_f)/\left(I_f^a/I_i^a\right)\right]}$ $x_i = \frac{K_f}{K_i}\cdot\frac{I_i^a}{I_f^a}\left(x_f^a+x_f\right)$	x_f^a is mass fraction added to the incremental phase as a scavenger I_f^a/I_i^a is the intensity ratio of the f phase to the I phase of the sample after increment I_f/I_i is the intensity ratio of the f phase to the I phase without increment	I_f^a, I_i^a is the intensity of i and f phases in the incremental sample. At 1:1:1… Calculation of K value ratio of external standard samples based on

Table 2.2 Results of a group of samples determined by different standard methods (%)

Determinant methods	Sample number to be tested	1			2			3		
	The phase in the sample to be measured	Cu	Ni	Si	Cu	Ni	Si	Cu	Ni	Si
	Proportioning ratio (%)	30	20	50	20	50	30	50	30	20
Popovic Increment method *c*	Increment phase and weight fraction	15	10		10		15		15	10
	Measurement result	31.6	20.9	47.5	19.9	50.8	30.3	55.9	27.9	19.2
Karlak external standard method		29.65	20.36	50.00	19.87	51.78	29.35	51.47	30.94	17.6
Sampling external standard method		29.65	20.37	50.00	19.89	51.76	29.37	51.47	30.93	17.6
Scavenging method of matrix absorption	$x_f = 25\%$ SiO_2	25.9	17.1	45.5	20.0	50.4	35.9	47.2	25.4	17.6
	$x_f = 25\%$ W	15.3	15.8	50.9	8.4	39.9	27.9	25.0	25.9	15.6

(1) When calculating the weight fraction of a phase in the multiphase sample, the increment method of Bezjak is the simplest, and the matrix elimination method is the second. The former only needs to measure two samples of the original sample and the new sample after increment, while the latter also measures two samples, but both samples need to be made, and the phase which does not exist in the original sample should be used as the elimination agent.
(2) When the weight fraction of each phase in the multiphase system is solved, and the selected diffraction lines do not overlap, the simplified external standard method is the simplest, the Popovic method is the second, and work amount of the matrix elimination method is the largest. The first two methods only need to measure two samples, while the last method needs to measure $(n + 1)$ sample, but the sample can contain amorphous phase, and the total amount of amorphous phase can be measured.
(3) When there is an overlap of diffraction lines in the composite phase, the external standard method of Karlak is generally adopted, and the work amount of experiment and calculation is large.
(4) From the measurement results of the two kinds of scavengers in Table 2.2, we can see that if the scavengers are not selected properly, the results are not credible. It is generally believed that the absorption coefficient, density, particle size, and X_f of the scavenger have an important influence on the measurement results. However, the method of using the incremental phase as the eliminator is especially suitable for the determination of trace phase. Of course, attention should be paid to the selection of the incremental phase as the eliminator in practical work.

2.2.3 Non-standard Method for Quantitative Phase Analysis and Its Comparison

1. **Non-standard sample method**
   Table 2.3 gives the working equations and key points of various non-standard methods in a comparative way.
2. **Experimental comparison of non-standard sample method [6]**
   When the K value is obtained by calculation or experiment, the simplified external standard method only needs to measure the intensity of the sample to obtain the weight fraction of each phase in the sample to be tested, so in this case, the simplified external standard method is also a non-standard method. The results of three non-standard sample methods and simplified external standard methods are listed in Table 2.4. By carefully comparing the items and data of this table, the following conclusions can be obtained:

   (1) The direct comparison method is the most convenient, a sample can give the result, but the K value needs theoretical calculation. The structure factor

Table 2.3 Comparison of non-standard methods for X-ray diffraction quantitative phase analysis

Methods	Working equation	Symbol description	Explain the main points of the method
Direct comparison method	$\begin{cases} \frac{I_m}{I_i} = \frac{K_m}{K_i} \cdot \frac{f_m}{f_i} \\ \sum f_i = 1 \end{cases}$ $\begin{cases} \frac{I_m}{I_i} = \frac{K_m}{K_i} \cdot \frac{\rho_i}{\rho_m} \cdot \frac{x_m}{x_i} \\ \sum f_i = 1 \end{cases}$	The value of K needs to be calculated theoretically $K = N^2 P F^2$ $\frac{1+\cos^2 2\theta}{\sin^2\theta\cos\theta} e^{-2M}$	Only one sample is needed to solve the problem In addition to the theoretical calculation of K value, we also need to know the density
Adiabatic method	$x_i = \left[\frac{K_i}{I_i\rho_i}\sum_{i=1}^{n}\frac{I_i\rho_i}{K_i}\right]^{-1}$	(1) The value of K needs to be calculated theoretically (2) $K_i\rho_i^{-1}$ obtained by experiment	K_i or $K_i\rho_i^{-1}$ can also be obtained from theoretical calculation and experiment
Zevin method: (1) The mass absorption coefficient of each phase is known	$\begin{cases} \sum_i^n \left[1 - \frac{I_{iJ}}{I_{iK}}\right] x_{iK}\mu_{im} = 0 \\ \sum_i^n x_{iK} = 1 \end{cases}$	The subscript lowercase letter is the phase number The subscript big letter is the sample number	All the n samples contain different contents of n A picture μ_{im} needs to be calculated according to the chemical composition of each phase

(continued)

Table 2.3 (continued)

Methods	Working equation	Symbol description	Explain the main points of the method
(2) The mass absorption coefficient of each sample is known	$\begin{cases} \sum_i^n \left[\frac{I_{iJ}}{I_{iK}} \cdot \frac{\overline{\mu}_{mJ}}{\overline{\mu}_{mK}}\right] x_{iK} = 1 \\ \sum_i^n x_{iK} = 1 \end{cases}$	The subscript lowercase letter is the phase number The subscript big letter is the sample number	μ_{mJ} needs to be calculated according to the chemical composition of each sample
(3) Absorption coefficient is unknown	$\begin{cases} \sum_i^n \left[\frac{I_{iJ}}{I_{iK}} \cdot \frac{I_{iJ} - I_{i(K+J)}}{I_{i(K+J)} - I_{iK}} \cdot x_{iK}\right] = 1 \\ \sum_i^n x_{iK} = 1 \end{cases}$	The subscript lowercase letter is the phase number. The subscript big letter is the sample number. The subscript $(K + J)$ indicates that samples K and J will be mixed at 1:1	All the n samples contain n phases with different contents
The improvement of Guo Chang-Lin et al.			N samples, the number of phases in each sample $\leq n$

can only be calculated by knowing the number and coordinate position of atoms in the single cell of the phase, and the temperature factor e^{-2M} can only be calculated by knowing the Debye temperature Θ, which is difficult to do in many cases, so it is often used in systems with simple structure, for example, the determination of α-phase and γ-phase in Fe-based or Fe–Ni-based alloys, and the determination of α-phase and β phase in titanium alloys.

(2) Zevin method is a good method without standard sample, which only involves the calculation of phase or sample mass absorption coefficient. The chemical composition of the channel phase or sample can be calculated by consulting the absorption coefficient. Obviously, this is not difficult to do, so it has a wide application prospect. However, it requires n samples to contain n phases with different weight fractions. This is different from the improved method proposed by Guo Chang-Lin, whose samples can be out of phase or multiphase. It is worth noting that, in principle, although the third method of Zevin is feasible, when the value of $\left[I_{iK} - I_{i(K+J)}\right]/\left[I_{i(K+J)} - I_{iJ}\right]$ is

Table 2.4 Results of a group of samples measured by different non-standard methods

Sample			1			2			3		
Phase			Cu	Ni	GaAs	Cu	Ni	GaAs	Cu	Ni	GaAs
Original ratio (%)			20.0	50.0	30.0	15.7	35.3	50.0	50.0	15.7	35.3
Determination results %	Direct comparison method		20.6	55.3	25.1	19.2	39.2	45.6	55.5	17.6	25.9
	Adiabatic method	Theoretical calculation of K value	20.6	55.3	25.1	19.2	39.2	45.6	55.5	17.6	25.9
		Experimental determination of K value	19.7	51.5	28.8	15.3	35.7	49.9	55.8	15.7	31.5
	Zevin method	Known $\overline{\mu}_{mi}$	25.8	45.4	29.9	19.3	29.9	50.8	57.8	15.6	29.6
		Known $\overline{\mu}_{mJ}$	25.2	45.7	30.1	19.8	30.0	51.1	55.8	15.0	30.2
	Simplified external standard method		19.7	51.5	28.8	15.3	35.7	49.9	55.8	15.7	31.5

within the range of statistical error of the integral strength measurement, $\left[I_{iK} - I_{i(K+J)}\right]/\left[I_{i(K+J)} - I_{iJ}\right] < 0$ there may be no solution.

The universal regression method and Lu Jin-Sheng's optimization calculation method are good improvements [3] and are expected to be widely used.

(3) Although the simplified external standard method belongs to the standard sample method, once the *K* value is measured by the experiment, it is a simple and easy non-standard sample. The weight fraction of each phase can also be obtained from the intensity measurement of a sample. When the *K* value is calculated theoretically, the adiabatic method is consistent with the direct method, and when the *K* value is obtained by experiments, the adiabatic method is consistent with the simplified external standard method. In the latter case, the simplified external standard method is actually a non-standard method, and after only the weight fraction of one phase is required, the other phases are related to the magnification of the phase, so the calculation is simple.

(4) It can be seen from Table 2.4 that the accuracy of the experimental determination is the highest by the simplified external standard method and the adiabatic method measured by the experiment *K* value. The Zevin method takes the second place, and the adiabatic method calculated by direct comparison method and *K* value theory is the worst. It can be seen that the accuracy of the method obtained from the experiment is high, which involves the use of external standard samples; the accuracy of the theoretical method of calculating constants is poor, and the more theoretical data (i.e., the data given in relevant books) are used in the calculation, the greater the error is, so the Zevin method is the best in the completely non-standard method, that is, there is no standard sample, and the theoretical data used are the least.

2.3 Accurate Determination of Lattice Parameters of Polycrystalline Samples

The determination of lattice parameters is always carried out on the samples with known crystal structure, so the correct indexing of each diffraction line in the diffraction pattern of the sample and the accurate diffraction angle data are the key. With these two sets of data, the lattice parameters can be obtained.

The active materials, crystal structure, and space group of Ni/MH and Li-ion batteries are listed in Table 2.5. It can be seen that there are not many varieties involved, and the crystal structure is limited to four crystal systems: cubic, hexagonal, rhombohedral, and orthorhombic, and the diffraction patterns of rhombohedral structure are often indexed by hexagonal structure. The following is a brief introduction to the method of accurate determination of lattice parameters.

Table 2.5 Electrode active materials and crystal structure of active materials for Ni/MH and lithium-ion batteries

Active material for Ni/MH battery				Active materials for lithium-ion battery			
Phase	Crystal structure	Card number of PDF	Space group	Phase	Structure	Card number of PDF	Space group
β-$Ni(OH)_2$	Hexagonal	03-0177	P-3m1	$LiNiO_2$	Rhombohedral	09-0063	R-3m
β-NiOOH	Hexagonal	06-0141		$Li(Ni,Co,Mn)O_2$	Rhombohedral	09-0063	R-3m
γ-NiOOH	Hexagonal	06-0075		$LiFePO_4$	Orthogonal	40-1499	Pnma
$LaNi_5$	Hexagonal	42-1191	P6/mmm	$FePO_4$	Orthogonal	30-0659	Cmcm
				2H-graphite	Hexagonal	41-1487	P63/mmc
				3R-graphite	Rhombohedral	26-1079	3R
				$Li_4Ti_5O_{12}$	Cubic		Fd3m
Diaphragm material							
Polypropylene (PP)	Monoclinic		$P2_1/c$	Polyethylene (PE)	Orthogonal		Pnam

2.3.1 *The Method of Solving Lattice Parameters by Jade Program Based on Simultaneous Equation Method*

According to the Bragg formula, the crystal plane spacing can be calculated and the lattice parameters can be calculated. For the cubic system,

$$a = \sqrt{h^2 + k^2 + l^2} \times d \tag{2.8}$$

It can be seen that each line can calculate a, and theoretically, these a should be equal, but because there are always errors in the determination, the values of each a are different. For non-cubic systems, it is necessary to solve the simultaneous equations. The following is the most general case, that is, for the Triclinic system:

$$\frac{1}{d_i^2} = h_i^2 A + k_i^2 B + l_i^2 C + h_i k_i D + k_i l_i E + l_i h_i F \tag{2.9}$$

There are six unknowns, and at least six diffraction lines are needed to form a set of simultaneous equations to obtain A, B, C, D, E, and F; thus, the reciprocal lattice parameters a^*, b^*, c^*, α^*, β^*, and γ^* are obtained. Then, the crystal lattice parameters a, b, c, α, β, and γ are obtained according to the relationship between crystal lattice and reciprocal lattice parameters. In order to obtain more accurate lattice parameters, for the pattern with n diffraction strips, $N =$ equations can be formed, N groups of lattice parameters can be obtained, and the final spherical average value or further processing, where m represents the number of lattice parameters. For Triclinic system, mole 6. As for other crystal systems, similar treatment can be made, just pay attention to:

Cubic system	$m = 1$	$A = B = C$	$D = E = F = 0$
Tetragonal system	$m = 2$	$A = B$	$D = E = F = 0$
Hexagonal	$m = 2$	$A = B$,	$D = A/2$, $E = F = 0$
Orthorhombic	$m = 3$	A, B, C	$D = E = F = 0$
Monoclinic	$m = 4$	A, B, C, D	$E = F = 0$
Triclinic system	$m = 6$	A, B, C, D, E, F	

The simultaneous equation can be used to solve the lattice parameters in the Jade program. The copying steps are as follows: selecting the data pattern of analysis, searching for peak, report, print, exit, option, Lattice calculate, determining the correct crystal system, carefully checking the *hkl* of the five lines listed must be completely correct, obtaining the lattice parameters and average lattice parameters of the corresponding lines, exit, Option, Cell refine, Calculate, and finally obtaining more accurate lattice parameters.

2.3.2 *Least Square Method for Accurate Determination of Lattice Parameters*

According to the principle of least square method, the best straight line is the one that minimizes the sum of squares of errors. Let us start with the general situation. It is known that the two physical quantities x and y have the following linear relationship.

$$Y = A + BX \tag{2.10}$$

The x is measured for n times, in order to obtain the intercept A and slope B of the best straight line. From the measured x value, the y value is calculated according to the Formula (2.10), and the error is

$$\Delta y_i = (A + Bx_i) - y_i \tag{2.11}$$

The sum of squares of the error is

$$\sum_i (\Delta y_i^2) = (A + Bx_i - y_i) + (A + Bx_i - y_i)^2 + \cdots \tag{2.12}$$

The necessary and sufficient condition for $\sum_i (\Delta y_i^2)$ is minimizing

$$\frac{\partial \sum_i (\Delta y_i^2)}{\partial A} = 0 \quad \frac{\partial \sum_i (\Delta y_i^2)}{\partial B} \tag{2.13}$$

So have

$$\begin{matrix} nA + B\sum x = \sum y \\ A\sum x + B\sum x^2 = \sum xy \end{matrix} \quad \text{Or} \quad \begin{pmatrix} n & \sum x \\ \sum x & \sum x^2 \end{pmatrix} \begin{pmatrix} A \\ B \end{pmatrix} = \begin{pmatrix} \sum y \\ \sum xy \end{pmatrix} \tag{2.14}$$

This is the canonical equation, and the maximum values of A and B can be obtained by solving the simultaneous equation. To sum up, the steps for writing the canonical equation are as follows: (1) replace the experimental values of x and y into (2.10), and there are n equations for n measurements; (2) multiply the coefficients of A by n equations and then sum them; and (3) multiply the coefficients of B by n equations and then sum them.

It is deduced that the cubic system is similar to the Formula (2.11).

$$\frac{\lambda^2}{2a^2}(h^2 + k^2 + l^2) + D\sin^2 2\theta = \sin^2\theta \tag{2.15}$$

Obtain the system of canonical equations

$$\begin{pmatrix} A\sum N^2 & D'\sum N\delta \\ A\sum N\delta & D'\sum \delta^2 \end{pmatrix}\begin{pmatrix} A \\ D' \end{pmatrix} = \begin{pmatrix} \sum N\sin^2\theta \\ \sum \delta\sin^2\theta \end{pmatrix} \tag{2.16}$$

here $A = \frac{\lambda^2}{4a^2}$, $N = h^2 + k^2 + l^2$, $\delta = 10\sin^2 2\theta$, $D' = D/10$

$$a = \frac{\lambda}{2\sqrt{A}} = \frac{\lambda}{2}\cdot\sqrt{\frac{\sum N^2\sum\delta^2 - \left(\sum N\delta\right)^2}{\sum\sin^2\theta\sum\delta^2 - \sum\delta\sin^2\theta\cdot\sum N\delta}} \tag{2.17}$$

The canonical equations of tetragonal and hexagonal systems are:

$$\begin{pmatrix} \sum N^2 & \sum NR & \sum N\delta \\ \sum NR & \sum R^2 & \sum R\delta \\ \sum N\delta & \sum R\delta & \sum\delta^2 \end{pmatrix}\begin{pmatrix} A \\ B \\ D' \end{pmatrix} = \begin{pmatrix} \sum N\sin^2\theta \\ \sum R\sin^2\theta \\ \sum\delta\sin^2\theta \end{pmatrix} \tag{2.18}$$

Although the canonical equations of tetragonal and hexagonal systems are the same, there are some differences in symbolic meaning.

Tetragonal system

$$N = h^2 + k^2,\ R = l^2,\ A = \frac{\lambda^2}{4a^2},\quad B = \frac{\lambda^2}{4c^2} \tag{2.19}$$

Hexagonal systems

$$N = h^2 + hk + k^2,\ R = l^2,\ A = \frac{\lambda^2}{3a^2},\quad B = \frac{\lambda^2}{4c^2} \tag{2.20}$$

For the rhombohedral system, it is generally indexed according to the hexagonal system, so the lattice constant is first calculated according to the hexagonal system, and then, the lattice constant a_R and α_R of the rhomboid cell is calculated according to the following formula.

$$a_R = \frac{1}{3}\sqrt{3a_H^2 + c_H},\quad \alpha_R = 2\sin^{-1}\frac{3}{2\sqrt{3 + (c_H/a_H)^2}} \tag{2.21}$$

The subscript H represents the hexagonal system.

The canonical equations of orthorhombic system are

$$\begin{pmatrix} \sum N^2 & \sum NR & \sum NS & \sum N\delta \\ \sum NR & \sum R^2 & \sum RS & \sum S\delta \\ \sum NS & \sum RS & \sum S^2 & \sum S\delta \\ \sum N\delta & \sum R\delta & \sum S\delta & \sum \delta^2 \end{pmatrix} \begin{pmatrix} A \\ B \\ C \\ D' \end{pmatrix} = \begin{pmatrix} \sum N \sin^2\theta \\ \sum R \sin^2\theta \\ \sum S \sin^2\theta \\ \sum \delta \sin^2\theta \end{pmatrix} \tag{2.22}$$

Among them

$$N = h^2, \quad R = k^2, \quad S = l^2, \quad A = \frac{\lambda^2}{4a^2}, \quad B = \frac{\lambda^2}{4b^2}, \quad C = \frac{\lambda^2}{4c^2}, \quad \delta = 10\sin^2 2\theta \tag{2.23}$$

2.4 Determination of Macroscopic Stress (Strain) [7]

It has been proved theoretically and experimentally that in the charge–discharge process of battery, the stress is caused by the de-intercalation of conductive ions in the positive and negative active materials, including the macroscopic stress of diffraction line displacement and the microscopic stress of line broadening. These two kinds of stresses need to be measured and analyzed by diffraction experimental method. This section introduces the measurement techniques of macroscopic stress, and the measurement of microscopic stress will be introduced in Sect. 2.5.

Macroscopic stress is a complex problem, which can be divided into three states: triaxial, biaxial (plane), and uniaxial according to the distribution state of stress, and the corresponding principal stress state. The measuring methods are also obviously different. For details, please refer to the relevant chapters of the book "Internal stress diffraction Analysis" [7] published by Beijing Science Press in 2013.

2.4.1 *Plane Stress State*

Because the stress state caused by atomic de-intercalation should belong to the plane stress state or the plane principal stress state, whether it is a wound cylindrical battery, a flat plate battery, or a button battery. When it is in the state of plane stress, the surface of the electrode is free, and the strain perpendicular to the surface is not zero. There are the following types:

$$\sigma^{2D} = \begin{pmatrix} \sigma_{11} & \sigma_{12} \\ \sigma_{21} & \sigma_{22} \end{pmatrix} \tag{2.24}$$

$$\varepsilon^{2D} = \begin{pmatrix} \varepsilon_{11} & \varepsilon_{12} & 0 \\ \varepsilon_{21} & \varepsilon_{22} & 0 \\ 0 & 0 & \varepsilon_{33} \end{pmatrix} \tag{2.25}$$

The relationship between ε and σ:

$$\begin{cases} \varepsilon_{11} = \dfrac{\sigma_{11} - \nu\sigma_{22}}{E} \\ \varepsilon_{22} = \dfrac{\sigma_{22} - \nu\sigma_{11}}{E} \\ \varepsilon_{33} = \dfrac{-\nu(\sigma_{11} + \sigma_{22})}{E} \end{cases} \tag{2.26}$$

$$\begin{cases} \sigma_{11} = \varepsilon_{11}E + \dfrac{\nu\ E(\nu\varepsilon_{11} + \varepsilon_{22})}{1 - \nu^2} \\ \sigma_{22} = \dfrac{E(\varepsilon_{11} + \varepsilon_{22})}{1 - \nu^2} \end{cases} \tag{2.27}$$

Among them, σ is stress, ε is strain, and the numbers of their subscript are the same as principal stress or principal strain, and the different numbers are shear stress or shear strain.

In the case of biaxial principal stress, there are:

$$\sigma^{2D} = \begin{pmatrix} \sigma_{11} & 0 \\ 0 & \sigma_{22} \end{pmatrix};\ \varepsilon^{2D} = \begin{pmatrix} \varepsilon_{11} & 0 & 0 \\ 0 & \varepsilon_{22} & 0 \\ 0 & 0 & \varepsilon_{33} \end{pmatrix} \tag{2.28}$$

It can be seen that it is much easier to solve the principal stress.

2.4.2 Method for Measuring Plane Stress

In the case of plane (two-dimensional) stress, the general formula of strain–stress relationship is

$$\varepsilon_{\phi,\psi}^{hkl} = \frac{1+\nu}{E}\left[\sigma_{11}\cos^2\phi + \sigma_{12}\sin(2\phi) + \sigma_{22}\sin^2\phi\right]\sin^2\psi - \frac{\nu}{E}(\sigma_{11} + \sigma_{22}) \tag{2.29}$$

Two-dimensional principal stress state

$$\varepsilon_{\phi,\psi}^{hkl} = \frac{1+\nu}{E}\left[\sigma_{11}\cos^2\phi + \sigma_{22}\sin^2\phi\right]\sin^2\psi - \frac{\nu}{E}(\sigma_{11} + \sigma_{22}) \tag{2.30}$$

The ψ angle is t the angle between o determine the stress direction and the surface normal Z, and ϕ to determine the intersection angle between the projection of the stress direction on the surface and the reference axis X.

The specific method of two-dimensional stress measurement is introduced below.

(1) **0°–45° method**

For the principal stress state, when $\psi = 0°$ and 45°, (2.14) becomes,

$$\varepsilon_{\phi,\,0}^{hkl} = \frac{d_{\phi\,0}^{hkl} - d_0^{\mathrm{hkl}}}{d_0^{hkl}} = -\frac{\nu}{E}(\sigma_{11} + \sigma_{22}) \tag{2.31}$$

$$\varepsilon_{\phi,\,45}^{hkl} = \frac{d_{\phi\,45}^{hkl} - d_0^{hkl}}{d_0^{hkl}} = \frac{1+\nu}{2E}\left[\sigma_{11}\cos^2\phi + \sigma_{22}\sin^2\phi\right] - \frac{\nu}{E}(\sigma_{11} + \sigma_{22}) \tag{2.32}$$

Because we know ϕ, we can be obtained simultaneously σ_{11} and σ_{22}.

(2) $\sin^2\psi$ **method**

In known ϕ cases, changing the ψ angle requires at least four points. Then draw the graph, and its slope M.

$$M = \frac{1+\nu}{E}\left[\sigma_{11}\cos^2\phi + \sigma_{22}\sin^2\phi\right] \tag{2.33}$$

The intercept I is

$$I = \frac{\nu}{E}(\sigma_{11} + \sigma_{22}) \tag{2.34}$$

Because we know ϕ, we can be obtained simultaneously σ_{11} and σ_{22}. Order

$$\sigma_\phi = \sigma_{11}\cos^2\phi + \sigma_{22}\sin^2\phi \tag{2.35}$$

If $\varepsilon_{0,\psi} = 0$, there is

$$\frac{1+\nu}{E}\left(\sigma_{11}\cos^2\phi + \sigma_{22}\sin^2\phi\right) = \frac{\nu}{E}(\sigma_{11} + \sigma_{22}) \tag{2.36}$$

Since ϕ is known and σ_{11} and σ_{22} obtained, the Besson ratio ν can be obtained. In the case of tensile stress, $\phi = 0, \sigma_{11} = \sigma, \sigma_{22} = 0$, there are

$$\varepsilon_{0\psi} = \frac{1+\nu}{E}\sigma \cdot \sin^2\psi - \frac{\nu}{E}\sigma \tag{2.37}$$

When $\varepsilon_{0,\psi} = 0$, there are

$$\begin{aligned} \sin^2\psi &= \frac{\nu}{1+\nu} \\ \frac{\partial}{\partial\sigma}\left[\frac{\partial\varepsilon_{\phi,\psi}}{\partial\sin^2\psi}\right] &= \frac{1+\nu}{E} \end{aligned} \tag{2.38}$$

Therefore, the elastic constant E and ν of the elastic constant can be obtained by the simultaneous Formula (2.38).

In some professional literature and some practical measurements, the Formula (2.30) and the Formula (2.29) are often written as follows:

$$\varepsilon_{\phi\psi}^{hkl} = \frac{1+\nu}{E}\sigma_\phi \sin^2\psi - \frac{\nu}{E}(\sigma_{11} + \sigma_{22}) \tag{2.39}$$

here

$$\sigma_\phi = \sigma_{11}\cos^2\phi + \sigma_{22}\sin^2\phi \text{ (two-dimensional principal stress state)} \tag{2.40}$$

$$\sigma_\phi = \sigma_{11}\cos^2\phi + \sigma_{12}\sin(2\phi) + \sigma_{22}\sin^2\psi \text{ (general two-dimensional stress state)} \tag{2.41}$$

Then, the determination of the above-mentioned 0°–45° method and the determination of the $\sin^2\psi$ method are more simple.

2.5 Diffraction Line Broadening Effect Caused by Microstructure [8, 9]

2.5.1 Broadening Effect of Microcrystals and Micro-stress

The grain size broadening effect of X-ray diffraction is expressed by Scherrer formula:

$$\beta_{hkl} = \frac{0.89\lambda}{D_{hkl}\cos\theta_{hkl}} \tag{2.42}$$

It is worth noting that D_{hkl} refers to the grain size along the normal direction of the *hkl* plane.

Broadening caused by micro-strain

$$\varepsilon_{\text{average}} = \frac{\beta_{hkl}}{4} ctg\theta_{hkl} \tag{2.43}$$

The β_{hkl} unit in the formula is radians. Equation (2.43) combines the average strain $(\varepsilon_{\text{average}})$ or stress $(\sigma_{\text{average}})$ with the FWHM β_{hkl} of the diffraction line.

2.5.2 Broadening Effect of Stacking Fault

The X-ray diffraction effect caused by stacking faults has different expressions for different structures.

For the dense hexagonal structure, there is no stacking fault effect when $h - k = 3n$ or *hko*; when $h - k = 3n \pm 1$, when l is even, the diffraction line is seriously broadened, and when l is odd, the diffraction line width is small. Its contribution to the half width of the diffraction line is

$$h - k = 3n \pm 1 \begin{cases} l = \text{even number} & \beta_f = \frac{2l}{\pi} tg\theta \left(\frac{d}{c}\right)^2 (3f_{\text{D}} + 3f_{\text{T}}) \\ l = \text{odd number} & \beta_f = \frac{2l}{\pi} tg\theta \left(\frac{d}{c}\right)^2 (3f_{\text{D}} + f_{\text{T}}) \end{cases} \tag{2.44}$$

β_f is measured in radians, d is the distance between crystal planes, and c is the lattice parameter of hexagonal c-axis. F_{D} and f_{T} are deformation stacking fault and twin stacking fault probability, respectively.

The opposite Face Centroid Cube (FCC) structure is as follows:

$$\beta_f = \frac{2}{\pi a} \sum \frac{|L_o|}{h_o(u+b)} tg\theta(1.5f_{\text{D}} + f_{\text{T}}) \tag{2.45}$$

The unit of β_f is in radians. The $\sum \frac{|L_o|}{h_o(u+b)}$ values of each *hkl* diffraction line are listed in Table 2.6.

For body-centered cubic (BCC) structures, there are

$$\beta_f = \frac{2}{\pi a} \frac{\sum |L|}{h_o(u+b)} tg\theta(1.5f_{\text{D}} + f_{\text{T}}) \tag{2.46}$$

The unit of β_f is in radians. The $\frac{\sum |L|}{h_o(u+b)}$ values of each *hkl* diffraction line are listed in Table 2.7.

Table 2.6 Several data related to the powder diffraction pattern of FCC structure with stacking faults

hkl	$\sum \frac{(\pm)L_o}{h_o^2(u+b)}$	$\sum \frac{\lvert L_o\rvert}{h_o(u+b)}$	$\Delta(2\theta)^o = \frac{90}{\pi^2} \frac{\sum(\pm)L_o}{h_o^2(u+b)} tg\theta\sqrt{3} f_D$
111	$\frac{1}{4}$	$\sqrt{\frac{3}{4}}$	$\frac{90}{\pi^2}\sqrt{3} f_D tg\theta_{111}\left(\frac{1}{4}\right)$
200	$-\frac{1}{2}$	1	$\frac{90}{\pi^2}\sqrt{3} f_D tg\theta_{200}\left(-\frac{1}{2}\right)$
220	$\frac{1}{4}$	$\frac{1}{\sqrt{2}}$	
311	$-\frac{1}{11}$	$\frac{3}{2}\sqrt{11}$	
222	$-\frac{1}{8}$	$\frac{\sqrt{3}}{4}$	$\frac{90}{\pi^2}\sqrt{3} f_D tg\theta_{222}\left(-\frac{1}{8}\right)$
400	$\frac{1}{4}$	1	$\frac{90}{\pi^2}\sqrt{3} f_D tg\theta_{400}\left(\frac{1}{4}\right)$

Table 2.7 A $\frac{\sum \lvert L\rvert}{h_o(u+b)}$ values of diffraction lines of BCC structure powder diffraction with stacking fault

hkl	110	200	211	220	310	222	321	400
$\frac{\sum \lvert L\rvert}{h_o(u+b)}$	$\frac{2}{3}\sqrt{2}$	$\frac{4}{3}$	$\frac{2}{\sqrt{6}}$	$\frac{2}{3}\sqrt{2}$	$4\sqrt{10}$	$2\sqrt{3}$	$\frac{5}{2}\sqrt{14}$	$\frac{4}{3}$

2.6 Least Square Method for Separating the Multiple Broadening Effect of XRD Line [7–15]

2.6.1 *Least Square Method for Separating Microcrystal and Micro-stress Broadening Effect*

When the effects of microcrystal and micro-strain exist separately, the crystallite size (D) and micro-strain (ε) can be obtained by Formulas (2.42) and (2.43), respectively.

Among them, $\beta_{1/2} = B_{1/2} - \beta_{1/2}^0$ the $B_{1/2}$ measured FWHM deducts the instrument broadening $\beta_{1/2}^0$, and the unit is radians. The angle marker c represents microcrystal, and s represents micro-strain. When these two broadening effects exist alone, they can be distinguished according to whether the $\beta\cos\theta$ and $\beta ctg\theta$ of each line are approximately the same.

When both exist at the same time, so there is

$$\beta = \beta_c + \beta_s \tag{2.47}$$

Replace (2.42) and (2.43) into (2.47) and multiply it by $\frac{\cos\theta}{\lambda}$

$$\frac{\beta\cos\theta}{\lambda} = \frac{0.89}{D} + \varepsilon \cdot \frac{4\sin\theta}{\lambda} \tag{2.48}$$

If the $\frac{\beta\cos\theta}{\lambda} \sim \frac{4\sin\theta}{\lambda}$ is plotted, the straight line can be obtained, and the micro-crystal scale D can be obtained from the intersection of the straight line and the ordinate, and the slope of the straight line is ε.

It is worth mentioning that $\frac{\beta\cos\theta}{\lambda} \sim \frac{4\sin\theta}{\lambda}$ is often difficult to make a straight-line chart because of the broadened anisotropy and measurement errors. So we use the least square method instead, and we order

$$\begin{cases} Y_i = \frac{\beta_i \cos\theta_i}{\lambda} & a = \frac{0.89}{D} \\ X_i = \frac{4\sin\theta_i}{\lambda} & m = \varepsilon \end{cases} \tag{2.49}$$

(2.48) is rewritten as

$$Y = \alpha + mX \tag{2.50}$$

The canonical equations of the least square are

$$\begin{cases} \sum^{n} Y_i = an + m\sum^{n} X_i \\ \sum^{n} X_i Y_i = a\sum^{n} X_i + m\sum^{n} X_i^2 \end{cases} \tag{2.51}$$

Written in matrix form (omitting the subscript) as

$$\begin{pmatrix} n & \sum X \\ \sum X & \sum X^2 \end{pmatrix}\begin{pmatrix} a \\ m \end{pmatrix} = \begin{pmatrix} \sum Y \\ \sum XY \end{pmatrix} \tag{2.52}$$

This is a typical system of bivariate first order equations, and its discriminant is

$$\Delta = \begin{vmatrix} n & \sum X \\ \sum X & \sum X^2 \end{vmatrix} \tag{2.53}$$

When $\Delta \neq 0$, there can be a unique solution:

$$\begin{cases} a = \frac{\Delta_a}{\Delta} = \frac{\begin{vmatrix} \sum Y & \sum X \\ \sum XY & \sum X^2 \end{vmatrix}}{\Delta} = \frac{\sum Y \sum X^2 - \sum X \sum XY}{n\sum X^2 - \left(\sum X\right)^2} \\ m = \frac{\Delta_m}{\Delta} = \frac{\begin{vmatrix} n & \sum Y \\ \sum X & \sum XY \end{vmatrix}}{\Delta} = \frac{n\sum XY - \sum X \sum Y}{n\sum X^2 - \left(\sum X\right)^2} \end{cases} \tag{2.54}$$

The least square method of separating microcrystal-micro-strain broadening effect is suitable for different crystal systems and different structures. There are two broadening effects of microcrystal and micro-strain in the XRD pattern of hydrogen storage alloy for negative electrode of MH–Ni battery, so this method is used to study the microstructure of hydrogen storage alloy. For the closely packed hexagons with stacking faults, only the lines independent of stacking faults (i.e., h–$k = 3n$ or hk0) can be calculated.

2.6.2 Least Square Method for Separating X-Ray Diffraction Broadening Effect Caused by Microcrystal-Stacking Fault

Using the Lorentzian approximation, affected by both microcrystals and stacking faults, the total FWHM β is the sum of microcrystalline broadening β_c and stacking fault broadening β_f, that is,

$$\beta = \beta_c + \beta_f \tag{2.55}$$

For the close-packed hexagonal structure, replace (2.42) and (2.44) into (2.55) and multiply it by $\frac{\cos\theta}{\lambda}$

$$h - k = 3n \pm 1 \begin{cases} l = \text{even} & \frac{\beta\cos\theta}{\lambda} = \frac{2l}{\pi}\left(\frac{d}{c}\right)^2 \frac{\sin\theta}{\lambda}(3f_{\mathrm{D}} + 3f_{\mathrm{T}}) + \frac{0.89}{D} \\ l = \text{odd} & \frac{\beta\cos\theta}{\lambda} = \frac{2l}{\pi}\left(\frac{d}{c}\right)^2 \frac{\sin\theta}{\lambda}(3f_{\mathrm{D}} + f_{\mathrm{T}}) + \frac{0.89}{D} \end{cases} \tag{2.56}$$

Order

$$\begin{cases} Y = \frac{\beta\cos\theta}{\lambda} & f = 3f_{\mathrm{D}} + 3f_{\mathrm{T}}(l = \text{even}) \\ & f = 3f_{\mathrm{D}} + f_{\mathrm{T}}(l = \text{odd}) \\ X = \frac{2l}{\pi}\left(\frac{d}{c}\right)^2 \frac{\sin\theta}{\lambda} & A = \frac{0.89}{D} \end{cases} \tag{2.57}$$

Rewrite (2.56) as

$$Y = fX + A \tag{2.58}$$

The derivation similar to formula (2.50)–(2.54) is as follows:

$$\begin{cases} A = \dfrac{\Delta_A}{\Delta} = \dfrac{\begin{vmatrix} \sum Y & \sum X \\ \sum XY & \sum X^2 \end{vmatrix}}{\Delta} = \dfrac{\sum Y \sum X^2 - \sum X \sum XY}{n \sum X^2 - (\sum X)^2} \\ f = \dfrac{\Delta_f}{\Delta} = \dfrac{\begin{vmatrix} n & \sum Y \\ \sum X & \sum XY \end{vmatrix}}{\Delta} = \dfrac{n \sum XY - \sum X \sum Y}{n \sum X^2 - (\sum X)^2} \end{cases} \tag{2.59}$$

After finding out $D_{\text{even}}, f_{\text{even}}, D_{\text{odd}}, f_{\text{odd}}$, using the Formula (2.60)

$$\begin{cases} f_{\text{even}} = 3f_{\text{D}} + 3f_{\text{T}} \\ f_{\text{odd}} = 3f_{\text{D}} + f_{\text{T}} \end{cases} \tag{2.60}$$

f_{D} and f_{T} were obtained simultaneously.

2.6.3 *Least Square Method for Separating the Double Broadening Effect of X-Ray Diffraction Caused by Micro-strain-Stacking Fault*

For closely packed hexagonal materials, $h - k = 3n \pm 1$, using Lorentzian approximation, there are.

$$\beta = \beta_f + \beta_s \tag{2.61}$$

(1) **Method 1**

Replace (2.43) and (2.44) into (2.61), and multiply it by $\frac{\cos\theta}{\lambda}$

$$\begin{cases} l = \text{even} & \frac{\beta\cos\theta}{\lambda} = \frac{2l}{\pi}\left(\frac{d}{c}\right)^2 \frac{\sin\theta}{\lambda}(3f_{\text{D}} + 3f_{\text{T}}) + \varepsilon\frac{4\sin\theta}{\lambda} \\ l = \text{odd} & \frac{\beta\cos\theta}{\lambda} = \frac{2l}{\pi}\left(\frac{d}{c}\right)^2 \frac{\sin\theta}{\lambda}(3f_{\text{D}} + f_{\text{T}}) + \varepsilon\frac{4\sin\theta}{\lambda} \end{cases} \tag{2.62}$$

Order

$$\begin{cases} Y = \frac{\beta\cos\theta}{\lambda} & f = 3f_{\text{D}} + 3f_{\text{T}}(l = \text{even}) \\ X = \frac{2l}{\pi}\left(\frac{d}{c}\right)^2 \frac{\sin\theta}{\lambda} & f = 3f_{\text{D}} + f_{\text{T}}(l = \text{odd}) \\ Z = \frac{4\sin\theta}{\lambda} & A = \varepsilon \end{cases} \tag{2.63}$$

Rewrite (2.62) as

$$Y = fX + A \tag{2.64}$$

The derivation similar to Formulas (2.50)–(2.54) is as follows:

$$\begin{cases} f = \dfrac{\Delta_f}{\Delta} = \dfrac{\begin{vmatrix} \sum YZ & \sum Z^2 \\ \sum XY & \sum XZ \end{vmatrix}}{\Delta} = \dfrac{\sum YZ \sum XZ - \sum Z^2 \sum XY}{\left(\sum XZ\right)^2 \sum X^2 \sum Z^2} \\ A = \dfrac{\Delta_A}{\Delta} = \dfrac{\begin{vmatrix} \sum XZ & \sum YZ \\ \sum X^2 & \sum XY \end{vmatrix}}{\Delta} = \dfrac{\sum XZ \sum XY - \sum X^2 \sum YZ}{\left(\sum XZ\right)^2 \sum X^2 \sum Z^2} \end{cases} \tag{2.65}$$

(2) **Method 2**

Replace the expressions (2.42) and (2.44) into (2.61) and divide them by $4tg\theta$ to get

$$\begin{cases} l = \text{even} \ \frac{\beta ctg\theta}{4} = \frac{l}{2\pi}\left(\frac{d}{c}\right)^2 (3f_{\mathrm{D}} + 3f_{\mathrm{T}}) + \varepsilon \\ l = \text{odd} \ \frac{\beta ctg\theta}{4} = \frac{l}{2\pi}\left(\frac{d}{c}\right)^2 (3f_{\mathrm{D}} + f_{\mathrm{T}}) + \varepsilon \end{cases} \tag{2.66}$$

Oder

$$\begin{cases} Y = \frac{\beta ctg\theta}{4} \quad \begin{matrix} f = 3f_{\mathrm{D}} + 3f_{\mathrm{T}}(l = \text{even}) \\ f = 3f_{\mathrm{D}} + f_{\mathrm{T}}(l = \text{odd}) \end{matrix} \\ X = \frac{2l}{\pi}\left(\frac{d}{c}\right)^2 \ A = \varepsilon \end{cases} \tag{2.67}$$

Rewrite (2.66) as

$$Y = fX + A \tag{2.68}$$

The derivation similar to Formulas (2.50)–(2.54) is as follows:

$$\begin{cases} A = \dfrac{\Delta_A}{\Delta} = \dfrac{\begin{vmatrix} \sum Y & \sum X \\ \sum XY & \sum X^2 \end{vmatrix}}{\Delta} = \dfrac{\sum Y \sum X^2 - \sum X \sum XY}{n \sum X^2 - (\sum X)^2} \\ f = \dfrac{\Delta_f}{\Delta} = \dfrac{\begin{vmatrix} n & \sum Y \\ \sum X & \sum XY \end{vmatrix}}{\Delta} = \dfrac{n \sum XY - \sum X \sum Y}{n \sum X^2 - (\sum X)^2} \end{cases} \tag{2.69}$$

2.6.3.1 Least Square Method for Separating the Triple Broadening Effect of X-Ray Diffraction Caused by Microcrystal-Micro-strain-Stacking Fault

For the closely packed hexagonal samples, when $h - k = 3n \pm 1$, the Lorentzian approximation is still used, and the total FWHM β of the diffraction line is

$$\beta = \beta_f + \beta_c + \beta_s \tag{2.70}$$

Replace (2.42)–(2.44) into (2.70) and multiply it by $\frac{\cos\theta}{\lambda}$

$$h - k = 3n \pm 1 \begin{cases} l = \text{even} & \frac{\beta\cos\theta}{\lambda} = \frac{2l}{\pi}\left(\frac{d}{c}\right)^2 \frac{\sin\theta}{\lambda}(3f_{\mathrm{D}} + 3f_{\mathrm{T}}) + \frac{0.89}{D} + \varepsilon\frac{4\sin\theta}{\lambda} \\ l = \text{odd} & \frac{\beta\cos\theta}{\lambda} = \frac{2l}{\pi}\left(\frac{d}{c}\right)^2 \frac{\sin\theta}{\lambda}(3f_{\mathrm{D}} + f_{\mathrm{T}}) + \frac{0.89}{D} + \varepsilon\frac{4\sin\theta}{\lambda} \end{cases} \tag{2.71}$$

Order

$$\begin{cases} Y = \frac{\beta\cos\theta}{\lambda} & f = 3f_{\mathrm{D}} + 3f_{\mathrm{T}}(l = \text{even}) \\ & f = 3f_{\mathrm{D}} + f_{\mathrm{T}}(l = \text{odd}) \\ X = \frac{2l}{\pi}\left(\frac{d}{c}\right)^2 \frac{\sin\theta}{\lambda} & A = \frac{0.89}{D} \\ Z = \frac{4\sin\theta}{\lambda} & B = \varepsilon \end{cases} \tag{2.72}$$

The matrix form of the canonical equation of the least square is

$$\begin{pmatrix} \sum X^2 & \sum X & \sum XZ \\ \sum X & n & \sum Z \\ \sum XZ & \sum Z & \sum Z^2 \end{pmatrix} \begin{pmatrix} f \\ A \\ B \end{pmatrix} = \begin{pmatrix} \sum XY \\ \sum Y \\ \sum YZ \end{pmatrix} \tag{2.73}$$

When the discriminant and unique solution of the system of ternary first-order equations:

$$\Delta = \begin{vmatrix} \sum X^2 & \sum X & \sum XZ \\ \sum X & n & \sum Z \\ \sum XZ & \sum Z & \sum Z^2 \end{vmatrix} \neq 0 \tag{2.74}$$

To have only a solution:

$$
\begin{aligned}
f &= \frac{\Delta_f}{\Delta} = \frac{\begin{vmatrix} \sum XY & \sum X & \sum XZ \\ \sum Y & n & \sum Z \\ \sum YZ & \sum Z & \sum Z^2 \end{vmatrix}}{\Delta} \\
A &= \frac{\Delta_A}{\Delta} = \frac{\begin{vmatrix} \sum X^2 & \sum XY & \sum XZ \\ \sum X & \sum Y & \sum Z \\ \sum XZ & \sum YZ & \sum Z^2 \end{vmatrix}}{\Delta} \\
B &= \frac{\Delta_B}{\Delta} = \frac{\begin{vmatrix} \sum X^2 & \sum X & \sum XY \\ \sum X & n & \sum XY \\ \sum XZ & \sum Z & \sum YZ \end{vmatrix}}{\Delta}
\end{aligned}
\tag{2.75}
$$

From the derivation of the formula, it can be known that only when the number of diffraction lines m_{even} and m_{odd} of $h - k = 3n \pm 1$, $l =$ even and $l =$ odd satisfy ≥ 3, it can be solved.

The above methods for separating the double broadening effect of microcrystal-stacking fault and micro-strain-stacking fault and the triple broadening effect of microcrystal-micro-strain-stacking fault are only deduced for dense hexagonal structures. However, the derived results are also applicable to face-centered and body-centered cubic structures.

Then use the diagonal method of solving the third-order determinant shown below.

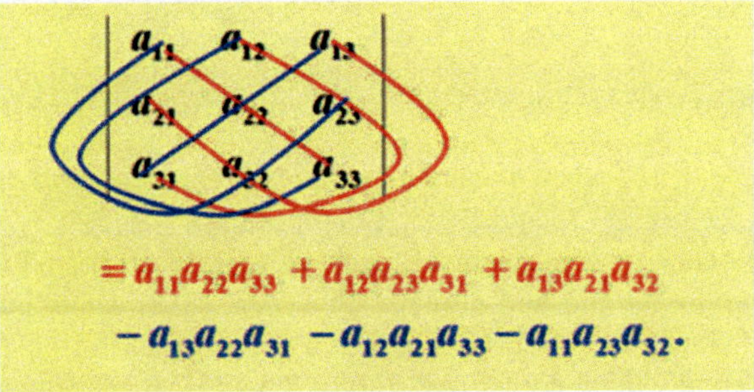

The determinant values of each numerator and denominator are obtained, and then divided by each other, f, A, and B are obtained, and then f, D, and D are obtained.

From the derivation of the formula, it is known that only when the number of diffraction lines m_{even} and m_{odd} of $h - k = 3n \pm 1$ minus $l =$ even and $l =$ odd satisfy ≥ 2 (double effect) and ≥ 3. It can only be solved when the triple effect.

2.6.4 The Structure of Series Calculation Programs

1. **Comparison of stacking fault broadening effects of dense heap hexagonal, face-centered cubic and body-centered cube**

For the purpose of comparison, the formulas related to the triple broadening effect of the three structures are rewritten as follows:

For CPH, $h - k = 3n \pm 1$

$$\begin{cases} l = \text{even} \ \frac{\beta\cos\theta}{\lambda} = \frac{2l}{\pi}\left(\frac{d}{c}\right)^2\frac{\sin\theta}{\lambda}(3f_D + 3f_T) + \frac{0.89}{D} + \varepsilon\frac{4\sin\theta}{\lambda} \\ l = \text{odd} \ \frac{\beta\cos\theta}{\lambda} = \frac{2l}{\pi}\left(\frac{d}{c}\right)^2\frac{\sin\theta}{\lambda}(3f_{\mathrm{D}} + f_{\mathrm{T}}) + \frac{0.89}{D} + \varepsilon\frac{4\sin\theta}{\lambda} \end{cases} \tag{2.76}$$

For FCC,

$$\frac{\beta\cos\theta}{\lambda} = \frac{1}{2\pi a}\sum\frac{|L_o|}{h_o(u+b)}\frac{\sin\theta}{\lambda}(1.5f_{\mathrm{D}} + f_{\mathrm{T}}) + \frac{0.89}{D} + \varepsilon\frac{4\sin\theta}{\lambda} \tag{2.77}$$

For BCC,

$$\frac{\beta\cos\theta}{\lambda} = \frac{1}{2\pi a}\frac{\sum|L|}{h_o(u+b)}\frac{\sin\theta}{\lambda}(1.5f_{\mathrm{D}} + f_{\mathrm{T}}) + \frac{0.89}{D} + \varepsilon\frac{4\sin\theta}{\lambda} \tag{2.78}$$

It can be seen that the expressions of the broadening effect caused by stacking faults of the three structures are similar, and the important differences are as follows:

(1) In the probability of stacking faults.

In relation, it has nothing to do with stacking fault for CPH, $h - k = 3n$ and $hk0$;

$f = 3f_{\mathrm{D}} + 3f_{\mathrm{T}}$ when $h - k = 3n \pm 1$, l is even, and falloff $3f_{\mathrm{D}} + f_{\mathrm{T}}$ when l is odd; for both FCC and BCC, it is fimble $1.5f_{\mathrm{D}} + f_{\mathrm{T}}$;

(2) The difference of coefficient of stacking fault term, for CPH, when l are the even number and odd number. The form is the same, but the value is different. However, for FCC and BCC, the different forms and values are also different, which come from Tables 2.5 and 2.6, respectively;

(3) For CPH, f_{D} and f_{T} can be obtained; for FCC, f_D can be obtained according to one of Formula (6.41), and then, f_{T} can be obtained; while for BCC, only $(1.5f_{\mathrm{D}} + f_{\mathrm{T}})$ can be obtained.

2. **Computing program structure**

The structure of the calculation program is shown in Fig. 2.13.

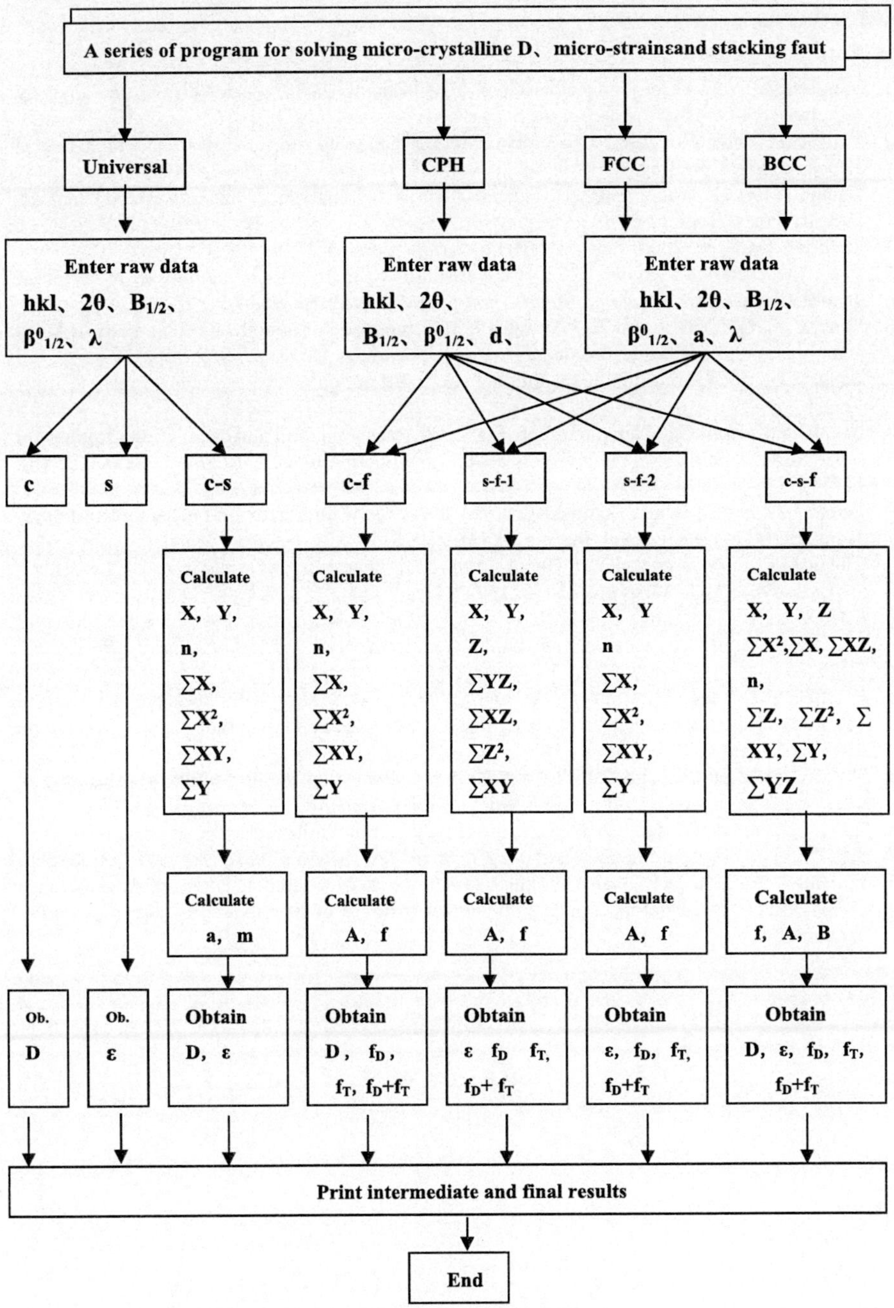

Fig. 2.13 Structure block diagram of computing program series

References

1. Yang, C.Z., D.C. Xie, G.Z. Chen, and F.M. Zhong. 1988. Phase diffraction analysis. Metallurgical Industry Press, Beijing.
2. Jiang, C.H., and C.Z. Yang. 2010. X-ray diffraction and scattering analysis of materials. Higher Education Press, Beijing.
3. Cheng, G.F., C.Z. Yang, and Y.H. Huang. 2010. X-ray analysis of nanomaterials. Chemical Industry Press, Beijing.
4. Materials Data Inc. 2004. Jade 6.0 XRD pattern processing. USA.
5. Yang, C.Z., and F.M. Zhong, 1983. Comparison of standard sample methods in X-ray phase quantitative analysis. *Shanghai Metals (non-ferrous Metals Fascicle)* 4 (3): 78.
6. Yang, C.Z., G.Z. Chen, and Z.Q. Wang. 1983. Comparison of non-standard methods in X-ray phase quantitative analysis. *Shanghai Metals (non-ferrous Metals fascicle)* 4 (4): 67.
7. Jiang, C.H., and C.Z. Yang. 2013. Internal stress diffraction analysis. Science Press, Beijing, 11.
8. Qin, P., Y.W. Lou, C.Z. Yang, and B.J. Xia. 2006. A new method and calculation program for separating multiple broadening effects of X-ray diffraction lines. *Acta Physica Sinica* 55 (3): 1325–1335.
9. Yang, C.Z., and J. Zhang. Some advances in the study of microstructure of nanomaterials by X-ray diffraction. *Progress in Physics* 28 (3): 280–313.
10. Cheng, G.F., C.Z. Yang, and Y.H. Huang. 2008. XRD characterization and study of Nano-ZnO. *Journal of Inorganic Materials* 23 (1): 199–202.
11. Cheng, L.F., C.Z. Yang, and Z.H. Pu et al. 2008. X-ray diffraction characterization of microstructure in face-centered cubic nanomaterials. *Monthly Journal of Scientific Research* 8 (8): 54–57.
12. Pu, Z.H., C.Z. Yang, F.L. Chen, Q. Pei, and Y.W. Lou. 2008. X-ray diffraction characterization and study of the microstructure in hexagonal close-packed nano material. *Powder Diffraction* 23 (3): 213–223.
13. Yang, C.Z., Z.H. Pu, and Z.L. Li. 2009. X-ray diffraction study on the preparation and microstructure of nano-sized nickel powder. *Nano-Science and Technology* 6 (2): 2–7.
14. Yang, C.Z., and C.H. Jiang. 2014. Line profile analysis and microstructure characterization of diffraction line broadening. *Physical and Chemical Test (Physical Fascicles)* 50 (9): 658–667.
15. Lou, Yu-Wan. 2007. XRD study on microstructure of positive and negative active substances in Ni–MH battery. Doctoral paper of Shanghai Institute of Microsystems and Information Technology.

Chapter 3
XRD Characterization of Active Material β-$Ni(OH)_2$ and AB_5 Alloy

Check for updates

MH-Ni and Cd-Ni batteries have been mass produced and widely used, and their positive active materials are both β-$Ni(OH)_2$ [1]. The negative material of the former is mainly AB_5 hydrogen storage alloy. β-$Ni(OH)_2$ belongs to hexagonal structure, P-3m1 (No. 164) space group.

Domestic production of β-$Ni(OH)_2$ has approached or reached the level of international high-quality materials. The technical parameters provided by the manufacturer include chemical composition, loose packing density and vibrating compaction density, particle size, mesh number, specific surface area and (101) half width of diffraction lines, etc. However, for the MH–Ni battery industry, the grain shape, grain size, and stacking fault probability of the initial β-$Ni(OH)_2$ are more important. Because in the process of activation and cycle life testing, the grain shape changes from short and fat column to near-equiaxed crystal, the grain will be refined, and the stacking fault probability will change [2], and these changes are closely related to the electrochemical performance of the material. Therefore, the choice of β-$Ni(OH)_2$ is one of the key problems for battery manufacturers. It is known that β-$Ni(OH)_2$ is prepared by co-deposition method, and its chemical reaction formula is as follows:

$$\begin{aligned} &NiSO_4 + 2NaOH \rightarrow \beta - Ni(OH)_2 \downarrow + Na_2SO_4 \\ \text{or}\quad &Ni(NO_3)_2 + 2NaOH \rightarrow \beta - Ni(OH)_2 \downarrow + 2NaNO_3 \end{aligned}$$

The ratio of the two solutions, in what way the mixed additives are injected, the co-deposition temperature, the stirring process during deposition, and the drying temperature and time of the precipitates all have important effects on the electrochemical performance of the products [3]. Most of these are done through the performance test of the battery. Wang et al. [4] studied the effect of preparation conditions on the structural parameters of β-$Ni(OH)_2$ microcrystals.

C. Yang et al., *Materials and Working Mechanisms of Secondary Batteries*,
https://doi.org/10.1007/978-981-19-5955-4_3

The electrochemical performance of positive active materials is sensitive to structure, so it is necessary to explore the correct and comprehensive characterization and comprehensive evaluation of β-$Ni(OH)_2$ from structure, as well as to establish testing standards for β-$Ni(OH)_2$ manufacturers and users.

AB_5 hydrogen storage alloy is mainly composed of $LaNi_5$ and belongs to hexagonal structure, P6/mmm (No. 191) space group.

This chapter is divided into two parts. On the basis of summarizing the previous work, the first part puts forward the principle and method of comprehensive and correct characterization and evaluation of β-$Ni(OH)_2$ by using conventional X-ray diffraction methods and comprehensively evaluates the grain shape and shape factor, average grain size, and stacking fault probability of β-$Ni(OH)_2$ based on the test and analysis of several β-$Ni(OH)_2$ products. The second part introduces the characterization of AB_5 alloy for MH/Ni battery.

3.1 The Principle and Method of Characterizing and Evaluating β-$Ni(OH)_2$

3.1.1 Two X-Ray Diffraction Analysis Methods for Characterizing β-$Ni(OH)_2$

Delmas and Tessier [5] consider the existence of growth stacking faults and deformation stacking faults in β-$Ni(OH)_2$ and use the method proposed by Treacy et al. and the DIFFax program [6] to simulate the diffraction pattern to establish the structural model and estimate the number of stacking faults. Thus, the simulation patterns with 10% deformation stacking fault probability (f_D) and 8% growth stacking fault probability (f_T) are given, which are in agreement with the diffraction pattern produced β-$Ni(OH)_2$ with good performance by typical process.

Langford et al. [8] applied pattern decomposition to the study of ZnO microcrystal size and stacking fault compound diffraction effect. Wang et al. [7] borrowed the method of Langford et al. [8] to deal with the microcrystal size and stacking layer in hexagonal ZnO and obtained that β-$Ni(OH)_2$ with an overall stacking fault probability of 14.9% has a higher discharge specific capacity (270 mAh/g).

The above two methods are very troublesome and not easy to popularize.

3.1.2 *A New Method for Characterizing the Microstructure of β-$Ni(OH)_2$*

The X-ray patterns of two kinds of β-$Ni(OH)_2$ are shown in Fig. 3.1. The diffraction data are given in Table 3.1. It can be seen from Fig. 3.1 and Table 3.1 that the broadening effect of each hkl diffraction plane is anisotropic.

(1) **A least square method for separating the double broadening effect of microcrystals and stacking faults** [9, 10]
Practice has proved that it is not enough to identify the crop phase of β-$Ni(OH)_2$ and determine the lattice parameters roughly, that is, to determine whether the product is β-$Ni(OH)_2$ from the *d* (crystal plane spacing) and I/I_1 (relative strength) data of XRD patterns, and to calculate the lattice parameters *a* and *c* from the *d* values of (001) and (100) crystal planes. Even if the above analysis results are completely consistent, the broadening of each line in the diffraction pattern is obviously different, which cannot be ignored, because these differences reflect the great differences in grain shape, grain size, and the number of stacking faults contained in the sample, and these differences have a

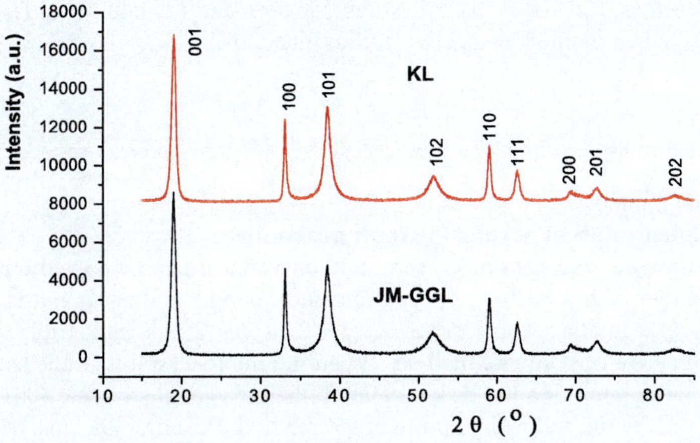

Fig. 3.1 Diffraction patterns of β-$Ni(OH)_2$ produced by different companies CuK_α

Table 3.1 Three diffraction data of β-$Ni(OH)_2$ produced from different companies

Sample	FWHM (°)							
	001	100	101	102	110	111	200	202
KL	0.625	0.326	1.093	2.330	0.395	0.622	1.607	1.182
JM-GGL	0.554	0.326	0.925	1.702	0.394	0.627	1.729	1.036
OMG	0.600	0.299	0.984	1.843	0.361	0.566	1.562	1.062

significant Influence on the electrochemical performance of β-$Ni(OH)_2$ and the charge–discharge performance and cycle life of Cd–Ni and MH–Ni batteries.

The methods for separating the microcrystal-stacking fault broadening effect described in Chap. 2 are as follows:

$$A = \frac{\Delta_A}{\Delta} = \frac{\begin{vmatrix} \sum XY & \sum X^2 \\ \sum Y & \sum X \end{vmatrix}}{\begin{vmatrix} \sum X & n \\ \sum X^2 & \sum XY \end{vmatrix}} = \frac{\sum XY \sum X - \sum X^2 \sum Y}{\left(\sum X\right)^2 - n \sum X^2} \tag{3.1}$$

$$f = \frac{\Delta_f}{\Delta} = \frac{\begin{vmatrix} \sum X & \sum XY \\ n & \sum Y \end{vmatrix}}{\begin{vmatrix} \sum X & n \\ \sum X^2 & \sum X \end{vmatrix}} = \frac{\sum X \sum Y - n \sum XY}{\left(\sum X\right)^2 - n \sum X^2} \tag{3.2}$$

$$\text{here} \begin{cases} Y = \frac{\beta \cos\theta}{\lambda} & f = 3f_D + 3f_T (l = \text{even}) \\ & f = 3f_D + f_T (l = \text{odd}) \\ X = \frac{2l}{\pi}\left(\frac{d}{c}\right)^2 \frac{\sin\theta}{\lambda} & A = \frac{0.89}{D} \end{cases} \tag{3.3}$$

Therefore, for h–$k = 3n \pm 1$, when l is even and l is odd, f_{even}, f_{odd}, D_{even}, and D_{odd} are obtained, respectively. Simultaneously

$$\begin{cases} f_{\text{even}} = 3f_D + 3f_T \\ f_{\text{odd}} = 3f_D + f_T \end{cases} \tag{3.4}$$

Get f_D and f_T.

(2) **Simplified solution of stacking fault probability**

From the above discussion, it can be known that it can be solved only when the number of h–$k = 3n \pm 1$, l is even, and l is odd, and m_{even} and m_{odd} must satisfy ≥ 2; in other words, when $m_{\text{even}} < 2$, $m_{\text{odd}} < 2$, it is impossible to obtain f_D and f_T by least square method. When the microcrystal is in the shape of a short, fat column or a flat disk parallel to the hexagonal structure (001), it can be solved by the following approximate method. Based on the analysis of the microcrystal profile (see Fig. 3.2), the following results are obtained:

$$\begin{cases} D_{101} = D_{001}/\cos\varphi_{z101} = 1.973 D_{001} \\ D_{102} = D_{001}/\cos\varphi_{z102} = 1.313 D_{001} \end{cases} \tag{3.5}$$

On behalf of others

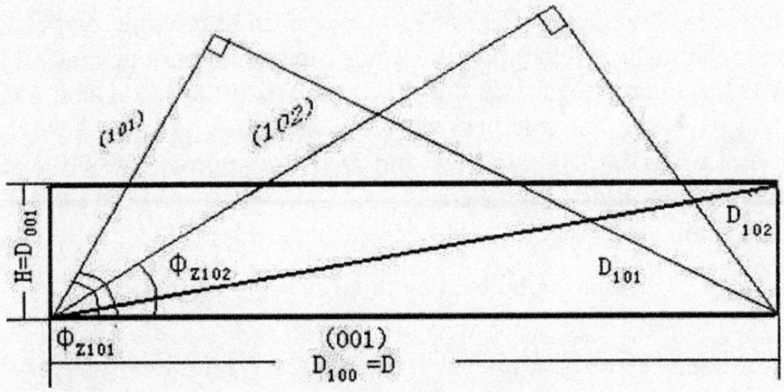

Fig. 3.2 Profile analysis of microcrystals

$$\begin{cases} \dfrac{\beta_{101} cos\theta_{101}}{\lambda} = \dfrac{2}{\pi}\left(\dfrac{d_{101}}{c}\right)^2 \dfrac{\sin\theta_{101}}{\lambda}(3f_\text{D} + f_\text{T}) + \dfrac{0.89}{D_{101}} \\ \dfrac{\beta_{102} cos\theta_{102}}{\lambda} = \dfrac{4}{\pi}\left(\dfrac{d_{102}}{c}\right)^2 \dfrac{\sin\theta_{102}}{\lambda}(3f_\text{D} + 3f_\text{T}) + \dfrac{0.89}{D_{102}} \end{cases} \tag{3.6}$$

If the peak shift caused by the difference of lattice parameters between different β-$Ni(OH)_2$ is ignored, the relevant data can be substituted into:

$$\begin{cases} 0.6122\beta_{101} = 34.4074 \times 10^{-3}(3f_D + f_T) + \dfrac{0.89}{1.973D_{001}} \\ 0.5827\beta_{102} = 52.4028 \times 10^{-3}(3f_D + 3f_T) + \dfrac{0.89}{1.313D_{001}} \end{cases} \tag{3.7}$$

The units of β_{101} and β_{102} are radians, and the units of D_{001} are Å, all are known, so f_D, f_T, and $f_D + f_T$ can be obtained.

3.1.3 *Judgment of Microcrystal Shape and Average Grain Size*

(1) **Judgment of grain shape**
It can be seen from the foregoing that in the polycrystalline XRD pattern of β-$Ni(OH)_2$, when $h–k = 3n$ or hk0, the diffraction line broadening has nothing to do with stacking faults, so D_{001}, D_{100}, D_{110}, and D_{111} can be calculated from the true half width of 001, 100, 110, and 111 diffraction according to Scheler's formula, which characterize the normal dimensions of microcrystals in (001), (100), (110), and (111) planes, respectively. D_{100}/D_{001} or $\arctan(D_{100}/D_{001})$ is defined as the shape factor of microcrystals.

When $D_{100} \approx D_{110}$ and $D_{100} > D_{001}$ several times, the microcrystal is in the shape of a short-fat column parallel to the hexagonal base plane (001); if $D_{100} \gg D_{001}$, the microcrystal is a flat disk parallel to the hexagonal base plane; when $D_{001} \gg D_{100}$, the microcrystal is acicular parallel to the hexagonal base plane; and when D_{001}, D_{100}, D_{110}, and D_{111} are approximately the same, the microcrystals are nearly equiaxed.

(2) **Average grain size**

The average grain size is obtained by the following methods:

$$\begin{cases} D_{\text{average}} = (D_{001} + D_{100} + D_{110} + D_{101} + D_{102} + \cdots + D_n)/n \\ D_{\text{average}} = (D_{001} + D_{100} + D_{110} + D_{\text{even}} + D_{\text{odd}} + \cdots + D_n)/n \end{cases} \tag{3.8}$$

$$D = D_{\text{average}} \pm \sigma$$
$$\sigma = \sqrt{\frac{\sum_1^n \left(D_{\text{average}} - D_i\right)^2}{n-1}} \tag{3.9}$$

σ is called standard error. The average σ/D is defined as the anisotropy factor of grain shape.

3.2 The Relationship Between Microstructure Parameters and Properties

3.2.1 Test Results of Several β-Ni(OH)$_2$ Samples

The X-ray diffraction data of 8 samples from different sources are analyzed as follows:

(1) **Grain shape and grain shape factor**
Table 3.2 shows the grain shape and shape factors of 8 samples. The grain is also in the shape of a short, fat column, along with the shape.

If the shape factor is different, the process of conversion to near-equiaxed axis in the process of battery charge, discharge, or cycle is also different.

(2) **The result of the determination of the probability of stacking fault**
In the above processing data, it should be explained that No. 8 sample because of nearly equiaxed crystal $D_{101} \approx D_{102} \approx (D_{001} + D_{100} + D_{110} + D_{111})/4$, but sample 6 because $0.89/D_{101} \ge \beta_{101}\cos\theta_{101}/\lambda$ or $0.89/D_{102} \ge \beta_{102}\cos\theta_{102}/\lambda$. So it can only be calculated by the least square method of decomposing microcrystal-stacking fault effect. The results are listed in Table 3.3.

It can be seen from the data that there are obvious differences in the deformation stacking fault probability f_D, the twin stacking fault probability f_T, and the sum of them, which may lead to different processes of stacking fault transition

Table 3.2 Grain shape and shape factors of different β-Ni(OH)$_2$

No	D_{001}	D_{100}	D_{110}	D_{100}/D_{001}	Grain shape
	nm				
1	16.8	81.6	97.6	4.857	A short-fat cylinder with a base parallel to the hexagonal base plane (001)
2	14.7	65.8	53.2	4.476	A short-fat cylinder with a base parallel to the hexagonal base plane (001)
3	19.2	112.4	69.5	6.176	A short-fat cylinder with a base parallel to the hexagonal base plane (001)
4	19.9	91.2	94.9	4.825	A short-fat cylinder with a base parallel to the hexagonal base plane (001)
5	15.1	72.2	42.5	4.768	A short-fat cylinder with a base parallel to the hexagonal base plane (001)
6	3.4	60.6	9.4	19.820	The shape of a disk with a base parallel to the hexagonal base (001)
7	25.1	82.3	79.9	3.279	A short-fat cylinder with a base parallel to the hexagonal base plane (001)
8	20.2	19.2	12.3	11.400	Near polyhedron or nearly equiaxed crystal

Table 3.3 Stacking fault probability of different β-Ni(OH)$_2$

	D_{101}	D_{102}	D_{even}	D_{odd}	f_D	f_T	$f_D + f_T$
	nm				%		
1	33.1	21.9	–	–	4.7	2.7	7.4
2	29.0	19.3	–	–	4.2	1.8	6.0
3	35.9	23.9	–	–	5.1	−0.2	4.9
4	37.3	24.8	–	–	2.6	1.7	4.3
5	29.8	19.8	–	–	5.2	1.8	7.0
6	–	–	11.1	14.1	8.8	−4.2	4.6
7	49.5	32.9	–	–	1.2	2.1	3.3
8	15.8	15.8	–	–	5.57	1.80	7.37

and stacking fault probability reduction during battery charge/discharge and cycle. In addition, it seems to suggest that the partial replacement of Ni by Co will reduce the stacking fault probability, and if there is too much Co to replace Ni, the total stacking fault probability will be lower, which is disadvantageous to the battery performance and increases the cost.

(3) **Average grain size**

The average grain size and shape anisotropy factors obtained according to formulas (3.8) and (3.9) are listed Table 3.4. It can be seen from the data that the larger the standard error is, the greater the relative error $\sigma/D_{Aver.}$ is and the greater the grain shape anisotropy factor is. Even if the bottom surface is parallel

Table 3.4 Average grain size of different β-Ni(OH)$_2$

Grain size	D_{001}	D_{100}	D_{101}	D_{102}	D_{110}	D_{111}	$D_{平均} \pm \sigma$	$\sigma/D_{Aver.}$
	nm							%
1	16.8	81.6	33.1	21.9	97.6	49.3	50.1 ± 32.0	63.900
2	14.7	65.8	29.0	19.3	53.2	27.5	34.9 ± 20.1	57.600
3	19.2	112.4	35.9	23.9	69.5	34.6	49.1 ± 34.7	70.700
4	19.9	91.2	37.3	24.8	94.9	39.9	51.2 ± 33.4	65.200
5	15.1	72.0	29.8	19.8	42.5	27.3	34.7 ± 20.7	59.700
6	3.4	60.6	14.1	11.1	9.4	14.1	19.9 ± 19.7	104.200
7	25.1	82.3	49.5	32.9	79.9	34.2	50.7 ± 24.9	49.100
8	20.2	19.2	15.8	15.8	12.3	11.4	15.8 ± 3.5	0.960

to the hexagonal base plane (001), the grain shape anisotropy factor $\sigma/D_{Aver.}$ is obviously different.

3.2.2 *The Relationship Between Microstructure Parameters and Properties of β-Ni(OH)$_2$*

(1) **The relationship between the structure of β-Ni(OH)$_2$ raw materials and battery performance**
The properties and microstructure parameters of β-Ni(OH)$_2$ prepared by different methods are shown in Table 3.5. It can be seen that the lattice parameters of four kinds of β-Ni(OH)$_2$ vary with different doping, and among the important electrochemical properties, the room temperature 1 °C capacity is related to the crystal shape, grain size, and total stacking fault probability, and the one with maximum stacking fault probability and moderate grain size has higher capacity.

The performance and microstructure data of β-Ni(OH)$_2$ from different sources are listed in Table 3.6. Although the lattice parameters are almost the same, their capacities are still different and correspond to the microstructure parameters.

Therefore, it can be concluded that the grain shape and size of positive active material β-Ni(OH)$_2$, especially the total stacking fault probability, have a good corresponding relationship with the charge–discharge capacity of MH/Ni battery. The materials with appropriate grain size and larger stacking fault probability correspond to larger charge–discharge capacity. This is consistent with the conclusions of the relevant literature. Considering the existence of growth stacking faults and deformation stacking faults in β-Ni(OH)$_2$ by Delmas and Tessier [5], the method proposed by Treacy et al. and the DIFFax program are used to simulate the diffraction pattern to establish the structural model

Table 3.5 Performance and microstructure data of β-$Ni(OH)_2$ produced by some units

	Ni-PTX	Ni-$Zn_4Co_{1.5}$	Ni-Cd_3Co	Ni-KY
Temperature 1 °C-1 capacity/mAh	592	571	591	578
600 °C, 1 °C-1 capacity/mAh	356.7	436.4	476.9	351.3
a/nm	0.31080	0.31098	0.31118	0.31157
c/nm	0.45849	0.45863	0.45996	0.46073
D_{001}/nm	13.1	15.5	14.6	11.9
D_{100}/nm	26.3	34.4	28.0	32.3
D_{100}/D_{001}	2.00	2.22	1.92	2.71
f_D/%	9.57	7.04	9.19	7.93
f_T/%	4.48	4.53	3.67	1.95
$(f_D + f_T)$/%	14.05	11.57	13.86	9.88

and estimate the number of stacking faults. It is consistent with the good electrochemical performance of β-$Ni(OH)_2$ with 10% deformation stacking fault probability (f_D) and 8% growth stacking fault probability (f_T) β-$Ni(OH)_2$. It is consistent with the conclusion that β-$Ni(OH)_2$ with a total stacking error probability of 14.9% has a higher discharge specific capacity (270 mAh/g), which is obtained by Wang et al. [7] by using Langford et al. to treat the microcrystal size and stacking layer in hexagonal ZnO.

(2) **Relationship between structure of β-$Ni(OH)_2$ raw materials and battery performance under high current charging**

The fast charge ability of high current is an urgent problem to be solved in the practical process of hydrogen-nickel battery. This section studies the relationship between the structure of β-$Ni(OH)_2$ raw materials and battery performance under high current charging. The diffraction patterns of several original β-$Ni(OH)_2$ samples are shown in Fig. 3.3. Its chemical composition is listed in Table 3.7.

Only from the X-ray diffraction patterns of β-$Ni(OH)_2$ samples, we cannot see the difference of three kinds of β-$Ni(OH)_2$ samples. By calculating the grain size and grain shape factor, as well as measuring the stacking fault probability, we can see the difference of microstructure parameters of No. 1, No. 2, and No. 3. The results are listed in Table 3.7. The corresponding battery performance is also listed in Table 3.8.

The comparison of Tables 3.7 and 3.8 shows that although the capacity of 1 °C increases with the decrease of stacking fault probability and shape factor, the utilization rate of active material, internal resistance, and the maximum voltage at different charging rates all increase with the increase of stacking fault probability and shape factor, which corresponds to the increase of Co content and the decrease of Ni content in β-(Ni, Co)$(OH)_2$. No.3 has the best fast charging performance.

Table 3.6 Performance and microstructure data of β-$Ni(OH)_2$ from different sources

Sample No.	a (Å)	c (Å)	Specific capacity (mAh/g)	Internal resistance (mΩ)	Utilization rate of β-$Ni(OH)_2$ (%)	$\frac{D_{100}}{D_{001}}$	D (nm)	f_D (%)	f_T (%)	$f_D + f_T$ (%)
1	4.118	4.607	271.9	14.6	92.6	4.076	21.81	4.34	1.47	6.81
2	4.118	4.602	271.0	14.3	92.9	4.475	27.14	7.23	−1.71	4.52
3	4.118	4.607	262.3	14.9	92.8	2.946	28.26	4.94	1.39	4.08

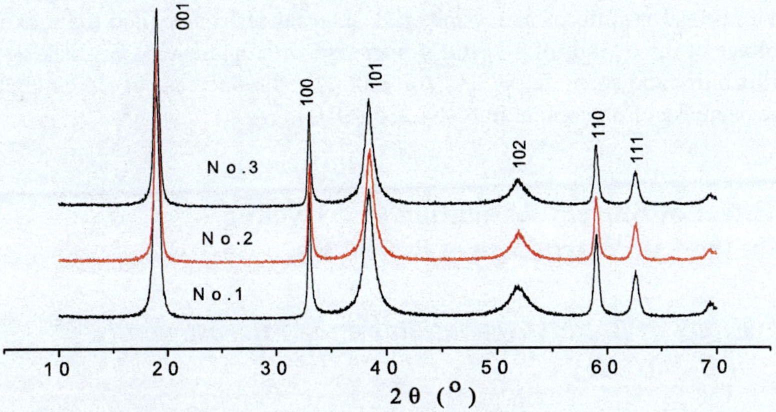

Fig. 3.3 Diffraction patterns of three original β-$Ni(OH)_2$ samples

Table 3.7 Chemical composition, grain shape, shape factor (D_{100}/D_{001}), and stacking fault probability of three original β-$Ni(OH)_2$ samples

No.	Ni (%)	Co (%)	Zn (%)	*a* (Å)	*c* (Å)	D_{001} (nm)	D_{100} (nm)	D_{100}/D_{001}	$f_T + f_D$(%)
1	57.2 ± 2.0	1.5 ± 0.5	4.8 ± 0.5	4.1269	4.6629	16.32	54.6	4.407	10.04
2	54.4 ± 2.0	4.5 ± 0.5	4.4 ± 0.5	4.1273	4.642	20.53	56.37	2.746	9.09
3	54.2 ± 2.0	5.5 ± 0.5	4.5 ± 0.5	4.1265	4.6563	20.17	54.49	2.702	9.04

Table 3.8 Performance of batteries corresponding to different β-$Ni(OH)_2$

	D_{100}/D_{001}	$f_T + f_D$ (%)	1 °C capacity (mAh)	Utilization rate of active material (%)	Internal resistance (mΩ)	The maximum voltage at different rates of charging		
						3 °C/V	4 °C/V	5 °C/V
No. 1	4.407	10.04	901.78	97.8	10.75	1.6505	1.7015	1.727
No. 2	2.746	9.09	934.17	97.6	10.5	1.6055	1.637	1.666
No. 3	2.702	9.04	949.08	97.1	9.95	1.6025	1.631	1.660

From the results of the 3.1 and 3.2 two sections, we can draw the following conclusions:

(1) The grain shape and size of positive active material β-$Ni(OH)_2$, especially the total stacking fault probability, have a good corresponding relationship with the charge–discharge capacity of MH/Ni battery. The material with appropriate grain size and larger stacking fault probability corresponds to the larger charge–discharge capacity.

(2) The utilization ratio of active material, internal resistance, and the maximum voltage of three kinds of β-Ni(OH)$_2$ increase with the increase of stacking fault probability and shape factor and correspond to the increase of Co content and the decrease of Ni content in β-(Ni, Co)(OH)$_2$.

3.3 Effect of Battery Activation and Cycling on the Microstructure of β-Ni(OH)$_2$

3.3.1 Effect of Battery Activation on the Microstructure of β-Ni(OH)$_2$

For comparative study, the diffraction patterns of the original β-Ni(OH)$_2$, activated, and cycled 100, 400 cycles are shown in Fig. 3.4. After activation and circulation, there is micro-stress, so it is the triple broadening effect of microcrystal-micro-strain-stacking fault (D-ε-f). The number of diffraction lines m_{even} and m_{odd} does not satisfy ≥ 3, so the data cannot be processed by the general method, but the simplified method of Formula (3.7) can be used to process the data, and f_D, f_T, and $f_D + f_T$ can be obtained. The calculated results are shown in Table 3.9.

It can be seen from the table:

(1) The grain is obviously refined by activation, especially the size in the direction of vertical *c*-axis is greatly reduced, so that the shape of microcrystal is transformed from chubby columnar to nearly equiaxed crystal. The cycle life test did not change this situation.
(2) Due to the charge and discharge of the battery, there is micro-strain (micro-stress) in β-Ni(OH)$_2$ after activation and circulation.

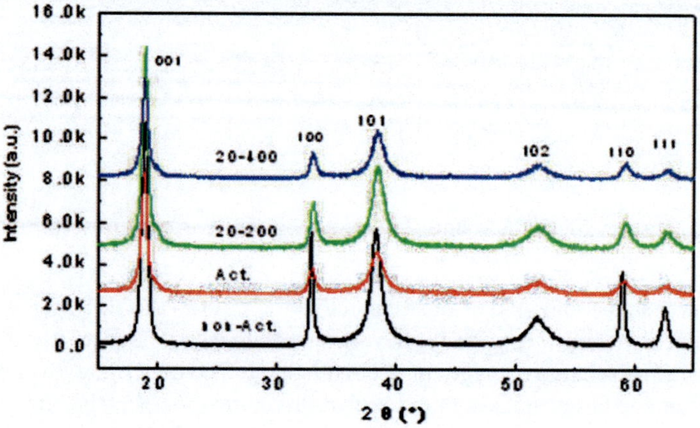

Fig. 3.4 XRD pattern of β-Ni(OH)$_2$ after inactivation, activation, and cycle 200, 400 cycles

Table 3.9 Effect of battery activation on microstructure parameters

Sample number		D_{002} (nm)	D_{100}	$D_{Average}$	$\varepsilon_{平均}$ ($\times 10^{-3}$)	f_D (%)	f_T	$f_D + f_T$
1	No active	16.50	29.40	27.20	–	9.86	1.59	11.45
	Only active	19.98	17.67	57.30	5.96	4.28	4.66	8.94
2	No active	15.33	41.00	21.2	–	5.90	15.70	9.80
	Only active	15.33	12.71	20.9	3.71	7.88	0.01	7.87

(3) After activation, the structure and probability of stacking fault also change, and the total probability of stacking fault becomes smaller.
(4) These changes are not completely reversible, indicating that the function of MN-Ni battery may be related to this irreversible change.

3.3.2 *Effect of Cycle on the Microstructure of β-Ni(OH)$_2$*

The effect of cycling on the structure of β-Ni(OH)$_2$ and the experimental results of the two groups of samples are given in Table 3.10. If the data are obtained from Table 3.10, the following results can be obtained:

(1) The grain size of β-Ni(OH)$_2$ is further refined by cycle life test.
(2) Repeated charge and discharge reduce the micro-strain and the residual strain with more cycles.
(3) The stacking fault after activation and cycle life test is dominated by deformation stacking fault, and the deformation stacking fault probability and total stacking fault probability vary with different cyclic conditions.

Table 3.10 Effect of battery cycle on microstructure parameters

		D_{002} (nm)	D_{100}	$D_{平均}$	$\varepsilon_{平均}$ ($\times 10^{-3}$)	f_D (%)	f_T	$f_D + f_T$
1	Only active	19.98	17.67	57.30	5.96	4.28	4.66	8.94
	100 period	19.78	17.82	54.90	5.57	5.12	2.83	7.95
	400 period	16.71	19.46	34.20	4.62	4.13	3.10	7.23
2	Only active	19.68	17.23	44.10	5.01	5.99	−1.12	4.87
	1C-1C-436 period	17.11	17.23	26.7	3.41	4.70	−0.62	4.08
	1C-10C-117 period	19.68	13.73	30.0	4.28	3.68	0.39	4.09
	3C-10C-95 period	17.11	15.27	40.2	5.86	9.02	−5.23	3.79

3.4 Comprehensive Evaluation of β-$Ni(OH)_2$

3.4.1 The Necessity of Comprehensive Evaluation

Summing up the experiment results in Sects. 3.2 and 3.3, the following facts can be obtained:

(1) The grain shape, size, and stacking fault probability of β-$Ni(OH)_2$ obtained by different preparation methods and different processes are obviously different.
(2) There is a certain relationship between the microstructure parameters of initial β-$Ni(OH)_2$ and the charge–discharge performance.
(3) The microstructure parameters of initial β-$Ni(OH)_2$ change during charge–discharge and cycle. The process of this change is also affected by charge–discharge conditions and cycle conditions.
(4) It can also be inferred from the above three that under the same charge and discharge conditions and the same cycle conditions, the changes described in (3) are also affected by microstructure parameters of initial β-$Ni(OH)_2$.

From the above research results, it is clear that in order to obtain good battery performance and long service life, it is very important for the research, development, and producers of MH–Ni and Cd–Ni batteries to choose β-$Ni(OH)_2$ with good grain shape, grain size, and higher layer error probability as the initial positive active material. On the other hand, it is also very important for trial producers or producers of β-$Ni(OH)_2$ to adopt what kind of process, especially the key processes to control grain shape, grain size, and stacking fault probability, in order to meet the different needs of the battery industry. In these two aspects, it is very important to comprehensively and correctly characterize and comprehensively evaluate β-$Ni(OH)_2$. Therefore, the establishment of industry (ministerial or national) β-$Ni(OH)_2$ test standard also shows its necessity and importance.

3.4.2 The Content of Comprehensive Evaluating Initial β-$Ni(OH)_2$

The contents, methods, and standards of comprehensive evaluation of β-$Ni(OH)_2$ need to be studied, discussed, and tested repeatedly by colleagues. What is proposed here is only for reference at the beginning. The content of comprehensive evaluating β-$Ni(OH)_2$ can be divided into two aspects: macro and micro.

(1) **Macro-performance parameters include:**
Content of main cationic elements: Ni and doped Co, Zn, etc.
Main impurity elements: content of impurity elements.
Particle shape, average particle size, and specific surface area.
Loose packing density and loose packing coefficient, vibrating packing density and vibrating packing coefficient.

(2) **Micro-performance parameters include:**
Whether the product belongs to β-$Ni(OH)_2$ with hexagonal (*P*-3m space group) structure, there should be no miscellaneous phase.

Grain shape and shape factor D_{100}/D_{001}, average grain size $D_{\text{average}} \pm \sigma$ and grain anisotropy factor $\sigma/D_{\text{average}}$, and stacking fault probability f_D, f_T, and $(f_D + f_T)$.

Except for whether it belongs to β-$Ni(OH)_2$, which has been provided by the production unit, it is necessary to find a more suitable and popularizing test method for these micro-parameters. Although the method proposed in this chapter is basically applicable and easy to popularize, it needs to be further studied and confirmed that the establishment of testing standards requires the joint efforts of peers and the organization of the relevant competent departments.

3.5 Characterization Method of AB_5 Alloy as Active Material [11]

$LaNi_5$-based hydrogen storage alloys have been studied for more than 30 years, and a series of AB_5 hydrogen storage alloys based on mixed rare earths have been developed, in which the *A* side is mixed rare earth and the *B* side is Ni, Co, Mn, and other elements. Because of its excellent cycle performance, it has been widely used in hydrogen purification, hydrogen storage, and MH–Ni batteries, and its usage is increasing year by year, and it has become a new industry.

The test contents of the microstructure of the alloy include:

(1) The phase structure of the alloy should be single-phase AB_5, hexagonal structure, space group P6/mmm, and there should be no A_2B_7, Ni, and other impurity phases. The change of phase structure of alloy powder before and after use.
(2) Determination of alloy lattice parameters.
(3) The microstructure of the alloy, that is, the determination of the residual stress (strain) and grain size of the alloy powder.

The above three microstructure parameters are measured by diffraction experiments and data processing on modern X-ray diffractometers. Although the operation and data processing of modern X-ray diffractometers use computers, different diffraction experimental conditions and parameters as well as different data processing methods will have an important impact on the test results. On the basis of summarizing the effects of Japanese Rigaku modern X-ray diffractometer, diffraction experimental conditions and parameters, and different data processing methods on diffraction peak position, intensity, and FWHM [12], an X-ray diffraction (XRD) method for characterizing the microstructure parameters of AB_5 hydrogen storage alloy is proposed, and these methods are used to characterize the structural characteristics of the initial AB_5 powder. The microstructure changes of AB_5 before and after activation and cycling were studied by static method, and significant results were obtained. These studies include: (1) the relationship between phase structure, lattice parameters, and weighted atomic radius of alloys with different composition and (2) microstructure of alloys with the same composition and different preparation methods. (3) The microstructure changes of AB$_5$ before and after charge–discharge activation and cycle life test of MH–Ni battery and the effect of Ca additive of β-Ni(OH)$_2$ on the microstructure of AB$_5$ alloy were also studied.

3.5.1 *Phase Structure Analysis*

Using Cu target, DS = 1°, SS = 1°, RS = 0.30 or 0.15 mm, diffracted light Monochromator, and 2θ scanning speed of 4°–8°/min, the power of X-ray source can be higher, for example, 50 kV, 100 mA, 40 kV, 40 mA, 4°/min. The scanning range is $2\,\theta =$ 15°–50°.

Single-phase identification method:

(1) Only the lines of AB$_5$ phase appear, and the diffraction data are as follows: hkl 100 001 101 110 200 111 002 201 ~ 2θ(°) 20.30 21.90 30.10 35.078 41.54 42.35 44.62 47.50
(2) There are no miscellaneous phases, such as A$_2$B$_7$ and Ni phase diffraction lines. Since the strongest line $d = 2.152$, 116 of A$_2$B$_7$ is very close to the strongest line 111 of AB$_5$, special attention should be paid to whether the second strong line 107 ($d = 2.74$, $2\theta \approx 33.9$ of A$_2$B$_7$) appears or not, and the second is to observe whether there are diffraction peaks between 200 ($d = 2.172$, $2\theta \approx 41.54°$) and 111 ($d = 2.12$, $2\theta = 42.52°$). Whether Ni exists or not, it is mainly observed whether there is a diffraction peak at $2\theta = 44.3°$ on the low corner side of AB$_5$, and if so, it belongs to the diffraction peak of Ni, as shown in Fig. 3.5.

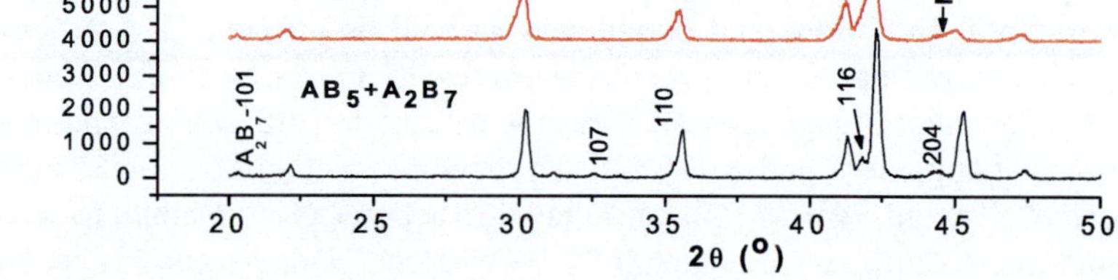

Fig. 3.5 Comparison of diffraction patterns with impurity phase Ni (middle), A$_2$B$_7$ (bottom) and non-impurity phase (top)

3.5.2 Accurate Determination Method of Lattice Parameters

In order to improve the measurement accuracy, the radiation is used, the scanning range of 2θ is 55°–85°, and the scanning speed is 4°/min. The $K\alpha_2$ diffraction peaks of all lines are removed by the program of removing $K\alpha_2$, and the parabola method is used to find the peak automatically. If necessary, the zero position of 2θ can be corrected to eliminate system errors. By using the least square method, the matrix form of regular equations of hexagonal crystal system is written.

$$\begin{bmatrix} \sum N & \sum NR & \sum N\delta \\ \sum NR & \sum R^2 & \sum R\delta \\ \sum N\delta & \sum R\delta & \sum \delta^2 \end{bmatrix} \begin{bmatrix} A \\ B \\ D' \end{bmatrix} = \begin{bmatrix} \sum N \sin^2 \sin\theta \\ \sum R \sin^2 \sin\theta \\ \sum \delta \sin^2 \sin\theta \end{bmatrix} \tag{3.10}$$

here

$$\begin{cases} A = \frac{\lambda^2}{3a^2}, & N = h^2 + hk + k^2 \\ B = \frac{\lambda^2}{4c^2}, & R = l^2 \\ D' = \frac{D}{10}, & \delta = 10\sin^2 2\theta \end{cases} \tag{3.11}$$

Finally, the following results are obtained:

$$\begin{aligned} &a = \sqrt{\lambda^2/3A}, \quad c = \sqrt{\lambda^2/4B} \\ &V = 0.866a^2c \end{aligned} \tag{3.12}$$

The actual operation is as follows: (1) press the format

$$hkl \qquad 2\theta\,(°)$$

and make a readable data set: AB_5-LP. txt and save it; (2) execute the program in dialog and finally output the intermediate and final results in the form of an array.

3.5.3 Measurement of Full Width at Half Maximum (FWHM) and Solution of Grain Size and Micro-strain by Least Square Method

1. **Measurement of full width at half maximum (FWHM) in diffraction experiment.**
   Using CuKα radiation, DS = 1°, SS = 1°, RS = 0.15 mm, Pyrolytic Graphite Monochromator, 2θ scanning range 20°–50°, scanning speed 2° or 4°/min, using a deeper sample frame. After obtaining the diffraction pattern:

(1) All the diffraction lines are treated by removing the $K\alpha_2$ component, and the parabolic method is used to find the peak automatically. Click Report, Print, and the data of all lines $B_{1/2}$ can be obtained.
(2) Create an array AB_5-XCS .txt in the following format:

$$Hkl \quad 2\theta(^\circ) \quad B_{1/2}(^\circ) \quad \beta^0_{1/2}(^\circ)$$

and it is preserved, in which $\beta^0{}_{1/2}$ is the broadening of the instrument.
(3) When the diffraction lines are seriously broadened, the FWHM data given by automatic peak finding are not reliable, and the FWHM data can be obtained by Fit all Line or Refine operation.

2. **Separate solution of double effect of Grain size and Micro-strain**

The crystallite size D and strain (ε) are derived from Sect. 2.6.1.

$$\begin{cases} a = \dfrac{\Delta_a}{\Delta} = \dfrac{\begin{vmatrix} \sum Y & \sum X \\ \sum XY & \sum X^2 \end{vmatrix}}{\Delta} = \dfrac{\sum Y \sum X^2 - \sum X \sum XY}{n\sum X^2 - \left(\sum X\right)^2} \\ m = \dfrac{\Delta_m}{\Delta} = \dfrac{\begin{vmatrix} n & \sum Y \\ \sum X & \sum XY \end{vmatrix}}{\Delta} = \dfrac{n\sum XY - \sum X \sum Y}{n\sum X^2 - \left(\sum X\right)^2} \end{cases} \tag{3.13}$$

here

$$\begin{cases} Y_i = \frac{\beta_i \cos\theta_i}{\lambda} & a = \frac{0.89}{D} \\ X_i = \frac{4\sin\theta_i}{\lambda} & m = \varepsilon \end{cases} \tag{3.14}$$

Finally, the following results are obtained:

$$\begin{aligned} D &= 0.89/a \\ \varepsilon &= m \end{aligned} \tag{3.15}$$

The above calculation can be carried out in dialog by using the solving program introduced by Qing et al. [9], Yan and Zhang [10]. The intermediate results and the final results can be given.

3.6 Characterization of AB_5 Alloy Prepared by Different Compositions and different methods [11]

3.6.1 Structure and Lattice Parameters of AB_5 Alloys with Different Compositions

The alloys with different compositions are melted by magnetic levitation method, annealed at 950 °C, then mechanically crushed, ball milled, and screened with 200 meshes. The composition and atomic radius of the seven alloys are listed in Table 3.11. For crop phase identification and lattice parameter determination, scanning is divided into two sections: $2\theta = 20°–50°$and 55°–85°, and the phase analysis results of the seven samples are listed in Table 3.13. The diffraction angle 2θ of each hkl crystal to $K\alpha_1$ radiation obtained from the pattern of $2\theta = 55°–85°$ is listed in Table 3.12. The lattice parameters *a* and *c* and cell volume *V* calculated by least square calculation are listed in Table 3.13. In order to explore the relationship between *a*, *c*, and *V* and alloy composition, the concept of weighted atomic radius *R* weight is introduced, which is defined as follows:

$$R_w = \frac{\sum_A a_{iA} R_{iA} + \sum_B a_{iB} R_{iB}}{\sum_A a_{iA} + \sum_B a_{iB}} \tag{3.16}$$

where a_i and R_i are the atomic fraction and atomic radius of the *i* element, and the subscript *A* and *B* represent the *A*-side and *B*-side elements, respectively, and their R_0 values are also listed in Table 3.13. It can be seen that except for the samples with a small amount of impurity phases, the weighted atomic radius (R_w) is linear with *a* and *c*, and *a* and *c* increase with the increase of the weighted atomic radius, as shown in Fig. 3.6.

Table 3.11 Composition atomic ratio of seven AB_5 alloys

	La	Ce	Pr	Nd	Er	Ni	Co	Mn	Al	Zr
1	0.620	0.270	0.050	0.080		3.690	0.720	0.400	0.190	
2	0.640	0.260	0.024	0.073		3.600	0.760	0.350	0.280	
3	0.625	0.270	0.026	0.077		3.650	0.766	0.340	0.243	
4	0.354	0.460	0.051	0.130		3.580	0.736	0.410	0.275	
5	0.315	0.490	0.052	0.141		3.720	0.700	0.355	0.230	
6	0.770	0.120	0.060	0.042	0.010	3.940	0.500	0.200	0.360	0.001
7	0.780	0.120	0.060	0.042		3.940	0.500	0.200	0.360	0.001
Atomic radius (nm)	0.1877	0.1824	0.1828	0.1822	0.1759	0.1246	0.1253	0.1366	0.1432	0.1599

Table 3.12 Diffraction data of seven AB_5 alloys, 2θ (°)

	112	211	202	300	301	220	221	113	311
1	58.597	60.841	62.760	64.241	68.877	75.782	81.082	81.786	83.558
2	58.536	60.857	62.679	64.358	68.878	75.879	80.862	81.590	83.640
3	58.520	60.922	62.702	64.441	68.900	75.940	80.800	81.821	83.861
4	58.600	60.982	62.681	64.558	69.041	76.160	80.860	81.878	84.040
5	58.583	61.098	62.819	64.658	69.158	76.204	80.780	81.999	84.219
6	58.560	60.720	62.660	64.160	68.642	75.638	80.245	81.583	83.539
7	58.600	60.721	62.759	64.157	68.697	75.622	80.879	81.697	83.522

Table 3.13 Calculation results of seven AB_5 alloys

	Main phase	Small amount	Trace amount	Weighted atomic radius (Å)	*a* (Å)	*c* (Å)	*V* ($Å^3$)
1	AB_5		A_2B_7	1.3623	4.9248	3.9450	82.817
2	AB_5			1.3647	4.9969	3.9992	86.475
3	AB_5			1.3664	4.9120	3.9294	82.312
4	AB_5			1.3629	4.9014	3.9373	81.914
5	AB_5			1.3616	4.8506	3.8902	79.265
6	AB_5	A_2B_7		1.3652	4.9599	3.9649	84.469
7	AB_5		α-Ni	1.3648	4.9310	3.9429	83.024

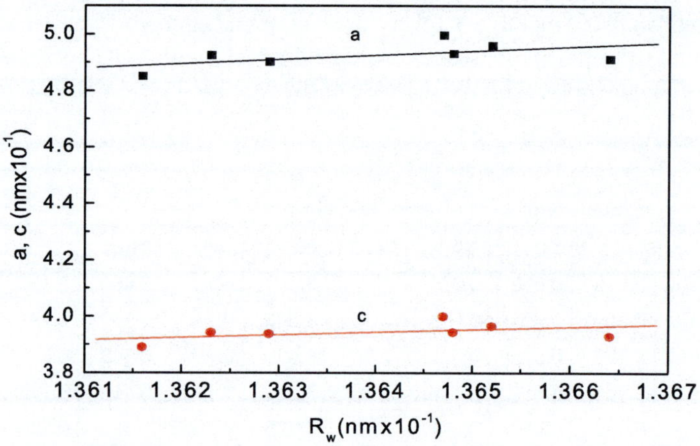

Fig. 3.6 Relationship between alloy weighted atomic radii R_w and *a* and *c* for different compositions

3.6.2 *Microstructure of Alloys Prepared by Different Methods*

The as-cast and annealed alloys of the same composition and the sheets obtained by single roll quenching with different linear speeds ($\varphi = 190$ mm) were mechanically ground and sifted over 200 meshes. The XRD patterns of as-cast, annealed (950 °C, 6 h), and five alloys with linear speeds of 20, 30, and 40 m/s are shown in Fig. 3.7. The phase structure is the same, but the linear shape is obviously different. The original data and the microcrystal size D and micro-strain ε obtained by the least square method are listed in Table 3.14. The results show that the grain size of the as-cast alloy is about 85 nm, there is a compressive strain of 0.75×10^{-4}, the grain size increases obviously after annealing, the micro-stress is not eliminated, and it may also be caused by grinding. It is easy to understand that there is tensile stress in the rapidly quenched alloy, which increases with the increase of cooling rate, but the grain size of 40 m/s quenched alloy is larger than that of 30 m/s and is similar to that of as-cast, which may be related to the different distribution of microcrystal and strain broadening between the two alloys.

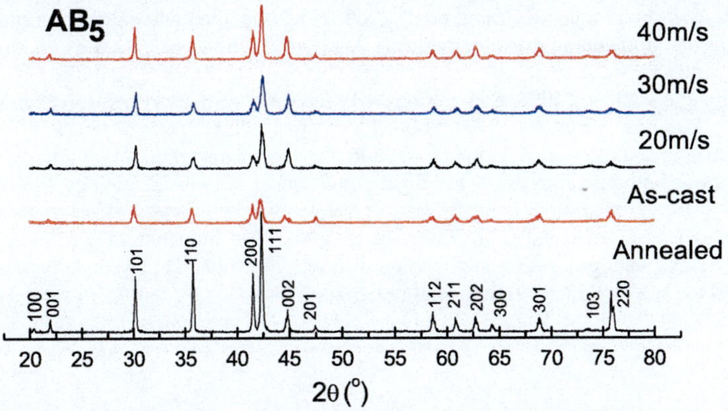

Fig. 3.7 XRD patterns of AB_5 alloys prepared by different methods as-cast

Table 3.14 XRD data of AB_5 alloys prepared by different methods

hkl		101	110	200	111	002	D (nm)	ε (10^{-4})
$\beta'_{1/2}$(°)		0.100	0.110	0.120	0.130	0.140		
$B'_{1/2}$(°)	As-cast	0.182	0.206	0.218	0.279	0.231	85	0.75
	Annealing	0.170	0.146	0.146	0.170	0.206	262	1.31
	20 m/s	0.194	0.243	0.267	0.279	0.261	48	0.69
	30 m/s	0.243	0.231	0.315	0.279	0.267	36	3.41
	40 m/s	0.170	0.206	0.231	0.218	0.255	87	4.20

References

1. Bode, H., J. Witte, et al. 1966. Nickel hydroxide electrode (I) Uber dur Nickel (II) Hydroxide hydrot. *Electrochem Acta* 11 (8): 1079–1087.
2. Lou, Yu-Wan, Chuan-Zheng Yang, and Xi-Gui Zhang, et al. 2006. Comparative study on microstructure of β-$Ni(OH)_2$ as cathode material for Ni–MH battery. *Science in China: Series E Technological Sciences* 36 (5): 467–482; 49 (3): 297–312.
3. Tessier, C. 1999. The structuir of $Ni(OH)_2$: from the ideal material to the electrochemically active one. *Journal of the Electrochemical Society* 146: 2059–2067.
4. Wang, C.Q., X.H. Ren, and W.Q. Jiang, et al. 1999. Effect of preparation conditions on structural parameters of $Ni(OH)_2$ microcrystals. *Chinese Journal of Nonferrous Metals* 9 (3): 504–509.
5. Delmas, C., and C. Tessier. 1997. Stacking faults in the structure of nickel hydroxide: a rationale of its high electrochemical activity. *Journal of Materials Chemistry* 7 (8): 1439–1443.
6. Treacy, M., J. Newsam, and M. Deem. 1991. A general recursion method for calculation diffraction intensities from crystals containing planar defaults. *Proceedings of the Royal Society of London. Series A: Mathematical and Physical Sciences* 43: 499–520.
7. Wang, C.Q., N. Wang, and N.N. Li, et al. 2002. Characterization of stacking fault structure of nickel hydroxide electrode materials. *Chinese Journal of Nonferrous Metals* 12 (3): 496–500.
8. Longford, J., and I.A. Bouluf. 1993. The use of pattern decomposition to study the combined X-ray diffraction effects of crystallite size and stacking faults in Ex-oxalate zine in oxide. *Journal of Applied Crystallography* 26 (1): 22–23.
9. Qing, P., Y.W. Lou, and C.Z. Yang, et al. 2006. A new method and calculation program for separating the broadening effect of multiple X-ray diffraction lines. *Acta Physica Sinica* 55 (3): 1325–1335.
10. Yan, C.Z., and J. Zhang. 2008. Some progress in the study of microstructure of nanomaterials by X-ray. *Progress in Physics* 28 (3): 280–313.
11. Wang, B.G., Z.L. Li, and C.Z. Yang, et al. 2006. Correct characterization of microstructure of AB_5 hydrogen storage alloy by X-ray diffraction. *Power Supply Technology* 30 (12): 1013–1016.
12. C.Z. Yang, L.F. Cheng, and B.G. Wang, et al. 2008. The influence of XRD experimental parameters and data processing methods on diffraction results. Chinese user papers of Neo-Science, No. 2, X-ray diffraction Special issue, 70–81.

Chapter 4
Solid-State Reaction and Formation Mechanism in the Process of $LiMeO_2$ Synthesis

Lithium-ion battery has been widely used. In order to obtain better charge and discharge performance and longer service life, the selection of battery positive materials is very important. Now, it has developed from $LiCoO_2$ to $LiNiO_2$, $Li(Ni_{1-x}Co_x)O_2$. In recent years, there are many researches on $Li(Ni,Co,Mn)O_2$ system, especially $Li(Ni_{1/3}Co_{1/3}Mn_{1/3})O_2$ was developed earlier [1–3]. In the past two years, Li $(Ni_{0.6}Co_{0.2}Mn_{0.2})O_2$ and $Li(Ni_{0.4}Co_{0.2}Mn_{0.4})O_2$ systems have also been reported [4, 5]. As far as the synthesis method is concerned, Sun, Quyang, and Wang et al. [1] prepared precursors using $Ni(NO_3)$ $2H_2O$, $Co(NO_3)$ $2H_2O$, and $Mn(NO_3)_2{\cdot}H_2O$ as raw materials and NaOH aqueous solution to adjust the pH value. Then, Li $(Ni_{1/3}Co_{1/3}Mn_{1/3})O_2$ was synthesized by mixing $LiOH{\cdot}H_2O$ with the precursor and calcining at 900C. In addition, some people use sulfate or acetate as raw materials to prepare precursors by co-precipitation method.

When our team in the synthesis of $Li(Ni_{0.6}Co_{0.2}Mn_{0.2})O_2$ and $Li(Ni_{0.4}Co_{0.2}Mn_{0.4})O_2$,

using nitrate as raw material, the precursor was prepared by co-deposition method adjusted PH value by NaOH or Na_2CO_3, respectively. Then, it was mixed with LiOH H_2O or Li_2CO_3 and synthesized by two-stage method at medium and high temperature. For more information on the synthesis process and electrochemical properties, please see two dissertations [4, 5].

The purpose of this chapter is to analyze the products of each stage of synthesis by X-ray diffraction (XRD) and to make a comprehensive analysis and comparison of the results, in order to find out the solid-state reaction, structure evolution in each stage, and the formation mechanism of $Li(Ni,Co,Mn)O_2$.

C. Yang et al., *Materials and Working Mechanisms of Secondary Batteries*,
https://doi.org/10.1007/978-981-19-5955-4_4

4.1 Experimental Steps to Study the Mechanism of $LiMeO_2$ Synthesis

4.1.1 Material Synthesis Strategy for the Study of the Formation Mechanism of $LiMeO_2$

The synthesis methods of $LiCoO_2$, $LiNiO_2$, Li $(Ni_{1-x}Co_x)O_2$, and $Li(Ni_xCo_yMn_{1-x-y})O_2$ are about the same. The detailed process can be found in the reference [1–5]. Generally, the precursor is prepared by co-deposition method, and then, the precursor is mixed with Li source and made by two steps roasting at medium and high temperature.

In order to facilitate the study of the structural evolution of the synthesis process and the synthesis mechanism of Li(Ni, Co, Mn)O_2, the following experimental steps were adopted.

(1) The precursors were prepared by co-deposition of pH value adjusted by NaOH and Na_2CO_3 aqueous solution, respectively, using nitrate as raw materials.
(2) The precursor was thermally decomposed at different medium temperatures.
(3) Two-stage synthesis of precursor + mixed with $LiOH{\cdot}H_2O$ or Li_2CO_3 at medium temperature or medium temperature + high temperature.
(4) In-situ X-ray diffraction observation and research during above variable temperature (room temperature ~1000°C) were carried out after mixing the precursor with $LiOH{\cdot}H_2O$.

4.1.2 X-Ray Diffraction Analysis

In addition to the phase identification and X-ray on-line study of the precursors obtained by co-deposition, the XRD analysis of the products in each stage is carried out in a quasi-dynamic manner, which involves not only the phase identification, but also the determination of the (FWHM) and the integral intensity ratio of the main diffraction lines, so that the Scherrer formula can be used.

$$D_{hkl} = \frac{0.89\lambda}{\beta_{hkl}\cos\theta_{hkl}} \tag{4.1}$$

The variation of grain size with calcination temperature and time is solved. Among $\beta = \beta_{1/2} - \beta_0$, that is, the measured FWHM minus the instrument broadening, the integral intensity ratio I_{003}/I_{104} and I_{101}/I_{104}, and for $Li(Ni_{0.6}Co_{0.2}Mn_{0.2})O_2$ and z = 0.24. According to the linear equation,

$$\sqrt{\frac{I_{003}}{I_{104}}} = 1.1968 - 2.2214X \tag{4.2}$$

$$\sqrt{\frac{I_{101}}{I_{104}}} = 0.7527 - 1.6295X \tag{4.3}$$

The changes of mixed occupation parameter X of Li and Ni at 3a and 3b sites with calcination temperature and time were quantitatively studied by linear equation [6, 7]. For the question of mixed occupancy, see Chapter 5.

4.2 XRD Analysis of Precursors and Thermal Decomposition Products

4.2.1 XRD Analysis of Precursors

When x:y:z = 0.6:0.2:0.2 or 0.4:0.2:0.4, the pH is adjusted by NaOH, and the XRD pattern of the obtained co-deposited precursor is shown in Fig. 4.1. It is identified as α-$(Ni_xCo_yMn_z)(OH)_2$ $0.75H_2O$ (reference PDF card number is 38–0715, $Ni(OH)_2$ $0.75H_2O$) or β-$(Ni_xCo_yMn_z)(OH)_2$ (reference PDF card number is 14–0117, β-$Ni(OH)_2$). Sometimes, α phase and β phase coexist, maybe α phase is more than β phase, or vice versa; it all depends on the specific process of co-deposition.

When x:y:z = 0.6:0.2:0.2, the pH value is adjusted by Na_2CO_3, and its precursor is alkaline carbonate (reference card number: 38–0714 $NiCO_3(OH)_2{\cdot}H_2O$), and the

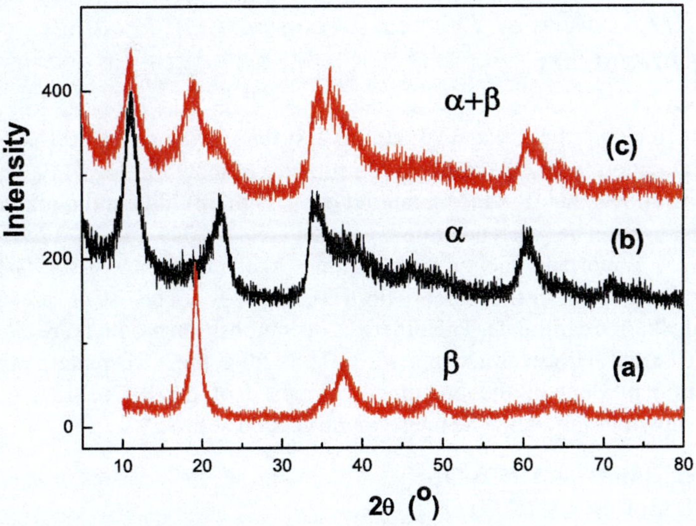

Fig. 4.1 XRD pattern of the precursor was co-precipitated by adjusting pH value with NaOH, **a** Pure β-$(Ni_xCo_yMn_z)(OH)_2$, **b** pure α-$(Ni_xCo_yMn_z)(OH)_2{\cdot}0.75H_2O$, and **c** coexistence of α phase and β phase

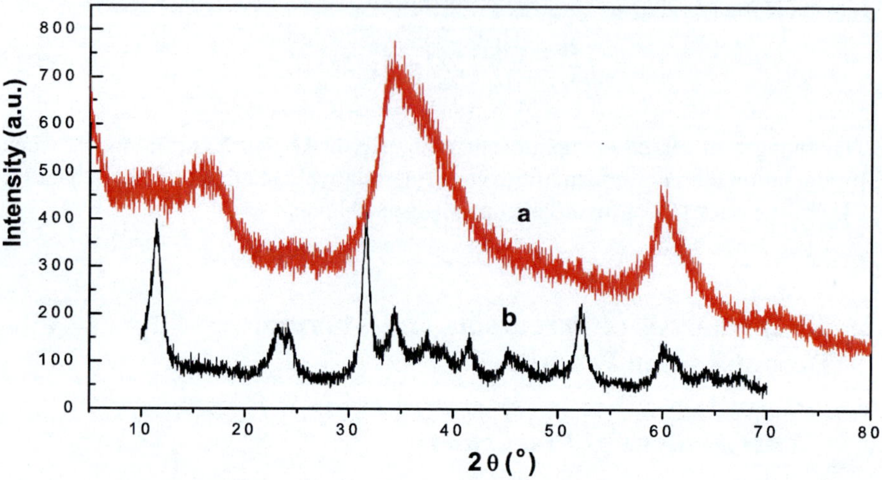

Fig. 4.2 XRD patterns of precursors were co-deposited with Na_2CO_3 to adjust pH value, **a** alkaline carbonate (poor crystallization, x:y:z = 0.6:0.2:0.2), **b** β-($Ni_xCo_yMn_z$)$(OH)_2$ + $MnCO_3$ (x:y:z: = 0.4:0.2:0.4)

crystallization condition is very poor, as shown in Fig. 4.2a. When the pH value is adjusted by Na_2CO_3, the precursor is β-($Ni_xCo_yMn_z$)$(OH)_2$ and $MnCO_3$ coexists, as shown in Fig. 4.2b.

4.2.2 XRD Analysis of Thermal Decomposition Products of Precursors

For the precursor of the x:y:z = 0.6:0.2:0.2 system calcined at 500–720°C, the typical XRD pattern of the decomposition product is shown in Fig. 4.3. The phase identification shows that it is face-centered cubic (Fm3m) NiO and diamond cubic (Fd3m) $MnCo_2O_4$. The reference number cards are 47–1049 and 23–1237, respectively. A very important phenomenon is that whether the precursor is pure α-($Ni_xCo_yMn_z$)$(OH)_2$, or β-($Ni_xCo_yMn_z$)$(OH)_2$, or α + β coexisting precursor or precursor of alkaline carbonate. Their thermal decomposition products are all mixture of these two oxides. However, when x:y:z = 0.4:0.2:0.4, the XRD pattern of thermal decomposition product of the precursor composed of β-($Ni_xCo_yMn_z$) $(OH)_2$ + $MnCO_3$ is shown in Fig. 4.4. The phase identification shows:

at 500 °C $NiMnO_3$ + $MnCo_2O_4$
at 600 °C $MnCo_2O_4$ + $NiMnO_3$
at 700 °C $MnCo_2O_4$ + NiO

Among them, $NiMnO_3$ belongs to the space group (No.148). It can be seen.

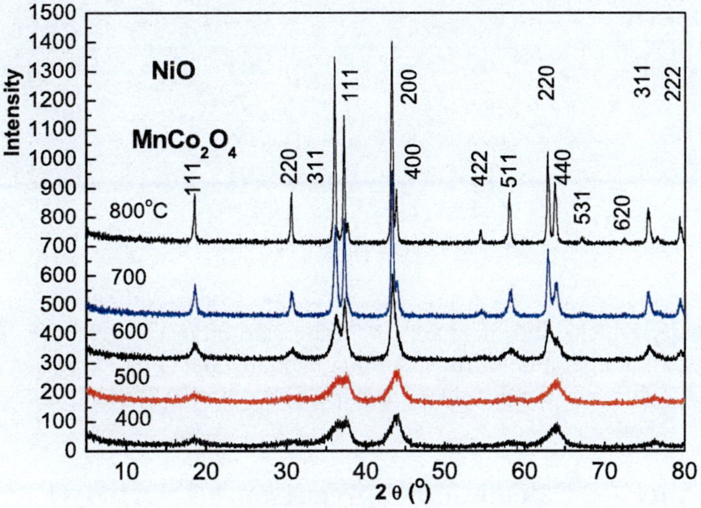

Fig. 4.3 XRD pattern of thermal decomposition products of precursors when $x : y : z = 0.6 : 0.2 : 0.2$

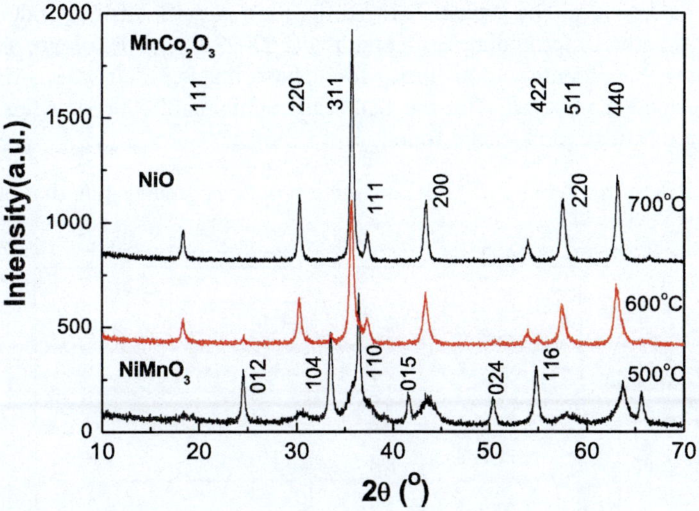

Fig. 4.4 Diffraction pattern of thermal decomposition products of precursors composed of β-$(Ni_xCo_yMn_z)(OH)_3 + MnCO_3$ when $x : y : z = 0.4 : 0.2 : 0.4$

(1) With the increase of temperature, the cubic phase increases and the rhombic phase decreases. While x:y:z = 0.6:0.2:0.2, with the increase of temperature, the proportion of Fd3m and Fm3m phase becomes smaller; in other words, the NiO phase increases rapidly with the increase of temperature.

Table 4.1 Data table of 200 diffraction lines of NiO at different temperatures for 3 h

Temperature and time	HWHM of (200) line (°)	D_{200}(nm)
400 °C 3 h	1.411	6.6
500 °C 3 h	1.227	7.7
600 °C 3 h	0.866	11.4
700 °C 3 h	0.271	56.1
800 °C 3 h	0.198	108.6

*β_o=0.120 2θ=43.98°

(2) No matter which kind of phase, the grain size increases gradually with the increase of temperature or time. Among them, the data of 200 diffraction lines of NiO at different temperatures for 3 h are shown in Table 4.1.

4.3 Products of Calcination of precursor + $LiOH{\cdot}H_2O$ or Li_2CO_3 at Medium and High Temperature

When x:y:z = 0.6:0.2:0.2, after the different precursors are uniformly mixed with $LiOH{\cdot}H_2O$ or Li_2CO_3. The typical X-ray diffraction patterns of the products after sintered at different intermediate temperatures (560–720 °C) for 6 h are shown in Fig. 4.5. After phase identification, the $LiMeO_2$ phase, that is, $Li(Ni_{0.6}Co_{0.2}Mn_{0.2})O_2$, has been formed (compared with the PDF card number: 09–0063). After careful observation, its characteristics are found.

(1) The diffraction intensity of (104) diffraction peak is greater than that of (003).

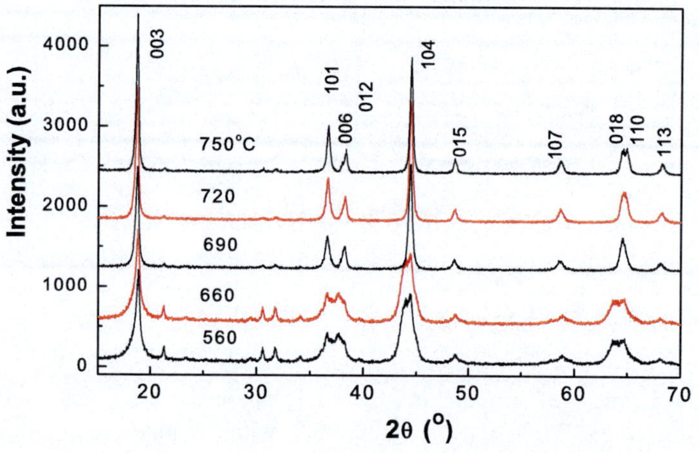

Fig. 4.5 XRD patterns of products sintered at different medium temperatures for 6 h

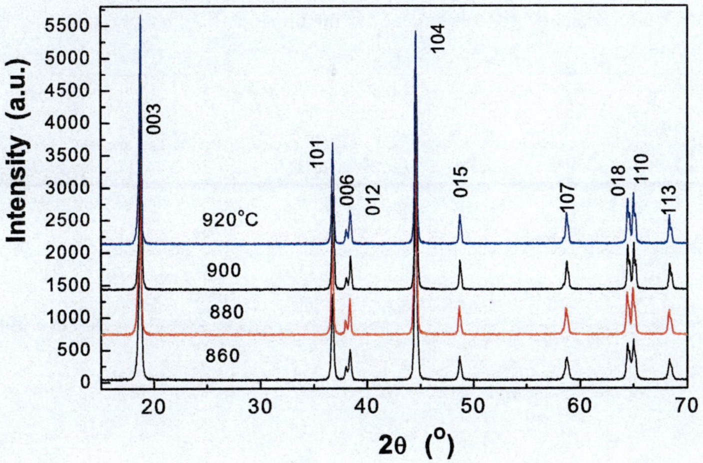

Fig. 4.6 XRD patterns of products sintered at 560 °C for 6 h at different high temperatures for 8 h

(2) The diffraction peaks were broadened obviously, and the two pairs of diffraction lines (006)/(012) and (018)/(110) could not be separated.

Because the diffraction pattern has the above characteristics, this semi-finished product is called distorted $Li(Ni_{0.6}Co_{0.2}Mn_{0.2})O_2$.

After the samples were calcined at 560 °C for 6 h, the XRD patterns of the products were shown in Fig. 4.6 after being calcined at 860, 880, 900, and 920 °C for 8 h. It was identified as a typical $Li(Ni_{0.6}Co_{0.2}Mn_{0.2})O_2$ by phase analysis, and the characteristics of its pattern are as follows:

(1) (003) the intensity of the diffraction peak is greater than that of (104), the broadening effect of the diffraction line decreases, and the line is sharp.
(2) The two pairs of diffraction lines (006)/(102) and (018)/(110) can also be obviously separated.

Comparative observation Figs. 4.5 and 4.6. It can be known that the characteristics have changed obviously, which implies that the fine structure and microstructure of $Li(Ni_{0.6}Co_{0.2}Mn_{0.2})O_2$ have changed during the two-stage temperature synthesis.

In order to further explain the changes of fine structure and microstructure, the. original data are listed in Table 4.2. The results calculated according to formula Eq. (4.1)–(4.3) are listed in Table 4.3. The lattice parameters a and c are obtained by Jade6.0 program Calculate Lattice and Cell Refinement processing. Comparing these data, we can see:

(1) In the middle temperature range (560–750 °C), the lattice parameter a decreases with the increase of temperature, but c increases, while the grain sizes D_{003}, D_{101}, and D_{104} grow up with the increase of temperature; generally speaking,

Table 4.2 Main XRD data of products calcined at medium and high temperature

	003			101			104		
	2θ(°)	HWHM(°)	I	2θ(°)	HWHM(°)	I	2θ(°)	HWHM(°)	I
560 °C 6 h	19.827	0.681	83				44.117	1.379	100
600 °C 6 h	19.735	0.496	96	36.549	0.616	37	44.278	0.580	100
660 °C 6 h	19.726	0.474	100	36.554	0.592	39	44.303	0.559	99
690 °C 6 h	19.702	0.348	96	36.555	0.467	38	44.337	0.365	100
720 °C 6 h	19.698	0.318	100	36.600	0.359	39	44.376	0.308	92
750 °C 6 h	19.677	0.280	100	36.631	0.318	38	44.406	0.290	83
$\beta_{1/2}$(°)		0.150			0.123			0.120	
560 °C 6 h + 860 °C8 h	19.759	0.255	100	36.751	0.209	38	44.506	0.232	90
560 °C 6 h + 880 °C 6 h	19.741	0.199	100	36.720	0.165	40	44.478	0.183	92
560 °C 6 h + 900 °C 6 h	19.835	0.203	100	36.809	0.172	41	44.554	0.178	90
560 °C 6 h + 920 °C 6 h	19.810	0.221	100	36.776	0.173	39	44.534	0.167	90

Table 4.3 Analysis results of XRD data at medium and high temperatures

	Lattice parameter (Å)		D_{003}	D_{101}	D_{104}	$\sqrt{\frac{I_{003}}{I_{104}}}$	X (%)	$\sqrt{\frac{I_{101}}{I_{104}}}$	X (%)
	a	c	(nm)	(nm)	(nm)				
560 °C 6 h	2.8985	14.1743	16.0		6.7	0.912	12.8		
600 °C 6 h	2.8804	14.1984	23.0	16.8	19.4	0.980	9.6	0.611	8.7
660 °C 6 h	2.8788	14.2047	24.6	17.6	19.3	1.006	8.6	0.632	7.4
690 °C 6 h	2.8784	14.2196	40.2	24.1	34.6	0.981	9.7	0.614	8.5
720 °C 6 h	2.8736	14.2250	47.4	35.1	45.1	1.042	7.0	0.648	6.4
750 °C 6 h	2.8697	14.2407	61.3	42.4	50.0	1.101	4.3	0.681	4.4
560 °C6 h + 860 °C 8 h	2.8663	14.1823	75.8	96.2	75.7	1.053	4.7	0.651	6.2
560 °C 6 h + 880 °C6 h	2.8685	14.1953	162.5	179.0	134.6	1.043	6.9	0.661	5.6
560 °C 6 h + 900 °C 6 h	2.8662	14.1315	168.5	168.8	146.2	1.052	6.5	0.675	4.8
560 °C 6 h + 920 °C 6 h	2.8676	14.1493	112.2	165.5	180.5	1.055	6.4	0.659	5.8

the mixed arrangement parameter X of Li and Ni at 3a and 3b sites decreases with the increase of temperature.

(2) In the high temperature range (860–920 °C), the lattice parameter a does not change much with the increase of temperature, but c decreases somewhat. The mixing parameter at 560 °C 6 h + 860 °C for 8 h is much lower than that at 560 °C for 6 h, but it does not change much in the high temperature section.

4.4 In-Situ XRD Study of Precursor + LiOH·H_2O at Variable Temperature

According to the element ratio requirement of $Li(Ni_{0.6}Co_{0.2}Mn_{0.2})O_2$, the precursor of β-$(Ni_{0.6}Co_{0.2}Mn_{0.2})(OH)_2$ and LiOH·H_2O were fully mixed and ground and then placed on the sample frame of high temperature accessory of Rigaku D/max-2200PC X-ray diffractometer. θ/2θ was scanned for 1 h at 25, 200, 400, 500, 600, 700, 800, 850, 900, and 950 °C, respectively, and the scanning range was 10–80°. One scan takes about 10 min. The diffraction patterns for the middle and high temperature segments are shown in Fig. 4.7a and b, respectively. The main data and analysis results are listed in Table 4.4.

From the analysis results of Fig. 4.7 and Table 4.4, we can see that:

(1) Up to 700C, Li_2CO_3 and $NiCo_2O_4$ have not disappeared; that is, the LiOH·H_2O → LiOH → Li_2CO_3 process occurs, but the lattice parameters are much smaller.
(2) $Li(Ni_{0.6}Co_{0.2}Mn_{0.2})O_2$ phase has begun to appear from 400C, and Li $(Ni_{0.6}Co_{0.2}Mn_{0.2})O_2$ has basically formed at 800C.
(3) Comparing Fig. 4.7a and b, it can be seen clearly that the integral intensity of line (104) is greater than that of line (003) in the middle temperature range (500–700 °C). At 800 °C, the situation is just the opposite, which is consistent with the results of Sect. 4.3.

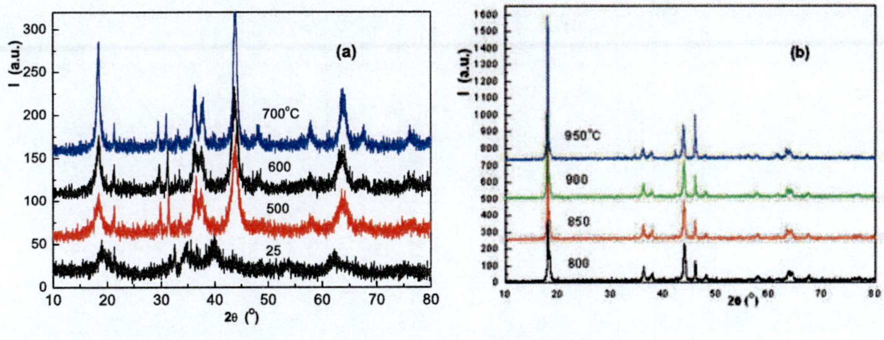

Fig. 4.7 Synthesis of in-situ XRD pattern, **a** XRD pattern obtained at medium temperature (400–700 °C), **b** XRD patterns obtained at high temperatures (800–950 °C)

Table 4.4 Main diffraction data and analysis results in Fig. 4.7

	003		104		Existing phase	D_{003}	D_{104}	I_{003}/I_{104}	X
	I	HWHM(°)	I	HWHM(°)		(nm)	(nm)		(%)
25 °C					β-(Me)$(OH)_2$ + LiOH·H2O				
400 °C 1 h	29	3.256	100	1.874	$LiMeO_2 + MeO_4 + Li_2CO_3$	2.5	2.8	0.290	29.6
500 °C 1 h	31	1.627	100	1.617	$LiMeO_2 + MeO_4 + Li_2CO_3$	5.3	5.4	0.314	28.7
600 °C 1 h	35	0.889	100	1.315	$LiMeO_2 + MeO_4 + Li_2CO_3$	10.4	6.8	0.355	21.4
700 °C 1 h	55	0.650	100	0.728	$LiMeO_2 + MeO_4 + Li_2CO_3$	16.1	13.4	0.552	20.4
800 °C 1 h	100	0.272	79	0.368	$LiMeO_2$	53.4	32.8	1.272	3.1
850 °C 1 h	100	0.213	63	0.337	$LiMeO_2$	88.5	34.5	1.597	-3.0
900 °C 1 h	100	0.154	36	0.358	$LiMeO_2$	257.8	34.1	2.755	-20.8
900 °C 1 h	100	0.155	34 ara>	0.354	$LiMeO_2$	248.9	34.7	2.907	-22.9
950 °C 1 h	100	0.164	39	0.322	$LiMeO_2$	194.2	40.3	2.564	-19.2
1000 °C 1 h	100	0.195	59	0.318	$LiMeO_2$	111.0	41	1.683	-4.5

(4) The data of D_{003} and D_{104} show that the grain size increases with the increase of temperature, but grows slowly in the high temperature section, and when the temperature reaches 850 °C, the grain size of $Li(Ni_{0.6}Co_{0.2}Mn_{0.2})O_2$ at [001] direction is several times of that of [104] direction. It shows that the grain shape is a columnar crystal perpendicular to (001).

(5) The mixed occupation parameter X of Li and Ni atoms at 3a and 3b decreases with the increase of temperature up to 800 °C. When the temperature is higher than 850 °C, because the diffraction lines of (003) are abnormally enhanced due to the preferred orientation of (001), X is negative value, which is not believed.

4.5 Solid-State Reaction in the Synthesis of Li(Ni,Co,Mn)O_2

From the results of Sects. 4.1, 4.2, 4.3 and 4.4, the solid-state reaction in the synthesis of Li (Ni,Co,Mn)O_2 can be summarized. For the samples of Ni:Co:Mn = 0.6:0.2:0.2, the synthesis process of $Li(Ni_{0.6}Co_{0.2}Mn_{0.2})O_2$ is as follows:

$0.6\ Ni(NO_3)_2 \cdot H_2O + 0.2\ Co(NO_3)_2 \cdot H_2O + 0.2\ Mn(NO_3)_2 \cdot H_2O$

\+ NaOH → α-$(Ni_{0.6}Co_{0.2}Mn_{0.2})(OH)_2$ or β-$(Ni_{0.6}Co_{0.2}Mn_{0.2})(OH)_2$ or $\alpha+\beta$

\+ Na_2CO_3 → $(Ni_{0.6}Co_{0.2}Mn_{0.2})(OH)_2CO_3 \cdot H_2O$

↓ thermal decomposition at medium temperature

$NiMnO_2 + NiO + NiMnO_3$

\+

$LiOH \cdot H_2O$ or Li_2CO_3

↓ Medium temperature roasting

Distorted $Li(Ni_{0.6}Co_{0.2}Mn_{0.2})O_2$

↓ High temperature roasting

Finished product $Li(Ni_{0.6}Co_{0.2}Mn_{0.2})O_2$

The main solid-state reactions are as follows:

$(Ni_{0.6}Co_{0.2}Mn_{0.2})(OH)_2CO_3{\cdot}H_2O \rightarrow NiO + MnCo_2O_4$
$(Ni_{0.6}Co_{0.2}Mn_{0.2})(OH)_2 \rightarrow NiO + MnCo_2O_4$
$Li_2CO_3{\cdot}H_2O \rightarrow Li_2CO_3 \rightarrow Li_2O + CO_2\uparrow$
$NiO + Li_2O \rightarrow LiNiO_2 + CO_2\uparrow$
$MnCo_2O_4 + Li_2O \rightarrow Li(Co,Mn)O_2.$
$LiNiO_2 + Li(Co,Mn)O_2 \rightarrow Li(Ni,Co,Mn)O_2$

For the samples of Ni:Co:Mn = 0.4:0.2:0.4, the synthesis process of $Li(Ni_{0.4}Co_{0.2}Mn_{0.4})O_2$ is as follows:

$0.4\ Ni(NO_3)_2{\cdot}H_2O + 0.2\ Co(NO_3)_2{\cdot}H_2O + 0.4\ Mn(NO_3)_2{\cdot}H_2O$

\+ +

NaOH Na$_2$CO$_3$

β-$(Ni_{0.4}Co_{0.2}Mn_{0.4})(OH)_2$ β-$(Ni,Co)(OH)_2 + MnCO_3$

↓ thermal decomposition at medium temperature

$NiO + CoMn_2O_4$ $NiMnO_3 + MnCo_2O_4 + NiO$

\+

$LiOH{\cdot}H_2O$ or Li_2CO_3

↓ Medium temperature roasting

Distorted $LiNi_{0.4}Co_{0.2}Mn_{0.4}O_2$

↓ High temperature roasting

Finished product: $LiNi_{0.4}Co_{0.2}Mn_{0.4}O_2$

The main solid-state reactions are as follows:

$(Ni_{0.4}Co_{0.2}Mn_{0.4})(OH)_2 \rightarrow NiO + CoMn_2O_4$
$Mn_2CO_3 + \beta\text{-}(Ni_{0.4}Co_{0.2}Mn_{0.4})(OH)_2 \rightarrow NiMnO_3 + MnCo_2O_4$
$NiMnO_3 + Li_2O \rightarrow Li(Ni,Mn)O_2$
$CoMn_2O_4 + Li_2O \rightarrow Li(Co,Mn)O_2$
$Li(Ni,Mn)O_2 + Li(Co,Mn)O_2 \rightarrow Li(Ni,Co,Mn)O_2$

4.6 Structural Evolution and Formation Mechanism During the Synthesis of $LiMeO_2$

The results of Sects. 4.2 and 4.4 are summarized, and the actual synthesis is that the precursor is fully mixed with the lithium source ($LiOH{\cdot}H_2O$ or Li_2CO_3) and placed in a rotary furnace crucible, which is divided into the following two processes:

(1) Heat up to a certain medium temperature (e.g., 650 °C) at a certain heating rate for several hours, then continue to rise to a high temperature (e.g., 900 °C) for several hours, and then come out of the furnace. The furnace body rotates slowly in the whole process.
(2) After the heat preservation of the medium temperature section is completed, the intermediate products are taken out, milled, and then directly heated to high temperature for several hours.

Therefore, the most substantial synthesis step is the solid–solid reaction of oxides of Ni, Co, and Mn (one or more) with Li_2O. Its formation mechanism is discussed as follows:

(1) NiO belongs to Fm3m (No.225) structure with a = 4.177 Å, and NiO may also belong to a structure (No.166), a = 2.955, c = 7.255 Å; $MnCo_2O_4$ belongs to a spinel (diamond cubic Fd3m) structure (No.227), a = 8.269 Å; and a compound $LiMeO_2$ belongs to R-3 m structure (No.166) a = 2.878 Å, c = 14.19 Å. Therefore, for the cubic oxides, such as NiO or $MnCo_2O_4$ or $NiCo_2O_4$, Li atoms not possible diffuse into this lattice at medium temperature, but to nucleate and grow again through solid–solid diffusion, Li atoms diffuse to the surface of cubic oxide particles and grains. When the composition reaches a certain level, it re-nucleates to form a new phase and then grows with the continued diffusion of Li atoms; that is, the formation of $LiNiO_2$ or Li $(Co,Mn)O_2$ has a nucleation-growth process.
(2) Both $NiMnO_3$ and $CoMnO_3$ belong to R-3 (No.148) structure, and the lattice parameters are as follows: a = 4.890 Å, c = 13.58 Å and a = 4.933 Å, c = 13.71 Å, respectively. If there are two molecules in the unit cell, then the two kinds of metal atoms occupy 3a and 3b sites, respectively, leaving one vacancy, and six oxygen atoms occupy 6c sites, which is similar to that of $LiMeO_2$(a = 2.86 Å, c = 14.19 Å) The difference of lattice parameter a is very large, but c not much different. If there are three molecules in the unit cell, atoms of two metals occupy 3a and 3b, respectively, and nine oxygen atoms occupy the 9d or 9e position. Therefore, it is possible that based on the R-3 lattice, the Li atom diffuses to the R-3 lattice, partially replacing the Ni atom 3a site. Partial Ni atoms diffuse into 3b site, thus forming a distorted $LiMeO_2$ with a certain mixed occupying parameter X, and the lattice parameters change accordingly.
(3) In addition, there is a direct heating process, through 500–700 °C especially when LiOH or Li_2CO_3 is in the liquid state (its melting point is 471C and 618C, respectively), and NiO or $MnCoO_2$ (or $MnCoO_3$) may be further oxidized to NiO_2 [or $(Ni,Co,Mn)O_2$] when it is in the melting state with the liquid lithium

source, which has the same structure as $LiMeO_2$ (No.166), a = 2.835 Å, c = 14.33 Å. The specific data are related to the proportion of Ni, Co, and Mn. In the NiO_2 structure with R-3 m, there are 3 molecules (z = 3) in the unit cell, 3 Ni atoms occupy 3a position, and 6 oxygen atoms occupy 6c position. Then, lithium atoms need to cross the smaller potential barrier and diffuse into the 3a and 3b sites of the $(Ni,Co,Mn)O_2$ lattice to directly form the distorted $Li(Ni,Co,Mn)O_2$ semi-finished product.

(4) The distorted $LiNiO_2$ and Li $(Co,Mn)O_2$ are formed by one or several of the above three mechanisms, and then, the distorted $Li(Ni,Co,Mn)O_2$ is formed by mutual diffusion, which has larger mixed occupying parameters and finer grain size, and the mixed occupying parameters decrease and the grain size increases with the increase of temperature.

(5) At high temperature, the atomic occupation of this distorted $Li(Ni,Co,Mn)O_2$ is adjusted with the increase of mutual diffusion coefficient, and the mixed occupying parameter X becomes smaller and smaller, which may be reduced to zero under certain conditions; that is, Li occupies 3a, Ni, Co, and Mn occupy 3b, and 6 oxygen atoms occupy 3c, achieving the ideal occupancy situation. At the same time, the grains grow gradually. Of course, the specific values of the mixed occupying parameters and grain size of the final product are also related to the process of the temperature section (such as temperature, holding time, and cooling rate).

Summing up the results of this chapter, we can know the physical process of $LiMeO_2$ synthesis.

(1) The core of $LiMeO_2$ synthesis is two steps:

(i) Oxides at medium temperature

$$\left.\begin{array}{l}\left.\begin{array}{l}\text{NiO face centered cube}\\ \text{MnCo}_2\text{O}_4\text{diamond cube}\end{array}\right\} + \begin{array}{l}\text{lithium source, interaction}\\ \text{re} - \text{nucleation and growth}\end{array}\\ \left.\begin{array}{l}\text{MnCoO}_3\text{, 3R.Rhombohedral}\\ \text{NiO}_2\text{, R} - \text{3mRhombohedral}\end{array}\right\} + \begin{array}{l}\text{lithium source interaction,}\\ \text{lithium atom diffusion to}\\ \text{oxide lattice}\end{array}\end{array}\right\}\begin{array}{l}\text{forms distorted}\\ \text{Li(Ni, Co, Mn)O}_2\end{array}$$

(ii) In the high temperature section, the distorted $Li(Ni,Co,Mn)O_2$ adjusts the atomic occupation and reduces the mixed occupation parameter X, and the grain size increases continuously. Finally, the final product $Li(Ni,Co,Mn)O_2$ with reasonable mixed space occupation and grain size and shape is formed.

The experimental results show that these processes have nothing to do with the specific phase to phase ratio of the precursor. Only to ensure that the atomic ratio of Ni, Co, and Mn in the precursor meets the requirements.

The experimental results show that these processes have nothing to do with the specific phase to phase ratio of the precursor. Only to ensure that the atomic ratio of Ni, Co, and Mn in the precursor meets the requirements.

From the above conclusion, it can be inferred that using nitrate or metal oxides, under the condition that the atomic ratios of Ni, Co, and Mn meet the requirements, the required $LiMeO_2$ can be directly synthesized by mixing with the lithium source ($LiOH{\cdot}H_2O$) and calcining at medium and high temperature. Our experiment has been successful, which will greatly simplify the process and reduce the cost. Metal composite oxides have been successfully used as intermediates to prepare multicomponent cathode materials for lithium-ion batteries [8].

(2) The formation mechanism of $LiMeO_2$ is controlled by diffusion, nucleation-growth of new phase, and/or structure of Li atoms unidirectional diffusion into R-3 or R-3 m. Then, it is completed by mutual diffusion, grain growth, and space occupation adjustment to improve the integrity.

References

1. Sun, Yncheng, Chuying Ouyang, Zhaoxing Wang, et al. 2004. Effect of Co content on rate performance of $Li(Mn_{0.5-x}Co_{2x}Ni_{0.5-x})O_2$ cathode materials for lithium-ion batteries. *Journal of the Electrochemical Society* 151 (4): A504–A508.
2. Chen, Y., G.X. Wang, K. Konstantinov, et al. 2003. Synthesis and characterization of $Li(Co_xMn_yNi_{1-x-y})O_2$ as a cathode material for secondary lithium batteries. *Journal of Power Sources* 119–121: 184–188.
3. Machheil, D.D., Z. Lu, and J.R. Dahn. 2002. Structure and electrochemistry of $Li[Ni_xCo_{1\text{-}2x}Mn_x]O_2(0{\le}x{\le}1/2)$. *Journal of the Electrochemical Society* 149 (10): A1332–A1336.
4. Zhang, Y. 2007. Preparation of lithium ion positive material $Li(Ni_xCo_yMn_{1\text{-}x\text{-}y})O_2$ by carbonate coprecipitation method. Master paper of Shanghai Institute of Microsystems and Information Technology, 5; Zhang, Y., H. Cao, J. Zhang, and B.J. Xia. 2006. Synthesis of $Li(Ni_{0.6}Co_{0.2}Mn_{0.2})O_2$ cathode material by a carbonate co-precipitation method and its electrochemical characterization. *Solid State Ionics* 177: 3303–3307.
5. Cao, H. 2006. Preparation and properties of lithium ion cathode material Li $(Ni_xCo_yMn_{1\text{-}x\text{-}y})$ O_2. Master paper of Shanghai Institute of Microsystems and Information Technology; Cao, H., B.J. Xia, et al. 2005. Synthesis and electrochemical characteristic of layered lithium ion batteries. *Solid State Ionics* 176: 1207–1211.
6. Zhang, X.G., J. Zhang, C.Z. Yang, and B.J. Xia. 2010. A new method for studying and analyzing the mixed occupation of lithium and nickel atoms in Li (Ni,Co,Mn) O_2. *Journal of Inorganic Material* 25 (1): 8–12.
7. Zhang, J., C.Z. Yang, and B.J. Xia. 2009. Simulation analysis and study on the structural changes of Li (Ni,Co,Mn) O_2 during synthesis. *Material Science and Engineering* 27 (6): 824–828.
8. Xia, B.J., J. Zhang, and X.W. Han. Preparation of multicomponent positive materials for lithium-ion batteries using transition metal composite oxides as intermediates. Patent number CN 100372774C; Xia, B.J., H.W. Han, and H. Cao. Multicomponent composite positive material for ion battery and preparation method. Invention patent CN 100342570C.

Chapter 5
Mixed Occupation of Ni/Li Atoms in $LiMeO_2$ Materials

In recent years, the research on the development of alternatives to $LiCoO_2$ as positive active materials for lithium-ion batteries is being accelerated [1, 2]. There have been three kinds of materials: $LiNiO_2$, $Li(Ni,Co)O_2$, and $Li(Ni,Co,Mn)O_2$, and the $Li(Ni,Co,Mn)O_2$ is one of the most widely studied and promising materials [2–4]. In the study, it is found that there is usually a mixed occupation of Li/Ni in these three kinds of positive active materials containing Ni, that is, the phenomenon of cationic mixed arrangement [5]. This phenomenon directly interferes with the detachment and re-intercalation of Li ion during charge and discharge, resulting in the decline of the electrochemical performance of the materials [6, 7]. Therefore, the study of crystallographic mixed occupation of Li/Ni atoms in $LiMeO_2$ materials has attracted wide attention of researchers.

The classical method to study the crystallographic space occupation is to carefully collect the diffraction pattern data of the powder sample, carry out the full-spectrum fitting calculation with the Rietveld method and calculation program, and finally obtain the atomic space occupation from the atomic occupation probability. Kin et al. [8] collected powder neutron diffraction data and confirmed by Rietveld refinement that there is a mixed occupation of 3a and 3b Li/Ni when $Li(Ni_{1/3}Co_{1/3}Mn_{1/3})O_2$ is charged to 4.7 V. Neutron diffraction is chosen because the difference of atomic scattering factor of X-ray by Ni, Co, and Mn is very small, but the difference of scattering by neutron is much larger. Liu et al. [9] tried to optimize the synthesis conditions of the materials by powder X-ray diffraction and Rietveld method of $CuK\alpha$ radiation. The author has also tried to use this method to study the atomic occupation of this kind of materials.

In the study, it is found that: (1) It takes several (5–8) h for each sample to collect experimental data with $CuK\alpha$ radiation; (2) it takes a lot of time to refine with GSAS et.al. other programs, especially when some finishing parameters need to be modified manually, sometimes even do not know how to modify the finishing parameters; and (3) the finishing method cannot establish the quantitative or semi-quantitative relationship between diffraction data and mixed space occupation.

C. Yang et al., *Materials and Working Mechanisms of Secondary Batteries*,
https://doi.org/10.1007/978-981-19-5955-4_5

Of course, because the Rietveld method is a full-spectrum fitting calculation, but also can make zero position and preferred orientation corrections, the results are much more accurate and reliable.

Based on the above reasons, it is particularly important to explore other simple and effective methods. For this reason, on the basis of a comprehensive study and analysis of the main diffraction intensity characteristics of the materials with this R-3 m structure, the author puts forward a set of simulation calculation method, that is, a new simulation calculation method based on the mixed space occupying model. This method is applied to the above three kinds of materials, and the quantitative relationship of mixed space occupying parameters is established by diffraction intensity ratio. Finally, experimental research and analysis are carried out.

5.1 Principles and Methods of Simulation Analysis [12–14]

5.1.1 The Principle of Simulation Analysis

In the case of symmetrical Bragg reflection geometry of powder X-ray diffractometer, the relative integral intensity of single-phase material is:

$$I_{hkl} = P_{hkl}|F_{hkl}|^2 \frac{1+\cos^2 2\theta_{hkl}}{\sin^2\theta_{hkl}\cos\theta_{hkl}} \tag{5.1}$$

where P_{hkl} is the multiplicity factor of (hkl) crystal plane, F_{hkl} is the structure factor, and the last term is the angle factor. The formula for calculating the structure factor is as follows:

$$F_{hkl} = \sum_{j=1}^{n} f_j e^{2\pi i(hx_j+ky_j+lz_j)}$$

$$= \sum_{j=1}^{n} f_j[\cos 2\pi(hx_j+ky_j+lz_j) + i\sin 2\pi(hx_j+ky_j+lz_j)] \tag{5.2}$$

where f_j is the atomic scattering factor of the j atom to the incident X-ray. x_j, y_j, z_j is the coordinate of the j atom in the unit cell, and the summation is carried out on all the atoms in the unit cell. For the $LiMeO_2$ of the No. 166 space group, it is the summation of three Li atoms, three Me atoms, and six oxygen atoms in the cell. The correct occupation of these 12 atoms in a unit cell is:

$$\left.\begin{array}{c} \text{Li3a}(0,0,0) \\ \text{Me3b}(0,0,1/2) \\ \text{O6c}(0,0,\pm z) \end{array}\right\} + (0,0,0),(1/3,2/3,2/3)\ \text{and}(2/3,1/3,1/3)\ \text{Symmetrical operation}$$

Generally, z is taken as 1/4. The 3a, 3b, and 6c position parameters of the 12 atoms of the main diffraction lines are listed in Table 5.1.

Special attention should be paid to the "addition" and "subtraction" relationship of the contribution of the 3a and 3b positions in the last rows of Table 1 to the intensity of the diffraction lines.

Let the atomic scattering factors at positions 3a, 3b, and 6c are f_{3a}, f_{3b}, and f_{6c}, for $Li_{1-X}Ni_X(Li_XNi_{0.6-X}Co_{0.2}Mn_{0.2})O_2$. They are obtained from the following formula:

$$\begin{cases} f_{3a} = (1 - X) f_{Li} + X f_{Ni} \\ f_{3b} = X f_{Li} + (0.6 - X) f_{Ni} + 0.2 f_{Co} + 0.2 f_{Ni} \\ f_{6c} = f_o \end{cases} \tag{5.3}$$

X is the mixed occupation parameter, and f_{Li}, f_{Ni}, f_{Co}, f_{Mn}, and f_o are the atomic scattering factors of Li, Ni, Co, Mn, and O. They are all functions of $\frac{\sin\theta}{\lambda}$, and decrease with the $\frac{\sin\theta}{\lambda}$ value increases. For the same diffraction pattern, λ is constant, which means that the atomic scattering factor decreases with the increase of 2θ; that is, the atomic scattering factor varies with different *hkl*, and the larger the diffraction angle 2θ is, the smaller the atomic scattering factor is.

Similarly, the angle factor $\frac{1+\cos^2 2\theta_{hkl}}{\sin^2\theta_{hkl}\cos\theta_{hkl}}$ of each diffraction line decreases with the increase of 2θ, and its specific value can be calculated or obtained by consulting the schedule of professional books.

So far, the main characteristics of the $LiMeO_2$ diffraction lines of this structure of R-3 m can be summarized as follows:

(1) In terms of the contribution of 6c oxygen atoms to the diffraction intensity, it can be divided into:

No contribution: such as 003, 101, 015, 009, 107, 018, and 113.
Addition contributions: such as 104 and 110.
Subtractive contributions: such as 006 and 012.

(2) In terms of the contribution of 3a and 3b atoms, it can be divided into:

Additive relationships: such as 006, 012, 104, 110, and 018.
Subtractive relationships: such as 003, 101, 015, 009, 107, and 113.

(3) The "addition" and "subtraction" relationship between 3a seat and 3b seat reveals a very important law. With 003 and 104 is an example

After calculating the F_{hkl} and merging the similar terms, the following results can be obtained:

$$\begin{cases} F_{003} = 3(1 - 2X) f_{Li}^{003} 3(0.62X) f_{i}^{003} - 0.6(f_{Co}^{003} + f_{Mn}^{003}) \\ F_{104} = 3 f_{Li}^{104} + 1.8 f_{Ni}^{104} + 0.6(f_{Co}^{104} + f_{Mn}^{104}) + 6 f_o^{104} \end{cases} \tag{5.4}$$

Table 5.1 (No. 166) atomic position parameters in the structure

hkl		003	101	006	012	104	015	009	102	018	110	113
Relative intensity I/I_1		100	40	12	15	95	18	9	20	25	25	13
position	Atomic coordinates	Position parameter $\cos 2\pi(hx_j + ky_j + lz_j) + i \sin 2\pi(hx_j + ky_j + lz_j)$										
3a	0 0 0	+ 1	+ 1	+ 1	+ 1	+ 1	+ 1	+ 1	+ 1	+ 1	+ 1	+ 1
	2/31/3 1/3	+ 1	+ 1	+ 1	+ 1	+ 1	+ 1	+ 1	+ 1	+ 1	+ 1	+ 1
	1/3 2/3 2/3	+ 1	+ 1	+ 1	+ 1	+ 1	+ 1	+ 1	+ 1	+ 1	+ 1	= 1
	Σ	+ 3	+ 3	+ 3	+ 3	+ 3	+ 3	+ 3	+ 3	+ 3	+ 3	+ 3
3b	0 0 1/2	–1	–1	+ 1	+ 1	+ 1	–1	–1	–1	+ 1	+ 1	–1
	2/3 1/3 5/6	–1	–1	+ 1	+ 1	+ 1	–1	–1	–1	+ 1	+ 1	–1
	1/3 2/3 7/6	–1	–1	+ 1	+ 1	+ 1	–1	–1	–1	+ 1	+ 1	–1
	Σ	–3	–3	+ 3	+ 3	+ 3	–3	–3	–3	+ 3	+ 3	–3
6c	0 0 1/4	–i	+ i	–1	–1	+ 1	–i	–i	+ i	–i	+ 1	–i
	2/3 1/3 7/12	–i	+ i	–1	–1	+ 1	–i	–i	+ i	–i	+ 1	––i
	1/3 2/3 11/12	–i	+ i	–1	–1	+ 1	–i	–i	+ i	–i	+ 1	–i
	0 0 – 1/4	+ i	–i	–1	–1	+ 1	+ i	+ i	–i	+ i	+ 1	+ i
	2/3 1/3 1/12	+ i	–i	–1	–1	+ 1	+ i	+ i	–i	+ i	+ 1	+ i
	1/3 2/3 5/12	+ i	–i	–1	–1	+ 1	+ i	+ i	–i	+ i	+ 1	+ i
	Σ	0	0	–6	–6	+ 6	0	0	0	0	+ 6	0
the total contribution in 3a and 3b		Subtract	Subtract	Add	Add	Add	Subtract	Subtract	Subtract	Add	Add	Subtract

The formula (5.4) shows that the diffraction intensity of the lines contributed by the additive relationship of 3a and 3b (such as 006, 012, 104, 110, and 018) has nothing to do with the mixed occupying parameter X, while the diffraction intensity of the subtractive lines (such as 003, 101, 015, 009, 017, and 113) varies obviously with the mixed occupying parameter X. If the intensity of the diffraction line of the subtraction relation can be measured absolutely, the mixed occupation of Li/Ni at 3a and 3b can be understood. But this is difficult to achieve in experiment. Therefore, the study of the diffraction intensity ratio of the addition relation and the subtraction relation becomes a feasible way for us to understand the mixed occupation of Li/Ni atoms.

When it comes to using the intensity ratio of diffraction pairs to study $LiNiO_2$-related problems, the following three points are worth explaining:

(1) When Dahn et al. [10] studied the diffraction intensity ratio of $Li_XNi_{2-X}O_2$, under the assumption that $Li_XNi_{2-X}O_2$ has the same structure as $LiNiO_2$ and that $Li_XNi_{2-X}O_2$ also has Li/Ni mixed occupancy, the relationship between $(I_{006} + I_{012})/I_{104} = R = 0.461 + 0.997(1-X) + 27.37\ (1-X)^2$ is given. According to the foregoing, it is known that the contributions of the three diffraction lines 3a and 3b of 006, 012, and 104 are all "additive", so it is impossible to reveal the problem of mixed occupation of Li/Ni, but only the relationship between $(I_{006} + I_{012})/I_{104}$ and component x in $Li_XNi_{2-X}O_2$.
(2) Ohzuku et al. [11] also used I_{003}/I_{104} to characterize the synthesis of $LiNiO_2$ when studying different synthesis processes, but did not mention the problem of Li/Ni mixed space occupation.
(3) Some people think that 003 is given by R-3 m layered structure, while 104 is given by R-3 m layered structure and cubic Fm-3 m structure, so I_{003}/I_{104} can be used to determine the mixing degree of "rock salt domain" of the sample, which is in good agreement with the ratio of n(Li)/m(Ni) obtained by chemical analysis. Although this seems to have no theoretical basis, it inadvertently touches on the problem of mixed occupation of Li/Ni.

5.1.2 The Method of Simulation Calculation

According to the principle introduced in Sect. 5.1.1 and the relevant data given, simulate and calculate the relationship of mixed occupying parameter X to ratio of diffraction intensity that is feasible, which can be calculated by hand or simulated by relevant programs. The author uses Powder Cell 2.4 program carries on the simulation calculation, and the input content of the data, take $Li_{1-x}Ni_X\ (Li_XNi_{1-X})O_2$, $X = 0.15$ as an example (see Table 5.2): space group No. 166, lattice parameter a, b, c, α, β, γ, and following content.

The output includes three: (1) crystal structure model and atomic bonding diagram; (2) diffraction pattern (spectrum figure), see Fig. 5.1; and (3) the digital form is shown in Table 5.3.

Table 5.2 Input content of the data of $Li_{1-x}Ni_x$ $(Li_xNi_{1-x})O_2$

Element name	Atom number	Ion	Location symbol wyck	Atomic coordinates			Occupancy probability SOF	Temperature factor B(Temp.)
				x	*y*	*z*		
Li	28	Li^{+1}	3a	0	0	0	0.85	0.66
Ni	3	Ni^{+2}	3a	0	0		0.15	0.22
Li	28	Li^{+1}	3b	0	0	1/2	0.15	0.66
Ni	3	Ni^{+2}	3b	0	0	1/2	0.85	0.22
O	8	O^{-2}	6c	0	0	0.25	1.00	0.44

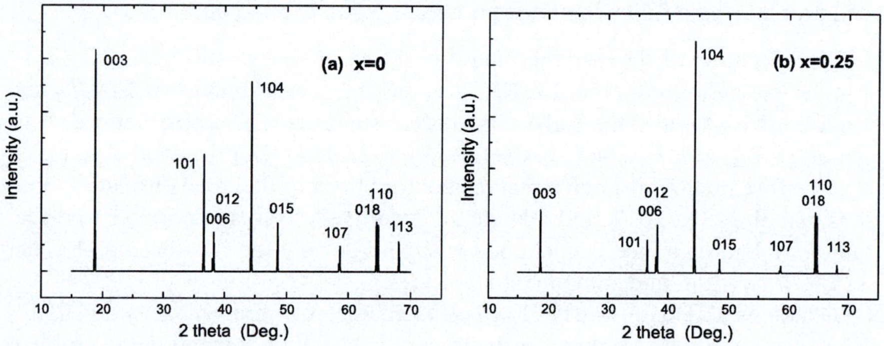

Fig. 5.1 Simulated diffraction pattern of $z = 0.25$ in $Li_{1-x}Ni_x(Li_xNi_{0.6-x}Co_{0.2}Mn_{0.2})O_2$ system

Table 5.3 Data format for analog output

h	k	L	2Theta/deg	d/(Å)	I/rel	|F(hkl)|	Mu	FWHM
0	0	3	19.153	4.63027	27.09	32.43	2	0.0707
1	0	1	37.307	2.40836	14.21	27.51	6	0.0707
0	0	6	38.868	2.31513	7.04	35.09	2	0.0707
0	1	2	39.018	2.30660	20.92	35.08	6	0.0707
1	0	4	45.320	1.99942	100.00	90.69	6	0.0707
0	1	5	49.625	1.83559	5.88	24.37	6	0.0707

Although the temperature factor is ignored in Eq. (5.1), it is found in the actual simulation that the change of B value between elements has a certain influence to the relative intensity. For example:

	(1) B	(2) B	(3) B
Li	0.66	0.44	0.22
Ni, Co, Mn	0.22	0.22	0.66
O	0.44	0.66	0.44
I_{003}/I_{104}	0.6684	0.6646	0.6545
I_{101}/I_{104}	0.5860	0.5894	0.5783

Therefore, all the following simulations use the first case, that is, keep Li, Ni (Co,Mn), and O at 0.66, 0.22,and 0.44, respectively.

Relative intensity ratio I_{h1k1l1}/I_{h2k2l2} can be obtained according to the calculation results of different materials and mixed occupying models (such as $Li_{1-x}Ni_x(Li_xNi_{0.6-x}Co_{0.2}Mn_{0.2})O_2$). Plots I_{h1k1l1}/I_{h2k2l2} [or $(I_{h1k1l1}/I_{h2k2l2})^{1/2}$] to X by Origin program, and finally obtains the relation equation by fitting method.

5.2 Simulation Results of Related Materials [12–14]

The chemical formula and Li/Ni mixed space occupying model of several related materials are shown in Table 5.4. In fact, as long as $u \neq 0$, a variety of materials with different components can be obtained, which are not listed here. As long as we master the simulation principles and calculation methods proposed in this book, all kinds of materials can be treated similarly.

5.2.1 *Material $Li(Ni_{0.6}Co_{0.2}Mn_{0.2})O_2$ Simulation Results*

The mixed occupation model of $Li(Ni_{0.6}Co_{0.2}Mn_{0.2})O_2$ is $Li_{1-x}Ni_x(Li_xNi_{0.6-x}Co_{0.2}Mn_{0.2})$ O_2. Set $X = 0.00$ and 0.25 and $Z = 0.25$.The diffraction pattern calculated by Powder Cell 2.4 program is shown in Fig. 5.1. It

Table 5.4 Chemical formulas of several related materials and their Li/Ni mixed occupying models

	Chemical formula of material	Li/Ni atomic mixed occupation model
General formula	$Li(Ni_uCo_vMn_{1-u-v})O_2$	$Li_{1-X}Ni_X(Li_XNi_{u-X}Co_vMn_{1-u-v})O_2$
$u = 0.0, v = 1.0$	$LiCoO_2$	
$u = 1.0, v = 0.0$	$LiNiO_2$	$Li_{1-X}Ni_X(Li_XNi_{1-X})O_2$
$u = 0.6, v = 0.4$	$Li(Ni_{0.6}Co_{0.4})O_2$	$Li_{1-X}Ni_X(Li_XNi_{0.6--X}Co_{0.4})O_2$
$u = 0.6, v = 0.0$	$Li(Ni_{0.6}Mn_{0.4})O_2$	$Li_{1-X}Ni_X(Li_XNi_{0.6-X}Mn_{0.4})O_2$
$u = 1/3, v = 1/3$	$Li(Ni_{1/3}Co_{1/3}Mn_{1/3})O_2$	$Li_{1-X}Ni_X(Li_XNi_{1/3-X}Co_{1/3}Mn_{1/3})O_2$
$u = 0.6, v = 0.2$	$Li(Ni_{0.6}Co_{0.2}Mn_{0.2})O_2$	$Li_{1-X}Ni_X(Li_XNi_{0.6-X}Co_{0.2}Mn_{0.2})O_2$
$u = 0.4, v = 0.2$	$Li(Ni_{0.4}Co_{0.2}Mn_{0.4})O_2$	$Li_{1-X}Ni_X(Li_XNi_{0.4-X}Co_{0.2}Mn_{0.4})O_2$

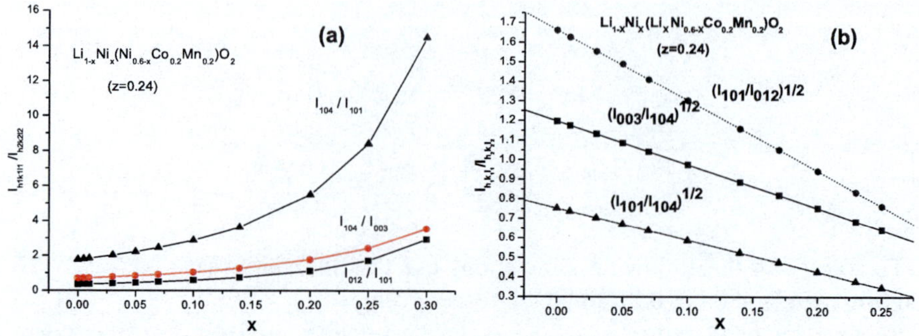

Fig. 5.2 Simulated results of $Li_{1-X}Ni_X(Li_XNi_{0.6-X}Co_{0.2}Mn_{0.2})O_2$. **a** $Y = (I_{h1k1l1}/I_{h2k2l2})$, **b** $Y = (I_{H1k1l1}/I_{H2k2l2})^{1/2}$

can be seen that the diffraction intensities of 003 and 104, 101 and 012 are reversed, that is, from $I_{003} > I_{104}$ and $I_{012} < I_{101}$ become to $I_{003} < I_{104}$ and $I_{012} > I_{101}$.

The relationship between the simulated diffraction intensity ratio I_{104}/I_{003}, I_{012}/I_{101}, and I_{101}/I_{104} of $Li_{1-x}Ni_x(Li_xNi_{0.6-x}Co_{0.2}Mn_{0.2})O_2$ and the mixed occupying parameter X is shown in Fig. 5.2a, which can be fitted by a "Cubic Quadrinomial". For example, for the $Li_{1-x}Ni_x(Li_xNi_{0.6-x}Co_{0.2}Mn_{0.2})O_2$ system $z = 0.24$, the relationship between I_{104}/I_{003} and the mixed occupying parameter X satisfies the following formula:

$$I_{104}/I_{003} = 0.693 + 3.270x - 5.074x^2 + 78.630x^3 \tag{5.5}$$

Write it in general formula:

$$y = a + bx + cx^2 + dx^3 \tag{5.6}$$

It is obviously impossible to find X with a numeral calculate method when y (I_{104}/I_{003}) is known. Therefore, you can only use the Origin program to draw and save. Then, using the intensity ratio data measured by the experiment, on the preserved Origin diagram, the mixed occupying parameter X is read on the curve by Screen Reader.

If the ordinate is changed to $(I_{h1k1l1}/I_{h2k2l2})^{1/2}$, the curve of the relationship between $(I_{h1k1l1}/I_{h2k2l2})^{1/2}$ and X is shown in Fig. 5.2b. There is a linear relationship between $(I_{003}/I_{104})^{1/2}$, $(I_{101}/I_{012})^{1/2}$, and $(I_{101}/I_{104})^{1/2}$ and X. Here, we only give when $z = 0.25$.

$$\begin{cases} (I_{003}/I_{104})^{1/2} = 1.06076 - 2.20433X \\ (I_{101}/I_{012})^{1/2} = 1.73537 - 3.62027X \\ (I_{101}/I_{104})^{1/2} = 0.77255 - 1.61161X \end{cases} \tag{5.7}$$

$$\begin{cases} X = 0.4812 - 0.4536(I_{003}/I_{104})^{1/2} \\ X = 0.4793 - 0.2762(I_{101}/I_{012})^{1/2} \\ X = 0.4794 - 0.6205(I_{101}/I_{104})^{1/2} \end{cases} \quad (5.8)$$

The linear equations obtained from the simulation of other related materials are listed in Table 5.5. It is obvious that it is very convenient to use this linear relationship for numerical calculation.

5.2.2 Simulation Results of $LiNiO_2$

The mixed arrangement model of $LiNiO_2$ is $Li_{1-X}Ni_X$ $(Li_XNi_{1-X})O_2$. Figure 5.3a shows the relationship between the diffraction intensity ratio of $Li_{1-X}Ni_X$ $(Li_XNi_{1-X})O_2$ and the mixed arrangement parameter *X*, which can be seen from Fig. 5.3a.

(1) (I_{006}/I_{012}) does not change with the mixed parameter *X*, because the contributions of the 3a and 3b positions to the two lines are additive, and the intensity ratio of the two lines has nothing to do with *X*.
(2) The other three curves are symmetrically distributed along the longitudinal axis of $X = 0.5$, and $(I_{101}/I_{104}) - X$ is a parabolic equation. When Li and Ni are completely interchanged ($X = 0$ and $X = 1$), the intensity ratio of each diffraction peak is exactly the same.
(3) When $X \le 0.5$, I_{104}/I_{003} and I_{012}/I_{101} increase with the increase of mixed parameter X, but I_{101}/I_{104} decreases with the increase of mixed parameter *X*.

Figure 5.3b shows the diagrams of $(I_{003}/I_{104})^{1/2}$, $(I_{101}/I_{012})^{1/2}$, and $(I_{101}/I_{104})^{1/2}$ with the mixed parameter *X*, all of which decrease linearly with the increase of *X*, and their equations are listed in Table 5.5. What makes sense is that the three lines intersect at one point at $X = 0.5$. If the experimental results show that $(I_{003}/I_{104}) = (I_{101}/I_{012}) = (I_{101}/I_{104})$, it shows that the Li/Ni mixed arrangement parameter *X* of the sample is 0.5; that is, 50% of the cationic atoms have mixed occupation.

5.2.3 Simulation Results of Other Four Materials

The relationships between $(I_{003}/I_{104})^{1/2}$, $(I_{101}/I_{012})^{1/2}$, and $(I_{101}/I_{104})^{1/2}$ and mixing parameter *X* of $Li(Ni_{0.6}Co_{0.4})O_2$, $Li(Ni_{0.6}Mn_{0.4})O_2$, $Li(Ni_{1/3}Co_{1/3}Mn_{1/3})O_2$, and $Li(Ni_{0.4}Co_{0.2}Mn_{0.4})$ materials were also simulated. $Li(Ni_{1/3}Co_{1/3}Mn_{1/3})O_2$ and $Li(Ni_{0.4}Co_{0.2}Mn_{0.4})O_2$ are given in Fig. 5.4a and b, respectively.

The linear equations and some important parameters obtained when $X = 0$ and $I_{104}/I_{003} = 1$ from the above simulation of six kinds of materials are summarized in Tables 5.6 and 5.7, respectively.

Table 5.5 Linear relationship between diffraction intensity ratio root value $[(I_{003}/I_{104})^{1/2}, (I_{101}/I_{012})^{1/2}$, 和 $(I_{101}/I_{104})^{1/2}]$ and *x* of six kinds of $LiMeO_2$ materials

	Z = 0.24	Z = 0.25
$LiNiO_2$	$X = 0.5554 - 0.4604(I_{003}/I_{104})^{1/2}$	$X = 0.5000 - 0.4689(I_{003}/I_{104})^{1/2}$
	$X = 0.464 - 3.2783(I_{101}/I_{012})^{1/2}$	$X = 0.5000 - 0.2955(I_{101}/I_{012})^{1/2}$
	$X = 0.4642 - 0.5888(I_{101}/I_{104})^{1/2}$	$X = 0.5000 - 0.6355(I_{101}/I_{104})^{1/2}$
$Li(Ni_{0.6}Co_{0.4})O_2$	$X = 0.5534 - 0.4533(I_{003}/I_{104})^{1/2}$	$X = 0.4900 - 0.4610(I_{003}/I_{104})^{1/2}$
	$X = 0.4486 - 0.2730(I_{101}/I_{012})^{1/2}$	$X = 0.4896 - 0.2862(I_{101}/I_{012})^{1/2}$
	$X = 0.4725 - 0.6228(I_{101}/I_{104})^{1/2}$	$X = 0.4895 - 0.6244(I_{101}/I_{104})^{1/2}$
$Li(Ni_{0.6}Mn_{0.4})O_2$	$X = 0.5273 - 0.4332(I_{003}/I_{104})^{1/2}$	$X = 0.4938 - 0.4681(I_{003}/I_{104})^{1/2}$
	$X = 0.4528 - 0.2708(I_{101}/I_{012})^{1/2}$	$X = 0.4361 - 0.2444(I_{101}/I_{012})^{1/2}$
	$X = 0.4505 - 0.5895(I_{101}/I_{104})^{1/2}$	$X = 0.4900 - 0.6372(I_{101}/I_{104})^{1/2}$
$Li(Ni_{1/3}Co_{1/3}Mn_{1/3})O_2$	$X = 0.5269 - 0.4427(I_{003}/I_{104})^{1/2}$	$X = 0.4742 - 0.4496(I_{003}/I_{104})^{1/2}$
	$X = 0.4474 - 0.2638(I_{101}/I_{012})^{1/2}$	$X = 0.4656 - 0.2635(I_{101}/I_{012})^{1/2}$
	$X = 0.4485 - 0.6032(I_{101}/I_{104})^{1/2}$	$X = 0.4656 - 0.6098(I_{101}/I_{104})^{1/2}$
$Li(Ni_{0.6}Co_{0.2}Mn_{0.2})O_2$	$X = 0.5388 - 0.4501(I_{003}/I_{104})^{1/2}$	$X = 0.4812 - 0.4536(I_{003}/I_{104})^{1/2}$
	$X = 0.4618 - 0.2777(I_{101}/I_{012})^{1/2}$	$X = 0.4793 - 0.2762(I_{101}/I_{012})^{1/2}$
	$X = 0.4619 - 0.6137(I_{101}/I_{104})^{1/2}$	$X = 0.4794 - 0.6205(I_{101}/I_{104})^{1/2}$
$Li(Ni_{0.4}Co_{0.2}Mn_{0.4})O_2$	$X = 0.3933 - 0 .3574(I_{003}/I_{104})^{1/2}$	$X = 0.4681 - 0.4456(I_{003}/I_{104})^{1/2}$
	$X = 0.3248 - .1616(I_{101}/I_{012})^{1/2}$	$X = 0.4641 - 0.2616(I_{101}/I_{012})^{1/2}$
	$X = 0.3258 - .4901(I_{101}/I_{104})^{1/2}$	$X = 0.4640 - 0.6101(I_{101}/I_{104})^{1/2}$

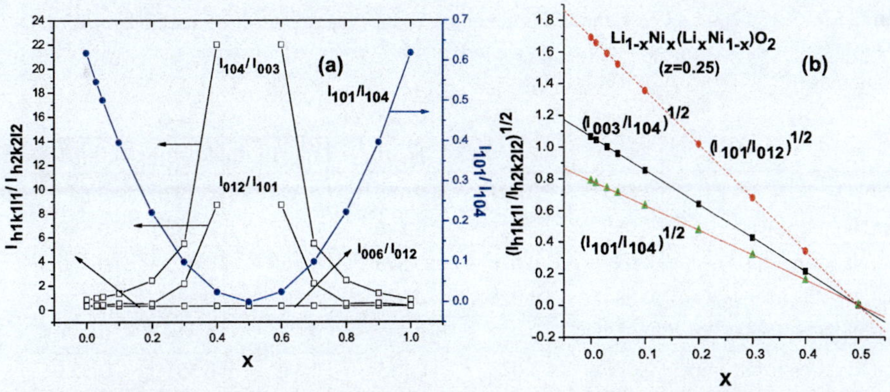

Fig. 5.3 Curve of the relationship between the diffraction intensity ratio of $Li_{1-x}Ni_x(Li_xNi_{1-x})O_2$ and the mixing parameter x. **a** I_{104}/I_{003}, $I_{012}/_{101}$, and I_{101}/I_{104}, **b** $(I_{003}/I_{104})^{1/2}$, $(I_{101}/_{012})^{1/2}$, and $(I_{101}/I_{104})^{1/2}$

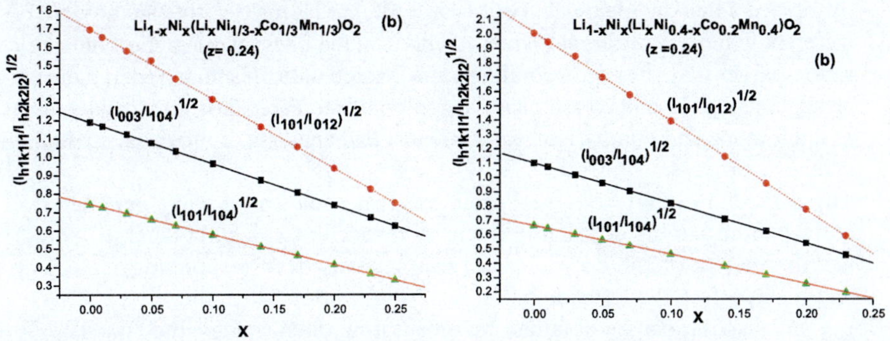

Fig. 5.4 $(I_{003}/I_{104})^{1/2}$, $(I_{101}/I_{012})^{1/2}$, and $(I_{101}/I_{104})^{1/2}$, and their relation curves with the mixed arrangement parameter x, **a** $Li(Ni_{1/3}Co_{1/3}Mn_{1/3})O_2$, **b** Li $(Ni_{0.4}Co_{0.2}Mn_{0.4})O_2$

Table 5.6 When $X = 0$ (i.e., no mixed arrangement), six kinds of $LiMeO_2$ materials mainly need to achieve the intensity ratio of diffraction lines

Materials	$x = 0$					
	$Z = 0.24$			$Z = 0.25$		
	I_{104}/I_{003}	I_{012}/I_{101}	I_{101}/I_{104}	I_{104}/I_{003}	I_{012}/I_{101}	I_{101}/I_{104}
$LiNiO_2$	0.6872	0.3594	0.6215	0.8797	0.3492	0.6192
$Li(Ni_{0.6}Co_{0.4})O_2$	0.6838	0.3702	0.5755	0.8838	0.3416	0.6147
$Li(Ni_{0.6}Mn_{0.4})O_2$	0.6748	0.3578	0.5840	0.8984	0.3141	0.5912
$Li(Ni_{1/3}Co_{1/3}Mn_{1/3})O_2$	0.7059	0.3477	0.5527	0.8988	0.3204	0.5830
$Li(Ni_{0.6}Co_{0.2}Mn_{0.2})O_2$	0.6981	0.3617	0.5666	0.8888	0.3322	0.5969
$Li(Ni_{0.4}Co_{0.2}Mn_{0.4})O_2$	0.8258	0.2473	0.4419	0.9060	0.3177	0.5785

Table 5.7 When $I_{104}/I_{003} = 1$, the Li/Ni mixed occupancy parameters X of six kinds of $LiMeO_2$ materials

Materials	When $I_{104}/I_{003} = 1$					
	X		Z = 0.24		Z = 0.25	
	Z = 0.24	Z = 0.25	I_{012}/I_{101}	I_{101}/I_{104}	I_{012}/I_{101}	I_{101}/I_{104}
$LiNiO_2$	0.095	0.031	0.568	0.393	0.397	0.545
$Li(Ni_{0.6}Co_{0.4})O_2$	0.095	0.029	0.596	0.367	0.386	0.544
$Li(Ni_{0.6}Mn_{0.4})O_2$	0.094	0.026	0.570	0.366	0.355	0.530
$Li(Ni_{1/3}Co_{1/3}Mn_{1/3})O_2$	0.084	0.025	0.527	0.365	0.356	0.522
$Li(Ni_{0.6}Co_{0.2}Mn_{0.2})O_2$	0.088	0.027	0.552	0.371	0.373	0.531
$Li(Ni_{0.4}Co_{0.2}Mn_{0.4})O_2$	0.036	0.023	0.315	0.349	0.352	0.523

From these data, we can see that:

(1) The diffraction intensity of materials with different composition has a certain difference when the ideal occupation ($X = 0$), but intensity ratios are obviously different when z is different. When z is known, the integral diffraction intensity ratio is larger than the corresponding value, which indicates that there is a mixed arrangement of Li and Ni atoms in the material [see Table 5.6], so it can be used as a reference and comparison for XRD test and analysis of the products in the synthesis process.

(2) The data of $I_{104}/I_{003} = 1$ [see Table 5.4 (b)] show the mixing parameter X corresponding to the diffraction intensity reversal of 104 and 003 ($I_{104} \geq I_{003}$) [see the left side of Table 5.7 left side] and the X values corresponding of I_{012}/I_{101} and I_{101}/I_{104} [see the left side of Table 5.7 right side]. Then, the values of I_{012} / I_{101} and I_{101}/I_{104} can be obtained by substituting the X values into $(I_{101}/I_{012})^{1/2} = a_2 - b_2X$ and $(I_{101}/I_{104})_{1/2} = a_3 - b_3X$. It can be seen that they are greatly influenced by z.

The numerical calculation can be carried out conveniently by using the linear relationship between (diffraction intensity ratio)$^{1/2}$ and X. Some practical examples are given below. Figure 5.5 shows the XRD pattern of $Li_{1-X}Ni_X(Li_XNi_{0.6-X}Co_{0.2}Mn_{0.2})O_2$ treated at medium temperature and high temperature, and it can be seen that $I_{104} > I_{003}$ of the semi-finished product at 660 °C for 6 h, while $I_{104} < I_{003}$ of the finished product.

at 660 °C 6 h + 900 °C for 8 h, that is, the intensity is reversed; the diffraction lines of the semi-finished product of 6 h at 550,660 °C are seriously broadened, 006 and 012, 018 and 110 cannot be separated, indicating that the grain is smaller.

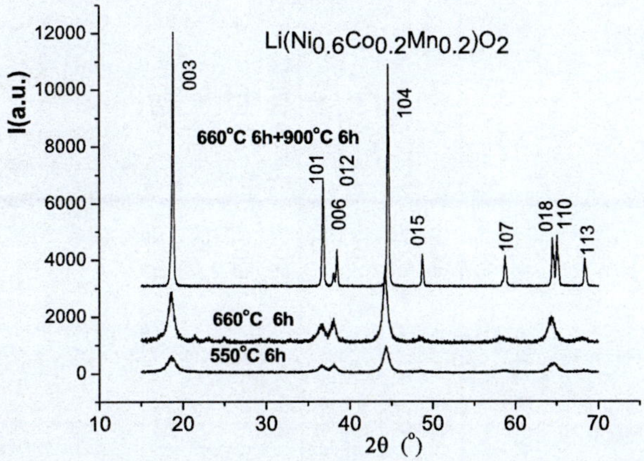

Fig. 5.5 $Li(Ni_{0.6}Co_{0.2}Mn_{0.2})O_2$ XRD patterns obtained at medium and high temperatures

5.3 The Mixed Occupying Parameter *X* Is Calculated from the Diffraction Integral Intensity Ratio

5.3.1 The Results of Experimental Determination

It can be seen from Sect. 5.1 that the relationship between the simulated diffraction intensity ratio I_{104}/I_{003}, I_{012}/I_{101}, and I_{101}/I_{104} and the mixed occupying parameter *X* can be fitted by "Cubic Quadrinomial", while the linear relationship between the mixed occupying parameter *X* and $(I_{104}/I_{003})^{1/2}$, $(I_{012}/I_{101})^{1/2}$, and $(I_{101}/I_{104})^{1/2}$ is shown. Therefore, the numerical calculation can be carried out by using the linear equation listed in Table 5.5. Table 5.8 shows some practical examples, that is, the determination results of semi-finished products and finished products calcined by $Li(Ni_{0.6}Co_{0.2}Mn_{0.2})O_2$ at medium and high temperatures.

From these experimental results, it can be seen that:

(1) The *X* obtained from the value of $(I_{104}/I_{003})^{1/2}$ is relatively sensitive to the *z* value, so it is necessary to confirm the *z* value in the 6c site of the oxygen atom, which is related to the preparation process of $LiNiO_2$ and the raw materials used [12]. According to our experimental results, $z = 0.24$ is more practical for $Li_{1-X}Ni_X(Li_XNi_{0.6-X}Co_{0.2}Mn_{0.2})O_2$.
(2) The *X* obtained from the value of $(I_{101}/I_{104})^{1/2}$ is not so sensitive to the *z* value; that is, when the *z* value of the 6c site cannot be determined, the measured data of $(I_{101}/I_{104})^{1/2}$ will be used to solve the *X*.
(3) The value of I_{012}/I_{101} is not sensitive to *z* value, but it is not included because the intensity of I_{012} is relatively too weak and interfered by 006, so the reliability of I_{012} is poor. It is not included.
(4) The results obtained by the two pairs of lines are inconsistent, which may be due to the experimental error caused by the abnormal enhancement of 003 preferred

Table 5.8 Semi-finished and finished $Li(Ni_{0.6}Co_{0.2}Mn_{0.2})O_2$ calcined at medium and high temperature

Temp(°C)/ Time (h)	$(I_{003}/I_{104})^{1/2}$	*X*		$(I_{101}/I_{104})^{1/2}$	*X*		$(I_{101}/I_{104})^{1/2}$	*X*	
		Z = 0.24	Z = 0.25		Z = 0.24	Z = 0.25		Z = 0.24	Z = 0.25
660/6	0.938	0.116	0.056	0.385	0.082	0.089	0.621	0.081	0.094
690/6	0.937	0.117	0.056	0.318	0.126	0.128	0.564	0.116	0.129
720/6	1.052	0.065	0.004	0.389	0.074	0.091	0.624	0.079	0.092
750/6	1.090	0.048	−0.013	0.439	0.043	0.059	0.663	0.055	0.068
860/8	1.005	0.064	0.026	0.417	0.057	0.063	0.646	0.066	0.078
880/8	1.047	0.067	0.006	0.434	0.045	0.057	0.659	0.058	0.071
900/8	1.052	0.065	0.004	0.442	0.037	0.050	0.665	0.054	0.067
920/8	1.078	0.053	−0.008	0.450	0.034	0.051	0.671	0.050	0.063

The intensity ratio of the main diffraction lines and the results of X-ray determinatio orientation, so the influence of 003 diffraction intensity due to the preferred orientation and the uncertainty of z value in 6c position should be avoided. Therefore, in the relativity measurement, the data of $(I_{101}/I_{104})^{1/2}$ with 101/104 pairs are more simple and more reliable, because it avoids not only the influence of texture but also the influence of z uncertainty.

However, it must be pointed out that the sample preparation method must be paid attention to when measuring the diffraction intensity ratio in order to avoid the 003 preferred orientation of grain distribution and increase the 003 diffraction intensity abnormally. For this reason, it is suggested to use a sample frame with a depth of 2 mm. After setting out the free fall, fill it with a piece of glass with a small cross-sectional area and, finally, gently scrape it flat with a piece of glass. The strength measured must be the integral strength after deducting the background. Therefore, it is suggested that the back and $K\alpha_2$ components should be removed by Jade program, and then, the data should be processed by Refine.

(5) As a relative measurement, the I_{104}/I_{101} line pair is relatively simple and avoids the influence of the enhancement of 003 diffraction intensity due to the abnormal preferred orientation and the uncertainty of z value in 6c position. For $Li(Ni_{0.6}Co_{0.2}Mn_{0.2})O_2$, when $z = 0.24$ and $X = 0.4619–0.6137(I_{101}/I_{104})^{1/2}$, the numerical calculation can be carried out, which is more convenient.

5.3.2 *The Effect of Medium Temperature in the Process of Synthesis*

The XRD pattern of the product sintered at medium temperature (650–750 °C, 6 h) is shown in Fig. 5.6. It can be seen that there is unreacted $LiCO_3$ when the temperature is lower than 700 °C, and the curve of I_{104}/I_{003} and I_{012}/I_{101} with low temperature sintering temperature is shown in Fig. 5.7.

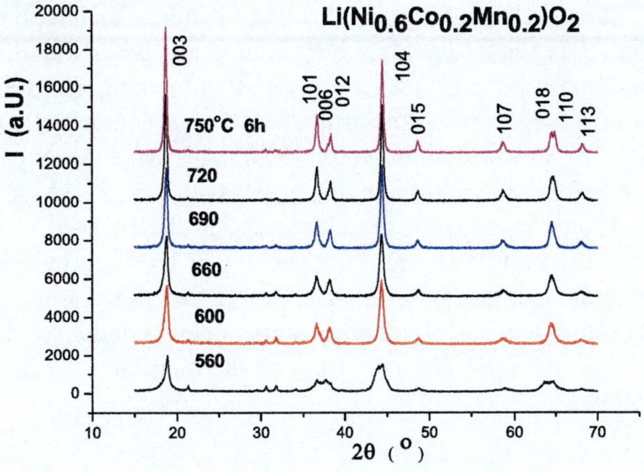

Fig. 5.6 XRD patterns of products sintered at different medium temperatures

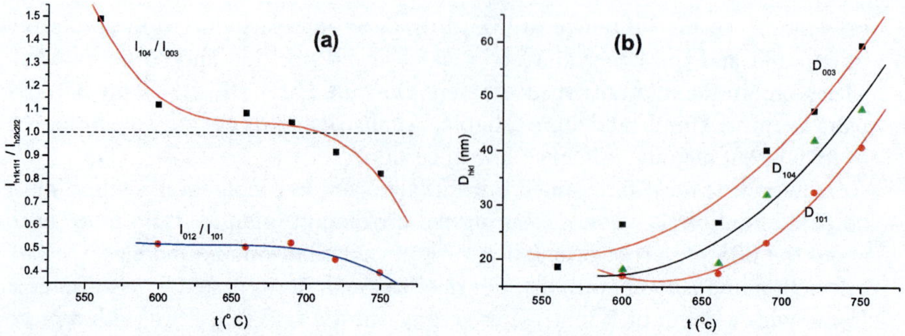

Fig. 5.7 Curves of I_{104}/I_{003} and I_{012}/I_{101} (**a**) and grain size D_{hkl} (**b**) varying with the temperature of medium temperature sintering

It can be seen from Fig. 5.7 that when the temperature is lower than 710 °C, the I_{104}/I_{003} is > 1, and when the temperature reaches 750C, the I_{104}/I_{003} is < 1, but the I_{012}/I_{101} is still greater than 0.697, and there is still a phenomenon of mixed arrangement. With the increase of temperature, the grains grow continuously, and when the temperature reaches 750 °C, 018 and 110 can be partially separated, as shown in Fig. 5.6.

5.3.3 *The Effect of High Temperature in the Process of Synthesis*

The XRD patterns of the materials sintered with high temperature range of 860, 880, 900, and 920 °C for 8 h are shown in Fig. 5.8. It can see that 018 and 110 better Fig. 5.9 variation curves of I_{104}/I_{003} (a) and D_{104} (b) at different high temperature the separation is with the increase of temperature. Figure 5.9 shows the variation curves of D_{104} and I_{104}/I_{003} at different high temperature.

It can be seen from Fig. 5.9 that the mixed arrangement of Li/Ni still exists in the materials sintered at high temperature, and when the mixed arrangement reaches the maximum at $t = 890$ °C, $x = 0.08$. An interesting phenomenon is that the large mixed arrangement hinders the growth of grains. The separation of 018 and 110 diffraction lines becomes more obvious with the increase of temperature, which is mainly due to the decrease of broadening effect with the growth of grains.

From the above research results, it can be seen that, on the whole, whether it is in the middle temperature section or the high temperature section, *X* decreases with the increase of temperature. Whether this is not in line with the general law of thermodynamics, high temperature is more conducive to the formation of disordered phase, and *X* should increase with the increase of temperature. The discussion is as follows.

Fig. 5.8 XRD patterns of products sintered at different high temperature

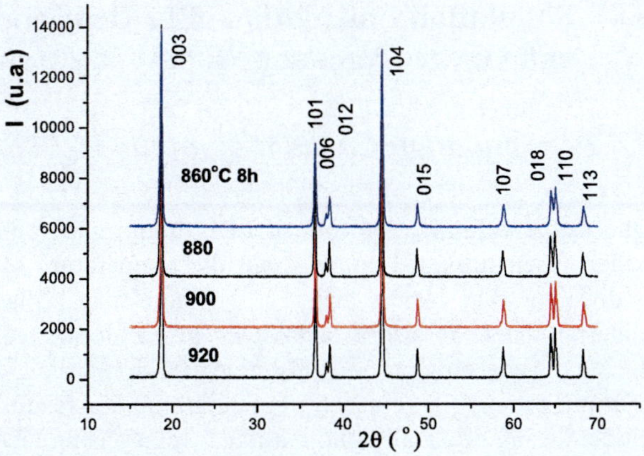

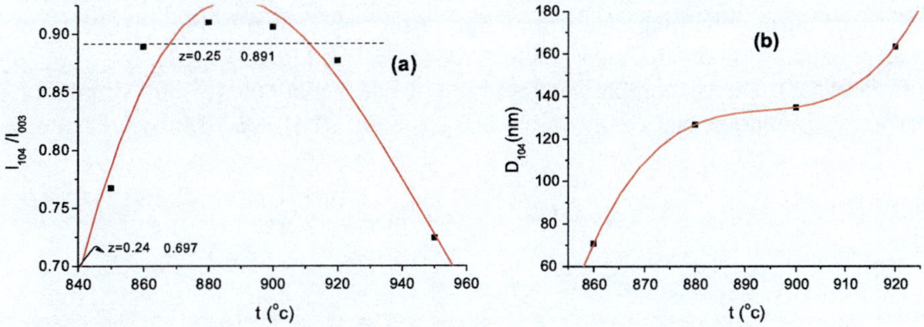

Fig. 5.9 Variation curves of I_{104}/I_{003} (**a**) and D_{104} (**b**) at different high temperature

In the order → disorder phase transition of solid solution, in the order temperature range, the order degree y generally decreases with the increase of temperature and increases with increase of time, and beyond this temperature range, it is different. This is the general law of thermodynamics. The synthesis process of $Li(Ni_{0.6}Co_{0.2}Mn_{0.2})$ O_2 is much more complicated. First of all, β-$(Ni_{0.6}Co_{0.2}Mn_{0.2})$ $(OH)_2$ is decomposed into oxides, and then, the oxides react with Li_2CO_3 (or LiOH) to form $Li(Ni_{0.6}Co_{0.2}Mn_{0.2})O_2$. It has not only a process of nucleation and growth, but also crystal has a process from imperfect to more perfect, that is, the process of mixed arrangement parameter X from large to small.

5.4 Simulation Calculation of Li Deficiency Model and Oxygen Vacancy Model

5.4.1 Simulation Calculation of the Li Deficiency Model

Due to the volatilization loss of Li in high temperature sintering, excessive Li is added when lithium is mixed, but the temperature of high temperature sintering is different. The loss is different, so there may be lack of Li and over-Li in the material. There is also a lack of Li in the de-intercalation-Li of lithium battery during the charging process. Figure 5.10a shows the assumption that all three types of atoms occupy sites correctly, but the simulation results under the lithium deficiency model during de-Li also show a linear relationship (see Table 5.9). The simulation results are shown in Fig. 5.10b assuming that raw material $Li(Ni_{0.6}Co_{0.2}Mn_{0.2})O_2$ has 0.1 mixed occupancy, that is, $Li_{1-X}Ni_{0.1}(Li_{0.1}Ni_{0.6-0.1}Co_{0.2}Mn_{0.2})O_2$, which remains unchanged during charging. The simulation results are shown in Fig. 5.10b. It can be seen that all of them are linear, which are summarized in Table 5.9.

Figure 5.11 shows comparison between experimental curve and simulation curve and simulation of Li removal of 2H-graphite/$Li(Ni_{1/3}Co_{1/3}Mn_{1/3})O_2$ battery

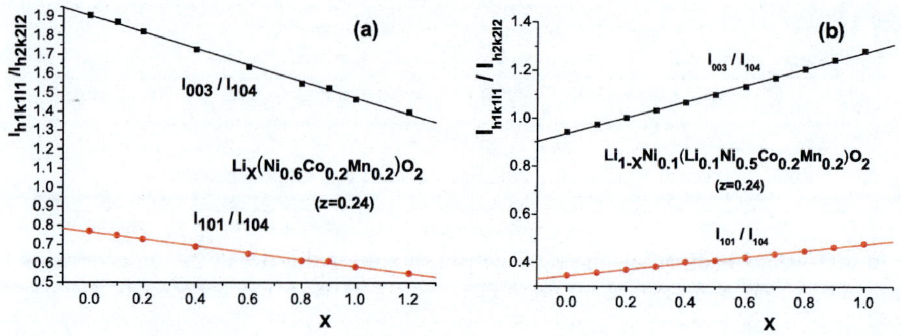

Fig. 5.10 $Li(Ni_{0.6}Co_{0.2}Mn_{0.2})O_2$ lithium deficiency model simulation results: I_{104}/I_{003} and $I_{012}/I_{101}-X$, **a** simulation results of lithium deficiency without mixed discharge, **b** simulation results of lithium deficiency without mixed discharge

Table 5.9 Corresponds to the linear equation obtained from the simulation results of the lithium deficiency model in Fig. 5.4

Materials and Li-deficient models	Linear equation
$Li_X(Ni_{0.6}Co_{0.2}Mn_{0.2})O_2$	$(I_{003}/I_{104}) = 1.904–0.432X$ $(I_{101}/I_{104}) = 0.765–0.181X$
$Li_{1-X}Ni_{0.1}(Li_{0.1}Ni_{0.6-0.1}Co_{0.2}Mn_{0.2})O_2$	$(I_{003}/I_{104}) = 0.935–0.339X$ $(I_{101}/I_{104}) = 0.343–0.133X$

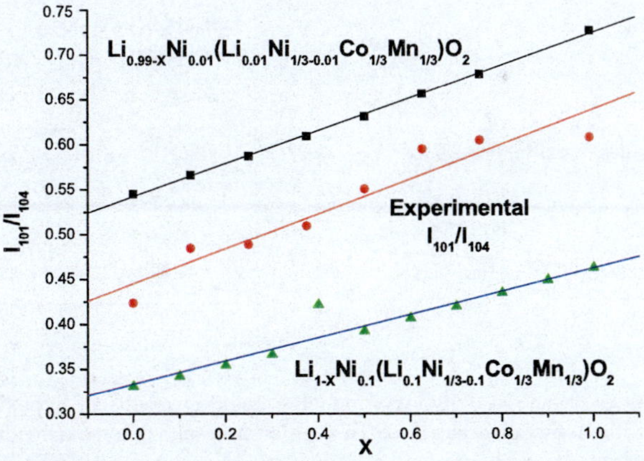

Fig. 5.11 Comparison of the curve of XRD intensity ratio I_{101}/I_{104} calculated by the model of Li removal during the charging process of 2H-carbon/$Li(Ni_{1/3}Co_{1/3}Mn_{1/3})O_2$ battery with the curve measured by experiment

during charging. The simulation curve assumes that there is a mixed arrangement of 0.01and 0.10 Li/Ni atoms in the raw materials, which can be simulated and calculated according to the $Li_{0.99-x}Ni_{0.01}(Li_{0.01}Ni_{0.324}Co_{1/3}Mn_{1/3})O_2$ and $Li_{1-X}Ni_{0.1}(Li_{0.1}Ni_{1/3-0.1}Co_{1/3}Mn_{1/3})O_2$ models, respectively. Compared with the experimental results can knew, the three straight lines are nearly parallel, and it can be estimated that there may be a mixed arrangement of 0.04 Li/Ni atoms in the raw materials.

5.4.2 Simulation Calculation of Oxygen Vacancy Model

The so-called oxygen vacancy model means that in the process of $LiMeO_2$ synthesis, for some reason, six oxygen atoms at 6c position are not filled, resulting in oxygen vacancy. Figure 5.12 shows the simulation results of hypoxia in two cases. The first case is that all three kinds of atoms occupy space correctly and only hypoxia occurs. The model can be written as $Li(Ni_{0.6}Co_{0.2}Mn_{0.2})O_{2-X}$. The simulation results are shown in Fig. 5.12a. It is worth mentioning that the occupation probability of oxygen atom is 1.0 at $X = 0$ and 0.8 at $X = 0.4$, and so on. In the second case, hypoxia occurs when the material has a mixed occupation of 0.10, and the model can be in Fig. 5.12b. It is known that the slope of the I_{101}/I_{003} line is very small; in other words, it hardly changes with the amount of hypoxia X, because the contribution of oxygen to 101 and 003 is zero. There was an approximate linear relationship between I_{101}/I_{104} and I_{104}/I_{003} and anoxic quantity X.

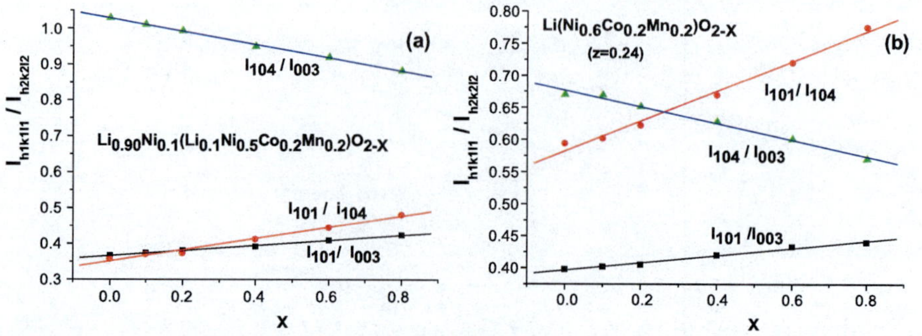

Fig. 5.12 Li ($Ni_{0.6}Co_{0.2}Mn_{0.2}$) O_2 hypoxia model simulation results of hypoxia: **a** simulation results of hypoxia without mixed arrangement, **b** simulation results of hypoxia with mixed arrangement written as follows. For $Li_{0.9}Ni_{0.1}$ ($Li_{0.1}Ni_{0.5}Co_{0.2}Mn_{0.2}$)$O_{2\text{-}X}$, the simulation results are shown

5.5 Summary of Mixed Occupation of Li and Ni Atoms

The main results in this chapter are summarized as follows:

(1) The positive active material $LiMeO_2$ for Li-ion battery belongs to the R-3 m space group (No. 166). Under ideal case, Li occupies 3a position, Me occupies 3b position, and O occupies 6c position. The mixed occupation of Li/Ni atoms in 3a and 3b often occurs in this material and can be expressed by the general formula $Li_{1-X}Ni_X$ ($Li_XNi_{u\text{-}X}Co_vMn_{1-u-v}$)$O_2$.

(2) In the analysis of the contribution of 3a and 3b atoms to the diffraction intensity of each (*hkl*) crystal plane, it is found that the intensity of the diffraction line of the addition relation of the two positions has nothing to do with the mixed occupation parameter *X*, while the intensity of the diffraction line of the subtraction relation is related to *X*. On this basis, it is proposed that it is feasible to use the intensity ratio method to study the mixed occupation of Ni/Li in this kind of materials. According to the mixed occupying model of $Li_{1-X}Ni_X(Li_XNi_{u\text{-}X}Co_vMn_{1-u-v})O_2$, the linear relations between the intensity ratio $(I_{003}/I_{104})^{1/2}$, $(I_{101}/I_{012})^{1/2}$, and $(I_{101}/I_{104})^{1/2}$ and the mixed occupying parameter *X* are obtained by means of Powder Cell program simulation, so the numerical calculation can be carried out. It has been applied in related materials: $LiNiO_2$, Li (Ni,Co)O_2, Li(Ni,Mn)O_2, and Li(Ni,Co,Mn)O_2.

(3) Using the intensity ratio measured by experiments, the mixed occupying parameter *X* can be obtained by numerical calculation. The practice shows that, as a relative measurement, the linear relationship between $(I_{101}/I_{104})^{1/2}$ and *X* is relatively simple and avoids the influence of the enhancement of 003 diffraction intensity due to the abnormal preferred orientation and the uncertainty of *z* value in 6c position. 101 and 104 are one of the three strong lines in the diffraction pattern (see the relative intensity data in the second row of Table 5.1).

(4) The fine structure changes of Li(Ni,Co,Mn)O_2 cathode materials synthesized by two-stage sintering method are explained: (1) There is mixed discharge of Li/Ni in the products at low temperature, and the mixing degree decreases and the grain size grows with the increase of temperature. (2) The mixing degree in the high temperature section decreases until it is close to zero, that is, the product that occupies space correctly, but when the temperature continues to rise, the mixed arrangement phenomenon begins to appear again and reaches a maximum. For $Li_{1-x}Ni_x(Li_xNi_{0.6-x}Co_{0.2}Mn_{0.2})O_2$ system, $X_{max} = 0.062$ at about 870 °C.

(5) The method of line-to-pair intensity ratio is also applied in the absence of Li (or de-Li during charging) and oxygen vacancy in the above six materials, and the quantitative relationship between intensity ratio and de-Li and anoxic quantity is obtained.

References

1. Luo, Z., N.L. Di, Z.Q. Kou, Z.H. Cheng, L.J. Liu, L.Q. Chen, and X.J. Huang. 2004. Măssbauer study and magnetic properties of electrochemical material $LiFePO_4$. *Chinese Physics* 13: 2158–2616.
2. Sun, Y.C., C.Y. Ouyang, Z.X. Wang, et al. 2004. Effect of Co content on rate performance of $LiMn_{0.5-x}Co_{2x}Ni_{0.5-x}O_2$ cathode materials for lithium-ion batteries. *Journal of Electrochemical Society* 151: A504–A508.
3. Li, D.C., M. Takahisa, and L.Q. Zhang. 2004. Effect of synthesis method on the electrochemical performance of $LiNi_{1/3}Mn_{1/3}Co_{1/3}O_2$. *Journal of Power Sources* 132: 150–155.
4. Yabuuchi, N., and T. Ohzuku. 2003. Novel lithium insertion material of $LiCo_{1/3}Ni_{1/3}Mn_{1/3}O_2$ for advanced lithium-ion batteries. *Journal of Power Sources* 119–121: 171–174.
5. Rougier, A., I. Saadoune, P. Gravereau, et al. 1996. Effect of cobalt substitution on cationic distribution in $LiNi_{1-y}Co_yO_2$ electrode materials. *Solid State Ionics* 90: 83–90.
6. Delmas, C., J.P. Pérès, A. Rougier, et al. 1997. On the behavior of the Li_xNiO_2 system: An electrochemical and structural overview. *Journal of Power Sources* 68: 120–125.
7. Hou, Z.F., H.Y. Liu, X.Z. Zhu, et al. 2003. Study on Li embedding performance of CuSn anode material for Li-ion battery. *Acta Physica Sinica* 52: 952.
8. Kim, J.-M., and H.-T. Chung. 2004. The first cycle characteristics of $Li[Ni_{1/3}Co_{1/3}Mn_{1/3}]O_2$ charged up to 4.7 V. *Electrochemica Acta* 49 (6): 937–944.
9. Liu, H.S., J. Li, Z.L. Gong, et al. 2004. Application of Rietveld method in the study of lithium nickel oxide electrode materials. *Power Supply Technology* 28: 612 gong.
10. Dahn, J.R., U. von Sacken, and C.A. Michal. 1990. Structure and electrochemistry of $Li_{1\pm y}NiO_2$ and a new Li_2NiO_2 phase with the Ni $(OH)_2$structure. *Solid State Ionics* 44: 87–97 [12].
11. Ohzuku, T., A. Ueda, and M. Nagoyama. 1993. Electrochemistry and structural chemistry of $LiNiO_2$ ($R\bar{3}m$) for 4 volt secondary lithium cells. *Journal of Electrochemical Social* 140 (7): 1862–1870.
12. Zhang, J., C.Z. Yang, X.G. Zhang, J. Li., and B.J. Xia. 2009. Simulation analysis and study of structural changes of Li (Ni,Co,Mn) O2 during synthesis. *Journal of Materials Science and Engineering* 27 (6): 824–828.
13. Zhang, X.G., J. Zhang, C.Z. Yang, and B.J. Xia. 2010. Simulation and experimental study on mixed arrangement of lithium and nickel in $LiMeO_2$ materials. *Journal of Inorganic Materials* 25 (1): 8–12.

14. Li, J., X.G. Zhang, J. Zhang, C.Z. Yang, and B.J. Xia. 2011. Study on the mixed occupation of Li/Ni atoms in Li (Ni,Me) O_2. New methods and applications. *Rare Metal Materials and Engineering* 40 (8): 1348–1354.

Chapter 6
Order–Disorder of Ni, Co, and Mn at (3b) Position in $Li(Ni_{1/3}Co_{1/3}Mn_{1/3})O_2$

6.1 Current Status of Superstructure Research in $Li(Ni_{1/3}Co_{1/3}Mn_{1/3})O_2$ [1–3]

Since the synthesis of $Li(Ni_{1/3}Co_{1/3}Mn_{1/3})O_2$, the arrangement of Ni, Co, and Mn in its 3b site has been the focus of debate. At present, there are three hypothetical models: The first is the $[\sqrt{3} \times \sqrt{3}]$R30 superlattice model calculated by Koyama et al. [1] through the first principle, that is, the three ions Ni, Co, and Mn are uniformly and regularly arranged in the 3b layer, as shown in Fig. 6.1a; the second is the stacking model with alternating arrangement of CoO_2, NiO_2, and MnO_2 layers, as shown in Fig. 6.1b; and the third is that Ni, Co, and Mn ions occupy the 3b site randomly and disorderly, that is, the disordered O3 structure. Table 6.1 shows the lattice parameters of the related materials and the average distance between transition metal ions and oxygen. It can be seen that the distance between Co–O in two kinds of superlattice $Li(Ni_{1/3}Co_{1/3}Mn_{1/3})O_2$ is almost the same as that in $LiCoO_2$, while the distance between Ni–O is larger than that in $LiNiO_2$. On the contrary, the distance of Mn–O is smaller than that of Mn–O in $LiMnO_2$. This shows that the local electronic structures of Co in two kinds of superlattice Li $(Ni_{1/3}Co_{1/3}Mn_{1/3})O_2$ are the same as those in $LiCoO_2$, while the local electronic structures of Ni and Mn are different from those in $LiNiO_2$ and $LiMnO_2$, respectively.

Yabuuchi et al. [2] calculated by the first principle that the unit cell formation energy of the $[\sqrt{3} \times \sqrt{3}]$ R30° superlattice is −0.17 eV, while the lattice formation energy of the stacking model is +0.06 eV. Therefore, the existence of the superlattice model is more reasonable and more likely. This crystal form can minimize the change of lattice volume in the process of charge and discharge, which is beneficial to the stability of the lattice. So as to improve the cycle performance of the material. Yabuuchi et al. [2] through the refinement of XRD structure found that the $[\sqrt{3} \times \sqrt{3}]$ R30° superlattice model is in good agreement with the actual diffraction spectrum of $LiNi_{1/3}Co_{1/3}Mn_{1/3}O_2$. This material is a $P3_1 12$ symmetric lattice rather than a simple R-3 m configuration. Further electron diffraction studies show that there is a $[\sqrt{3} \times$

C. Yang et al., *Materials and Working Mechanisms of Secondary Batteries*,
https://doi.org/10.1007/978-981-19-5955-4_6

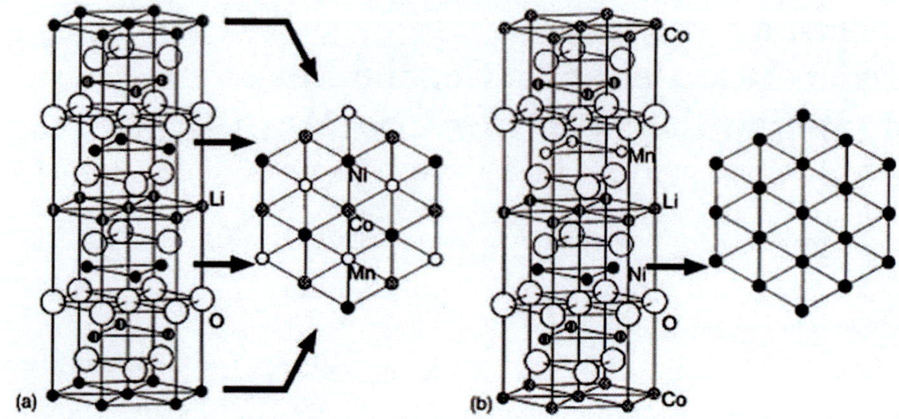

Fig. 6.1 Structural model of superstructure $Li(Ni_{1/3}Co_{1/3}Mn_{1/3})O_2$. **a** is considered to be composed of $(Ni_{1/3}Co_{1/3}Mn_{1/3})O_2$ sheets with $[\sqrt{3} \times \sqrt{3}]R30^{\circ}$ superstructure, **b** is composed of alternating NiO_2, CoO_2, and MnO_2 slices

Table 6.1 Lattice parameters of several materials and average distance between transition metal ions and oxygen

	Lattice parameter (nm)		Average distance between atoms (nm)		
Superstructure Li $(Ni_{1/3}Co_{1/3}Mn_{1/3})O_2$	a	c	Co–O	Ni–O	Mn–O
Calculated according to the model in Fig. 6.1a	0.2831	1.388	0.192	0.202	0.194
Calculated according to the model in Fig. 6.1b	0.2827	1.398	0.192	0.199	0.196
$LiCoO_2$ Calculate	0.2823	1.359	0.191		
Experiment	0.2814	1.4044			
$LiNiO_2$ Calculate	0.2837	1.321		0.194	
Experiment	0.2880	1.4187			
$LiMnO_2$ calculation	0.2932	1.416			0.203

$\sqrt{3}]$ R30° superlattice in the [11.0] crystal zone. As shown in Fig. 6.2, the spot index is R-3 m structure represented by hexagonal [2].

Whitfield et al. [3] studied the structure of $Li(Ni_{1/3}Co_{1/3}Mn_{1/3})O_2$ by neutron powder diffraction and anomalous dispersive powder diffraction. The interaction between elements was enhanced by resonance diffraction. The distance between three transition metal elements was measured by neutron diffraction. It is found that there is no superlattice structure shown in Fig. 6.1a. Moreover, the distribution of 3b site of Ni, Co, and Mn in the lattice is not random. The results of Rietveld structure refinement analysis show that about 2% of Ni^{2+} migrates from 3b site to 3a site

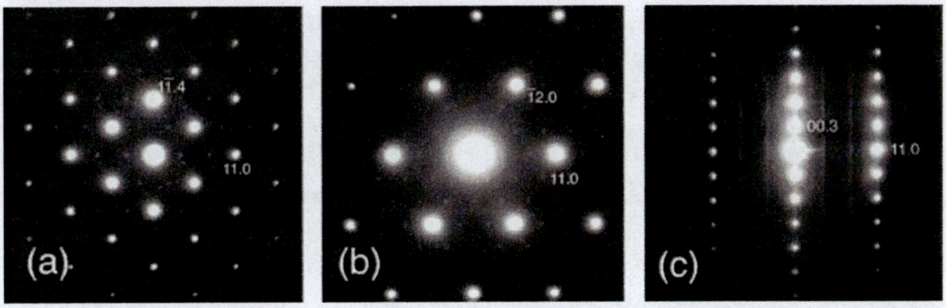

Fig. 6.2 Electron diffraction pattern of $LiNi_{1/3}Co_{1/3}Mn_{1/3}O_2$ **a** [221] crystal zone, **b** [001] crystal zone, **c** [110] crystal zone

instead of Li^+, that is, there is a mixed arrangement of cations. While Co and Mn only occupy the 3b site of transition metal in the lattice. Because the radius of Ni^{2+} ($r_{Ni}{}^{2+} = 0.69$ Å) is similar to that of Li^+ ($r_{Li}{}^{+} = 0.76$ Å), Ni^{2+} is easy to mixed arrangement with Li^+. The higher the concentration of Ni^{2+} in the lithium layer, the more difficult it is to de-intercalate Li^+ in the layered structure, and the more worse the electrochemical performance is. The results are shown in Fig. 6.3. It can be seen that it is of theoretical and practical significance to study the superstructure of $Li(Ni_{1/3}Co_{1/3}Mn_{1/3})O_2$.

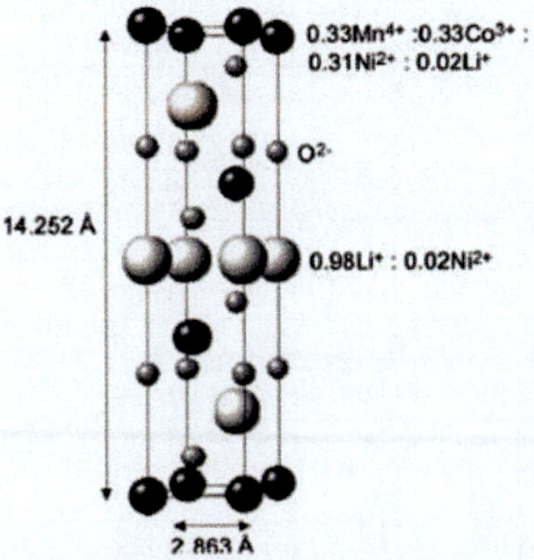

Ion	position	x y z	occupation probability	Thermal parameters
Li^{+1}	3a	000	0.9781	1.236
Ni^{+2}	3a	000	0.0213	0.429
Li^{+1}	3b	0 0 1/2	0.0242	1.236
Ni^{+2}	3b	0 0 1/2	0.3092	0.429
Co^{+2}	3b	0 0 1/2	0.3320	0.429
Mn^{+2}	3b	0 0 1/2	0.3346	0.429
O^{-2}	6c	0 0 0.259	1.000	0.663

Fig. 6.3 The structure of Li $(Ni_{1/3}Co_{1/3}Mn_{1/3})O_2$ after refinement. The occupation parameters of each ion are shown in the table on the right

6.2 Unit Cell Structure and Diffraction Pattern of Superstructure in Li $(Ni_{1/3}Co_{1/3}Mn_{1/3})O_2$ [4]

6.2.1 *Cell Structure of Disordered-Ordered in Li $(Ni_{1/3}Co_{1/3}Mn_{1/3})O_2$*

The following are the results of the simulation study using the Power.cell program. Let the long-range order degree $S = 0$, 0.5, and 1, and to input: Space group No.1 66, $R\overline{3}\frac{2}{m}$, $a = 2.963$, $c = 14.252$, and data in Table 6.2 when $S = 0$, that is, Ni, Co, and Mn occupy the three crystallographic positions 001/2, 1/3 2/3,7/6, and 2/3 1/3 5/6 according to their atomic ratios, so their SOF is 0.333; when $S = 1$, Ni, Co, and Mn occupy the three crystallographic positions 001/2, 1/3 2/3 7/6, and 2/3 1/3 5/6, respectively, and their SOF is 1.000; when $S = 0.5$, occupy is complicated, Ni, Co, and Mn occupy by $S = 0.5$ orderly three crystallographic positions of 001/2, 1/3 2/3 7/6, and 2/3 1/3 5/6, respectively, and at the same time, they occupy three (3b) positions with a probability of 0.5×0.333, respectively. The crystal structure models obtained are shown in Fig. 6.4a and b, respectively, and the angle between a and b is 120°. It should be noted that the Ni atoms shown in Fig. 6.4a are all 1/3(Ni,Co,Mn) atom.

Table 6.2 Crystallographic occupation of various elements with different occupying order degree

S	Element	Z	Ion	Position	x	y	z	SOF	B(temp.)
0.0	Li	3	Li^{+1}	3a	0	0	0	1.00	1.236
	Ni	28	Ni^{+2}	3b	0	0	1/2	1/3	0.429
	Co	27	Co^{+2}	3b	0	0	1/2	1/3	0.429
	Mn	25	Mn^{+2}	3b	0	0	1/2	1/3	0.429
	O	8	O^{-2}	6c	0	0	1/4	1.00	0.663
0.5	Li	3	Li^{+1}	3a	0	0	0	1.00	1.236
	Ni	28	Ni^{+2}		0	0	1/2	1/2	0.429
	Co	27	Co^{+2}		1/3	2/3	7/6	1/2	0.429
	Mn	25	Mn^{+2}		2/3	1/3	5/6	1/2	0.429
	Ni	28	Ni^{+2}	3b	0	0	1/2	1/6	0.429
	Co	27	Co^{+2}	3b	0	0	1/2	1/6	0.429
	Mn	25	Mn^{+2}	3b	0	0	1/2	1/6	0.429
	O	8	O^{-2}	6c	0	0	1/4	1.00	0.663
1.0	Li	3	Li^{+1}	3a	0	0	0	1.00	1.236
	Ni	28	Ni^{+2}	3b	0	0	1/2	1.00	0.429
	Co	27	Co^{+2}	3b	0	0	1/2	1.00	0.429
	Mn	25	Mn^{+2}	3b	0	0	1/2	1.00	0.429
	O	8	O^{-2}	6c	0	0	1/4	1.00	0.663

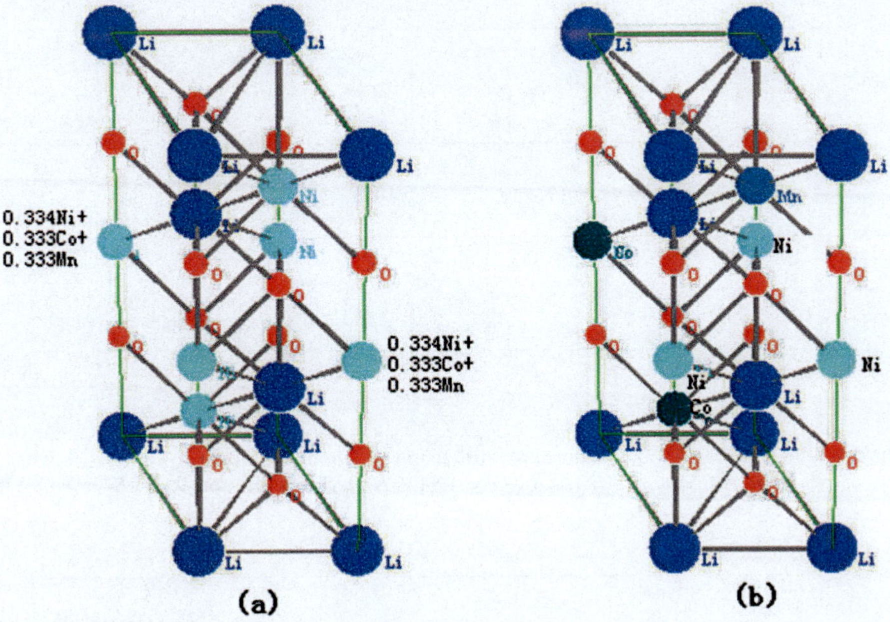

Fig. 6.4 Crystal structure model of long-range order in Li $(Ni_{1/3}Co_{1/3}Mn_{1/3})$ O_2 when $S = 0$ (**a**) and $S = 1$ (**b**). In Figure (a), the atoms shown with Ni are all 1/3 Ni,Co,Mn atomic occupancy

6.2.2 Comparison of X-Ray and Neutron Diffraction Patterns with Wavelengths of 1.5406

For general X-ray wavelengths, such as CuKα, the difference of scattering factors of Ni, Co, and Mn atoms is very small, while the difference of scattering.

factors of three transition atoms to neutron rays is much larger, so neutron diffraction is often used to study the ordered occupancy problem of the nearest neighbor atoms. Figures 6.5 and 6.6 show the X-ray and neutron diffraction patterns of $S =$ 0 and S$S =$ 1, respectively, and there are obvious differences. In order to further study the differences between them, the relative intensity data of X-ray and neutron diffraction with wavelengths of 1.54056 are listed in Table 6.3. Comparing Figs. 6.4 and 6.5 with the data in Table 6.3, we can see that:

(1) There is a very obvious difference in the relative intensity distribution between the X-ray diffraction pattern and the neutron diffraction pattern, no matter $S = 0.0$ or 1.0. The first is the difference of the three strong lines in the pattern, which is marked in boldface in the table, and the most obvious is the difference of the strongest lines. These differences are caused by different atomic scattering factors.

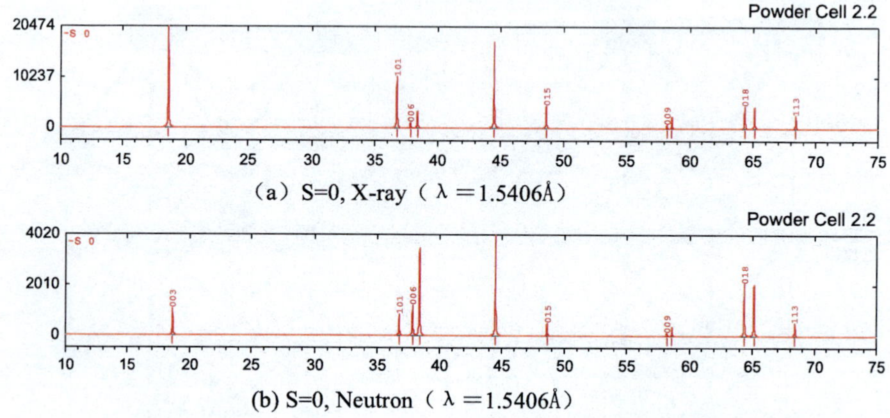

Fig. 6.5 When, $S = 0$, X-ray **a** and neutron diffraction pattern of $Li(Ni_{1/3}Co_{1/3}Mn_{1/3})O_2$ **b** with both wavelengths of 1.5406, **a** $S = 1$, X-ray($\lambda = 1.5406$ Å), **b** $S = 1$, Neutron($\lambda = 1.5406$ Å)

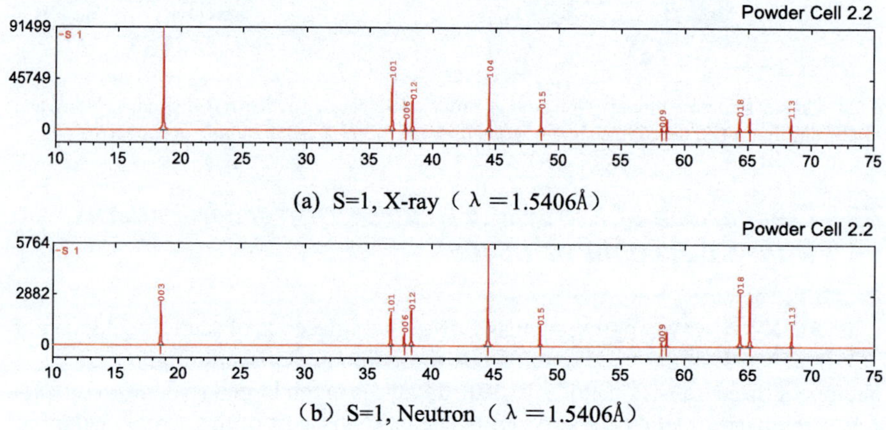

Fig. 6.6 When, $S = 1$, X-ray **a** and neutron diffraction pattern, **b** of $Li(Ni_{1/3}Co_{1/3}Mn_{1/3})O_2$ with both wavelengths of 1.5406

$S = 0$				$S = 1.0$			
X - ray		Neutron		X - ray		Neutron	
hkl	I/I_1	*hkl*	I/I_1	*hkl*	I/I_1	*hkl*	I/I_1
003	100	104	100	003	100	104	100
104	85	012	90	104	51.2	018	52
101	51	018	52	101	50.8	114	51

Table 6.3 Comparison of relative intensity data of X-ray and neutron diffraction at wavelengths of 1.5406Å and $S = 1.0$

hkl	d (Å)	2θ (°)	relative intensity (%)			
			$S = 0$		$S = 1$	
			X-ray	Neutron	X-ray	Neutron
003	4.75067	19.663	**100.00**	26.69	**100.00**	42.42
101	2.44274	36.763	**51.14**	21.04	**50.78**	33.65
006	2.37533	37.845	5.66	29.82	10.21	11.93
012	2.34173	38.409	16.38	**89.90**	29.41	34.74
104	2.03516	44.481	**84.84**	**100.00**	**51.24**	**100.00**
015	1.887073	48.632	21.47	12.30	21.24	19.80
009	1.58356	58.213	3.89	2.94	2.87	4.76
107	1.57348	58.622	11.40	8.71	11.32	14.10
018	1.44677	64.339	22.18	**51.93**	13.90	**52.02**
110	1.43150	65.110	21.20	50.91	13.32	**51.02**
113	1.37063	68.1389	13.32	13.31	13.22	21.69

The three sets of numbers in blackbody represent the three strongest lines in the diffraction pattern

(2) For X-ray diffraction, the relative intensity of the secondary strong line decreases with the increase of S, and the relative intensity of the third strong line is almost unchanged with the increase of S; for neutron diffraction, the relative intensity of the secondary strong line decreases with the increase of S, and the relative intensity of the third strong line almost does not change with the increase of *S*.

6.2.3 Comparison of Abnormal Diffraction Patterns

The so-called abnormal diffraction is to carry on the diffraction experiment by using the radiation of the absorption limit wavelength of a certain element in the sample, which is called abnormal scattering or abnormal diffraction or resonance scattering, sometimes also called selective element diffraction.

When there is abnormal diffraction, the total atomic scattering factor f is given by the following formula.

$$f_{(S,\lambda)} = f_{0(S)} + f'_{(S,\lambda)} + if''_{S,\lambda} \tag{6.1}$$

Among them, f' and f'' are, respectively, the real part and the imaginary part of the abnormal scattering correction. The phase of f' is opposite to phase of f_0, and the phase of f'' shows a 90 displacement. The f is

$$f = \sqrt{(f_0 + f')^2 + (f'')^2} \tag{6.2}$$

And can be approximately written as

$$f = f_0 + f' + \frac{1}{2}\frac{(f'')^2}{f_0 + f'} \tag{6.3}$$

The effect of abnormal scattering on the total factor scattering factor can be summarized as follows:

(1) The abnormal scattering correction component varies with the wavelength of the incoming ray. Near the absorption limit, great changes will take place.
(2) When the wavelength of the incident ray is far from the absorption limit, the abnormal scattering correction can be ignored; near the absorption limit, the abnormal scattering correction component is quite large and increases with the increase of the scattering angle, so the abnormal scattering correction cannot be ignored; when the incident ray wavelength is consistent with the absorption limit, the abnormal scattering influence is very great, and it can play a dominant role in the high diffraction angle.
(3) Because of the characteristic absorption limit of each element, the abnormal scattering correction is related to the atomic number of the scatterer. Therefore, generally speaking, it is advantageous to use abnormal diffraction to study the long-range order of neighboring atoms. Table 6.4 shows the main diffraction data (2θ and relative intensity) of the simulated X-ray diffraction spectrum in which the absorption limits of Ni, CuKα and Co, and Mn are incident rays. It can be seen that the position of the diffraction lines moves to a large angle with the increase of the wavelength, while the relative intensity has no obvious change. Does this indicate that the abnormal diffraction effect does not play an important role here, or the Power.cell program cannot automatically correct the abnormal scattering.

6.2.4 Comparison of Different Positions Occupied by Ni Co Mn (3b)

In the R-3 m (No. 166) structure, the positions of (3b) have three positions of 0 0 1/2, 1/3 2/3 7/6, and 2/3 1/3 5/6, respectively. Ni, Co, and Mn can occupy three different positions in turn. Figure 6.7 shows the crystal cell structure of the three cases. It can be seen that their cell structures are the same in form, but the same positions are occupied by different atoms.

The diffraction pattern data (2θ, relative intensity, and structure amplitude) of the three occupying materials are summarized in Table 6.5. The strongest lines are the same, all 003, and the intensity data of the second strong line and the third strong

Table 6.4 Comparison of abnormal diffraction data (2θ, relative intensity) when $S = 1$

hkl	*d* (Å)	Absorption limit wavelength (Å)				Absorption limit wavelength (Å)			
Elements		Ni	Cu	Co	Mn	Ni	Cu	Co	Mn
Wavelength (Å)		1.48801	1.54056	1.6082	1.8964	1.48801	1.54056	1.6082	1.8964
		2θ(°)				Relative intensity, I/I_1			
003	4.75067	19.021	19.663	19.489	23.027	100.00	100.00	100.00	100.00
101	2.44274	35.592	36.893	38.573	45.846	51.12	50.78	50.33	48.35
006	2.37533	36.508	37.845	39.572	47.055	10.28	10.21	10.11	9.69
012 012	2.34173	37.171	38.536	40.295	47.931	29.63	29.41	29.12	27.87
104	2.03516	42.994	44.592	46.660	55.682	51.76	51.24	50.58	47.84
009	1.87073	46.971	48.735	51.021	61.046	21.58	21.24	21.01	19.79
015	1.58356	56.049	58.213	61.030	73.566	3.92	3.87	3.80	3.64

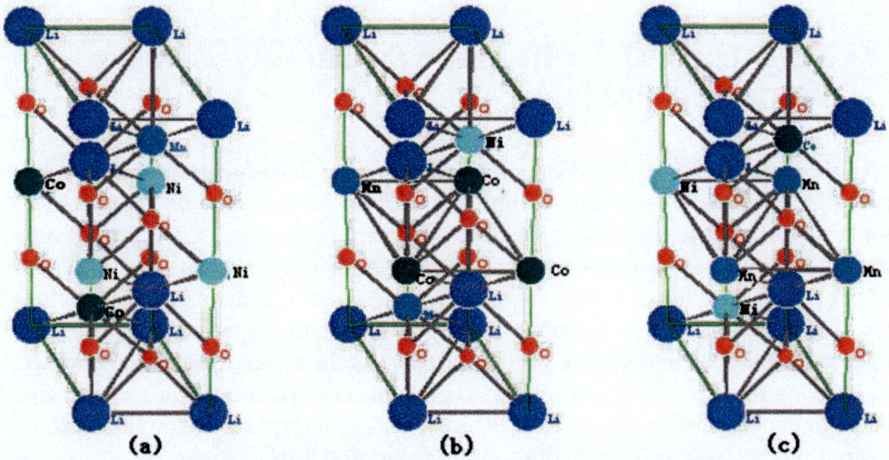

Fig. 6.7 Comparison of unit cell structure with different positions occupied by Ni, Co, and Mn, respectively, **a** Ni 0 0 1/2; Co 1/3 2/3 7/6; Mn 2/3 1/3 5/6, **b** Co 0 0 1/2: Mn 1/3 2/3 7/6; Ni 2/3 1/3 5/6, **c** Mn 0 0 1/2; Ni 1/3 2/3 7/6; Co 2/3 1/3 5/6

line change little. The structural amplitudes of the three strong lines do not change much.

Table 6.5 Comparison of diffraction data (2θ, relative intensity, and structure amplitude) at different positions occupied by Ni, Co, and Mn, respectively

hkl	d (Å)	2θ (°)	Ni* 0 0 1/2; Co 1/3 2/3 7/6 Mn 2/3 1/3 5/6		Co 0 0 1/2: Mn 1/3 2/3 7/6 Ni 2/3 1/3 5/6		Mn 0 0 1/2 Ni 1/3 2/3 7/6 Co 2/3 1/3 5/6	
			Relative intensity	|*F*(*hkl*)|	Relative intensity	|*F*(*hkl*)|	Relative intensity	|*F*(*hkl*)|
003	4.75067	19.663	**100.00**	131.94	**100.00**	125.55	**100.00**	135.13
101	2.44274	36.893	**58.25**	119.22	**57.93**	113.12	**58.46**	122.31
006	2.37533	37.845	11.80	95.94	11.42	89.88	11.98	99.02
012	2.34173	38.536	32.85	93.96	31.76	87.91	33.41	97.04
104	2.03516	44.592	**56.44**	145.09	**57.39**	139.22	**56.06**	148.09
015	1.87073	48.735	27.15	111.32	26.97	105.58	27.27	114.27
009	1.58356	58.213	5.33	104.76	5.29	99.37	5.35	107.57

The three sets of numbers in blackbody represent the three strongest lines in the diffraction pattern

6.3 The Effect of Order Degree on the Intensity of Matrix Diffraction Lines [4]

From the introduction of "the General principle of the determination of long-range ordering degree in ordered Phase" and the ordered phase based on "face center" and "body center" and the example of A-15 structure in Sect. 5.2–6.6 of 《Phase diffraction Analysis [5], it is known that superlattice diffraction lines may occur after ordering of some solid solutions, of course, the degree of ordering has an obvious effect on the intensity of the superlattice lines. However, the superlattice line may not appear after the ordering of some solid solutions. But the effect of the order degree on the intensity of different diffraction lines is obviously different. So the order degree can be obtained through the study of the intensity of matrix diffraction lines, which depends on the crystal structure of the solid solution.

From the previous research, it is known that the superlattice diffraction line does not appear after the ordering of Li $(Ni_{1/3}Co_{1/3}Mn_{1/3})O_2$ with rhombohedral structure of R-3 m (No. 166). Let us study the effect of ordering degree on the intensity of matrix diffraction lines.

Let the atomic scattering factors at positions 3a, 3b, and 6c be f_{3a}, f_{3b}, and f_{6c}. For $Li(Ni_{1/3}Co_{1/3}Mn_{1/3})O_2$, they are obtained from the following formula:

$$\begin{aligned} f_{3a} &= f_{(3a)}1 + f_{(3a)2} + f_{(3a)3} = 3f_{Li} \\ f_{3a} &= f_{(6c)1} + f_{(6c)2} + \ldots + f_{(3c)6} = 6f_o \end{aligned} \quad (6.4)$$

f_{3b} is more complicated, when $S = 0$.

$$f_{3b} = \frac{1}{3}(f_{Ni} + f_{Co} + f_{Mn}) \times 3 \quad (6.5)$$

When $S = 1$

$$f_{3b} = f_{Ni} + f_{Co} + f_{Mn} \tag{6.6}$$

When $0 < S < 1$

$$\begin{aligned} f_{3a} = &\frac{1}{3}(f_{Ni} + f_{Co} + f_{Mn}) \times f_{3b} = S \times (f_{Ni} + f_{Co} + f_{Mn}) \\ &+ (1 - S) \times \frac{1}{3}(f_{Ni} + f_{Co} + f_{Mn}) \times 3 \end{aligned} \tag{6.7}$$

From the formula (6.7), it is known that when $0 < S < 1$, the atomic scattering factor of the (3b) position is divided into two terms, the first term describes the ordered occupation part, and the second term describes the contribution of the disordered occupation of the remaining atoms after the ordered occupation. f_{Li}, f_{Ni}, f_{Co}, f_{Mn}, and f_o are atomic scattering factors of Li, Ni, Co, Mn, and O. They are all functions of $\frac{\sin\theta}{\lambda}$ and decrease with the $\frac{\sin\theta}{\lambda}$ value increases. For the same diffraction pattern, λ is constant, which means that the atomic scattering factor decreases with the increase of 2θ, that is, the atomic scattering factor varies with different hkl, and the larger the diffraction angle 2θ is, the smaller the atomic scattering factor is.

Similarly, the angle factor $\frac{1+\cos^2 2\theta_{hkl}}{\sin^2\theta_{hkl}\cos\theta_{hkl}}$ of each diffraction line decreases with the increase of 2θ, and its specific value can be calculated or obtained in the table in the appendix of professional books.

The "addition" and "subtraction" relationship between 3a site and 3b site reveals a very important law. Taking the three diffraction lines of 003, 101, and 104 as examples, the F_{hkl} is calculated and the similar terms are combined to obtain:

$$\begin{cases} F_{003} = 3f_{Li}^{003} - S \times (f_{Ni}^{003} + f_{Co}^{003} + f_{Mn}^{003}) - (1 - S) \times \frac{1}{3}(f_{Ni}^{003} + f_{Co}^{003} + f_{Mn}^{003}) \times 3 \\ F_{101} = 3f_{Li}^{101} - S \times (f_{Ni}^{101} + f_{Co}^{101} + f_{Mn}^{101}) - (1 - S) \times \frac{1}{3}(f_{Ni}^{101} + f_{Co}^{101} + f_{Mn}^{101}) \times 3 \\ F_{104} = 3f_{Li}^{104} + S \times (f_{Ni}^{104} + f_{Co}^{104} + f_{Mn}^{104}) + (1 - S) \times \frac{1}{3}(f_{Ni}^{104} + f_{Co}^{104} + f_{Mn}^{104}) \times 3 + 6f_{O}^{104} \end{cases} \tag{6.8}$$

It can be seen that the intensity of the diffraction line of (104) is stronger than that of (101). In order to calculate the intensity of 003, 101, and 104 diffraction lines, the data in Table 6.6 are required. The results of careful calculation and analysis of formula (6.8) show that the order degree S has no effect on the diffraction intensity of the matrix, in other words, the diffraction intensity has nothing to do with the order and disorder of Ni, Co, and Mn at 3b position. So ordered degree S cannot be obtained from the intensity data of the matrix line.

Table 6.6 Data of three diffraction lines: 003, 101, and 104

hkl	θ (°)	Angle factor	Multiplicity factor	sinθ/λ	Atomic scattering factor (f_i^{hkl})				
					Li	Ni	Co	Mn	O
003	9.331	72.533	2	0.105	2.215	25.600	24.650	22.570	7.250
101	19.446	17.255	6	0.205	1.741	21.370	20.540	19.000	5.634
104	22.296	11.310	6	0.246	1.627	19.590	19.810	17.359	4.814

6.4 Superlattice Lines of $Li(Ni_{1/3}Co_{1/3}Mn_{1/3})O_2$ [4]

6.4.1 *$Li(Ni_{1/3}Co_{1/3}Mn_{1/3})O_2$ Superlattice Lines Whether to Appear or Not*

From the previous research, it is known that the superlattice diffraction line does not appear after the ordering of $Li(Ni_{1/3}Co_{1/3}Mn_{1/3})O_2$ with rhombohedral structure of R-3 m (No. 166). In order to verify whether this is correct, let us first study the diffraction pattern of the classical $AuCu_3$ superstructure (Fig. 6.8a, b, and c). On base of the space group and crystallographic position given in Table 6.7, the diffraction patterns can obtain. It can be seen that the superlattice lines can be displayed only when the *x y z* of the crystallographic position is the same, but the space groups are different, that is, the space groups with different extinction rules are used to calculate the diffraction patterns. From this conclusion, it is inferred that there is no superlattice line in the previous Li $(Ni_{1/3}Co_{1/3}Mn_{1/3})O_2$-simulated diffraction pattern with the space group No.166, which does not prove that the superstructure will not appear superlattice line.

In the rhombohedral crystal system space group (No. 143–No. 167), the author does not find the same *x y z* coordinate as the (3b) position of No.166, but the space group with different extinction conditions corresponds to it. It can be seen that it is true to change No.166 into $P3_112$ (No. 151) symmetry after ordering at (3b) position that is worth discussing. Because the atomic position and extinction law of No.151 are:

(3a)	$x\ -x\ 1/3$	$x\ 2x\ 2/3$	$-2x\ -x\ 0$	General
(3b)	$x\ -x\ 5/6;$	$x\ 2x\ 2/3;$	$-2x\ -x\ 1/2$	$00l = 3n$
(6c)	xyz	$-yx-yz+1/3$	$-x+y\ -xz+2/3$	Special condition
	$-y-x\ -z+2/3$	$-x+yy\ z+1/3$	$x\ x-y\ -z$	No extra conditions

For this reason, we calculate the crystallographic position parameters of the possible superlattice lines: 001, 002, 004, 111, as shown in Table 6.8. By analyzing this table, we can see that

(1) The atoms of (3a) and (6c) sites have not contribution to 001, 002, 004, 111

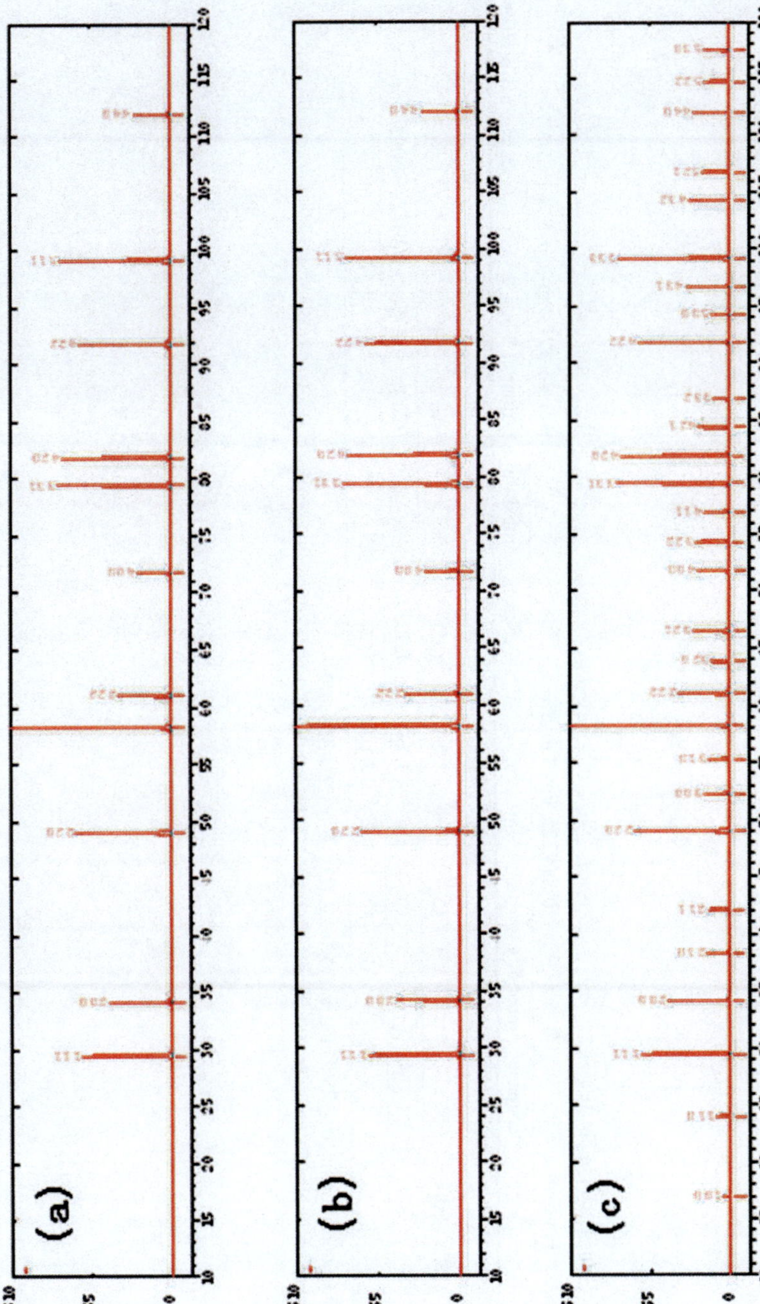

Fig. 6.8 Simulated diffraction patterns of $AuCu_3$ disordered (**a**) and ordered ($S = 1$, *b*, *c*). The input data are as follows in Table 6.7

Table 6.7 Input data in Fig. 6.7

(a) No. 225, Fm-3 m, $S = 0$				(b) No.225, Fm-3 m, $S = 1$				(c) No. 221, Pm-3 m, $S = 1$			
Atom	Site	*x y z*	Occupancy probability	Atom	Site	*x y z*	Occupancy probability	Atom	Site	*x y z*	Occupancy probability
				Au	1a	000	1.00				
Au	1a	000	0.25	Cu	3c	1/2 1/2 0	1.00	Au	1a	000	1.00
Cu	1a	000	0.25	Cu		1/2 0 1/2	1.00	Cu	3c	0 1/2 1/2	1.00
				Cu		0 1/2 1/2	1.00				

Note: The 4a position of No. 225 includes 000, 0 1/2 1/2, 1/2 0 1/2, 0 1/2 , 1/2
The 1a position of No.221 includes 000, 3c include 0 1/2 1/2, 1/2 1/2 0, 1/2 0 1/2

Table 6.8 The possible superpoint front of the No. 166 space group: crystallographic position parameters of 001, 002, 004, and 111

hkl		001	002	004	111
Position	Atomic coordinate	Position parameter $\cos 2\pi(hx_j + ky_j + lz_j) + i \sin 2\pi(hx_j + ky_j + lz_j)$			
3a	000	1	1	1	1
	2/3 1/3 1/3	−0.5 + 0.866*i*	−0.5 − 0.866*i*	−0.5 + 0.866*i*	−0.5 + 0.866i
	1/3 2/3 2/3	−0.5 − 0.866*i*	−0.5 + 0.866*i*	−0.5 − 0.866*i*	−0.5 − 0.866i
	Σ	0	0	0	0
3b	00 1/2	−1	1	1	−1
	2/3 1/3 5/6	0.5 − 0.866*i*	−0.5 − 0.866*i*	−0.5 + 0.866*i*	0.5 − 0.866*i*
	1/3 2/3 7/6	0.5 + 0.866*i*	−0.5 + 0.866*i*	-0.5 − 0.866*i*	0.5 + 0.866*i*
	Σ	0	0	0	0
6c	00 1/4	*i*	−1	1	*i*
	2/3 1/3 7/12	−0.866 − 0.5*i*	0.5 + 0.866*i*	−0.5 + 0.866*i*	−0.866 − 0.5*i*
	1/3 1/3 11/12	0.866 − 0.5*i*	0.5 − 0.866*i*	−0.5 − 0.866i	0.866 − 0.5*i*
	00 −1/4	−1	−1	1	−1
	2/3 1/3 1/12	0.866 + 0.5*i*	0.5 + 0.866*i*	−0.5 + 0.866*i*	0.866 + 0.5*i*
	1/3 2/3 5/12	−0.866 + 0.5*i*	0.5 − 0.866*i*	−0.5 − 0.866*i*	−0.5 + 0.866*i*
	Σ	0	0	0	0

(2) For the case of (3b) site, it is more complicated; if the atomic scattering factor of each position is the same, when the position is disordered, the factor scattering factor of each position is $\frac{f_{Ni}+f_{Co}+f_{Mn}}{3}$, and the sum of the contribution of the three positions is zero, then the superlattice lines will not appear. When the factor scattering factors of the three positions are different and the sum of the contributions of the three positions is not zero, the superlattice lines appear, and the greater the difference of the atomic scattering factors of the three positions, the stronger the intensity of the superlattice lines. Therefore, the use of abnormal diffraction technique and neutron diffraction is beneficial to the study of superlattice structure.

6.4.2 The Effect of Ordered Degree S on the Intensity of Superlattice Lines

In order to calculate the effect of ordering degree S on diffraction intensity, the required data are given in Table 6.9. Let us calculate the effect of ordering S on the intensity of diffraction lines. Because 104 and 111 overlap, only 002, 003, 004, and

Table 6.9 Related parameters of several base lines and superlattice lines, λ = 1.54056 Å

hkl	001	002	003	004	101	104	111
Multiple factors	2	2	2	2	6	6	6
d	14.2520	7.1260	4.75067	3.5630	2.44274	2.03516	2.0288
2θ	6.196	12.412	19.663	24.970	36.893	44.592	44.627
Angle factor	680.9	168.5	73.499	39.82	17.15	11.31	11.31
Sinθ/λ	0.035	0.070	0.105	0.140	0.205	0.246	0.246
f_{Li}			2.215		1.741	1.627	
f_{Ni}	27.631	26.638	25.600	23.008	21.370	19.590	19.590
f_{Co}	26.538	25.664	24.650	22.970	20.540	19.810	19.810
f_{Mn}	24.149	24.271	22.570	21.142	19.000	17.359	17.359
f_O			7.250			4.814	

101 diffraction lines are calculated. Table 6.10 shows the calculation results, which can be seen as follows:

(1) The intensity of matrix diffraction lines does not change with order degree *S*; in other words, the intensity of matrix diffraction lines is not affected by *S*.
(2) The intensity of the superlattice diffraction line increases with the increase of *S*. It can use given the intensity data of superlattice line intensity or the intensity ratio of the superlattice line to the matrix diffraction lines to solve the ordered degree *S*.
(3) However, the relative intensity of the superlattice line is very weak, even though the relative intensity of the superlattice lines of 002 and 004 is very weak, even at the time of $S = 1.0$. The relative intensity is only 0.225 and 0.043%. Therefore, it is difficult to detect superlattice lines in the experiment.

6.5 Summary and Prospect of the Study on the Superstructure of $Li(Ni_{1/3}Co_{1/3}Mn_{1/3})O_2$ [4]

For $Li(Ni_{1/3}Co_{1/3}Mn_{1/3})O_2$ when Ni, Co and Mn orderly occupy the 3b site, the order degree is not effect on the intensity of matrix diffraction lines, but the intensity of the superlattice diffraction line increases with the increase of the order degree. Therefore, theoretically, the order degree S can be solved according to this relation, but because the intensity of superlattice diffraction lines is very weak, it is difficult to observe superlattice diffraction lines in general diffraction experiments. Therefore, whether there is order occupying in 3b site and the formation of superstructure of $Li(Ni_{1/3}Co_{1/3}Mn_{1/3})O_2$ needs to be further studied in the following aspects:

(1) To explore the best process of ordering treatment of $Li(Ni_{1/3}Co_{1/3}Mn_{1/3})O_2$ materials at appropriate medium temperature.

Table 6.10 The influence of the calculated intensity and relative intensity of several matrix lines 003, 101 and superlattice lines 002, 004 and the degree of order S, $\lambda = 1.54056$ Å

S		0.0	0.2	0.4	0.6	0.8	1.0
Intensity	002	0	19.164	228.926	515.1063	1216.3021	1431.104
	003	635,262.963	635,262.963	635,262.963	635,262.963	635,262.963	635,262.963
	004	0	10.910	43.476	97.8314	173.7932	271.747
	101	321,050.875	321,050.875	321,050.875	321,050.875	321,050.875	321,050.875
Relative intensity	002	0.0	0.0029	0.0360	0.0811	0.1915	0.2253
	003	100.00	100.00	100.00	100.00	100.00	100.00
	004	0.0	0.0017	0.0068	0.0154	0.0274	0.0428
	101	50.54	50.54	50.54	50.54	50.54	50.54

The three sets of numbers in blackbody represent the three strongest lines in the diffraction pattern

(2) To improve the detection ability of diffraction experiments, using ultra-strong X-ray sources, such as synchrotron radiation X-ray sources, and high-efficiency super-energy detectors, finely experimental operations (such as slow speed scanning), and abnormal diffraction techniques or/and neutron diffraction techniques. In short, it is necessary to greatly improve the ability to detect ultra-weak lines.
(3) According to the relationship that the diffraction intensity increases with the increase of the order degree S, the order degree of the material can be determined and can establish the relationship between ordered treatment process and material (battery) performance.

References

1. Koyama, Y., I. Tanaka, H. Adachi, et al. 2003. Crystal and electronic structures of superstructural $Li_{1-x}[Co_{1/3}Ni_{1/3}Mn_{1/3}]O_2$ (0 < X< 1). *Journal of Power Sources* 119–121: 644–648.
2. Yabuuchi, N., Y. Koyama, N. Nakayama, et al. 2005. Solid-state chemistry and electrochemistry of $LiCo_{1/3}Ni_{1/3}Mn_{1/3}O_2$ for advanced lithium-ion batteries. *Journal of the Electrochemical Society* 152 (7): 1434–1440.
3. Whitfield, P., I. Davidson, L. Cranswick, et al. 2005. Investigation of possible superstructure and cation disorder in the lithium battery cathode material $LiMn_{1/3}Ni_{1/3}Co_{1/3}O_2$ using neutron and anomalous dispersion powder diffraction. *Solid State Ionics* 176: 463–471.
4. Zhang, J., C.Z. Yang, X.W. Han, et al. 2014. Diffraction research on the order-disorder effect of transition metal in $Li(Ni_{1/3}Co_{1/3}Mn_{1/3})O_2$ (2014, will be published).
5. Yang, C.Z., D.C. Xie, G.Z. Chen, and F.M. Zhong. 1989. *Phase diffraction analysis*, Chap. 5, pp. 179–184. Metallurgical Industry Press.

Chapter 7
Preparation and X-Ray Diffraction Characterization of $LiFePO_4$

In 1997, Professor Goodenough [1] of the University of Texas in the USA first reported the research results of $LiFePO_4$ as positive materials for lithium-ion batteries, which became a milestone in the research of $LiFePO_4$ lithium-ion positive materials. Due to the advantages of a wide range of raw materials, low price, friendly environment, good thermal stability and high safety performance of the prepared battery, $LiFePO_4$ has a great market prospect in the field of movable power supply, especially in the field of large power supply for electric vehicles and static energy storage. This large power supply requires low volume specific capacity of materials, but high requirements for material price, safety, and environmental protection. As a result, $LiFePO_4$ has become a new generation of positive materials for lithium-ion batteries with the most potential for development and application.

Polyanionic phosphate materials represented by $LiFePO_4$ have attracted wide attention because of their outstanding characteristics such as high safety, high power, ultra-long cycle life, wide electrochemical window, low cost, and so on. $LiFePO_4$ has become the most likely cathode material to replace Li-ion secondary battery, so it has become a hot spot of development and research in the battery industry. The main contents are as follows: (1) the synthesis and modification of $LiFePO_4$ materials; (2) the research and development of 2H-graphite/$LiFePO_4$ battery; (3) the study of charge and discharge performance and working mechanism of this kind of battery; (4) the industrialization of 2H-graphite/$LiFePO_4$ battery to meet the (HEV) requirements of hybrid electric vehicles.

C. Yang et al., *Materials and Working Mechanisms of Secondary Batteries*,
https://doi.org/10.1007/978-981-19-5955-4_7

7.1 Preparation Method of $LiFePO_4$ and Carbon-Coated Nano-$LiFePO_4$ [2, 3]

7.1.1 Synthesis Methods of $LiFePO_4$

(1) **High-temperature solid-state method** [4, 5]

The basic process of preparing $LiFePO_4$ by high-temperature solid-state method is to mix iron salt, lithium salt, and phosphate. The mixture is heated at a lower temperature (300 °C) to remove volatile matter and then sintered at a higher temperature to obtain a complete $LiFePO_4$ crystallization powder. The heat treatment process is usually completed in N_2 or Ar atmosphere to prevent Fe^{2+} from oxidizing to Fe^{3+}. The operation flow of high-temperature solid-state method is simple, and it is a mature method to prepare positive electrode materials for Li-ion batteries. The advantage of high-temperature solid-state method is that it is easy to operate and easy to realize batch industrialization. However, because it needs to be sintered at high temperature for a long time, the energy consumption is high and the cost is relatively higher, and the $LiFePO_4$ particles obtained by high-temperature solid-state method are generally larger. The results show that the temperature has a great influence on the particle size of $LiFePO_4$ product during heat treatment. If the temperature is too high, it is easy to cause sintering caking. Lowering the sintering temperature and adding carbon particles to $LiFePO_4$ can reduce the particle size of the product.

Because the price of ferrous salt for preparing $LiFePO_4$ is higher than that of trivalent iron salt, it is considered to choose cheap Fe^{3+} salt as iron source to prepare $LiFePO_4$, such as Fe_2O_3, $FePO_4$, and so on. $LiFePO_4$ can be prepared by adding a certain amount of reducing agent in the preparation process, but the conditions must be well-controlled; otherwise, Fe^{3+} salt is easy to be left in the product, which affects the performance of $LiFePO_4$.

Although the process of preparing $LiFePO_4$ by high-temperature solid-state method is simple, there are often impurity phases, such as Li_3PO_4, Fe_2O_3, and so on. The mechanism of impurity phases produced by raw materials of different systems may be different. At present, there is a lack of systematic research in this field. Reducing atmosphere can be used to avoid impurity phase, or a little reductive substance, such as carbon, can be added to the raw material.

(2) **Carbon-thermal reduction method** [6]

A novel carbon-thermal reduction method was used to prepare $LiFePO_4$ materials. In this method, cheap Fe_2O_3 is used as raw material, and carbon reduction reaction is used to prepare $LiFePO_4$ material, which is based on the following reaction:

$$2LiH_2PO_4 + Fe_2O_3 + C = 2LiFePO_4 + CO + 2H_2O \tag{7.1}$$

$$LiH_2PO_4 + 0.45Fe_2O_3 + 0.1Mg\,(OH)_2 + 0.45C$$

$$= LiFe_{0.9}Mg_{0.1}PO_4 + 0.45CO + 1.1H_2 \quad (7.2)$$

Using Fe_2O_3 instead of expensive FeC_2O_4 salt as raw material can greatly reduce the cost, but, as it is difficult to ensure that the Fe^{3+} in the raw material is fully reduced, it may be difficult to avoid impurity phase in the product.

(3) **Mechanochemical method** [7]

The $Fe_3(PO_4)_2\ 5H_2O$ and Li_3PO_4 were mixed in a certain proportion, and the conductive additive was added, and the $LiFePO_4$ crystal powder was obtained by ball milling for 24 h. However, the crystal integrity of $LiFePO_4$ obtained in this way is poor, and it needs to be treated by heating. The $LiFePO_4$ powder with good integrity can be obtained by heating to 550 °C in nitrogen atmosphere and holding for 15 min. The basic principle of high-energy ball milling is to use the energy generated by high-speed rotating steel balls to promote the reaction between raw materials. The actual discharge specific capacity of $LiFePO_4$ materials can reach 150 mAh/g under 0.2 C charge–discharge conditions. Because the reactants are uniformly mixed during ball milling, the product prepared by high-energy ball milling has high activity. Compared with the high-temperature solid-state method, the heating temperature of the high-energy ball milling method is lower, the heating time is shorter, and the energy consumption can be saved. However, due to the high requirement of ball milling tank, it is difficult to get large-scale application.

(4) **Sol–gel method** [8, 9]

The gel method is to mix Fe^{2+}, Li^+, and PO_4^{3-} into a solution, add a certain amount of complexing agent to make the product form a sol, and then heat up the gel-like substance. The product prepared by sol–gel method has uniform particles, good properties, and low heat treatment temperature, which can save a lot of energy consumption. The sol–gel method has the advantages of low gel heat treatment temperature, small particle size, narrow particle size distribution, simple operation, and easy control. The disadvantage is that the gelation process needs to be completed for a long time under specific pH conditions, and the precursor shrinks greatly after drying. Firstly, the idea of preparing $LiFePO_4$ by sol–gel method was put forward, in which ascorbic acid was added into LiOH and $Fe(NO_3)_3$, and then phosphoric acid was added. The pH value was adjusted by ammonia, and the gel obtained at 60 °C was heat-treated to obtain pure phase $LiFePO_4$. Liu [17] mainly uses the special reduction ability of ascorbic acid to reduce Fe^{3+} to Fe^{2+}, which not only avoids the use of expensive Fe^{2+} salt as raw material, reduces the cost, but also solves the atmosphere requirements of precursors. $LiFePO_4$ was prepared by sol–gel method. $CH_3COOLi\ 2H_2O$, $Fe(CH_3COO)_2$, and H_3PO_4 were mixed in N-N-dimethylformamide with atomic ratio of Li:Fe:P = 1.05:1:1 and then heated in air. The obtained solids were heated to 350 °C and kept for two hours. Then uniform and fine $LiFePO_4$ products were obtained by heating in Ar (5%H_2) atmosphere between 700 and 800 °C for 10 h.

(5) **Co-precipitation method** [10]

The co-precipitation method uses the soluble salt of Fe^{2+}, Li^+ and as raw materials to precipitate $LiFePO_4$ from the solution by controlling the pH value of the solution; then the precipitate is filtered, washed, and dried; finally, the precipitate is treated at high temperature to get the $LiFePO_4$ product. Generally speaking, the high-temperature treatment time is shorter than that of the pure high-temperature solid-state method, and the temperature can also be slightly lower.

Amorphous $FePO_4$ and $LiFePO_4$ particles with nanometer size were prepared by co-precipitation method. The specific method is to add $Fe(NH_4)_2$ $(SO_4)_2$ $6H_2O$ to the continuously stirred $NH_4H_2PO_4$ solution at the ratio of 1:1, and then add oxidant H_2O_2 to form precipitation and stop stirring, rest for 3 h, filter out the precipitation and dry, and the amorphous $FePO_4$ can be obtained.

Co-precipitation-based reaction.

$$\begin{aligned} & Fe(NH_4)_2(SO_4)_2 + NH_4H_2PO_4 + 0.5H_2O_2 \\ & \quad = FePO_4 + 2(NH_4)HSO_4 + H_2O + NH_3 \end{aligned} \tag{7.3}$$

By adding the obtained $FePO_4$ to the acetonitrile solution of LiI, the reaction based on $LiFePO_4$, can be obtained by "lithiation".

$$FePO_4 + LiI = LiFePO_4 + 0.5I_2 \tag{7.4}$$

Finally, crystalline $LiFePO_4$ can be obtained by heat treatment of amorphous $LiFePO_4$. The $FePO_4$ and $LiFePO_4$ particles obtained by co-precipitation method can reach nanometer size and have good electrochemical performance, especially the high current charge–discharge performance has been greatly improved. However, the process is complex, energy consumption is high, and it is difficult to scale up.

(6) **Hydrothermal method** [11–13]

Hydrothermal method refers to the general name of chemical reactions carried out in fluids such as water or steam under high temperature and high pressure. The hydrothermal method has the advantages of simple operation and controllable process. The $LiFePO_4$ powder prepared by this method has uniform phase and fine particle size, which is suitable for cathode materials of lithium-ion batteries. However, the hydrothermal method needs high-temperature and high-pressure equipment, so it is difficult to industrialized production. $LiFePO_4$ particles were synthesized by hydrothermal method using $FeSO_4{\cdot}7H_2O$, H_3PO_4, and LiOH as raw materials and ascorbic acid as reducing agent. $LiFePO_4$ particles were synthesized by hydrothermal method at 150–220 °C. The first discharge capacity of 0.3 mA cm^2 is 169 mAh/g at room temperature. In order to avoid the oxidation of Fe^{2+} to Fe^{3+}, during the mixing process, ferrous salt and phosphoric acid solution were mixed first and then stirred by adding LiOH solution [because $Fe(OH)_2$ was easily oxidized to

Fe^{3+}], $LiFePO_4$, was synthesized in a short time (5 h) and low temperature (120 °C). The average particle size of $LiFePO_4$ is only 3 μm, and about 60% lithium ion is de-intercalated at 0.14 mA cm^2 current density.

7.1.2 *Methods to Improve the Performance of $LiFePO_4$ [2, 14]*

The main methods to improve the electrical conductivity of lithium ferrous phosphate materials include: (1) particle nanocrystallization; (2) surface coating of conductive layer, such as nanocarbon layer; (3) bulk doping of lithium ferrous phosphate; (4) formation of Fe_2P, Fe_3P and $Fe_{15}P_3C_2$ phases with good electronic conductivity on the surface of lithium ferrous phosphate during synthesis; (5) improvement of surface morphology of lithium ferrous phosphate. The comparison of electronic and ionic conductivities of Mg, Ni and undoped electrons and ions is given in Table 7.1 for reference.

7.2 X-Ray Characterization of $LiFePO_4$ Material

The X-ray characterization of $LiFePO_4$ materials should include: (1) Dynamic or quasi-dynamic observation and analysis in each stage of various synthesis methods, especially the structural identification of the intermediates in each stage. If the test and analysis in this aspect is done well, the various chemical reactions and physical mechanisms in each stage can be further understood, and the structural changes and synthesis mechanism in each stage of material synthesis can be revealed as described in Chaps. 4, 5, and 6. Explore the best synthesis process. The work in this area is not very good [2, 3, 14], mainly because the material synthesis preparation personnel are not familiar with the X-ray analysis, do not make good use of the relevant analysis tools, and what is important is that there is no close cooperation and participation of X-ray analysis experts; (2) The comprehensive characterization

Table 7.1 Electronic and ionic conductivity (s/cm) of $LiFe_{0.95}Mg_{0.05}PO_4$, $LiFe_{0.95}Ni_{0.05}PO_4$, and $LiFePO_4$

Temperature (°C)	$LiFe_{0.95}Mg_{0.05}PO_4$		$LiFe_{0.95}Ni_{0.05}PO_4$		$LiFePO_4$	
	σ_e	σ_i	σ_e	σ_i	σ_e	σ_i
−30	8.31×10^{-5}	5.91×10^{-5}	3.90×10^{-5}			
25	1.65×10^{-4}	1.79×10^{-4}	6.40×10^{-4}	5.04×10^{-5}	3.75×10^{-9}	5.0×10^{-5}
60	2.21×10^{-4}	2.83×10^{-4}	9.00×10^{-4}		2.04×10^{-8}	1.06×10^{-4}

of the finished product $LiFePO_4$; (3) The changes of structure and fine structure of $LiFePO_4$ during each electrode (activation, charge–discharge, cycle, storage), as well as the relationship between them and properties, will be introduced in Chaps. 12, 16, and 19, respectively.

The specific contents of X-ray analysis should be the phase identification, the determination of lattice parameters, the determination and analysis of microstructure parameters, and the X-ray energy spectrum analysis of the valence states of the main positive ions.

The comprehensive characterization of the finished product $LiFePO_4$ includes:

(1) **Macroperformance parameters**

Contents of main cationic elements: Li, Fe, and dopants;

Main impurity elements: content of impurity elements;

Particle shape, average particle size, and specific surface area;

Loose packing density and loose packing coefficient, vibrating packing density and vibrating packing coefficient.

(2) **Microscopic performance parameters**

There should be no miscellaneous phases in the $LiFePO_4$, of whether the final product belongs to orthogonal Pnma (No. 62) structure;

Lattice parameters and the difference between the lattice parameters and the standard value $a = 0.60157$, $b = 1.03915$, $c = 0.47207$ nm;

Grain shape and shape factor $D_{100}/D_{010}/D_{001}$, average grain size $D_{\text{aver.}} \pm \sigma$.

Except for whether the $LiFePO_4$, production unit has provided these micro-parameters, it is necessary to find a more suitable and popularizing test method. As for the establishment of testing standards, it requires the joint efforts of peers, as well as the relevant competent departments to come forward and organize.

The following sections introduce the influence of synthesis process parameters on the structure of nano-$LiFePO_4$/C compound material [2].

7.3 Synthesis of Nanometer $LiFePO_4$/C-Composites Material [2, 3, 14–16]

In order to improve the electronic or ionic conductivity of lithium ferrous phosphate, there are three commonly used technical routes: carbon coating, bulk doping, and nanocrystallization. By reducing the particle size of lithium ferrous phosphate and shortening the effective diffusion stroke of lithium ion, the ionic conductivity of the material can be effectively improved, and the electronic conductivity of the material can be improved by in-situ carbon coating, and the nanoparticle size of lithium ferrous phosphate can be further controlled.

Based on the carbon coating technology [16], the formation of nanocarbon network distribution on the surface of $LiFePO_4$ is the key to optimize and improve the properties of materials. There are two elements to construct this structure: one is the introduction mode of conductive carbon, and the other is how to make nanocarbon uniformly dispersed between the grain boundaries and particles of $LiFePO_4$ to form a conductive network connection to reduce the electron transfer resistance without hindering the transport of lithium ions. In this study, a new sol–gel process was used. Based on the dispersed structure of the gel network, introduced carbon source as precursor components together, and in-situ pyrolysis reaction to form carbon-coated $LiFePO_4$ nanocomposites. Because the gel itself is a flowable component and has a network-like solid component, the reaction component can be uniformly dispersed and fixed. Selecting a suitable medium to ensure the solubility and dispersion between the components of the precursor and the medium is the key to the sol–gel method.

At present, it has been reported that there are two main kinds of media used in the preparation of $LiFePO_4$ by sol–gel method. One is the sol–gel method based on citric acid system, the main raw materials are ferric citrate and citric acid, this method can only be realized in a suitable range of solution pH, citrate as a carbon source at the same time; it must be a large excess to meet the carbon content required in $LiFePO_4$/C composites. The other is alcohol system, mainly ethanol, ethylene glycol, and polyol. $LiFePO_4$ materials are prepared by this system, and the selected iron source is generally organic divalent iron salt or trivalent iron salt with pH regulator. The cost of raw materials is high, the sol–gel process is long, and the synthesis method tends to be complex.

7.3.1 Structural Design of Nano-$LiFePO_4$/C Composite Material

Firstly, the design idea of the synthesis of nano-$LiFePO_4$/C compound material is considered. In the traditional carbon-coated modification of $LiFePO_4$, the introduction of conductive additive carbon mainly plays an action in three aspects:

(1) as a reducing agent to reduce Fe^{3+} to Fe^{2+};
(2) as a conductive agent to improve the overall electrical conductivity of the material;
(3) as a crystal growth inhibitor to hinder the growth and aggregation of $LiFePO_4$ grains.

The mechanism of carbon coating to improve the electrical conductivity of the materials can be explained by a simple model proposed by Dominko et al., as shown in Fig. 7.1a, b. In the process of charge and discharge, lithium ion and electron need to enter and exit or intercalate at the same active site on the surface of active material at the same time. If lithium ion and electron are not at the same active site on the surface of active material, then electrons need to be transmitted through the surface of active material to reach the position of lithium ion, or lithium ion needs to be

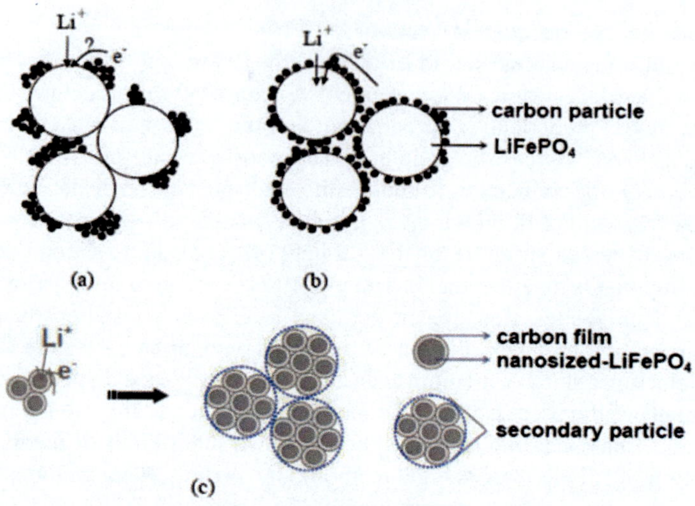

Fig. 7.1 Schematic diagram of structural design of $LiFePO_4$/C composites. The traditional $LiFePO_4$/C composite route of **a**, **b**; **c** "nanomicron" composite route

diffused to the position of electron. When the active material is a semiconductor or insulator, it is difficult for electrons to conduct. On the other hand, the diffusion of lithium ion in $LiFePO_4$ is limited by one-dimensional diffusion channel, and the diffusion rate is slow. If a layer of carbon is uniformly coated on the surface of the active material, electrons can be transmitted quickly through the carbon layer on the surface. In the traditional carbon-doped modification methods, carbon is generally dispersed on the surface of $LiFePO_4$ particles in the form of nanocarbon particles or discontinuous carbon coating, and the electron transport path is discontinuous; in addition, the $LiFePO_4$ particles obtained by the traditional preparation method are generally thicker and the diffusion path of Li^+ is longer, so the modification effect of the material is not very ideal. Based on this, we propose the design idea of "nanomicron" structure of carbon-coated $LiFePO_4$ from two aspects: reducing the particle size of the product and improving the coating effect of carbon, as shown in Fig. 7.1c.

7.3.2 *Synthesis of Nano-$LiFePO_4$/C Composite Material*

According to this design, the design process of the composite is as follows: Firstly, Li, Fe, P components and organic carbon sources were dispersed in the precursor medium at molecular level by liquid-phase method, and nanocrystalline $LiFePO_4$, organic compounds were synthesized by heat treatment to decompose in situ and uniformly disperse on the interface and surface of nanocrystals to form a uniformly coated thin carbon film. The agglomerated primary particles are treated into micron uniform secondary particles by crushing and sieving methods. The secondary particles are clusters of nanocrystalline $LiFePO_4$, and the particles are connected with each other by surface carbon to form a conductive network. Using the above material design ideas, the primary particle nanocrystallization can greatly shorten the diffusion path of Li^+ in $LiFePO_4$ and increase the apparent diffusion coefficient; the formation of conductive network can effectively improve the electrical conductivity of the material; in addition, the micrometerization of secondary particles is beneficial to the improvement of material vibrating density and bulk energy density.

Liu [2] is a commonly used raw material for the preparation of $LiFePO_4$, using iron nitrate ($Fe(NO_3)_3 \cdot 9H_2O$) containing nine crystalline water as iron source, lithium dihydrogen phosphate (LiH_2PO_4) as lithium source, sucrose as carbon source, and using the mixed phase of organic solvent ethylene glycol monomethyl ether and water as medium. Nano-$LiFePO_4$/C composite cathode materials with carbon nanonetwork distribution were prepared. The method is mainly based on the sol–gel conversion process of hydrated ferric nitrate and ethylene glycol methyl ether, which is used as a gel network reaction matrix to realize the molecular mixing of precursor raw materials. Nitric acid, the preparation of nano-$LiFePO_4$/C composites produced by the two reactions, can play an adaptive role in the regulation of solution pH. This method can not only reduce the particle size of the product, but also improve the purity and electrical conductivity of the product. The process is simple, and the raw material cost is low and has the prospect of large-scale production. For the synthesis system, the following issues will be discussed:

(1) The effects of different sintering temperature, sintering time and carbon content on the physical and electrochemical properties of the products;
(2) Study on the high-temperature properties of the composites;
(3) Preliminary analysis of the structure of carbon in the composites;
(4) Combined with the analysis of liquid phase precursor synthesis chemistry and heat treatment reaction process, the formation mechanism of nano-$LiFePO_4$/C composites is discussed.

7.4 X-Ray Study on the Synthesis Process of Nano-$LiFePO_4$/C Composite Material [2]

7.4.1 Effect of Heat Treatment Temperature on Fine Structure of Nano-$LiFePO_4$/C Composite Material

From the above analysis, it can also be seen that the continuous decomposition of sucrose can make $LiFePO_4$ crystal growth and carbon coating at the same time, and the subsequent reaction will also be helpful to inhibit the oxidation of Fe^{2+} and improve the purity of $LiFePO_4$ products.

Since the obtained precursors are dry gel-like substances containing Fe, P, Li, carbon sources, and excess ethers, it is necessary to undergo subsequent heat treatment. Heat treatment temperature is one of the key parameters affecting the properties of $LiFePO_4$ cathode materials. First of all, it is discussed in detail. Figure 7.2 shows the XRD spectra of the products obtained at different heat treatment temperatures (600, 650, 700, and 750 °C), holding for 10 h and carbon excess of 3% (calculated according to the theoretical content of residual carbon accounting for 3% of the $LiFePO_4$ products). Compared with PDF card (40-1499), the main crystal phase of the products obtained at four heat treatment temperatures is $LiFePO_4$, space group Pmnb with olivine structure. Except for a small amount of impurity phase containing Fe^{3+} in the products obtained at 600 °C, the products corresponding to the other three heat treatment temperatures are pure $LiFePO_4$ materials, which is basically consistent with the above thermal analysis results. The appearance of impurity phase at 600 °C indicates that some raw materials at lower temperature are not fully involved in the formation of $LiFePO_4$. In addition, the corresponding diffraction peak of residual carbon cannot be seen from the XRD spectrum, which may be due to the existence of amorphous carbon and low content.

The crystal structure cell parameters and average grain size of the four samples are given in Table 7.2. The calculation is based on the Scherrer formula, and the diffraction peak corresponding to the (311) crystal plane is selected for calculation. It can be seen that the lattice constant of each sample increases with the increase of heat treatment temperature, but the change is very small, indicating that the primary

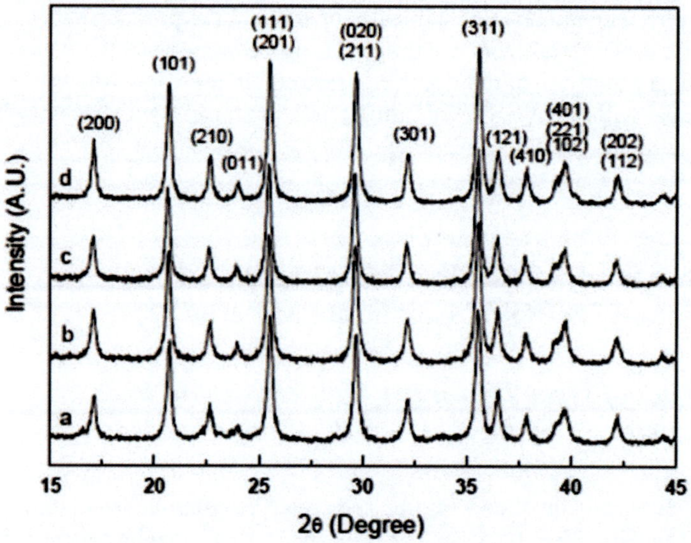

Fig. 7.2 XRD patterns of $LiFePO_4/C$ composites obtained at different heat treatment temperatures. **a** 600 °C, **b** 650 °C, **c** 700 °C, **d** 750 °C

Table 7.2 Unit cell parameters of $LiFePO_4$ samples obtained at different heat treatment temperatures

Treatment temperature (°C)	Lattice parameters (Å)			Average grain size *D* (nm)
	a	*b*	*c*	
600	10.341	6.014	4.702	32.1
650	10.346	6.015	4.711	33.1
700	10.349	6.020	4.709	33.6
750	10.354	6.023	4.712	35.1
PDF standard value	10.347	6.019	4.704	

grain size of $LiFePO_4$ synthesized by sol–gel method is not significantly affected by the heat treatment temperature.

7.4.2 *Effect of Carbon Coating Amount on the Structure of Nano-$LiFePO_4$/C Composite Material*

Based on the previous material design ideas, the composite mode of carbon plays an important role in the modification of $LiFePO_4$ materials. According to the study of the structure and properties of carbon with different heat treatment temperature and time, it is found that nanograins with narrow particle size distribution can be prepared by in-situ cracking of carbon with ethylene glycol methyl ether and water as reactants. In order to further analyze the influence of the morphology and structure of surface carbon on the properties of $LiFePO_4$ materials, we also investigated the effect of carbon coating amount on the structure and properties of nano-$LiFePO_4$/C composites. Figure 7.3 shows the XRD pattern of the product without carbon source, sucrose excess and excess 3% and 6%, respectively. It can be seen that all the products are $LiFePO_4$, with orthorhombic olivine structure and have no obvious impurity phase. With the increase of carbon addition, the XRD diffraction peak shifts slightly to a low angle, indicating that more pyrolytic carbon has an obvious inhibitory effect on the growth of $LiFePO_4$ grains during heat treatment, which can also be seen from the calculation results of lattice constant and average grain size in Table 7.3.

7.4.3 *Effect of Heat Treatment Time on the Structure of Nano-$LiFePO_4$/C Composite Materials*

In order to investigate the effect of heat treatment time on the structure of the product, the XRD spectra of the products obtained by heat treatment at 700 °C for 5, 10, 15, and 20 h were given in Fig. 7.4. It can be seen that the products obtained at each heat treatment time are $LiFePO_4$ with intact crystal, no obvious impurity phase, and

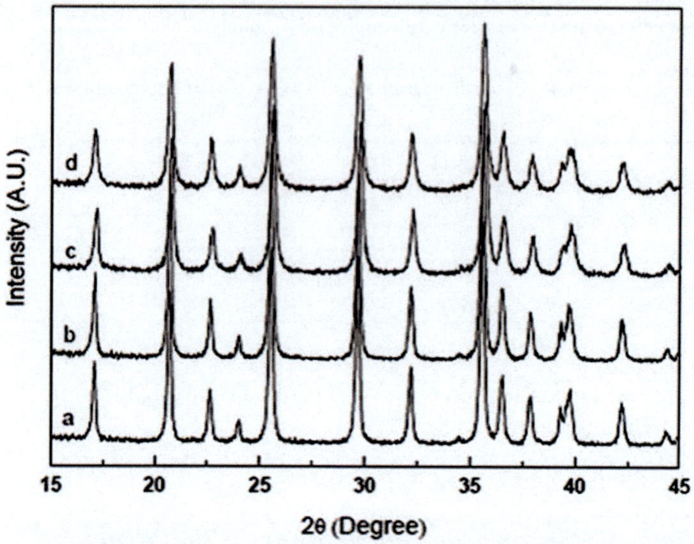

Fig. 7.3 XRD diagram of $LiFePO_4$/C composites with different carbon content, **a** sugar-free; **b** 0% excessive; **c** 3% excessive; **d** 6% excessive

Table 7.3 Cell parameters of $LiFePO_4$ samples prepared with different carbon content

Sample	Lattice parameters (Å)			Average grain size *D* (nm)
	a	*b*	*c*	
No sugar	10.356	6.028	4.716	58.6
0%	10.351	6.023	4.711	51.2
3%	10.349	6.020	4.709	33.1
6%	10.332	6.014	4.702	32.0

olivine structure. With the increase of heat treatment time, the XRD diffraction peak shifts slightly to a lower angle, and the intensity of the diffraction peak increases slightly, which indicates that the grain size and crystallinity of the product increase slightly with the increase of heat treatment time. The corresponding lattice constant and average grain size are listed in Table 7.4 after calculation, and the corresponding lattice constant increases slightly with the extension of heat treatment time, but the increase is not obvious. It shows that the heat treatment time has little effect on the primary particles of $LiFePO_4$ prepared by sol–gel method.

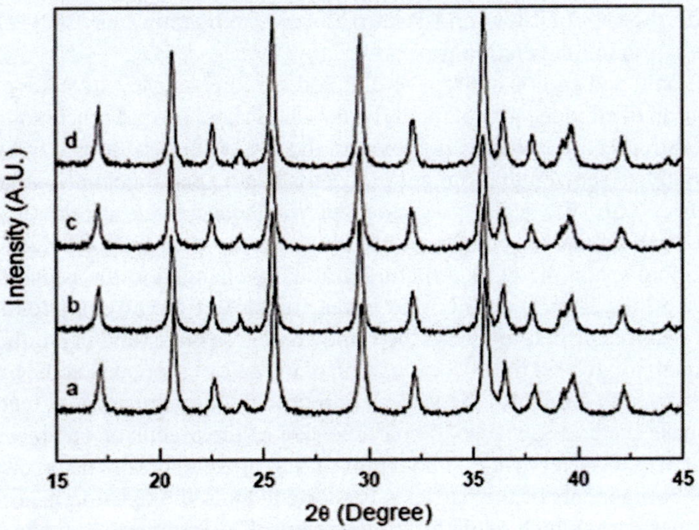

Fig. 7.4 XRD patterns of primary particles of $LiFePO_4$/C composites obtained at different heat treatment times. **a** 5 h, **b** 10 h, **c** 15 h, **d** 20 h

Table 7.4 Unit cell parameters of $LiFePO_4$ samples obtained with different heat treatment time

Heat treatment time (h)	Lattice parameters (Å)			Average grain size *D* (nm)
	a	*b*	*c*	
5	10.316	6.013	4.696	31.0
10	10.349	6.020	4.709	33.1
15	10.353	6.017	4.710	33.9
20	10.358	6.021	4.715	34.6

7.5 Analysis of Heat Treatment Reaction of Precursor During Synthesis of Nano-$LiFePO_4$/C Materials [2]

The hydrolyzed product forms a gel by polymerization reaction:

$$-Fe - OH + HO - Fe \rightarrow -Fe - O - Fe - + H_2O \tag{7.5}$$

$$-Fe - OR + RO - Fe \rightarrow -Fe - O - Fe - + ROH \tag{7.6}$$

In this way, the Fe–O bond formed after polymerization is obviously weaker than Fe–O in inorganic salts, and when the aqueous solution of LiH_2PO_4 and sucrose is mixed with the solution of $Fe(NO_3)_3$ $9H_2O$ and ethylene glycol methyl ether, the corresponding peaks of Li–O and P–O still exist, indicating that LiH_2PO_4 is not involved reaction in the gelation process.

P, Li, Fe salt, and carbon source were mixed at molecular level by sol–gel in the mixed solution of ethylene glycol methyl ether and water. The gel precursor obtained after evaporation is amorphous, as shown in the XRD diagram of Fig. 7.5. Further FE-SEM analysis showed that the gel precursor had a spatial network structure, as shown in Fig. 7.5b. When the precursor gel was heat-treated at 300 °C, no new diffraction peak appeared, but the morphology of the precursor changed greatly, and the original spatial network structure gradually changed to the island structure of nanocrystallites. This is mainly due to the fracture of the original continuously distributed network structure due to the removal of free water and crystalline water from the precursor gel and the decomposition of organic compounds in the precursor. This process is very similar to the granulation process of precursor raw materials in the solid phase method. Because the particle size of the precursor obtained by heat treatment at 300 °C is very small, about tens of nanometers, the diffusion distance of the internal components of the particles is shortened and the reaction is accelerated at high temperature, which leads to the decrease of the synthesis temperature and reaction time of $LiFePO_4$. When the heat treatment temperature is 400 °C, there is a weak peak near $2\theta = 23°$, which is confirmed to be the characteristic peak of $LiHPO_4$. It shows that $LiHPO_4$, is first formed during LiH_2PO_4 heat treatment, while there seem to be a few diffraction peaks of $LiFePO_4$ crystal plane near $2\theta = 35°$, indicating that the crystallization temperature of $LiFePO_4$ may be lower in differential thermal curve analysis. Because the precursor is a molecular mixing process, the decrease of crystallization temperature is very possible. The obvious olivine $LiFePO_4$ diffraction peak appears when the heat treatment temperature is 500 °C. With the increase of heat treatment temperature, the intensity of the diffraction peak increases and the crystallization tends to be complete. Through the above analysis of the heat treatment process of the precursor and the TG-DTA analysis of the junction and the precursor, the reaction mechanism of the heat treatment process can be discussed as follows.

The following reactions will occur during the heat treatment at 150–450 °C for the $(–Fe–O–Fe–)_n$ formed in (7.5) and (7.6).

$$(-Fe - O - Fe-)_n \rightarrow nFe_2O_3 \tag{7.7}$$

With the decomposition of sucrose and excess ethylene glycol methyl ether to carbon and a small amount of active hydrogen, the Fe_2O_3 of nanocrystals changes rapidly to Fe_3O_4 and FeO. Taking carbothermal reduction as an example, the reaction equation can be expressed as follows:

$$3nFe_2O_3 + nC \rightarrow 2nFe_3O_4 + nCO \tag{7.8}$$

$$2nFe_3O_4 + 2nC \rightarrow 6nFeO + 2nCO \tag{7.9}$$

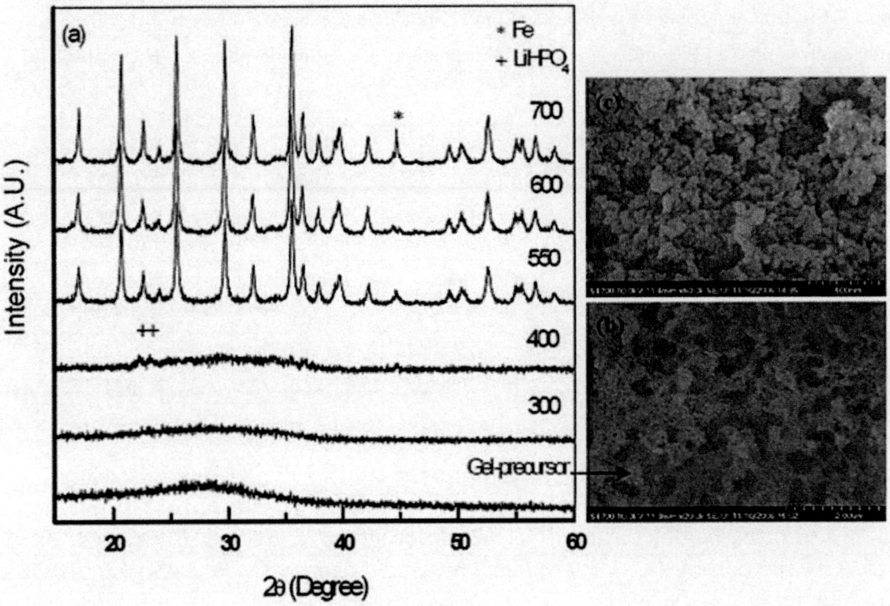

Fig. 7.5 **a** XRD patterns of precursor condensation and its heat treatment at different temperatures; **b, c** FE-SEM diagrams of precursors and their products obtained by heat treatment at 300 °C for 5 h, respectively

Combined with the results of TG-DTA analysis of precursors and XRD analysis of the above precursors during heat treatment, the crystal of $LiFePO_4$ began to precipitate when the temperature reached 500 °C, and the following reactions occurred.

$$6n\text{LiHPO}_4 + 6n\text{FeO} + 3n\text{C} \rightarrow 6n\text{LiFePO}_4 + 3n\text{H}_2\text{O} + 3n\text{CO} \qquad (7.10)$$

Through the above analysis, the formation mechanism of nano-$LiFePO_4$ can be summarized as follows:

$$\text{LiH}_2\text{PO}_4 + 0.5\text{Fe}_2\text{O}_3 + 0.5\text{C} \rightarrow \text{LiFePO}_4 + \text{H}_2\text{O} + 0.5\text{CO} \qquad (7.11)$$

7.6 Phase Structure and Powder Diffraction Data of $LiFePO_4$ and $FePO_4$

The material and structure data related to $FePO_4$ and $LiFePO_4$ can be found in the database of the International Centre for Diffraction Data (ICDD) as shown in

Table 7.5 Structural data of $LiFePO_4$- and $FePO_4$-related substances

Substance name	Lithium phosphate iron ore	Isophosphate iron (manganese) ore	Iron phosphate	Iron phosphate
Chemical formula	$LiFePO_4$	$(Fe,Mn)PO_4$	$FePO_4$	$FePO_4$
PDF card number	40-1499	34-0134	30-0659	31-0647
Space group	Pnma (No. 62)	Pnma (No. 62)	Cmcm (No. 63)	$P_{22}2222$ (No. 180)
Crystal system	Orthogonal	Orthogonal	Orthogonal	Hexagonal
Lattice parameter	$a = 6.0157$	$a = 5.83$	$a = 5.227$	$a = 5.170$
	$b = 10.3915$	$b = 9.79$	$b = 7.770$	
	$c = 4.7207$	$c = 4.769$	$c = 6.322$	$c = 11.37$

Table 7.5 for comparison and application in the study. $LiFePO_4$ was discovered in mineral research, so its mineralogical name is lithium phosphate iron ore (triphylite), belongs to orthorhombic, Pnma (No. 62) space group, $a = 7.08$, $b = 10.44$, $c = 4.75$ Å. X-ray density $D_x = 3.47$ g/cm^3, measured density $D_m = 3.45$ g/cm^3, PDF card number is 40-1499.

It can be seen from Table 7.5 that almost the same structure (Pnma) as $LiFePO_4$ is $(Fe,Mn)PO_4$, its mineralogical name is iron isophosphate (manganese) ore (heterosite), PDF card number is 34-0134. $FePO_4$ has two structures: orthogonal [Cmcm (No. 63)] and hexagonal [$P_{22}2222$ (No. 180)]. The powder diffraction data of related materials are listed in Table 7.6 for comparison and application in the research work.

Table 7.6 Powder diffraction data CuKα of the structure of $LiFePO_4$- and $FePO_4$-related substances

Material chemical formula	$LiFePO_4$			$(Fe,Mn)PO_4$			$FePO_4$			$FePO_4$		
PDF No.	40-1499			34-0134			30-0659			31-0647		
Structure	Orthogonal			Orthogonal			Orthogonal			Hexagonal		
Space group	Pnma (No. 62)			Pnma (No. 62)			Cmcm (No. 63)			$P_{22}2222$ (No. 180)		
Lattice parameter	$a = 6.0157$			$a = 5.83$			$a = 5.227$			$a = 5.170$		
	$b = 10.3915$			$b = 9.79$			$b = 7.770$					
	$c = 4.7207$			$c = 4.769$			$c = 6.322$			$c = 11.37$		
2θ (°), CuKα	θ (°)	*hkl*	I/I_1	θ (°)	*hkl*	I/I_1	θ (°)	*hkl*	I/I_1	θ (°)	*hkl*	I/I_1
17.121	17.121	020	34									
17.956				17.956	020	60						
19.845										**19.845**	**100**	**20**
20.511							**20.511**	**110**	**60**			
20.607				20.607	011	70						
20.751	**20.751**	**011**	**76**									
21.340										21.340	101	1
22.648	22.648	120	26									
22.861							22.861	020	40			
23.515										23.515	003	1
23.662				23.662	120	25						

(continued)

Table 7.6 (continued)

23.986				23.986	101	6						
24.000	24.00	101	10									
24.877							**24.877**	**111**	**80**			
25.280										**25.280**	**102**	**100**
25.525	**25.525**	**111,021**	**70**									
25.748				**25.748**	**111**	**85**						
25.900				25.900	021	20						
26.927							26.927	021	40			
29.676	**29.676**	**121,200**	**100**									
30.192				30.192	121	35						
30.616				**30.616**	**200**	**85**						
32.161	32.161	031	34									
33.117				33.117	031	15						
34.291							34.291	200	60			
34.440	34.440	220	1									

(continued)

Table 7.6 (continued)

34.741										34.741	110	5
35.122							**35.122**	**112**	**100**			
35.525	35.525	131	81									
35.758				35.758	200	4						
36.465	36.465	211	25									
36.649				**36.649**	**131**	**100**						
37.279				37.279	211	40						
37.618										**37.618**	**104**	**10**
37.834	37.834	140	16									
38.216										38.216	112	10
38.235	38.235	002	1									
38.728				38.728	012	16						
38.799							38.799	130	60			

Bold to represent the three strong lines in the diffraction pattern

References

1. Padhi, A.K., K.S. Nanjundaswamy, and J.B. Goodenough. 1997. Phospho-olivines as positive-electrode materials for rechargeable lithium batteries. *Journal of the Electrochemical Society* 144 (4): 1188–1192.
2. Liu, H. 2008. Study of synthesis and performance of cathode material $LiFePO_4$ for Li-ion battery. Doctoral paper of Shanghai Institute of Microsystems and Information Technology.
3. Mi, C.F. 2005. Study on olivine $LiFePO_4/C$ composite positive material. Doctoral thesis of Zhejiang University, Hangzhou, 5.
4. Prince, A.A.M., S. Mylswamy, T.S. Chan, R.S. Liu, B. Hannoyer, M. Jean, C.H. Shen, S.M. Huang, J.F. Lee, and G.X. Wang. 2004. Investigation of Fe valence in $LiFePO_4$ by Mossbauer and XANES spectroscopic techniques. *Solid State Communications* 132 (7): 455–458.
5. Takahashi, M., S. Tobishima, K. Takei, and Y. Sakurai. 2001. Characterization of $LiFePO_4$ as the cathode material for rechargeable lithium batteries. *Journal of Power Sources* 97–98 (1–2): 508–511.
6. Barker, J., M.Y. Saidi, and J.L. Swoyer. 2003. Lithium iron(II) phospho-olivines prepared by a novel carbothermal reduction method. *Electrochemical and Solid-State Letters* 6 (3): A53–A55.
7. Franger, S., F.L. Cras, C. Bourbon, and H. Rouault. 2003. Comparison between different $LiFePO_4$ synthesis routes and their influence on its physico-chemical properties. *Journal of Power Sources* 119–121 (1–2): 252–257.
8. Croce, F., A.D. Epifanio, J. Hassoun, et al. 2002. A novel concept for the synthesis of an improved $LiFePO_4$ lithium battery cathode. *Electrochemical and Solid-State Letters* 5 (3): A47–A50.
9. Iltchev, N., Y. Chen, S. Okada, and J. Yamaki. 2003. $LiFePO_4$ storage at room and elevate temperatures. *Journal of Power Sources* 119–121 (1–2): 749–754.
10. Scaccia, S., M. Carewska, P. Wisniewski, and P.P. Prosini. 2003. Morphological investigation of sub-micron $FePO_4$ and $LiFePO_4$ particles for rechargeable lithium batteries. *Materials Research Bulletin* 38 (7): 1155–1163.
11. Chen, J.J., and M. Stanley. 2006. Hydrothermal synthesis of lithium iron phosphate. *Electrochemistry Communications* 8: 855.
12. Yang, S.F., P.Y. Zavalij, and M.S. Whittingham. 2001. Hydrothermal synthesis lithium iron phosphate cathodes. *Electrochemistry Communications* 3: 505–508.
13. Lee, J., and A.S. Teja. 2005. Characteristics of lithium iron phosphate particles synthesized in subcritical and supercritical water. *The Journal of Supercritical Fluids* 35: 83–90.
14. Lai, C.Y. 2006. Study on preparation and modification of lithium ion battery material $LiFePO_4$. Doctoral paper of Shanghai Institute of Microsystems and Information Technology.
15. Liu, Hui, Jingying Xie, and Ke Wang. 2008. Synthesis and characterization of nano-$LiFePO_4$/carbon composite cathodes from 2-methoxyethanol-water system. *Journal of Alloys and Compounds*. In Press. Available online May 13, 2007.
16. Liu, Hui, Yi Feng, Zenghui Wang, Ke Wang, and Jingying Xie. 2008. A PVB-based rheological phase approach to nano-$LiFePO_4/C$ composite cathodes. *Powder Technology* 184 (3): 313–317.

Chapter 8
Preparation and XRD Analysis of Carbon Materials Used for Li-Ion Batteries

8.1 Preparation of Carbon Electrode Materials for Li-Ion Batteries [1–4]

Negative materials for lithium-ion batteries require the following characteristics: (1) As low as possible electrode potential; (2) Ion has high diffusivity in the negative electrode solid-state structure; (3) High de-embedding reversible; (4) Good conductivity and thermodynamic stability; (5) Good compatibility to electrolyte solvent; (6) Rich in resources and low in price; (7) Safety, no pollution.

8.1.1 Classification and Structure of Carbon Materials

Carbon materials have the characteristics of high electrical conductivity, good chemical stability, low price, and so on. There are many kinds of crystal forms in solid carbon materials, such as graphite, diamond, and amorphous carbon. The structural diversity of carbon materials comes from the diversity of chemical bonding between C–C atoms. According to the degree of graphitization, carbon materials are generally divided into graphite and non-graphite. There are natural graphite, artificial graphite, graphitized carbon materials, including carbon fiber, modified graphite, composite graphite, treated artificial graphite, graphitized mesophase carbon microspheres, and so on. According to their structural characteristics, non-graphite carbon materials can be divided into two categories: easy graphitization carbon and difficult graphitization carbon, that is, soft carbon and hard carbon. The following is the structure of several carbon materials.

(1) **Graphite**

In graphite crystal, the sp2 hybrid orbitals of carbon atoms form three coplanar σ bonds with strong interaction. Carbon atoms and carbon atoms form a huge six-ring network structure through continuous sp2 bonds, forming a two-dimensional

C. Yang et al., *Materials and Working Mechanisms of Secondary Batteries*,
https://doi.org/10.1007/978-981-19-5955-4_8

graphite layer. The hybrid electrons form large electron conjugated large π bonds on both sides of the network layer, and the Van der Waals force between the layers binds each layer together to form a layered structure called crystallization. Because the interaction between layers is weaker than that of chemical bonds, graphite is easy to dissociate, appear soft, and lubricating. However, due to the resonance of π electrons along the network plane, graphite shows great differences, such as the thermal expansion coefficient of the graphite layer in the plane direction (*a*-axis direction) is $-1.5 \times 10^{-6}\ K^{-1}$, and the vertical direction (*c*-axis direction) is $2.8 \times 10^{-6}\ K^{-1}$; the carbon atoms in the graphite crystal structure are arranged orderly into hexagonal carbon rings, the carbon atomic lamellae are quite large, and the layers are parallel and oriented arrangement to each other, and there is only a weak Van der Waals force between the layers. In hexagonal graphite, the usual arranged order of carbon layers is ABAB... In the other graphite, there is an unusual arranged order of ABCABC... and is called rhombohedral graphite. Many artificial carbon materials have different degrees of stacking defects due to spiral dislocations or lamination dislocations; as a result, the carbon atoms deviate from the normal position, and the periodic stacking dislocation structure cannot be maintained. Because these disordered graphite layers are nailed to each other, it is very difficult to eliminate this dislocation completely by high-temperature treatment. This type structure is called turbulent layer structure. In the turbulent layer structure, the value of the interlayer spacing is chaotic, but the average value is larger than that of graphite. With the increase of pyrolysis temperature, the interlayer spacing decreases, which is close to graphite. However, high pressure and high temperature are needed to reach the value of graphite.

Generally speaking, natural graphite is extracted from natural graphite ore and then coated or treated with some other treatments. It has a relatively high degree of graphitization and high capacity, which can be close to the theoretical capacity of graphite. The post-treatment process is no longer graphitized at high temperature, and of course, a very small number of high-end natural graphite is coated and then graphitized.

Artificial graphite refers to the treatment of other compounds containing C, which is generally graphitized at high temperature. Of course, some low-end artificial graphite may not be graphitized at high temperature. Generally speaking, we can roughly have the impression that the cycle capacity of artificial graphite is better than that of natural graphite. Because of poor circulation, the choice of electrolyte is more important, and natural graphite is softer. However, if the compaction is too high, the particles may be deformed, and the liquid absorption capacity will decrease sharply.

(2) **Amorphous carbon**

Amorphous carbon refers to the carbon material with low crystallinity (i.e., graphitization degree), small grain size, and large crystal plane spacing (d_{002}) and the most disordered structure, i.e., the so-called amorphous carbon. The arrangement of hexagonal carbon atoms in a plane is regular, but the number of carbon rings forming microcrystals is less than that of graphite; only a small number of plane layers are parallel to each other, and the other orientations are messy. According to the difficulty of graphitization, amorphous carbon is divided into soft carbon and hard carbon. **Soft carbon** is easily graphitized carbon, which refers to amorphous carbon that can only be graphitized at a high temperature above 2500 °C including coke, petroleum coke, carbon fiber, carbon microspheres, and so on. **Hard carbon** refers to the carbon material which is difficult to graphitize at high temperature above 2500 °C. It is usually the carbon material obtained by the pyrolysis of high molecular polymer. The common hard carbon is pyrolysis products of phenolic resin, epoxy resin, polyfurfuryl alcohol, etc. They do not have a clear XRD peak, but TEM observations show that their basic structure is dominated by a two-dimensional graphite layer. Because it is formed directly by solid phase pyrolysis, the movement of intermediate phase graphitization unit is strictly limited, which makes it difficult to graphitize at high temperature.

(3) **Fullerene**

The discovery of fullerenes and carbon nanotubes is a milestone in the history of world science. The fullerene family includes C60 and C70 et al. The structure of a C60 molecule is similar to that of a football. It is a sphere composed of 12 pentagons and 20 hexagons. It has five symmetrical axes and high symmetry and belongs to icosahedral point group (Ih point group). Each carbon atom is connected to the adjacent three carbon atoms by sp2 hybrid orbitals, and the remaining p orbitals form π bonds in the periphery and inner cavity of the C60 molecule. In addition to C60, the carbon clusters with closed cage structure may also have C28, C32, C50, C70, C84, and so on, which form a series of closed cage structure, and it be called as fullerene. At present, fullerenes are mainly prepared by physical methods: arc method, benzene flame combustion method, and DC arc method. With the rapid development of fullerene chemistry, the chemical synthesis of fullerene has become a research field that has attracted much attention. The synthesis of fullerenes by chemical method has the advantages of controllable product, high yield, and convenient reaction operation. At the same time, the synthesis of this spherical molecule will greatly enrich the research content of organic chemistry and effectively promote the research and development of organic chemistry.

8.1.2 Structural Defects of Carbon Materials

Due to the various hybrid forms of carbon atom bonding and the diversity of carbon material structure, carbon materials have a variety of structural defects, such as plane displacement, spiral dislocation, stacking defects, and so on. Structural defects in the carbon layer in the ideal graphite crystal structure and carbon atoms are bonded through sp2 hybrid orbitals to form a six-membered ring network structure; however, there may be sp, sp3 hybrid carbon atoms in practical carbon materials. With different hybrid forms, the distribution density of electron cloud is also different, which leads to the change of electron density in the carbon plane layer, the deformation of the carbon plane layer and the structural defects in the carbon plane layer. In addition, when there are other impurity atoms in the carbon plane structure, the size, and charge of the impurity atoms are different from those of the carbon atoms, which will also cause structural defects in the carbon plane layer.

In the carbon materials prepared by pyrolysis with organic compounds as precursors, the carbon atoms along the edges may still be connected with some functional groups (such as –OH, =O, –O–, $-CH_3$, and so on), which will also cause structural deformation of the carbon plane.

The carbon plane is irregularly arranged, forming the stacking defects in the carbon material. Pore defects are caused by pores left by volatilization of gaseous materials in the process of preparing carbon materials.

Whether artificial graphite or modified natural graphite, in order to meet the needs of high-power performance of power batteries, on the one hand, the size of graphite particles should be reduced, on the other hand, nanoscale pore channels should be introduced, which will help to reduce the distance of lithium-ion migration or diffusion and provide more lithium-ion transport channels.

Hard carbon is relative to soft carbon, which here refers to the carbon material obtained by heat treatment at low temperature, with a small amount of graphite structure. For example, sucrose is used for hydrothermal treatment to get carbon balls, and then heat treated in an inert atmosphere for further carbonization. Because it is carbonized at low temperature, the hard carbon material has more nanopores, which is convenient for the rapid intercalation and de-intercalation of lithium ions, has excellent fast charge and discharge ability, and can even be used as a capacitor electrode material. For hard carbon materials, although the reversible capacity is not high, the irreversible capacity is large for the first time, so in order to solve the practical application of hard carbon, the formation of a good solid electrolyte intermediate phase (interface) film (SEI) on the surface of the material may be one of the development directions of the material. In addition, in order to further improve the rapid charge–discharge ability of hard carbon, nanospherical hard carbon can also be prepared. In order to avoid the agglomeration of carbon spheres, the surface of polyacrylonitrile nanospheres was coated with a layer of oxide, which was carbonized at low temperature and then removed.

8.1.3 Study on the Preparation of Carbon Nanospheres Using Polyacrylonitrile as Precursor

Polyacrylonitrile (PAN) has high carbonation yield and is used in commercial carbon fiber production; it also plays an important role in the synthesis of highly ordered graphite nanosheets and other carbon nanomaterials. For example, the activated carbon fiber prepared by PAN has nitrogen-containing functional groups and is an excellent catalyst for the removal of hydrogen chloride and sulfur oxides. Because of its unique nanostructure and rich nitrogen-containing groups, polyacrylonitrile is an excellent catalyst for the removal of hydrogen chloride and sulfur oxides. Carbon

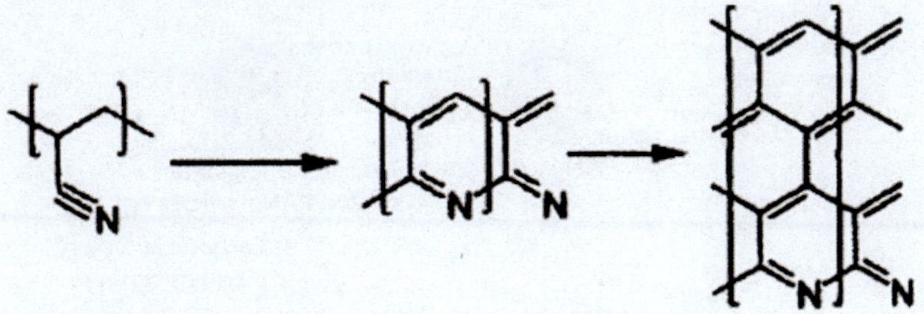

Fig. 8.1 Schematic diagram of PAN high-temperature carbonization process

nanospheres (CNSs) with PAN as precursor also have potential application value in the field of catalyst carrier electrode materials.

The carbonization process of PAN is shown in Fig. 8.1. Because PAN contains unsaturated –CN groups, cross-linking occurs in the process of carbonization, which leads to the aggregation and fusion of PAN nanospheres and the destruction of the desired nanostructure. In order to avoid reunion, core–shell copolymers were prepared, with cross-linked polyacrylic acid as shell and polyacrylonitrile as core micelle. Under the protection of shell, the core of polyacrylonitrile was constrained, and carbon nanospheres could be obtained by carbonization at high temperature; however, the preparation of core–shell copolymer micelles is complicated, the operation is tedious, and the cost is high. An improved synthesis method is proposed to limit the agglomeration of nanospheres by using inorganic shell layer. By coating a layer of inorganic/hard shell on the surface of nanospheres, the uniform dispersed carbon nanospheres can be obtained by preventing the cross-linking and agglomeration carbonization between nanospheres in the process of high-temperature carbonization and removing the inorganic shell layer by acid. In addition, the products were prepared as electrodes, and the properties of carbon nanospheres as anode materials for lithium-ion batteries were explored.

The preparation process of carbon nanospheres is shown in Fig. 8.2, which mainly includes four steps: (1) PAN latex particles are obtained by emulsion polymerization; (2) A layer of titanium phosphate is coated on the surface of the particles; (3) High-temperature carbonization; (4) The coating layer is removed, and the carbon nanospheres were obtained. The specific experimental steps of obtaining carbon nanospheres are as follows:

93.8 mL water, 1.0 g sodium dodecyl sulfate, 6.2 mL acrylonitrile monomer, and 0.10 g potassium persulfate were successively added to the three-neck flask, stirred in nitrogen atmosphere for 15 min, then the air was removed and the raw materials were mixed evenly, then stirred and reacted under 70 °C for 11 h. The emulsion obtained after polymerization was freeze-dried to obtain loose PAN powder. A certain amount of PAN powder was dispersed in phthalate diethylene glycol diacrylate (PDDA) 2.0 wt% solution and soaked for 20 min to make its surface positively charged. After centrifugation, the product was dispersed in 10 mol/L $Ti(SO_4)_2$ solution, dissolved in 0.1 mol/L H_2SO_4, pH 0.95, soaked for 10 min, and centrifuged again, and the product

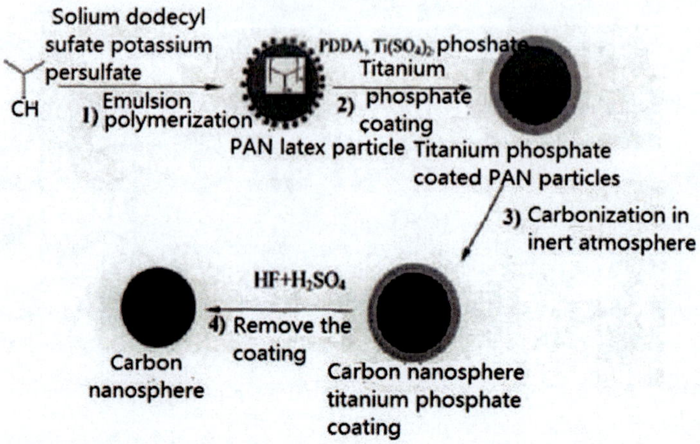

Fig. 8.2 Schematic diagram of the preparation process of carbon nanospheres

was transferred to phosphate solution (Na_2HPO_4 and NaH_2PO_4 solution, PH adjusted to 4.0) and soaked for 2 min. Then the PAN particles coated with titanium phosphate were obtained by centrifugation and drying. The coated PAN was calcined at 230 °C in air for 2 h to make the PAN particles cross-linked to stabilize its structure, and then 400 °C calcined in Ar atmosphere for 2 h. 800 °C roasting for 2 h, natural cooling to room temperature; the black powder obtained after roasting was placed in a mixed acid solution of HF (40 wt%) and H_2SO_4 (HF/H_2PO_4 volume ratio of 4:1), soaked for 4 h to remove the coating, then centrifuged, washed with a large amount of distilled water, and dried to get the final product.

In addition, the uncoated PAN powder was treated by the same heat treatment, and the product was obtained for comparison. Figure 8.3 shows the ETM images of the products obtained at different preparation stages. Figure 8.3a shows the morphology of PAN particles prepared by emulsion polymerization. It can be seen that PAN latex particles are nanoparticles with an average particle size of about 60 nm. The surface of the carbon ball is coated with titanium phosphate and carbonized, and the surface of the carbon ball is coated with a thickness of about 5 nm (as shown in Fig. 8.3b). The EDX results obtained on the coating layer show the signal peaks such as C, O, Cu, P, Ti, and so on. Because the tested substrate copper film contains C and Cu, it can be judged that the coating layer is a compound containing P Ti O. After the coating layer was removed by mixed acid, the pure carbon nanospheres with an average particle size of about 50 nm were obtained, as shown in Fig. 8.3c. The EDX results showed that only Cu and C existed, and there were no elements such as Ti, P, and O, which proved that the inorganic coating was removed completely. The average particle size of the carbon nanospheres is smaller than that of the precursor PAN latex particles, which is caused by the shrinkage of the polymer during dehydrogenation and denitrification at high temperature. Figure 8.3d shows the TEM image of the experimental products. There is no inorganic coating to protect the products of PAN particles calcined at the same high temperature. It can be seen that PAN agglomerates obviously and becomes a larger carbon block after carbonization. The comparative experiments clearly show that the titanium phosphate coating on the surface of PAN latex particles isolates the contact between PAN latex particles and limits their carbonization in a small space, thus preventing the agglomeration caused by –CN cross-linking between PAN particles. Therefore, this method is an effective method to synthesize carbon nanospheres with polyacrylonitrile as precursor and ensure its intact nanomorphology.

Figure 8.4 shows the XRD spectrum of the obtained carbon nanospheres, and two peaks, one strong, and one weak, corresponding to the two scattering peaks of amorphous carbon. Calculated from the Bragg formula $2d\sin\theta = \lambda$ can be observed that $d_{002} = 0.356$ nm. The crystal plane spacing value is larger than $d_{002} = 0.335$ nm of graphite, indicating that the graphitization degree of carbon nanospheres is not high, which is due to the lower carbonization temperature. At the same time, the diffraction peak is wide, which is not only related to the carbonization degree of the material, but also related to the nanometer size of carbon spheres. No other diffraction peaks were observed in the spectrum, indicating that the purity of the product was high and the inorganic shell had been removed by acid washing.

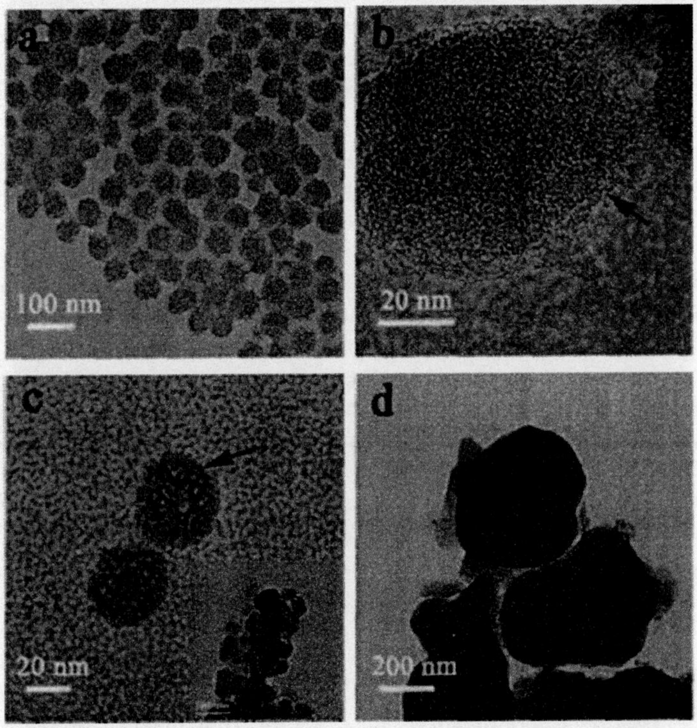

Fig. 8.3 **a** Agglomerated latex particles, **b** phosphoric acid-coated carbon nanospheres, **c** removing the coated carbon nanospheres, **d** compare the TEM photographs of the experimental products

Fig. 8.4 XRD pattern of carbon nanospheres

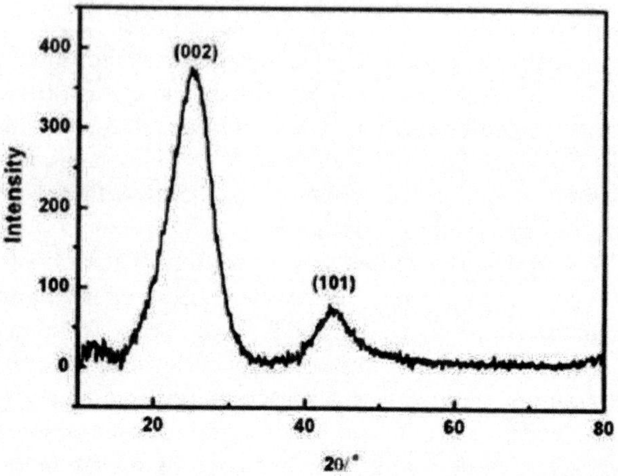

Carbon, the negative active material for lithium-ion batteries, can be divided into three types according to its structural characteristics: graphite, easy graphitization carbon, and difficult graphitization carbon (commonly known as soft carbon and hard carbon). Soft carbon mainly includes mesophase carbon microspheres, petroleum coke, needle coke, carbon fiber, etc., while hard carbon mainly includes resin carbon (such as phenolic resin, epoxy resin, and polyfurfuryl alcohol PFA-C), organic polymer pyrolytic carbon (including polyvinyl alcohol group, polyvinyl chloride group, polyacrylonitrile group, etc.), and carbon black. Because the crystallinity of soft carbon is similar to that of graphite, it is generally believed that it is easier to intercalate lithium than hard carbon; that is, it is easier to charge and the security is also better.

Graphite carbon material is a common negative material because of its mature application technology, outstanding performance in safety and cycle life, cheap and non-toxic. The negative materials of conventional Li-ion batteries include natural graphite, natural graphite-modified materials and mesophase carbon microspheres and petroleum coke artificial graphite. The price of natural graphite and natural graphite-modified materials is relatively low, but the charge–discharge efficiency and service life need to be further improved. Mesophase carbon microspheres have a special structure, spherical lamellar structure, and smooth surface, with a diameter of 5–40 μm. The unique morphology of the material makes it have significant advantages in specific capacitance (more than 330 mAh/g), safety, discharge efficiency, cycle life (more than 2000 cycles), but the cost needs to be reduced. The products of petroleum coke are outstanding in discharge efficiency and cycle life, but there are some problems such as high cost and complex preparation process. In recent years, with the deepening of the research work, researchers have found that through the surface modification and structure adjustment of graphite and all kinds of carbon materials, or the partial disorder of graphite, or the formation of nanoscale pores, holes and channels in all kinds of carbon materials, it is beneficial to the intercalation and de-intercalation of lithium. At present, due to the problems of low initial charging efficiency, low compaction density and immature process, hard carbon materials have not entered the stage of large-scale commercialization, and the related fields in China are still in the experimental stage. Among the researchers of domestic universities and research institutes, the most famous is the research group of Academician Chen Li-Quan. The spherical hard carbon specific capacity developed by them can reach 400 mAh/g, with the first effect reaching 80%. Domestic studies on hard carbon have not seen related reports on industrialization, and there are not many enterprises abroad to achieve industrial sales, the most famous is Japan's Wu Yu Chemical, but the hard carbon capacity announced by the company is less than 300 mAh/g, which is far from meeting the requirements of current power vehicles. The power market requires lithium-ion batteries to have high-rate discharge performance, high safety performance, high efficiency, and high cycle life. At the same time, in view of the industrialization prospect of electric vehicles, the materials to be developed should have the characteristics of low cost. The author believes that the most practical anode materials used in power batteries are carbon microsphere composites and natural graphite-modified materials.

8.2 Phase Structure and Conventional XRD Analysis of Carbon Materials

8.2.1 Structural Data of Important Phases of Carbon Materials

There are many isomers of carbon materials. The important and common ones are amorphous carbon, hexagonal graphite, rhombohedral graphite, diamond, face-centered carbon and face-centered cubic C60, and so on. The structural parameters are listed in Table 8.1. The diffraction data of carbon materials and C–Li compounds related to lithium-ion batteries are given in Table 8.2, for reference and application.

Table 8.1 Structural data of several important carbon materials

	Name	PDF card	Space group	No.	*a* (Å)	*c* (Å)
1	Diamond	06-0675	Fd3m	No. 227	3.566	
2	2H-graphite	41-1487	$P6_3mmc$	No. 194	2.470	6.724
3	3R-graphite	26-1079	3R	No. 146	2.456	10.4
4	FCC carbon		Fm3m	No. 225	3.552	
5	C60	9-1717	Fm3m	No. 225	12.38	
6	Amorphous carbon					

Table 8.2 Comparison of diffraction data of several important carbon materials in lithium-ion batteries, CuKα_1 radiation

2θ (°)		2H-graphite 41-1487		3R-graphite 26-1079		LiC_{24} 35-1047		LiC_{12} 35-1046		LiC_6 34-1320	
Space group lattice parameter (Å)		P6$_3$mmc $a = 2.470$ $c = 6.724$		3R $a = 2.456$ $c = 10.4$		P6$_3$mmc $a = 2.47$ $c = 6.711$		P6$_3$mmc $a = 2.288$ $c = 7.066$		P6$_3$mmc $a = 4.305$ $c = 7.036$	
		hkl	I/I_0	*hkl*	I/I_0	*hkl*	I/I_0	*hkl*	I/I_0	*hkl*	I/I_0
1	12.54							001	10		
2	17.04					002	40				
3	24.03									001	100
4	25.28							002	100		
5	25.65					003	100				
6	26.38	002	100								
7	26.62			003	100						
8	34.60					004	80				
9	38.27							003	20		
10	42.22	100	2								
11	43.49			101	11						
12	44.39	101	6								
13	46.37			012	9					002	100
14	49.21										

8.2.2 *Characteristics of X-Ray Diffraction Patterns of 2H-Graphite*

First of all, let's observe a X-ray diffraction of typical 2H-graphite (see Fig. 8.5). Referring to the characteristics of the fault broadening effect of dense hexagonal stacking layer in Chap. 2, we can see that the broadening of hexagonal-graphite diffraction lines is inconsistent, and in other words, there is a phenomenon of selective broadening. After being processed with Jade5.0, the diffraction data are listed in Table 8.3. The analysis of these data shows that:

(1) When $h - k = 3n$ or $l \neq 0$, such as 001, 002, 100, 004, 110, 112, 006, the $\beta \cos \theta$ values are similar, except 001 and 112, but the β ctg θ values are quite different, indicating that there is only microcrystalline broadening effect in these lines. The $\beta \cos \theta$ is somewhat different between the lines, indicating that the grain shape is not spherical, but polyhedral, which is explained by each *hkl* normal to the grain scale data D_{hkl} (see the last row of Table 8.3).

(2) When $h - k = 3n \pm 1$, the diffraction line broadening effect is much larger than that of $h - k = 3n$, which indicates that there is an obvious selective broadening effect in the 2H-graphite XRD pattern.

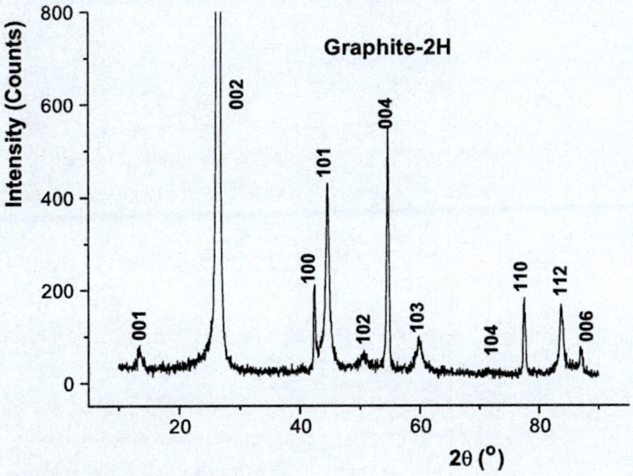

Fig. 8.5 Typical 2H-graphite X-ray diffraction flower

(3) When $h - k = 3n \pm 1$, the broadening effect of the diffraction line of l = even number (such as 102) is obviously larger than l = odd number (such as 101).

8.2.3 *Quantitative Analysis of 2H-Graphite and 3R-Graphite*

In the graphitization treatment of negative electrode active materials for lithium-ion batteries, it is sometimes not possible to obtain a single hexagonal graphite, but the coexistence of hexagonal graphite and rhombohedral (3R) graphite. Therefore, it is required to determine the relative content of 2H and 3R graphite.

The partial diffraction pattern of 2H and 3R graphite is shown in Fig. 8.6. As can be seen from the picture, in all the sample, there are four diffraction peaks of 101, 012 of 2H-graphite and 101, 012 of 3R-graphite that appear in the 2θ angle range from 40 to 50°. 3R content is calculated according to US Patent (US5554462) as follows:

$$W_{3R} = \frac{I_{3R\text{-}101} \times \frac{15}{12}}{\left[I_{3R\text{-}101} \times \frac{15}{12}\right] + I_{2H\text{-}101}} \times 100\% = \frac{1.25 I_{3R\text{-}101}}{1.25 I_{3R\text{-}101} + I_{2H\text{-}101}} \tag{8.1}$$

In the determination of 3R content, CuKα radiation was used FT step scanning with step length of 0.01°, recording time of 1.0 s per step, or sampling width of 0.01°, 0.5°/min continuous scanning, scanning range of 40–48°. By measuring the integral intensity of 3R-101 and 2H-101 diffraction peaks, respectively, the relative content of 3R graphite can be calculated according to formula 8.1. The determination results of 10 samples are given in Table 8.4.

Table 8.3 Corresponds to the XRD data and analysis results of Fig. 8.5

hkl	001	002	100	101	102	004	103	110	112	006
2θ (°)	13.323	26.518	42.291	44.510	50.555	54.616	59.807	72.424	83.546	86.9977
FWHM (°)	0.689	0.252	0.277	0.635	1.824	0.338	1.350	0.282	0.549	0.397
I/I_1	0.5	100.0	1.5	7.9	1.7	5.5	2.3	1.5	2.9	1.2
$\beta_{1/2}$ (°)	0.120	0.110	0.106	0.106	0.105	0.105	0.105	0.104	0.103	0.103
β ($\times 10^{-3}$) (rad)	9.931	2.478	2.985	9.233	30.002	4.067	21.729	3.107	7.784	5.131
$\beta \cos\theta$ ($\times 10^{-3}$)	9.861	2.412	2.784	8.545	27.129	3.614	19.836	2.424	5.805	3.723
$\beta\ \mathrm{ctg}\,\theta$ ($\times 10^{-3}$)	85.032	10.516	7.717	22.563	56.958	7.877	81.368	3.876	9.714	5.408
D_{hkl} (nm)	13.6	56.8	49.2	16.0	5.0	37.9	7.3	56.6	23.6	36.8

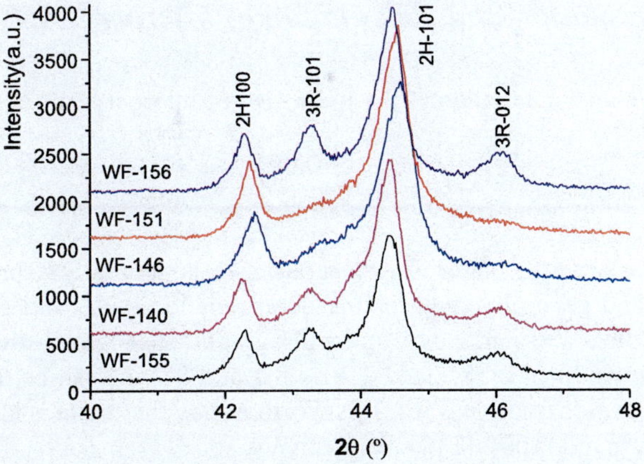

Fig. 8.6 2H and 3R graphite coexist partial diffraction pattern, CuKα radiation

Table 8.4 Relative content of 2H and 3R graphite in 10 samples

	Sample number	3R (101)	2H (101)	3R (%)	2H (%)
1	WF-140	20,371	44,831	36.2	63.8
2	WF-155	13,850	41,881	29.2	70.8
3	WF-150	16,246	52,937	27.7	72.3
4	NG-9	32,262	112,896	26.3	73.7
5	NG-10-2	18,265	71,255	24.3	75.7
6	NG-13	32,651	55,533	42.4	57.6
7	NG-14	26,158	27,767	54.3	45.9
8	NG-15	31,614	98,756	28.6	71.4
9	NG-16	25,200	63,142	33.3	66.7
10	NG-16	26,257	88,645	27.0	73.0

The key to the experimental determination is how to separate the diffraction peaks and how to measure the integral intensity of the diffraction peaks. If you have lines shape WF-155, WF-140, and WF-156 in Fig. 8.2, you can use *BG*, *Apply*, *Strip K-alpha2* in the Jade program. *Refine* also adjusts "Lorentzia" parameters until the difference between fitting and experimental values reaches the minimum, and *Report* can achieve the purpose of peak separation and integral intensity. Table 8.4 shows that the determination results of 10 samples are all obtained by this method. If it is linear like WF-146 and WF-151 in Fig. 8.6, if it is carried out according to the above method, the result is more than ten percent, but the actual result is only a few percent, so both peak separation and integral intensity need human intervention.

8.2.4 Determination and Calculation of Microcrystal Size

The general formula for calculating the grain size of microcrystals is:

$$D_{hkl} = \frac{0.89\lambda}{\beta_{hkl} \cos \theta_{hkl}} \tag{8.2}$$

In the books of carbon materials, the grain size is generally expressed by L_{hkl} instead of D_{hkl} and generally requires the determination of L_a and L_c, that is, the grain size along the *a*-axis and *c*-axis of the 2H-graphite hexagonal structure. As can be seen from the data in Fig. 8.1 and Tables 8.1 and 8.2, L_a can be determined by (100) and (110) two diffraction lines. It is worth noting that there is an angle of 30° and 60° between *a* and [100], [110], respectively, so

$$L_a = \frac{L_{100}}{\cos 30°} = \frac{0.89\lambda}{0.866\beta_{100} \cos \theta_{100}} = \frac{1.028\lambda}{\beta_{100} \cos \theta_{100}} \tag{8.3}$$

$$L_a = \frac{L_{110}}{\cos 60°} = \frac{0.89\lambda}{0.5\beta_{110} \cos \theta_{110}} = \frac{1.780\lambda}{\beta_{110} \cos \theta_{110}} \tag{8.4}$$

However, because the (100) diffraction lines of 2H-graphite are often interfered by 2H-graphite (101) and 3R-graphite (101), the (110) diffraction lines of 2H-graphite are often used to determine L_a.

In principle, (001), (002), and (004) can be used to determine L_c, but because the relative intensity of (001) and (004) is much lower than that of (002), it is often used (002) to determine L_c, directly.

$$L_c = \frac{K\lambda}{\beta_{002} \cos \theta_{002}} \tag{8.5}$$

When deriving the Scherrer formula from Chap. 4, we know that K should generally be 0.89. However, the advanced carbon materials laboratory of Institute of Metals, Chinese Academy of Sciences provides $K = 0.94$, which assumes that the linear shape is Gaussian type and the crystal is a small and uniform cube.

8.2.5 Determination of Alignment (Orientation Ratio) [5]

1. **The significance and characterization of orientation ratio**

The so-called orientation ratio, also called as alignment ratio, is the degree of preferred orientation in a polycrystalline sample in which a crystal plane is parallel to the sample surface (electrode surface). In an ideal polycrystalline powder sample, the orientation of any crystal plane is completely disordered; that is to say, there is no problem of preferred orientation. If we test this kind of sample by X-ray diffraction, the sample is filled into the sample rack in some way, and compacted and flattened to make the surface of the sample flat. During the test, the surface of the sample is strictly tangent to the focusing circle and passes through the X-ray powder diffractometer circle center. If the above sample preparation does not change the completely disordered distribution of grain orientation, then the relative intensity of each diffraction line in the diffraction pattern is inherent in this structure; if the sample preparation process changes the disordered distribution of grain orientation, then a preferred orientation occurs.

In the manufacturing process of graphite electrode, the distribution of grain orientation in the active material will be changed and the preferred orientation will be produced. The simplest way to characterize the preferred orientation is to use the diffraction intensity ratio of two crystal planes perpendicular to each other and its variation with the compaction density of the electrode.

In the X-ray diffraction test of the electrode, the surface of the electrode is directly used as the test surface. The graphite with hexagonal structure is often characterized by the diffraction intensity ratio I_{002}/I_{110} (I_{110}/I_{002}) or I_{004}/I_{110} (I_{110}/I_{004}) of its base plane (002) or (004) and its perpendicular hexagonal cylinder (110) or (100). Because the diffraction geometry of the powder X-ray diffractometer determines that the diffraction intensity is given by those crystal planes parallel to the sample surface, the physical meaning of the intensity ratio I_{002}/I_{110} is the ratio of the probability of the (002) parallel sample surface to that of the (110) parallel sample surface.

2. **The effect of orientation ratio and density of graphite electrode on the performance of the battery**

From the crystal model of 2H-graphite, it is known that there is a large space between the two hexagonal base planes (graphite layer spacing, that is, d_{002}), and the diffusion velocity of embedded atoms perpendicular to the cylinder is much larger than that perpendicular to the base plane (about 10^6 times). Therefore, theoretically, the higher the probability that the (001) plane of the 2H-graphite grain is perpendicular to the electrode surface, the more favorable for the intercalation of Li atoms, while the orientation of the (110) plane is of no great importance.

The manufacturing method of the negative electrode sheet is that the active material is 2H-graphite, which is coated on both sides of the current collector copper foil with water-based adhesive and then pressed into a pole sheet with a certain compaction density by roller. Generally speaking, the higher the compaction density of the active material 2H-graphite is, the more favorable it is to improve the capacity of the single cell. When the compaction density of the electrode made of 2H-graphite is at 1.58–1.68 g/cm^3, the orientation ratio of the electrode active material:

	I_{110}/I_{004}	I_{004}/I_{110}
Usually	~0.07 or >0.07	~14.3 or <14.3
Preferably	0.09 or >0.09	~11.1 or <11.1
The best is	0.18 or >0.18	~5.56 or <5.56

Therefore, some customers require that the alignment ratio ≤ 20, and the best is 5–8, which seems reasonable.

If the orientation ratio of the active material is lower than the above range, the prepared battery shows a large electrode expansion during charging, so it is difficult to increase the battery capacity per unit volume. On the other hand, if the orientation ratio exceeds the above range, that is, $0.18 < I_{110}/I_{004} < 0.07$ or $14.3 < I_{004}/I_{110} < 5.56$, it is difficult to increase the filling density and discharge capacity.

From powder to beating, coating, to rolling into the required poles, the orientation ratio of active materials will change obviously, which is closely related to the compaction density and the shape of the original 2H-graphite particles. Therefore, it is of great practical significance to study the orientation ratio of graphite electrode. The research in this area is almost blank.

3. **The method of determining the orientation ratio of 2H-graphite**

The X-ray diffraction patterns of 2H-graphite powder and 2H-graphite electrode are shown in Fig. 8.7. It can be seen from Fig. 8.7 that all the spectral lines of the powder appear, but except 002 is very strong, and the other lines are very weak. 004 is the stronger diffraction line in the weak, but the orientation ratio can still be characterized by measuring the diffraction intensity of (002) (004) and (100) (110). However, for the negative 2H-graphite electrode sheet, the (111) diffraction lines of the current collector Cu seriously interfere with the (100) and (101) of 2H-graphite, so the orientation ratio can only be characterized by measuring the diffraction intensities of (002), (004), and (110).

(1) **Preparation of negative electrode sheet**

The negative electrode active material 2H-graphite is mixed with aqueous solution of thickener carboxymethyl cellulose (CMC) and binder resin styrene butadiene rubber (SBR). The ratio is negative powder: CMC:SBR:SP = 94.5:2.25:2.25:1.

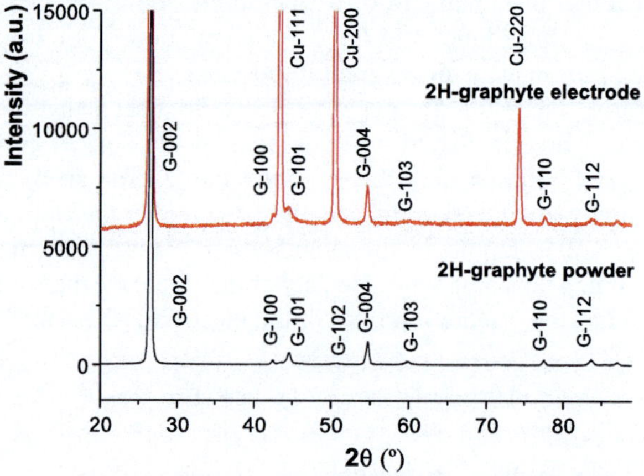

Fig. 8.7 Comparison of X-ray patterns between hexagonal graphite powder and electrode sheet

To stirs the mixture fully to make a uniform slurry. By adjusting the blade seam width, the paste is coated on 12 μm thick double-sided smooth copper foil, so that the polar one-sided density after drying is 20 mg/cm^2. After baking in a vacuum drying oven at 110 °C, cutting by a slicer, and rolling the poles according to the required compaction density, a group of poles sheet with a compaction density of 1.50–1.90 ± 0.05 g/cm^3 is prepared for use.

(2) **Determination of compaction density of electrode**

The compaction density of the electrode is measured on the electrode whose length and width, and total thickness are L, W, and T, respectively. Calculated according to formula (8.6)

$$\rho = \frac{G - g}{L \times W \times (T - t)} \quad (\text{g/cm}^3) \tag{8.6}$$

G and g are the total mass of the measured polar sheet and the mass of the Cu collector, respectively, and T and t are the total thickness of the polar plate and the thickness of the Cu collector foil, respectively.

(3) **Method of determination by X-ray powder diffractometer**

Because the determination of orientation ratio is to measure the intensity of two diffraction lines. Generally speaking, the greater the intensity of the measurement, the smaller the error. Therefore, under the condition that the resolution is allowed, a larger diaphragm system and a higher incident ray intensity should be adopted. DS = SS = 1°, RS = 0.30 mm, sampling width 0.01–0.02°, 2θ scanning speed 0.5–1°/min. Generally, 0.02 and 1°/min; 0.01 and 0.5°/min are used for more accurate measurement. The scanning range of the three lines is as follows:

hkl	2θ scanning range (°)
002	25.5–29
004	53.5–56
110	76.5–79

(4) **An example of measuring the result with 004–110 line pair**

Figure 8.8 shows the determination results of five samples, where A, B, C, D, and E represent different samples. From Fig. 8.8 Visible:

(i) When characterized by I_{004}/I_{110}, the density of the unpressed electrode is generally lower, and the ratio of I_{004}/I_{110} is also lower. When the density is 1.50 g/cm^3, the value of I_{004}/I_{110} increases rapidly, which seems to be a common phenomenon. But what value it increases to is closely related to the original material. Then with the increase of pressing density, different raw materials show obviously different rules of change.

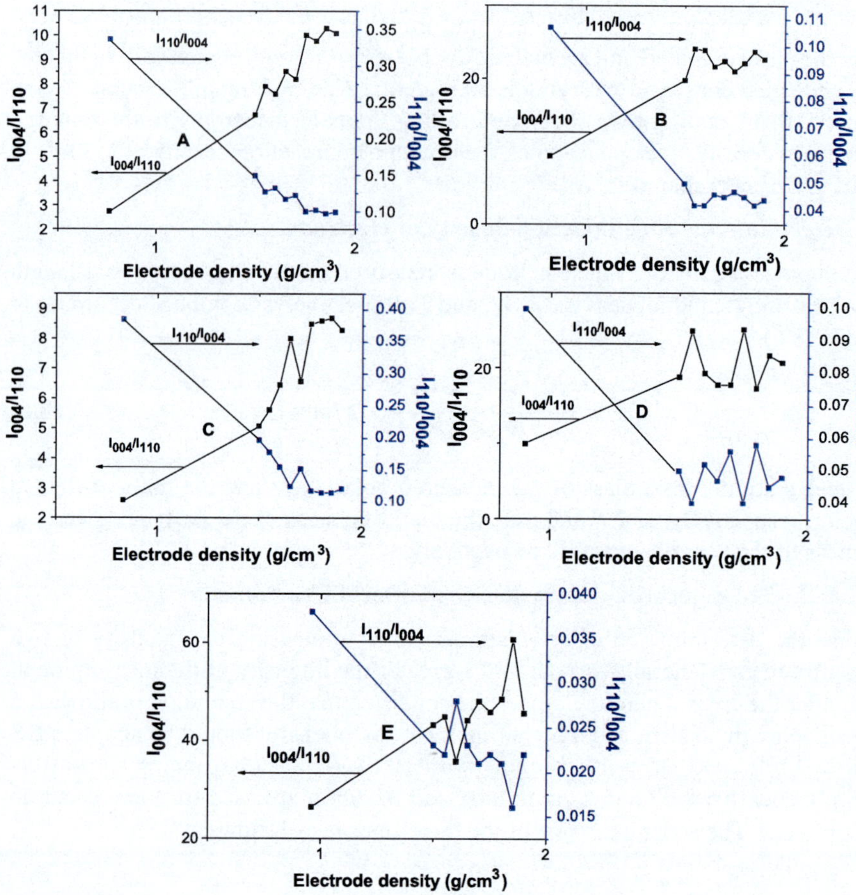

Fig. 8.8 Relationship between compaction density and orientation ratio of several typical graphite electrodes. A—PC-LF-2; B—LF; C—FT-2; D—new-FT-17; E—FT-21

(ii) When characterized by I_{110}/I_{004}, the density of the unpressed electrode is generally lower, and the value of I_{110}/I_{004} is also higher. When the density is 1.50 g/cm^3, the value of I_{004}/I_{110} decreases rapidly, which is also a common phenomenon. Then with the increase of pressing density, different raw materials show obviously different rules of change.

This shows that it is possible to characterize it in both ways. However, the physical meaning is more clearly represented by the value of I_{004}/I_{110}, and the larger the value is; that is shown that the greater the probability that (004) is parallel to the surface of the electrode, it is not beneficial to the intercalation of Li in graphite.

In order to compare the differences between different samples and remove the data of unpressed points, the results of Fig. 8.8 are redrawn and the curve of the relationship between the compaction density and orientation ratio of the electrode is shown in Fig. 8.9. As can be seen from the figure:

(i) FT-21 (E) has the highest degree of orientation, FT-2 (C) and PC-LF-2 (A) have the lowest degree of orientation, and new-FT-17 (D) and LF (B) have a medium degree of orientation.
(ii) E varies greater with the compaction density and fluctuates greater, which is disadvantageous to the battery performance, so it exceeds the "usual choice." The general trend of orientation degree of C and A increases with the increase of compaction density, but the range is not large and the "best choice" of attachment orientation ratio.
(iii) It seems that D and B have a medium degree of orientation, and the change with the compaction density is small, especially B is the best, and its fluctuation with the compaction density is the least, which accords with the "optimization."

(5) An example of measuring the result with 002–110 line pair

In addition, the ratio of I_{002}/I_{110} to I_{004}/I_{110} can also be used to characterize the orientation ratio. Figure 8.10a shows the relationship between the orientation ratio I_{002}/I_{110} of three kinds of 2H-graphite electrodes and the compaction density. For the convenient of comparison, the curve of the relationship between I_{004}/I_{110} and electrode compaction density is shown in Fig. 8.10b. They all show that the relationship between the orientation ratio of the ink electrode and the compaction density is similar, but the relative error of the I_{002}/I_{110} display is larger, Table 8.5 shows the determination results of the orientation ratio I_{110}/I_{004} of a group of powder samples. It should be noted that the measured result is the orientation ratio of the sample plane of the polycrystal X-ray diffractometer, which is different from the orientation ratio of the electrode sheet, and when necessary the electrode orientation ratio should be determined.

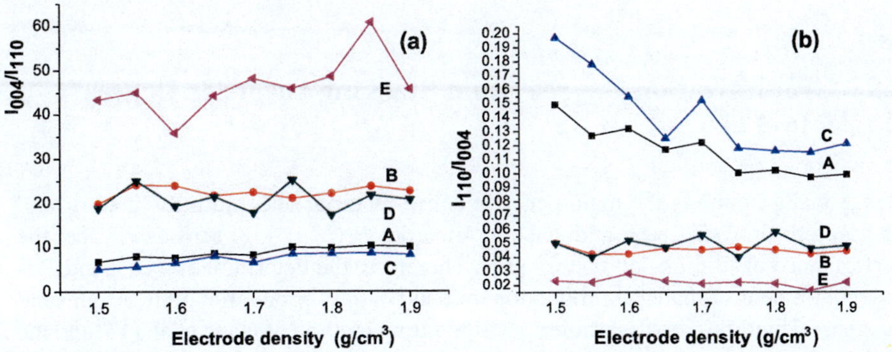

Fig. 8.9 Relationship between electrode density and orientation ratio of several typical graphite electrodes. **a** I_{004}/I_{110}, **b** I_{110}/I_{004}

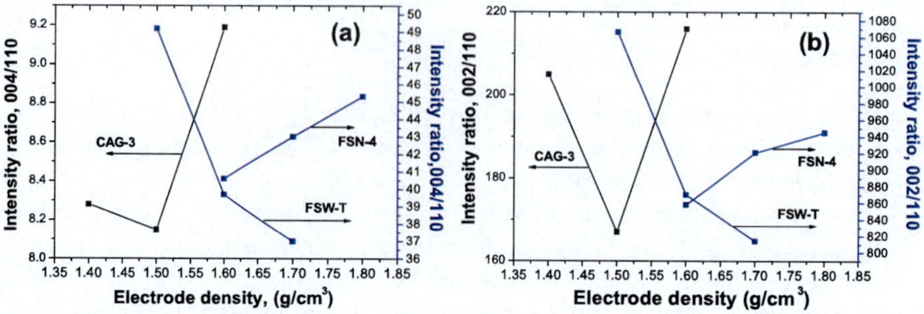

Fig. 8.10 Relationship between the ratio of I_{002}/I_{110} of graphite electrode and the compaction density of graphite electrode

Table 8.5 Determination of orientation ratio I_{110}/I_{004} of a group of powder samples

	FJ-38	FJ-39	FJ-50	FJ-51
I_{110}	47,377	40,375	59,010	57,082
I_{004}	191,456	139,298	161,330	153,161
I_{004}/I_{110}	4.049	3.460	2.732	3.676
I_{110}/I_{004}	0.247	0.289	0.366	0.272

8.2.6 *Comprehensive Determination Results of a Group of Powder Samples Containing 2H and 3R Phases*

Table 8.6 shows the comprehensive determination results of a group of samples containing 2H and 3R graphite powders. It can be seen that in the samples containing 2H and 3R graphite, the grain shape is not spherical, the content of rhombic graphite is different, and the orientation ratio is obviously different.

8.3 Graphitization Degree *g* and Stacking Disorder Degree *P* [6–11]

Hexagonal graphite is the main negative active material of lithium-ion battery, and its graphitization degree g and stacking disorder P (= 1 − g) seriously affect the performance of lithium-ion battery [12]. Therefore, the determination of g and P is the requirement of material production unit and battery production unit. At present, the general methods are the whole spectrum fitting method of Hang et al. [13] and the d_{002} method using the linear relationship between 2H-graphite d_{002} and P [13–15]. The former method requires a long time for data collection and a special calculation program, while the latter method generally uses silicon powder as the internal standard. The purpose of this experimental study is: (1) the influence of different

Table 8.6 Results of X-ray diffraction determination of a group of samples

	ZH-1001	ZH-1002	ZH-1003	ZH-1004	YF—WF-142	X50-080801	NG-24	NG-29
a (Å)	2.4658	2.4648	2.4591	2.4675	2.4686	2.4643	2.4655	2.4639
c (Å)	6.6975	6.7096	6.7027	6.7313	6.7269	6.7010	6.7149	6.6983
$FWHM_{110}$	0.363	0.368	0.322	0.268	0.274	0.319	0.280	0.272
L_a (nm)	76.3	77.8	92.4	122.8	119.5	93.7	114.5	99.7
$FWHM_{002}$	0.328	0.302	0.321	0.260	0.245	0.301	0.249	0.306
L_c (nm)	173.4	152.7	157.8	221.9	246.6	174.3	239.5	205.5
I_{3R101}	1112	1022	1108	871	1324	3503	3120	7118
I_{2H101}	8197	7418	9163	7773	19,683	11,056	12,840	12,416
W_{3R} (%)	14.50	14.69	13.10	21.30	7.76	28.37	23.30	41.74
I_{004}	15,424	18,470	19,438	10,952	8368	22,354	15,986	186.3
I_{110}	3104	2135	2124	2303	3044	3328	2999	3981
I_{110}/I_{004}	0.201	0.115	0.109	0.210	0.364	0.149	0.187	0.214

experimental conditions on the results of whole spectrum fitting, the influence of different solutions of d_{002}, and the comparison of d_{002} and θ_{002} methods with whole spectrum fitting method. (2) The best scheme of whole spectrum fitting method and D_{002} and θ_{002} diffraction experimental determination with Si powder as internal standard sample is put forward.

Different arrangements of carbon atoms form a variety of carbon materials with specific properties. If the carbon atoms are arranged in a straight-line and their length is several nm, this is the carbon nanotube, which belongs to a hexagonal graphite structure (see Sect. 8.7.5); but the carbon atoms are mostly arranged in a two-dimensional network, and if the grid size is smaller, there is a completely disordered arrangement between the grids; this is amorphous carbon, and its X-ray scattering pattern is shown in Fig. 8.11c. It can be said that the structure of all carbon materials is formed by the expansion or/and overlapping stacking of hexagonal network planes composed of carbon atoms. Because the interaction between overlapping layers is very weak, it is not always stacked according to certain rules. In ideal graphite, hexagonal network planes are marked ABAB... or ABCABC.... The former is hexagonal 2H-graphite, belonging to $P6_3mmc$ (No. 194) space group, and the latter is rhombohedral 3R-graphite, belonging to R3 (No. 146) space group. Their structural models are shown in Fig. 8.12a, b, respectively.

The hexagonal network surface of carbon except according to the aforementioned ABAB... and ABCABC.... In addition to the stacking formation of graphite, on the contrary, not only the hexagonal mesh surface is small, but also the stacking between the network surfaces is completely irregular, and it is only parallel stacking, which is called the chaotic layer structure of carbon. Because there is no periodicity along the so-called *c*-axis direction, the random layer structure belongs to a two-dimensional structure. The diffraction pattern is shown in Fig. 8.11b.

Compared with 2H-graphite (see Fig. 8.11a), the *hk*0 diffraction lines of graphite are asymmetrically broadened and degenerate from three-dimensional *hk*0 to two-dimensional structure of *hk* 10 and 11 shown in Fig. 8.11b are degenerated from 100 and 110, respectively. By analyzing these diffraction lines in detail, the probability g_1 of graphite structure on two parallel stacked network surfaces can be obtained.

There are several X-ray methods to determine the graphitization degree.

8.3.1 Fourier Analysis of (10) and (11) Diffraction Lines [13, 14]

Intandov et al. [8] wrote the corrected *hk* intensity distribution as

$$\frac{P'_{2\theta}(hk)4\sin\theta(\sin^2\theta - \sin^2\theta_0)^{1/2}}{KPF^2(1+\cos^2 2\theta)} = \sum_{-\infty}^{+\infty} A_{n(hk)}\cos 2\pi n h_3 \tag{8.7}$$

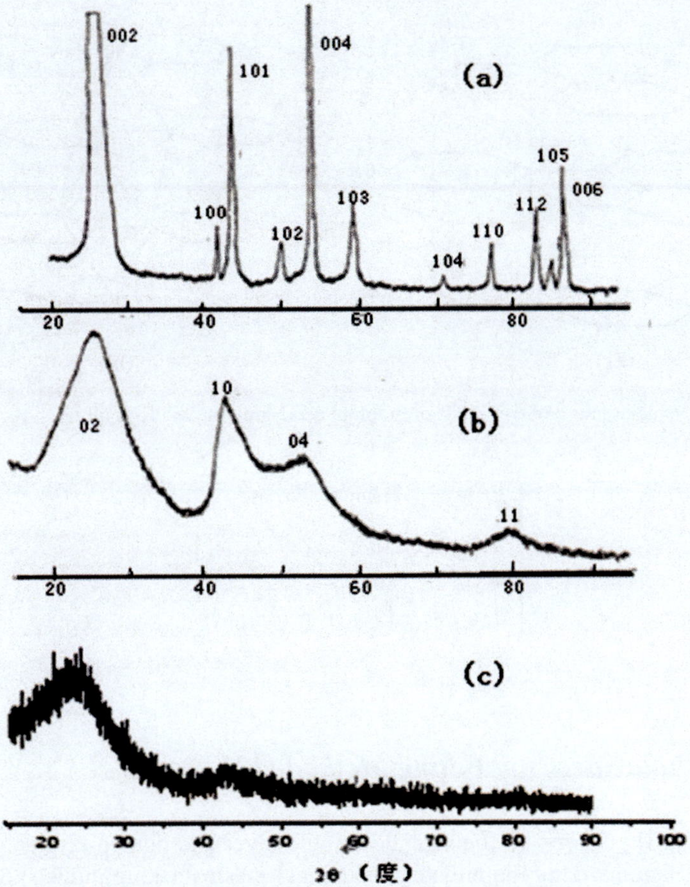

Fig. 8.11 X-ray powder diffraction pattern of carbon materials **a** hexagonal graphite, **b** amorphous carbon of coke, **c** after heat treatment at 1000 °C

where K is constant, P and F are multiplicity factor and structure factor, respectively. θ_0 is the Bragg angle corresponding to of hk, since the left side of the factor (8.7) can be measured. So the Fourier coefficient A_0 can be obtained by Fourier transformation.

$$A_{n(hk)} = \int_{-\infty}^{+\infty} \frac{P'_{2\theta(hk)} 4 \sin\theta \left(\sin^2\theta - \sin^2\theta_0\right)^{1/2}}{KPF^2\left(1 + \cos^2 2\theta\right)} \cos 2\pi n h_3 \mathrm{d}h_3 \tag{8.8}$$

When $n = 1$, $A_{n(hk)}$ is the probability of graphite stacking closest to the net surface, g_1 is obtained. When $n = 2$ can obtain stacking probability with the second net face the ABAB… is ABCABC…, that is the stacking probability g_{ABA} and g_{ABC} of 2H and 3R:

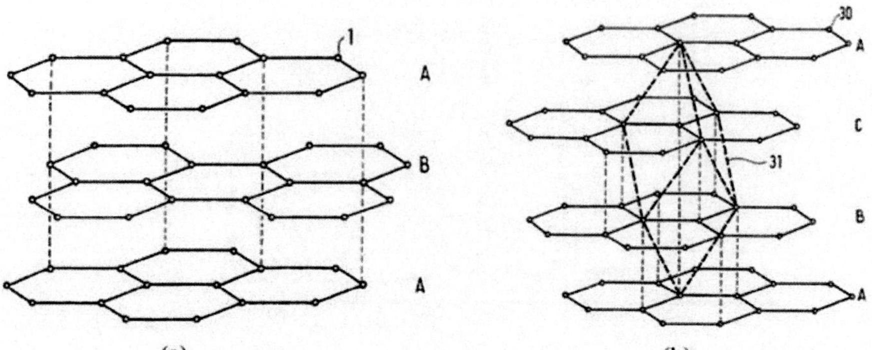

Fig. 8.12 Crystal structure of two kinds of graphite **a** 2H-graphite, **b** 3R-graphite

$$\begin{cases} g_1 = -2A_{1(10)} = A_{1(11)} \\ g_{\mathrm{ABA}} = \dfrac{1}{3}\left\{2A_{2(10)} + A_{2(11)}\right\} \\ g_{\mathrm{ABC}} = \dfrac{2}{3}\left\{A_{2(11)} - A_{2(10)}\right\} \end{cases} \tag{8.9}$$

8.3.2 *Whole Spectrum Fitting Method [13]*

Hang et al. [13] put forward the so-called single-layer and double-layer models on the basis of summarizing the previous hexagonal mesh stacking models of carbon atoms and established a whole spectrum fitting-refinement program.

The "single-layer" model thinks that if the carbon atom occupies the A position in the basic hexagonal network, the upper layer can be chosen as follows:

(1) The atom has a disordered shift with probability *P*.
(2) The probability of occupying the B position (1 – *P*)/2.
(3) The probability of occupying the C position is also (1 – *P*)/2.

This model shows that ABAB... stacking cannot be reproduced, but can describe the structure of disordered carbon.

The "two-tier model" assumes that the carbon infrastructure is a unit of two layers of stacking recognized by AB and then expands horizontally infinitely. This kind of double-layer stacking is selected as follows:

(1) The probability of disordered displacement between two adjacent double-layer units is *P*.
(2) The probability of the accepted displacement between two adjacent double-layer units is P_t, such that the local is AB/CA/BC order.

Table 8.7 Refinement results of Hang of three samples [2]

	S2000	S2850 "Single-layer" model	S2850 "Double-layer" model
Crystal plane spacing, d_{002} (Å)	3.430 (2)	3.382 (1)	3.382 (1)
Lattice parameter, a (Å)	2.459 (3)	2.463 (10)	2.461 (1)
Disordered shift probability, P	0.89	0.28 (1)	0.26 (1)
Displacement probability, p_t	0	0	0.128
Graphitization degree, g	0	0	0
Strain, ε	0.003 (1)	0	0
Root mean square displacement, $(\langle\sigma\rangle^2)^{1/2}$	0.108 (5)	0.065 (5)	0.091 (5)
Grain size along a-axis, L_a (nm)	10.5 (1)	19.8 (8)	19.6

(3) The probability that there is no displacement at all is $(1 - P - P_t)$, which gives a very clear AB/AB…stacking, if $P = 0$, $P_t = 0$ this is the ABAB…stacking in 2H-graphite; if $P_t = 1$ and $P = 0$, this is 3R-graphite ABCABC…stacking. Of course, may also include a small portion of 3R stacking in 2H-graphite.

From the above-mentioned "single-layer" model and "double-layer" model, it can be known that there can be no two-dimensional diffraction lines (hk) in the double-layer model when the P value is large. When the P value is small, the single-layer model cannot reproduce the 102, 103 diffraction peaks observed in the better crystalline graphite, but the double-layer model can and can accurately reproduce the peaks in the range of 100–101 in graphite.

They use Fortran 77 to program refinement. The refinement results of the three examples are shown in Table 8.7. Among them, ε is the strain, the $(\langle\sigma\rangle^2)^{1/2}$ is root mean square displacement, and the L_a is the grain size in the a-axis direction.

8.3.3 *The Relationship Between Stacking Disorder Degree* $\mathbf{P_{002}}$ *and* $\mathbf{d_{002}}$ *[14]*

Tandolf et al. [8] put Franklin, Bacon, Mering and Maire and Houska and Warren obtained the relationship between d_{002} and graphitization degree g is plotted relation curve as shown in Fig. 8.13. The relationship between the solid lines in the figure is

$$d_{002} = 3.440 - 0.086\left(1 - P_{002}^2\right) \tag{8.10}$$

The relationship between dotted lines is

$$g = \frac{3.440 - d_{002}}{3.440 - 3.354} \tag{8.11}$$

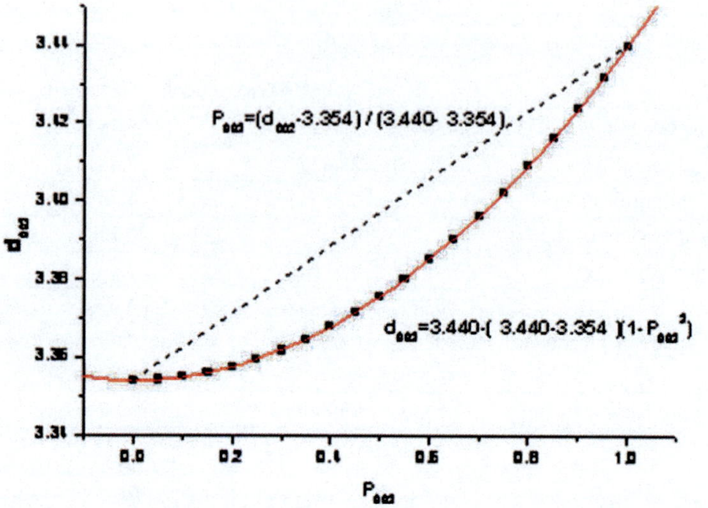

Fig. 8.13 Relationship between the degree disorder P_{002} and the average plane spacing d_{002}

$$P_{002} = 1 - g = \frac{d_{002} - 3.354}{3.440 - 3.354} \tag{8.12}$$

According to the relation of formula (8.10), it is assumed that $g = 0.0, 0.10, \ldots, 0.90, 1.00$, find out the corresponding d_{002}, draw the figure with Origin software, get the relation curve of $d_{002} \sim g$, and save it. Later, for unknown samples, the accurate d_{002} can be obtained by experiments, and the graphitization degree g can be obtained on the curve by using the Screen Reader function of Origin software, which is called curve method.

Sometimes for the sake of simplicity, in the relative determination of g (or) P, it is very convenient to solve the linear relationship of (8.11) or (8.12) formula.

Let $g = 0.0, 0.1, \ldots, 1.0$, d_{002} is obtained by formula (8.11) and substituted into Bragg formula, $\lambda = 0.154056$ nm, Bragg angle θ_{002} (°) is obtained. Then Fig. 8.14 is drawn by Origin program, and the linear equation is

$$g = -38.3007 + 2.96038\theta_{002}\ (°) \tag{8.13}$$

In this way, the Si standard sample can be used as the internal standard to accurately determine the Bragg angle θ_{002} of the 002 diffraction peak of graphite.

$$\theta_{\text{Gra}-002} = \theta^{\text{Exp.}}_{\text{Gra}-002} - \left(\theta^{\text{Exp.}}_{\text{Si}-111} - \theta^{\text{Stand.}}_{\text{Si}-111}\right) = \theta^{\text{Exp.}}_{\text{Gra}-002} - \left(\theta^{\text{Exp.}}_{\text{Si}-111} - 14.221°\right) \tag{8.14}$$

Taking the degree as the unit, the graphitization degree g can be obtained by introducing the formula (8.13). Compared with the d_{002} method, the error introduced

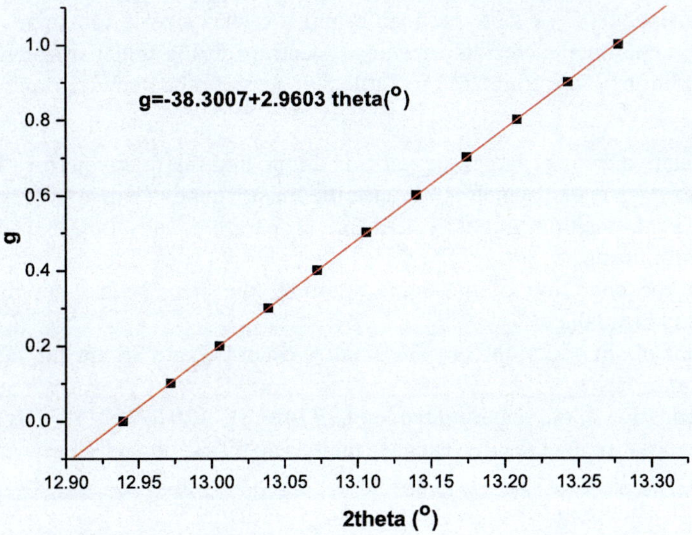

Fig. 8.14 Relationship between graphitization degree g and (002) semi-diffraction angle θ (°), $CuK\alpha_1$ radiation

by this approximation is in the range of 0.001–0.003, so it is not necessary to replace the Bragg formula to calculate d_{002} and then replace formula (8.11) to find g.

8.4 Experimental Determination of Graphitization Degree *g* and Disorder Degree *P* [11]

8.4.1 The Effect of Different Experimental Conditions on the Degree of Disorder P *Obtained by Whole Spectrum Fitting*

According to the requirements of Ref. [13] and the current experimental conditions of whole spectrum determination, see Table 8.8.

Using Japanese (Rigaku) polycrystal X-ray diffractometer, DS and SS are evenly divided into 2, 1, 1/2, 1/4, 1/6, 0.05° et al. and RS is 2.0, 1.0, 0.6, 0.3, 0.15, 0.05 mm. Obviously, DS and SS cannot meet the requirements. In addition, we think that

Table 8.8 Whole spectrum experimental conditions for current experimental determination

DS (°)	SS (°)	RS (mm)	Step size (°)	Time (s/step)	2θ scanning range (°)	Total time (min)
0.2	0.2	0.30	0.05	0.9	10–120	33

the step size of 0.05° is not reasonable and the step size is too large. For this reason, the experimental results of whole spectrum fitting under different experimental conditions *P* are compared in Table 8.9. As it can be known from the data in the table:

(1) The relative error of larger step size is larger, and the relative error of smaller step size is smaller, which is because the linear shape is worse than that of the latter. Peak position, intensity, and line shape are equally important in whole spectrum fitting.
(2) Under the condition of the same aperture, the error with large diffraction intensity is smaller.
(3) In terms of slit width, the error is smaller when DS and SS are 1/2° and RS = 0.30 mm.
(4) The reduction of the scanning range of 2θ from 10–120 to 20–110° has no effect on the measurement results, because there is no diffraction peak between 2θ and 110–120°. To sum up, it is reasonable to choose conditions in Table 8.10 for daily measurement.

8.4.2 *The Effect of Different Solution Methods of* $\mathbf{d_{002}}$ *on the Determination of* $\mathbf{g}$

The current method is to use Si powder as the internal standard sample to determine the two diffraction peak positions of 2H-graphite (002) and Si (111). The $2\theta_{002G}$ value is modified by $2\theta_{Si111exp.} - 2\theta_{Si111stand.} = \Delta 2\theta_{Si111}$, the d_{002G} is calculated by Bragg formula: $2d \sin \theta = \lambda$, and then the *g* value is calculated by substituting formula (8.10). Although this method is simple, it only considers the systematic error, but not the accidental error. Therefore, we use three methods to calculate d_{002}: (1) the current method; (2) using the modified *d* values of 002, 004, and 006 of Si standard sample to calculate *c* value, respectively, and then extrapolating to cos 2θ = 0 by extrapolating function cos 2θ to get c_0, then $d_{002G} = 0.5c_0$; (3) using whole spectrum (also adding Si sample). After calibrating the $2\theta_{hkl}$, of each *hkl*, the lattice parameters a_0 and c_0 are obtained by the least square method, and then the $d_{002G} = 0.5c_0$ is obtained. Finally, P_{002} is obtained according to the linear equation and the reading on the curve by Screen Reader of Origin software.

The original data and analysis results of the four samples are given in Tables 8.11 and 8.12, respectively. From the results of Table 8.12, it can be seen that the extrapolation of *c* by d_{002} and d_{004} is not credible, mainly because there are too few data points; only the results of d_{002} are consistent with those of full-spectrum d_{002}, and the results obtained by whole-spectrum least square method are credible.

Table 8.9 Whole spectrum fitting method to solve the graphitization degree under different experimental conditions

	DS (°)	SS (°)	RS (mm)	Step size (°)	Time (s)	2θ scanning range (°)	Total time (min)	2θ (°)	I (cps)	P (%)	Relative error (%)
WH-103	1/6	1/6	0.30	*0.05*	0.9	10–120	33	26.60		14.5	10.6
WH-103-0	1/6	1/6	0.30	*0.05*	0.9	10–120	33	26.60	3416	12.7	−3.8
1	1/4	1/4	0.15	0.02	0.4	10–120	36.7	26.62	6148	13.8	+4.5
2	1/4	1/4	0.30	0.02	0.4	10–120	36.7	26.60	10,290	13.3	+0.8
3	1/4	1/4	0.30	*0.05*	0.9	10–120	33	26.65	9996	12.8	−3.0
4	1/2	1/2	0.15	0.02	0.4	10–120	36.7	26.68	21,340	13.3	+0.8
5	1/2	1/2	0.30	0.02	0.4	20–110	30	26.7	39,603	13.3	+0.8
6	1/2	1/2	0.30	*0.05*	0.9	10–120	33	26.7	36,774	12.8	−3.0
7	1	1	0.15	0.02	0.4	10–120	36.7	26.7	50,010	12.5	−5.3
8	1	1	0.30	0.02	0.4	10–120	33.3	26.7	89,083	13.5	+2.3
9	1	1	0.30	*0.05*	0.9	10–120	33			12.8	−0.3
10	1	1	0.30	0.02	3°/min	10–120	36.7			13.1	−0.8
										$\overline{P} = 13.2$	

Italic values is the scores, 1/2 is a half degree
Note WH-103 and WH-103-N ($N = 0, 1, \ldots, 10$) for twice sample preparation

Table 8.10 Reasonable experimental parameters for whole spectrum measurement

Scan mode	DS (°)	SS (°)	RS (mm)	Step size (°)	Scanning parameters (s/step)	2θ scanning range (°)	Total time (min)
Timing counting	1/2	1/2	0.30	0.02	0.4–0.5	10–110	30–37.5
Continuous scan	1	1	0.30	0.02	2.5–3.0	10–110	36–30

8.4.3 d_{002} and θ_{002} Methods Are Compared with Whole Spectrum Fitting [11]

For comparison, the determination results of d_{002} method (straight-line and curve), θ_{002} method and whole spectrum fitting (WSF) method are summarized in Table 8.13, it can be seen that most of the WSF obtained by whole spectrum fitting (WSF) is between the d_{002} straight-line calculation and the curve reading.

The d_{002} straight-line method is in good agreement with the θ_{002} straight-line method, and there is only an error in the thousandth percentile.

Through theoretical analysis and XRD experimental study, the following conclusions are drawn:

(1) The recommended XRD experimental conditions for the determination P_{FWHM} by whole spectrum fitting method are as follows: DS = SS = 1/6°, RS = 0.30 mm or 0.15 mm Cu target, scanning range of 2θ = 20–110°, step width 0.02°, counting time 0.4–0.5 s/step, total 30–37.5 min, or sampling width 0.02°, continuous scan 2.5–3.0°/min, total of 30–36 min.
(2) The formulas for calculating graphitization degree g by using d_{002} and θ_{002} linear calculation methods with Si powder as internal standard are, respectively.

$$g_{002} = \frac{3.440 - d_{002}}{3.440 - 3.354}; \tag{8.15}$$

$$g_{002} = -37.3007 + 2.9603\theta_{002} \ (度) \tag{8.16}$$

$$\theta_{\text{Gra}-002} = \theta^{\text{Exp}}_{\text{Gra}-002} - \left(\theta^{\text{Exp}}_{\text{Si}-111} - \theta^{\text{Stan.}}_{\text{Si}-111}\right); \quad P = 1 - g \tag{8.17}$$

The recommended XRD experimental conditions are: DS = SS = 1°, RS = 0.30 mm, step width 0.02°, 2θ continuous scanning, range = 25.5–29.5°, the scanning speed is 10°/min.

Table 8.11 Raw data for determination of graphitization degree from four samples 2θ (°)

Graphite	Correct	002		100	101		004				110	102
Si			111			220		311	400	331		
Si_{stand}			28.442			47.302		56.122	49.130	76.376		
1024	Before	26.328	28.322									
	After	26.492	28.442									
	Before	26.478	28.412	42.373	44.477	47.273	54.570	56.096	69.097	76.353	77.465	83.561
	After	26.508	28.442	42.403	44.507	47.302	54.594	56.122	69.130	76.375	77.488	83.584
1108	Before	26.476	28.384									
	After	26.525	28.442									
	Before	26.419	28.329	42.281	44.410	47.196	54.545	56.016	69.062	76.262	77.402	83.491
	After	26.532	28.442	42.387	44.516	47.302	54.651	56.122	69.130	76.376	77.515	83.591
1111	Before	26.456	28.367									
	After	26.531	28.442									
	Before	26.438	28.339	42.286	44.421	47.223	54.549	56.035	69.046	76.294	77.397	83.555
	After	26.541	28.442	42.365	44.500	47.302	54.635	56.122	39.130	76.376	77.479	83.642
1207	Before	26.376	28.303									
	After	26.515	28.442									
	Before	26.439	28.363	42.310	44.451	47.230	54.553	56.051	69.042	76.285	77.417	83.499
	After	26.518	28.442	42.382	44.523	47.302	54.623	56.122	69.130	76.376	77.508	83.580

Note DS = SS = 1°, RS = 0.30 mm, 2θ = 25–30° 1°/min step size = 0.02°; DS = SS = 1°, RS = 0.30 mm, 2θ = 20–110° 4°/min step size = 0.02°

Table 8.12 Results of four sample calculations and curve readings *g*

No.	d_{002}			Whole spectrum analysis								
	Method			d_{002}			Extrapolation			Least square method		
	10.10	10.12	Curve	10.10	10.12	Curve	10.10	10.12	Curve	10.10	10.12	Curve
1024	0.910	0.911	0.707			0.748	0.967	0.970	0.831	0.948	0.950	0.778
1108	0.958	0.962	0.808	0.969	0.971	0.830	1.010	1.023		0.964	0.966	0.808
1111	0.966	0.969	0.823	0.981	0.985	0.868	0.891	0.891	0.667	1.061	1.067	
1207	0.943	0.945	0.778	0.949	0.951	0.775	1.035	1.039		1.005	1.008	0.685

Table 8.13 Comparison of determination results *P* of d_{002} method, θ_{002} method, and whole spectrum fitting method

Sample number	d_{002}	d_{002} method calculation	θ_{002} method calculation	d_{002} (curve)	Whole spectrum fitting, P_{FWHM}
SOFMP-051024	3.3671	0.152	0.153	0.239	0.106
SOFMP-051108	3.3876	0.391	0.393	0.192	0.068
SOFMP-051111	3.3569	0.033	0.031	0.177	0.065
SNHT-051207	3.3589	0.057	0.055	0.222	0.115
KS-44	3.3565	0.029	0.026	0.181	0.121
KC-44	3.3552	0.017	0.011	0.111	0.079
FSN-060401	3.3550	0.012	0.009	0.105	0.101
FSN-060402	3.3556	0.019	0.016	0.138	0.049
SCB-060401	3.3667	0.148	0.147	0.384	0.312
SOFMP-051124	3.3578	0.044	0.042	0.201	0.086
SOFMP-6D-A-2	3.3571	0.036	0.034	0.193	0.079
SOFMP-5D-A-1	3.3591	0.059	0.057	0.236	0.097
SOFMP-5D-A-2	3.3581	0.048	0.046	0.215	0.089
SOFMP-8D	3.3587	0.055	0.053	0.233	0.070

8.5 A New XRD Method for Determining the Stacking Disorder Degree of 2H-Graphite [9, 10]

8.5.1 *The Langford Method Improved by the Author*

Langford et al. [16] used Lorentzian approximation when studying hexagonal ZnO. It is concluded that the total line width β has the following relationship with microcrystal broadening β_C and stacking fault broadening β_P.

$$\beta = \beta_C + \beta_P \tag{8.18}$$

For isotropic nearly spherical microcrystals

$$\beta = \beta_C + A\cos\Phi_Z \tag{8.19}$$

$$\text{when}\begin{cases} h-k=3n \text{ or } l=0 & A=0 \\ h-k=3n\pm 1,\ l=\text{Even number} & A=3P/(2c) \\ h-k=3n\pm 1,\ l=\text{Odd number} & A=P/(2c) \end{cases} \tag{8.20}$$

where c is the lattice constant of the c-axis and Φ_Z is the angle between the diffraction plane and the hexagonal base plane (001). The Scherrer equation $\beta_C = \frac{0.89\lambda}{D\cos\theta}$ and formula (8.20) are substituted into (8.18) and multiplied by the $\frac{\cos\theta}{\lambda}$, obtain

$$\begin{matrix} h-k=3n\pm 1 & l=\text{even} \\ & l=\text{odd} \end{matrix}\begin{cases} \dfrac{\beta\cos\theta}{\lambda} = \dfrac{0.89}{D} + \dfrac{\cos\theta}{2c}\cdot\dfrac{\cos\phi_z}{\lambda}\cdot 3P \\ \dfrac{\beta\cos\theta}{\lambda} = \dfrac{0.89}{D} + \dfrac{\cos\theta}{2c}\cdot\dfrac{\cos\phi_z}{\lambda}\cdot P \end{cases} \tag{8.21}$$

From the expressions (8.19) and (8.20), it can be seen that (1) the $h-k=3n$, $l=0$, the line has no stacking fault broadening effect; (2) the line with an $h-k=3n\pm 1$, $l=$ even number, the diffraction line is seriously broadened; while $l=$ odd number the diffraction line is smaller broadened, which is consistent with the experimental results described in Sect. 8.1.2.

Order

$$Y = \frac{\beta\cos\theta}{\lambda} \qquad a = \frac{0.89}{D}$$

$$X = \frac{\cos\phi_Z}{2c}\cdot\frac{\cos\theta}{\lambda} \quad l=\text{even } F=3P;\quad l=\text{odd } F=P \tag{8.22}$$

Rewrite (8.21)

$$Y = a + FX \tag{8.23}$$

The matrix form of the canonical equations of the least square is

$$\begin{pmatrix} n & \sum X \\ \sum X & \sum X^2 \end{pmatrix}\begin{pmatrix} a \\ F \end{pmatrix} = \begin{pmatrix} \sum Y \\ \sum XY \end{pmatrix} \tag{8.24}$$

The results of simultaneous solution are as follows:

$$\begin{cases} D = 0.89 \times \dfrac{n\sum X^2 - \left(\sum X\right)^2}{\sum X^2 \sum Y - \sum X \sum XY} \\ F = \dfrac{n\sum XY - \sum X \sum Y}{n\sum X^2 - \left(\sum X\right)^2} \end{cases} \tag{8.25}$$

8.5.2 Least Square Method for Solving P_{AB} and P_{ABC}, Respectively [9, 10]

According to Warren [17], we deduce and replace f_D and f_T with P_{AB} and P_{ABC}.

$$h - k = 3n \pm 1 \begin{cases} l = \text{Even} \quad \beta_P = \dfrac{2l}{\pi}\mathrm{tg}\theta(3p_{AB} + 3p_{ABC}) \\ l = \text{Odd} \quad \beta_P = \dfrac{2l}{\pi}\mathrm{tg}\theta(3p_{AB} + p_{ABC}) \end{cases} \tag{8.26}$$

The Scherrer equation and formula (8.26) are substituted into formula (8.18) and multiplied by $\frac{\cos\theta}{\lambda}$, and the result is obtained.

$$h - k = 3n \pm 1 \begin{cases} l = \text{Even} \quad \dfrac{\beta\cos\theta}{\lambda} = \dfrac{2l}{\pi}\left(\dfrac{d}{c}\right)^2 \dfrac{\sin\theta}{\lambda}(3p_{AB} + 3p_{ABC}) + \dfrac{0.89}{D} \\ l = \text{Odd} \quad \dfrac{\beta\cos\theta}{\lambda} = \dfrac{2l}{\pi}\left(\dfrac{d}{c}\right)^2 \dfrac{\sin\theta}{\lambda}(3p_{AB} + p_{ABC}) + \dfrac{0.89}{D} \end{cases} \tag{8.27}$$

$$\text{order } Y = \frac{\cos\theta}{\lambda} \qquad A = \frac{0.89}{D}$$

$$X = \frac{2l}{\pi}\left(\frac{d}{c}\right)^2 \frac{\sin\theta}{\lambda} \qquad P = \begin{cases} l = \text{Even} 3p_{AB} + 3p_{ABC} \\ l = \text{Odd } 3p_{AB} + p_{ABC} \end{cases} \tag{8.28}$$

Rewrite (8.27):

$$Y = A + PX \tag{8.29}$$

The derivation of equations similar to (8.24) to (8.25)

$$\begin{cases} D = 0.89 \times \dfrac{n\sum X^2 - (\sum X)^2}{\sum X^2 \sum Y - \sum X \sum XY} \\ P = \dfrac{n\sum XY - \sum X \sum Y}{n\sum X^2 - (\sum X)^2} \end{cases} \tag{8.30}$$

When $k - k = 3n \pm 1, l =$ even, obtain P_{even}, $P_{\mathrm{even}} = 3P_{\mathrm{AB}} + 3P_{\mathrm{ABC}}$; when $k - k = 3n \pm 1, l =$ odd, obtain D_{odd}, $P_{\mathrm{odd}} = 3P_{\mathrm{AB}} + P_{\mathrm{ABC}}$, so as long as the diffraction lines of $l =$ even and $l =$ odd n_{even}, $n_{\mathrm{odd}} \geq 2$, we can get D, P; D_{even}; P_{even}, D_{odd}, P_{odd}. Simultaneous

$$\begin{cases} P_{\mathrm{even}} = 3P_{\mathrm{AB}} + 3P_{\mathrm{ABC}} \\ P_{\mathrm{odd}} = 3P_{\mathrm{AB}} + P_{\mathrm{ABC}} \end{cases} \tag{8.31}$$

Get P_{AB} and P_{ABC}.

It can be seen from Fig. 8.1 and Tables 8.1 and 8.2 that in 2H-graphite, when $h - k = 3n \pm 1$, 1 = odd, there are 101 and 103, when 1 = even, there are only 102, While 022 (or 202) does not appear because of extinction or too weak intensity.

Sometimes it is difficult to obtain credible half-width (FWHM) of 102 and 103 lines when the stacking disorder is large. In this case, if the shape of the microcrystal is polyhedral or nearly equiaxed, and D_{002}, D_{100}, and D_{004} are roughly the same, then:

$$\overline{D} = \frac{D_{002} + D_{100} + D_{004}}{3} \tag{8.32}$$

Use the following formula

$$\frac{\beta_{101}\cos\theta_{101}}{\lambda} = \frac{0.89}{\overline{D}} + \frac{\cos\phi_{Z101}}{2c}\frac{\cos\theta_{101}}{\lambda}P \tag{8.33}$$

$$\frac{\beta_{101}\cos\theta_{101}}{\lambda} = \frac{0.89}{\overline{D}} + \frac{2}{\pi}\left(\frac{d_{101}}{c}\right)^2\frac{\sin\theta_{101}}{\lambda}(3P_{\mathrm{AB}} + P_{\mathrm{ABC}}) \tag{8.34}$$

$$\frac{\beta_{102}\cos\theta_{102}}{\lambda} = \frac{0.89}{\overline{D}} + \frac{4}{\pi}\left(\frac{d_{102}}{c}\right)^2\frac{\sin\theta_{102}}{\lambda}(3P_{\mathrm{AB}} + 3P_{\mathrm{ABC}}) \tag{8.35}$$

P, $(3P_{\mathrm{AB}} + P_{\mathrm{ABC}})$ and $(P_{\mathrm{AB}} + P_{\mathrm{ABC}})$ can be obtained by using (8.33–8.34) and (8.35), respectively.

8.5.3 Experimental Determination of Stacking Disorder Degree of 2H-Graphite

In order to determine the stacking disorder in 2H-graphite experimentally and compare with the results measured by Hang et al. [13] by whole spectrum fitting method, the X-ray diffraction patterns and data were obtained by using the experimental conditions required by copper target and Hang program, that is, DS = SS = 1/6°, RS = 0.3 mm, step 0.05°, timing counting (FT) mode with recording time of 0.9 s per step, 2θ scanning range of 10–120°. Three typical patterns of $2\theta = 20$–90° are shown in Fig. 8.15. The obvious differences are as follows: (1) The linear shape of SOFMP-8D is the best, the broadening effect is the smallest, and the linear shape of SCB-060401 is the worst, and the broadening is the most serious. 100 and 101 cannot be separated, and 102 and 103 have been submerged in the background, indicating that the sample has the highest degree of disorder and the lowest degree of graphitization, and there may be interference from 3R graphite 101 line; (2) the corresponding (002) peak shifts to a low angle (see Fig. 8.15b). The experimental conditions determined by d_{002} method are as follows: the sampling width of DS = SS = 1° is 0.02°, 1°/min, the scanning range of 2θ is 25–29°, and the sample is prepared according to graphite:Si = 2:1.

In order to study and determine in detail the stacking disorder of actual samples, Table 8.14 lists the original data of 8 samples and the necessary data for calculation. Among them, β (°) is obtained by deducting the background Kα_2 from the Jade7.0 program and obtained by Refine, which is obtained by the β and 2θ relation interpolation of standard silicon under the same scanning condition, which is called instrument broadening.

The analysis and calculation results of eight samples are listed in Table 8.15; for comparison, the results of Hang et al. [13]. Whole spectrum fitting and d_{002} straight-line calculation are also listed in Table 8.15.

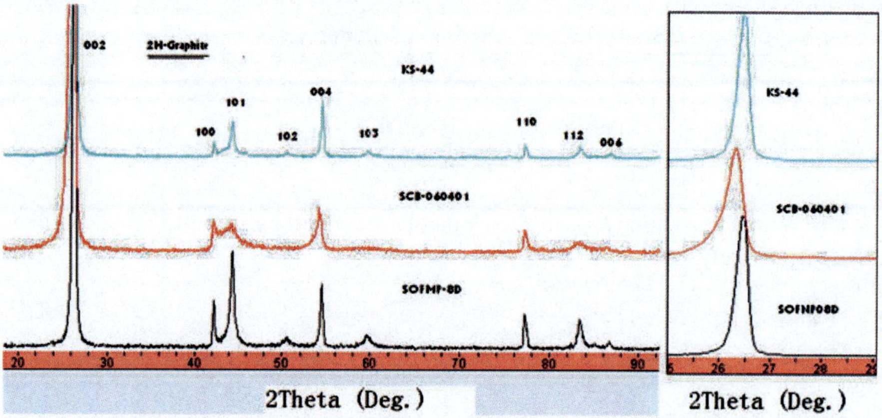

Fig. 8.15 X-ray diffraction flowers (left) and local magnification (right) of three typical 2H-graphite

Table 8.14 Original data of ten samples and relevant data necessary for calculation

hkl		002	100	101	102	004	103
d (Å)		3.3756	2.1386	2.0390	1.8073	1.6811	1.5478
2θ (°)		26.47	42.34	44.47	50.56	54.60	59.89
$\beta^0_{1/2}$ (°)		0.110	0.106	0.106	0.105	0.105	0.105
ϕ_z (°)		0.000	90.000	77.321	74.115	0.000	73.205
$\cos\theta$		0.9734	0.9325	0.9256	0.9042	0.8885	0.8665
$\cos\phi_z$				0.2195	0.2737		0.2889
$\frac{\cos\phi_z}{2C}\cdot\frac{\cos\theta}{\lambda}$ $(\times 10^{-3})$				9.8061	11.9447		12.0824
$\frac{2l}{\pi}\left(\frac{d}{c}\right)^2\frac{\sin\theta}{\lambda}$ $(\times 10^{-3})$				14.3776	28.3974		32.7863
		002	100	101	102	004	103
SCFMP-051024 No. 1	$\beta_{\frac{1}{2}}$ (°)	0.289	0.248	0.603	1.397	0.273	1.369
	D_{hkl} (Å)	450.9	593.3			467.6	
SCFMP-051108 No. 2	β (°)	0.242	0.239	0.444	1.296	0.233	0.903
	D_{hkl} (Å)	611.4	633.4			690.7	
SCFMP-051111 No. 3	$\beta_{\frac{1}{2}}$ (°)	0.242	0.218	0.384	0.812	0.240	0.891
	D_{hkl} (Å)	611.4	752.2			654.9	
SNHT-051207 No. 4	β (°)	0.267	0.214	0.572	1.303	0.311	
	D_{hkl} (Å)	514.0	780.0			429.2	
FSN-HE-060402 No. 5	β (°)	0.221	0.173	0.479		0.217	0.614
	D_{hkl} (Å)	727.1	1257.3			789.3	
SCFMP-8D No. 6	β (°)	0.248	0.245	0.477	1.072	0.252	1.06
	D_{hkl} (Å)	584.8	606.1			601.4	
FSN-HE-060401 No. 7	β (°)	0.274		0.395		0.263	0.390
	D_{hkl} (Å)	492.1				559.5	
KS-44 No. 8	β (°)	0.233	0.228	0.535		0.222	0.689
	D_{hkl} (Å)	656.1	690.5			755.6	

A careful comparison shows that:

(1) Except for a few samples, the calculated results of the improved Langford method P are basically consistent with those of P_{WSF}, such as No. 2, 3, 4, 6, 8. Only when the two diffraction peaks of 100 and 101 cannot be well separated, will there be a big difference.

(2) When the credible FWHM data of 101 and 102 can be obtained, according to the method proposed by the author, the result of $P_{AB} + P_{ABC}$ is also in good agreement with P_{WSF}, such as No. 1, 3, 4; but when there is only FWHM of 101 but not 102, only $3P_{AB} + P_{ABC}$ can be obtained, but the sum of disordered degree ($P_{AB} + P_{ABC}$) can not be obtained.

Table 8.15 Comparison of experimental data analysis results with P_{FWHM} and P_{d002}

Sample No.		$\overline{D}$ (nm)	Calculate by (8.29)	Calculate by (8.30) and (8.31)			FWHM method	d_{002} straight-line method
			P	P_{AB}	P_{ABC}	$P_{AB} + P_{ABC}$	P_{WSF}	P_{d002}
SCHMP-051024	1	50.39	0.0604	0.0821	0.0525	0.1346	0.106	0.091
SOFMP-051108	2	64.51	0.0736	0.1152	0.0118	0.1270	0.068	0.042
SCFMP-051111	3	67.28	0.0552	0.0478	0.0217	0.0695	0.065	0.033
SNHT-051207	4	57.44	0.1134	0.0728	0.0531	0.1259	0.115	0.099
			Calculate by (8.29)		Calculate by (8.30)			
			P		$3P_{AB} + P_{ABC}$			
FSN-HE-060402	5	92.46	0.1002		0.2051		0.049	0.019
SOFMP-8D	6	59.74	0.0816		0.1670		0.07	0.055
FSN-HE-060401	7	52.58	0.0458		0.0935		0.101	0.012
KS-44	8	70.07	0.1097		0.2245		0.121	0.029

(3) The calculated P, $P_{AB} + P_{ABC}$, and P_{WSF} can all reveal the difference of stacking disorder among different samples; in other words, the trends of the results of the three methods are consistent.

(4) Although the d_{002} method is simple, the result is not very reliable, especially when using linear relationship.

8.6 Discussion on the Method of Determining Graphitization Degree and Stacking Disorder Degree of 2H-Graphite

The main points of several methods for analyzing and determining the stacking disorder of hexagonal graphite are summarized in Table 8.16. It can be seen that each method has its own characteristics, but from the point of view of experimental procedures and data processing, the two methods proposed in this paper are much simpler than the full-spectrum fitting method used by Hang et al. [13], and the results are basically consistent with the full-spectrum fitting method. The d_{002} straight-line method is the simplest, and the result is not very reliable, but it is acceptable as a relative daily measurement.

Based on the above comparison, the following conclusions can be drawn:

(1) Although the hexagonal graphite structure is not dense hexagonal, its stacking square (ABAB ...) along the normal direction of the hexagonal grid plane is the same as the dense hexagonal. The selective broadening effect of stacking

Table 8.16 Comparison of several methods for analyzing and determining the stacking disorder of hexagonal graphite

Compare items	Improved Langford method	$(P_{AB} + P_{ABC})$ method	P_{WSF} method	D_{002} straight-line method
Principle	Based on the double broadening effect of stacking fault and microcrystal-stacking fault	Based on AB and ABC stacking model double broadening effect of microcrystal-stacking fault	Based on ABAB two-layer and A and B single-layer stacking model	Empirical approximate linear relationship based on stacking disorder degree and d_{002}
Experimental requirements	No special requirements, $\theta/2\theta$ continuous scan, 2°/min. The scanning range 2θ is 20–65°	No special requirements, $\theta/2\theta$ continuous scan, 2°/min. The scanning range 2θ is 20–65°	FT mode. The scanning range 2θ is 10–120° or 20–110°	Need to add silicon standard sample. The scanning range 2θ is 25–30°
Calculating method	Only the FWHM of 002, 100, 101, 004, and 103 need to $\overline{D}$ be solved by the substitution formula (8.28) from the FWHM of 002, 100, and 004	It only takes the data of 002, 100, 101, 102, and 004 to $\overline{D}$ obtain the simultaneous solution of the substitution formulas (8.30) and (8.31) from the FWHM of 002, 100, and 004	Using the whole spectrum fitting program compiled by Hang	$P_{002} = \frac{d_{002}-3.354}{3.440-3.354}$ $g = -38.3007 + 2.96030\theta$ $P_{002} = 1 - g$
Characteristics	(1) The double broadening effect of microcrystal and disorder is clearly considered based on the model (2) The experiment is simple (3) The physical meaning of the calculation process is clear and simple	(1) Based on the double broadening effect of microcrystal-stacking disorder (2) The experiment is simple (3) The physical meaning of the calculation process is clear and simple, and the two kinds of stacking disorder can be distinguished	(1) Although it is clear based on the model, it is based on the whole pattern (2) The requirement of the experiment is higher (3) The physical meaning of the process is not clear when calculated by a special program	(1) Based on the simple empirical relationship, the broadening effect of stacking disorder is ignored (2) The experiment is simple and the time is the shortest, but the standard sample is needed (3) The calculation is the simplest

(continued)

Table 8.16 (continued)

Compare items	Improved Langford method	($P_{AB} + P_{ABC}$) method	P_{WSF} method	D_{002} straight-line method
Comprehensive evaluation	The method is simple, and the result is reliable	The method is simpler, and the result is reliable	The method is troublesome, and the result is credible	The method is the simplest, and the result is not credible

disorder on XRD pattern is the same as that of dense hexagonal stacking disorder, that is, the broadening effect of $h - k = 3n$ and $hk0$, stacking disorder is the same as that of dense hexagonal stacking disorder, and the even broadening of $h - k = 3n \pm 1$ is serious, and the even broadening of $l =$ odd is smaller.

(2) The Langford method for dealing with isotropic near-equiaxed crystal ZnO has been improved and successfully applied to the determination of 2H-graphite. The disorder degree of graphite is in good agreement with the result of whole spectrum fitting, but both the experiment and data processing are much more convenient and easy to popularize.

(3) Two stacking disorder models and data mathematical methods proposed by the author, as long as 101 and 102 credible ones can be obtained. Two kinds of disorder can be obtained from FWHM data, and the results of P_{AB} and P_{ABC}, $P_{AB} + P_{ABC}$ are in good agreement with the results of whole spectrum fitting.

(4) The following experimental conditions are recommended: DS = SS = 0.25°, step size = 0.02°, 2θ scanning speed 2°/min. The scanning range of 2θ is 20–65°, continuous scanning mode.

8.7 X-Ray Analysis of Several Carbon Materials Used in Chemical Power Industry

8.7.1 Negative Material for Supercapacitors—Active Carbon

The X-ray diffraction pattern of active carbon is shown in Fig. 8.16. (a) is the active carbon used for the negative electrode of supercapacitor using β-$Ni(OH)_2$ as the positive active material, and (b) is generally active carbon. The two very wide peaks in Fig. 8.16a are 26.60° and 43.42°, respectively, with a very wide half-width, which shows that they are typical amorphous carbon materials. In addition to the wide scattering peaks mentioned above, there are five sharp peaks of seemingly face-centered cubic carbon in Fig. 8.16a, b, and the positions of each diffraction peak are 43.842°, 51.237°, 75.857°, 92.478°, and 97.270°, respectively. Figure 8.16b is particularly obvious.

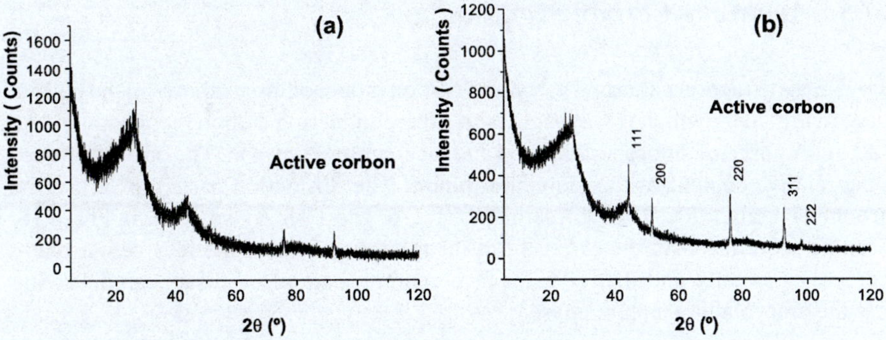

Fig. 8.16 X-ray diffraction pattern of activated carbon, **a** anode material, **b** general active carbon for supercapacitors

8.7.2 Superconductor Carbon Monosodium

The X-ray diffraction pattern of superconducting carbon monosodium is shown in Fig. 8.17. In addition to the above-broadened scattering pattern, its three peaks.

The sites are 26.88, 43.82, and 78.96°, corresponding to the scattering peaks of amorphous carbon. In addition, the sharp peaks of carbon with face-centered cubic structure appear, which can be indexed as 111, 200, 220, 311, 222, and the peaks are 43.821, 51.201, 75.779, 92.438, and 97.998°, respectively, and the linear shape is very good, indicating that the grain size reaches the order of μm.

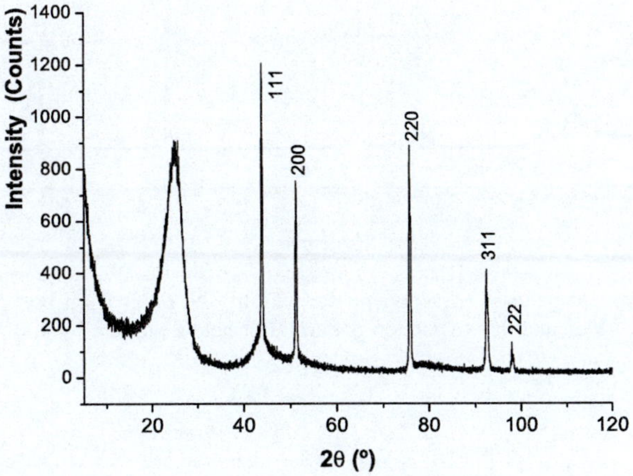

Fig. 8.17 X-ray diffraction pattern of superconducting carbon monosodium

8.7.3 Acetylene Carbon Monosodium

The X-ray diffraction pattern of acetylene carbon monosodium is shown in the bottom curve in Fig. 8.18, which is very similar to superconducting carbon monosodium and is also composed of amorphous carbon and face-centered carbon. The only difference is the relative content of face-to-center phase. The diffraction pattern of acetylene carbon black after 900 °C heat treatment for 2, 6, and 10 h is also given in Fig. 8.18. It can be seen that with the extension of heat treatment time, the face center phase decreases gradually, and after 10 h, the face-centered carbon has disappeared, leaving only the amorphous graphite phase.

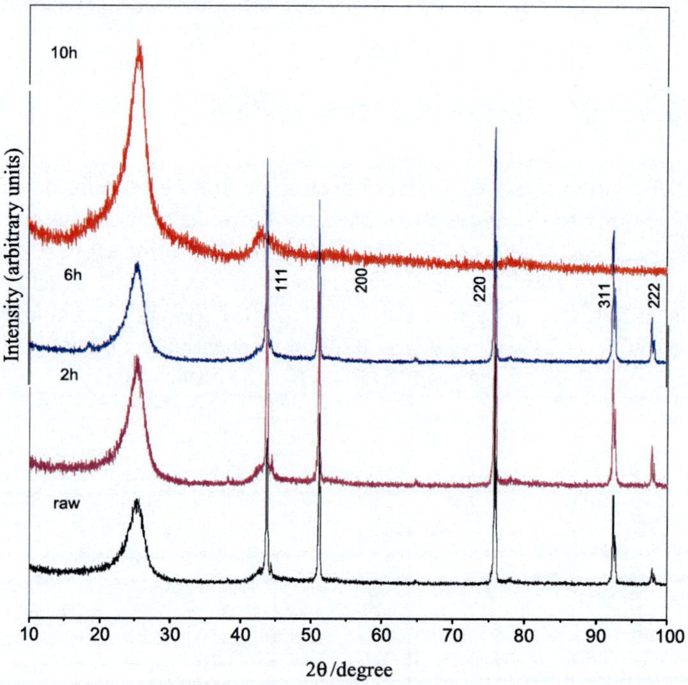

Fig. 8.18 X-ray element patterns of acetylene carbonium, the order from bottom to top is the original acetylene carbide and the diffraction pattern after heat treatment by 900 °C for 2, 6, and 10 h

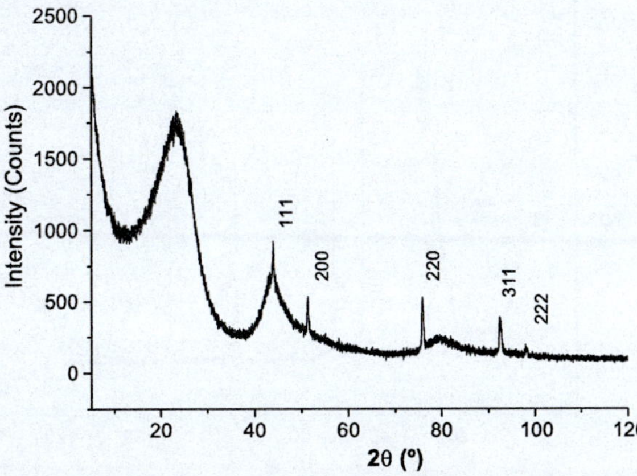

Fig. 8.19 X-ray diffraction pattern of hard carbon (Carbontron)

8.7.4 Hard Carbon

The X-ray diffraction pattern of hard carbon is shown in Fig. 8.19. It can be seen that the X-ray diffraction pattern of hard carbon is also similar to that of acetylene carbon and superconducting carbon. The only difference is that the amorphous graphite phase is the main phase, in other words, the content of the amorphous phase is much higher than that of the face-centered phase.

8.7.5 Carbon Nanotubes

The X-ray diffraction patterns of the so-called carbon nanotubes are shown in Fig. 8.20, and the corresponding peaks are 25.659, 43.806, 51.164, 75.800, 92.438, 97.996°. Using the search/matching results in the Jade program, Fig. 8.21 shows that the peak position is 25.659 corresponding to 002 of hexagonal graphite, which is asymmetric at the bottom of the face-centered 111 peak. The lines of the face-centered cubic phase are very sharp, indicating that the grain size is of the order of μm. Therefore, the nanowires in the material belong to hexagonal graphite.

8.7.6 Study on the Structure of Face-Centered Cubic Carbon

The face-centered cubic phase mentioned above is the result of visual observation, and its line distribution conforms to the law of its line distribution is consistent with the law of two densities and one looseness, which looks like a face-centered cubic structure. However, the careful determination of the peak positions of each

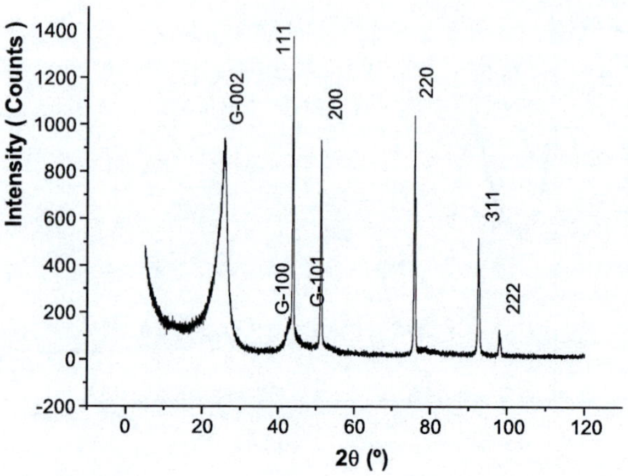

Fig. 8.20 Experimental result

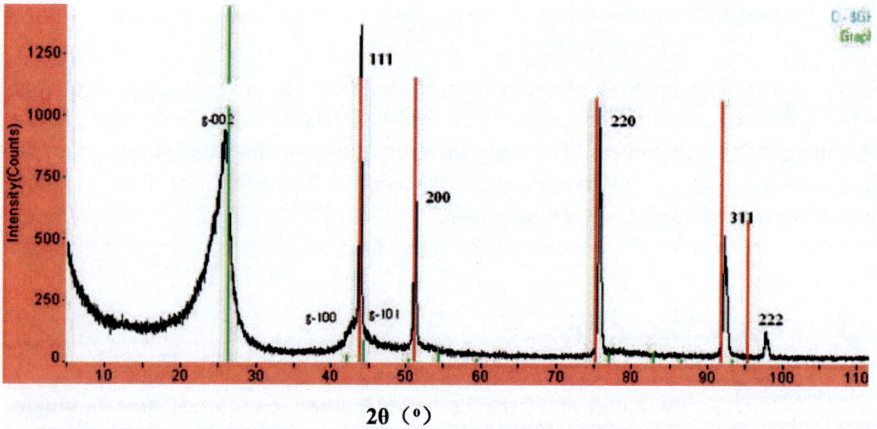

Fig. 8.21 Analysis result

diffraction peak is listed in Table 8.17. The "Pattern indexing" in the Jade program cannot be indexed by cubic, tetragonal, orthorhombic, and monoclinic systems. Two possibilities are obtained by hexagonal indexing: R3 (No. 148) and R32 (No. 155) space group. It can be seen that the indexing results are included in Table 8.17. It can be seen that it cannot be credible.

This kind of phase which looks like a face-centered cubic structure cannot be indexed by the cubic system, indicating that the displacement of each diffraction line is irregular. Then there are two possibilities:

1. There are deformation stacking faults in the face-centered cubic structure, which will shift 111 to the high angle and 200 to the low angle, while their second-order diffraction is the opposite, that is, 222 to the low angle and 400 to the high angle.

Table 8.17 Indexing results of space groups R3 (No. 148) and R32 (No. 155)

Experimental observation results		Calculation results of two hexagonal space groups	
2θ (°)	Relative strength	*hkl*	2θ (°)
43.821	100	101	43.863
		003	44.391
51.201	63.7	012	51.255
		110	75.041
75.779	79.7	104	75.800
		021	91.197
		113	91.564
92.438	35.0	015	92.299
		202	96.663
97.998	7.9	006	98.144
		024	119.774

2. There is the first kind of anisotropic residual stress in the face-centered cubic structure, which makes the displacement direction of each diffraction line irregular. Let's study this problem:

   (1) It may be thought that it is caused by measurement error. If the accurate lattice parameter a_0 can be obtained by extrapolation, then the diffraction angle of each crystal plane can be calculated by a_0 and then calculate $\Delta 2\theta$. The results are listed in the penultimate column of Table 8.18, all $\Delta 2\theta$ of which are more than 1°. It is obvious that it is not caused by the measurement error.

   (2) The peak positions of 220 and 311 are not affected by deformation stacking faults, so the average values $\overline{a}$ obtained by 220 and 311, when calculate the diffraction angle 2θ of each crystal plane, and compared with the actual measured value, that is, the calculation $\Delta 2\theta = 2\theta_{\text{exp.}} - 2\theta_{\text{cal.}}$ If it is negative, it indicates that the actual diffraction line is displaced to the low angle direction, otherwise the actual diffraction line shifts to the high angle direction, and the calculated results are listed in the most right side of Table 8.18. It can be seen that no matter 111, 200, or 222, the peak displacement caused by deformation stacking fault is not attached.

   (3) The average value of the lattice parameters calculated by each single line ($\overline{a} = 3.5528$ Å), and then the 2θ and $\Delta 2\theta = 2\theta_{\text{exp.}} - 2\theta_{\text{cal.}}$ of each crystal plane are calculated, and the results are listed in the penultimate column of Table 8.18.

The strain calculated according to the definition of strain $\varepsilon = (d_{\text{act.}} - d_0)/d_0$ is listed in Table 8.19. The positive sign is tensile strain, and the negative sign is compressive strain, which shows that the strain is anisotropic. It is worth mentioning

Table 8.18 Analysis of structural parameters of face-centered cubic carbon

	Experimental value		PDF 43-1104	FCC hkl	Single-line calculation of lattice parameters	Calculated on average = 3.5528 2θ (°)		Calculated according to cos 2θ extrapolation a_0 = 3.4946 2θ (°)		Average = 3.5429 to calculate 2θ (°)	
	2θ (°)	d (Å)	2θ (°)		a (Å)	2θ (°)	$\Delta 2\theta$	2θ (°)	$\Delta 2\theta$	2θ (°)	$\Delta 2\theta$
1	43.821	2.0642	43.914	111	3.5753	44.114	−0.293	44.888	−1.067	44.243	−0.422
2	51.201	1.7827	51.282	200	3.5674	51.395	−0.194	52.315	−1.004	51.549	−0.348
3	75.779	1.2542	75.367	220	5.5474	75.647	+0.132	77.136	−1.357	75.896	−0.117
4	92.438	1.0669	92.087	311	3.5385	91.956	+0.479	93.949	−1.511	92.288	+0.150
5	97.998	1.0206	95.570	222	3.5355	97.363	+0.635	99.558-	1.560	97.728	+0.270
$\bar{a}$ = 3.5528						Existence of anisotropic strain		Caused by non-measurement error		No stacking fault effect	

Table 8.19 Results of anisotropic strain measurement in face-centered cubic carbon

hkl	$d_{\text{exp.}}$	$d_{\text{cal.}}$	$d_{\text{exp.}} - d_{\text{cal.}}$	$\varepsilon = (d_{\text{exp.}} - d_{\text{cal.}})/d_{\text{cal.}}$ (10^{-3})
111	2.0642	2.0512	+0.0130	+6.338
200	1.7827	1.7764	+0.0063	+3.546
220	1.2542	1.2561	−0.0019	−1.513
311	1.0669	1.0712	−0.0043	−4.014
222	1.0206	1.0256	−0.0050	−4.875

that 111 and 222 are parallel crystal planes, and their normal strains should be the same. The data in Table 8.19 show that there is little difference in the absolute value of strain, but the direction of strain is opposite, which is unreasonable and may be caused by the error caused by the weak diffraction intensity of 222.

Based on the above discussion, the between the crystal planes does not conform to the rule of face-centered cubic. The so-called face-centered cubic carbon can be called distorted face-centered cubic carbon because of its anisotropic strain.

References

1. Invention patent. Preparation of artificial graphite carbon anode material and manufactured artificial graphite carbon anode material. 上海杉杉科技有限公司, 旺剑桥 洪苏宁 刘芳 于晓阳 于宪 脏志请, 公开号: CN1927707, 公开日: 2007.3.14 shimo.
2. Invention patent. Preparation method of artificial graphite anode material for lithium-ion power battery. 天津市贝特瑞新能源材料有限责任公司, 贾永平, 计同顺, 王成扬, 丁艳青, 代建国, 江 斌, 申请号: 200610014878, 2008.10.28.
3. Invention patent. Composite graphite anode materials and other preparation methods for lithium-ion secondary batteries. 深圳市贝特瑞电子材料有限公司, 岳敏 王桂林, 申请号: CN200510034329, 公开号: CN1702892, 公开日: 2005.11.30.
4. Invention patent. An enhanced preparation method of tin-graphite composite anode material for lithium-ion battery. 上海杉杉科技有限公司, 宁波杉杉新材料科技有限公司, 黄瑞安, 张殿浩, 吴敏昌, 申请号: 200810203232, 公开日: 2010.6.16.
5. Liu, F., X.G. Lu, and C.Z. Yang. 2011. Study on orientation ratio of graphite electrode for lithium-ion battery. *Battery Technology*.
6. Norio, I., and I. Michio. 1993. Relations between structural parameters obtained X-ray powder diffraction of various carbon materials. *Carbon* 31 (7): 1107–1113.
7. Li, C.J., B.X. Ma, and X.X. He. 1999. Characterization of graphitization degree of carbon/carbon composite material (I). *New Carbon Material* 14 (1): 19–25.
8. 稻坦稻夫, 白石念. 1991. 石墨化度的评价. 炭素 5 (118): 165–175 (in Japanese).
9. Li, H., C.Z. Yang, and F. Liu. 2008. A new method of X-ray diffraction method for determining the stacking disorder of hexagonal graphite. *Chinese Science, Series B Chemistry* 28 (9): 755–760.
10. Li, Hui, Chuanzheng Yang, and Fang Liu. 2009. Novel method of determing stacking disorder degree in hexagonal graphite by X-ray diffraction. *Science in China Series B: Chemistry* 52 (2): 174–180.
11. Li, H., C.Z. Yang, and F. Liu. 2009. Determination of graphitization degree and disorder degree of carbon electrode materials by X-ray diffraction. *Journal of Testing Technology* 23 (2): 161–167.

12. Don, J., X.F. Li, Lei Yan, and W.C. Shen. 1998. Research progress of carbon electrode materials for lithium ion batteries. *New Carbon Material* 13 (3): 55–62.
13. Hang, Shi, J.N. Reimers, and J.R. Dahn. 1993. Structure-refinement program for disordered carbons. *Journal of Applied Crystallography* 26: 827–836.
14. Ergun, S. 1976. Physics and chemistry of carbon. *Carbon* 14: 139–147. Edited by P.L. Walken, vol. 3, 211. New York: Marcel Deuber.
15. Drits, V.A., and C. Tchoubar. 1991. *X-ray diffraction by disordered lamellar structure*. Springer-Verlag.
16. Langford, J.I., A. Bouitif, et al. 1993. The use of pattern decomposition to study the combined X-ray diffraction effects of crystallite size and stacking faults in ex-oxalate zinc oxide. *Journal of Applied Crystallography* 26 (1): 22–32.
17. Warren, B.E. 1969. *X-ray diffraction*, 275–313. London: Addison-Wesley.

Chapter 9
Solid Electrolyte Interface Film on Graphite Surface of Li-Ion Battery

At present, there are mainly two kinds of negative active materials for lithium-ion batteries: lithium and hexagonal graphite, and $Li_4Ti_5O_{12}$ is newly developed. The former is mostly used in button batteries, while commercial lithium-ion batteries generally use hexagonal graphite. Therefore, the charge and discharge process of battery must involve the repeated intercalation of lithium ion in lithium and graphite electrodes, and the repeated intercalation and de-intercalation of lithium ion in lithium and graphite electrodes will inevitably involve the migration process of lithium ion in the surface film of lithium and graphite electrodes. Therefore, the properties of the electrode surface film directly determine the intercalation kinetics of lithium ion in lithium and graphite electrodes and the interface stability of the electrolyte, so it fundamentally determines the performance indexes of the battery, such as the first irreversible capacity, cycle life, storage performance, safety, and stability. Therefore, the in-depth study of the formation mechanism, chemical composition, structure, storage stability, and influencing factors of the electrode surface film of negative active materials has become one of the hotspots in the current lithium-ion battery research. The full name of the electrode surface film is the solid electrolyte interphase (interface) film, which is generally referred to as SEI film. In this chapter, we will introduce the SEI film on the surface of graphite electrode.

9.1 Overview of SEI Film on Graphite Surface [1]

As a kind of carbon element, graphite has very stable chemical properties and can exist in nature. Because of its low price and excellent performance, it has been widely used as a negative active material for commercial lithium-ion batteries.

In 1955, Hérold, a Frenchman, first discovered that graphite and metal lithium can form a series of lithium-graphite intercalated compounds (for short Li-GICs) [2] through common chemical reactions. When the amount of lithium embedded in graphite reaches the maximum, Li-GICs can be written as a compound with a standard

C. Yang et al., *Materials and Working Mechanisms of Secondary Batteries*,
https://doi.org/10.1007/978-981-19-5955-4_9

chemical formula LiC_6. The chemical activity of saturated lithium intercalation phase LiC_6 is similar to that of lithium metal, which can be obtained by direct reaction of graphite and molten lithium in inert atmosphere above 350 °C [2]. After discovering that graphite has lithium intercalation activity, people soon realized the potential application value of graphite as a negative lithium intercalation active material for lithium-ion batteries.

First, graphite has a very low electrochemical lithium intercalation potential of about 0.25 V versus Li/Li^+.

Second, the molar volume of graphite lithium intercalation phase and de-lithium phase is relatively close, and the difference is less than 10%.

Third, graphite has a wide range of sources, low price, and environmentally friendly.

Unfortunately, for a long time, people have not been able to find a suitable electrolyte to realize the reversible electrochemical intercalation of lithium in graphite. It was not until 1990 that Dahn et al. [3] found for the first time that lithium ion could be reversibly intercalated in graphite electrode in $LiAsF_6$'s mixed electrolyte of propylene carbonate (PC) and ethylene carbonate (EC). At the same time, they found that the graphite electrode used will be accompanied by the first irreversible capacity loss during the charge and discharge process, and the irreversible capacity loss is proportional to the surface area of the graphite electrode used. Therefore, they believe that a uniform protective film is likely to be formed on the surface of the graphite electrode during the first lithium intercalation. Because the lithium intercalation voltage of graphite electrode is very close to that of lithium metal electrode, the protective film formed on the surface of graphite electrode and the SEI film formed on the surface of lithium metal in the same electrolyte should have the same chemical composition. This conclusion has been confirmed by relevant experiments. It can be seen that the protective film on the surface of graphite electrode should also be the conductor of lithium ion and the insulator of electrons, so Dahn et al. [3] also named the protective film on the surface of graphite electrode as solid electrolyte mesophase (interface) film, that is, SEI film.

Before Dahn realized reversible electrochemical lithium intercalation of graphite electrode with mixed electrolyte based on PC and EC for the first time, people basically used non-aqueous electrolyte based on single PC to study the electrochemical lithium intercalation behavior of graphite electrode, but no valuable experimental results were obtained. The reason is that the reduction products of PC cannot form an effective SEI film on the surface of the graphite electrode, thus preventing the graphite layer peeling off (graphite exfoliation) caused by the co-embedding of PC and lithium ion in the graphite layer. As a homologue of PC, EC has a relatively high melting point (PC melting point is −48.8 °C, while EC melting point is 36.4 °C), so it rarely attracted attention in the early stage of lithium-ion battery research. However, since the discovery of the magical film-forming function of EC, EC has become an indispensable combination solvent for commercial lithium-ion battery electrolyte. Thus, it can be seen that the formation of an effective SEI film on the surface of graphite electrode plays a decisive role in the reversible electrochemical intercalation and de-lithium of graphite.

9.2 Formation Mechanism of SEI Film on Graphite Surface

Since it was confirmed that there is a SEI film on the graphite surface, a variety of models have been proposed to explain the formation process of SEI film on the graphite surface, among which the more famous models are Peled model and Besenhard model.

The Peled model was originally proposed by Peled to explain the formation process of SEI film on the surface of lithium metal [4]. Later, it was borrowed and improved by scholars to explain the formation process of SEI film on the surface of graphite [5]. However, strictly speaking, the formation process of SEI film on the surface of graphite is not exactly the same as that on the surface of lithium metal. The establishment of SEI film on the surface of lithium metal can be accomplished by placing lithium metal in the corresponding electrolyte, while the establishment of SEI film on the surface of graphite must polarize the graphite cathode immersed in the electrolyte to a sufficiently low voltage range. However, in view of the fact that the SEI film with the same chemical composition can be formed on the surface of graphite and lithium metal in the same electrolyte, some scholars have adopted the main contents of the Peled model which is used to describe the formation process of SEI film on the surface of lithium metal. It is considered that the formation process of SEI film on graphite surface can be divided into the following steps [5]:

(1) The organic solvent molecules in the electrolyte obtain electrons on the graphite electrode, undergo a two-dimensional electrochemical interfacial electron transfer reaction, and form an intermediate product, the so-called anion free radical (radical anion).
(2) The anion free radicals undergo a series of common chemical decomposition reactions in the electrolyte close to the graphite surface, and the final reduction product of the electrolyte is formed.
(3) Some insoluble substances in the reduction products of the electrolyte precipitate in the electrolyte close to the graphite surface and finally form a SEI film coated on the graphite surface.

In the original paper revealing the existence of SEI film on the surface of graphite, Dahn et al. used the Peled model to explain the formation process of SEI film on the surface of graphite [3]. In addition, based on the Peled model and by comparing the interaction forces between the reduction products and the graphite electrode, Ein-Eli successfully explained why the reduction products of EC can form an effective SEI film on the graphite surface, while the reduction products of PC cannot form an effective SEI film on the graphite surface. The existence of anion free radicals, which is the key intermediate in Peled model, has been confirmed by Endo et al. with the help of electron spin resonance (ESR). In addition, Wang et al. studied the reduction mechanism of EC in detail through the first-principles calculation method based on density functional theory, and the results also proved that EC molecules were indeed reduced to anion free radicals at first, and then the anion radicals were broken down to form the final reduction product.

Although the Peled model has its success in explaining the formation mechanism of SEI film on graphite surface, it still has many deficiencies. Besenhard et al. [6] measured the real-time change of the volume expansion of graphite electrode in the process of charge and discharge by accurate dilatometer (dilatometer). The results show that the end face of graphite, that is, the plane through which lithium ion is intercalated into graphite layer, has expanded by 150% at a certain lithium intercalation potential. Obviously, such a large amount of expansion can never be caused by the intercalation of lithium ions into the graphite layers alone. Therefore, they proposed another model that is different from the Peled model, which is now called the Besenhard model:

(1) Because the Van der Waals force connecting the graphite interlayer is very weak, the organic solvent molecules in the solvation layer in the electrolyte are easily intercalated with lithium ions into the graphite interlayer to form a ternary graphite intercalation compound $Li(Sol)_xC_y$.
(2) The ternary graphite-embedded compound takes place in continuous reduction and decomposition reaction in the graphite layer near the graphite surface or in the electrolyte and finally forms the reduction product of the electrolyte.
(3) Some insoluble substances in the reduction products of the electrolyte precipitate in the graphite layer near the graphite surface or in the electrolyte and finally form a SEI film which not only penetrates the graphite interlayer but also covers the graphite surface.

The biggest difference between the Besenhard model and the Peled model is that the former believes that the solvent molecules will be co-embedded in the graphite layer together with lithium ions. In order to verify the above conclusion, Chung et al. [7] systematically studied the electrochemical behavior of graphite in the electrolyte containing two different organic solvents. The results are shown in Fig. 9.1. As can be seen from the figure, graphite can be reversibly electrochemically intercalated and stripped of lithium in the electrolyte containing super-butylene carbonate (referred to as t-BC), but not in the electrolyte containing cis-butyl carbonate (abbreviated as c-BC). t-BC and c-BC belong to isomerism. In the t-BC molecule, the two methyl groups are located on both sides of the carbonate ion plane; in the c-BC molecule, the two methyl groups are located on the same side of the carbonate ion plane. According to the Peled model, although t-BC and c-BC have different molecular configurations, the reduction products formed by electrochemical reduction on the graphite electrode are the same. Therefore, the SEI film formed on the graphite surface in the two electrolytes containing t-BC and c-BC should have the same chemical composition. Therefore, the graphite electrode should show similar electrochemical lithium intercalation behavior in the above two electrolytes, but the experimental facts are completely opposite, which indicates that the Peled model needs to be improved. However, according to the Besenhard model, this differential behavior can be well explained. Because the two methyl groups in c-BC are located on the same side of the carbonate ion plane, the molecular volume of PC is almost the same as that of monomethyl. Therefore, c-BC is very easy to intercalate graphite layer with lithium ion like PC and cause graphite exfoliation. However, because the two methyl groups in t-BC are located on both sides of the carbonate ion plane, their molecular volume is larger than that of PC. Therefore, t-BC is not as easy as PC to intercalate into the interlayer of graphite with lithium ion and cause graphite exfoliation. Therefore, graphite can electrochemically intercalate lithium reversibly in the electrolyte containing t-BC, but not in the electrolyte containing c-BC. Figure 9.2 illustrates this explanation [7].

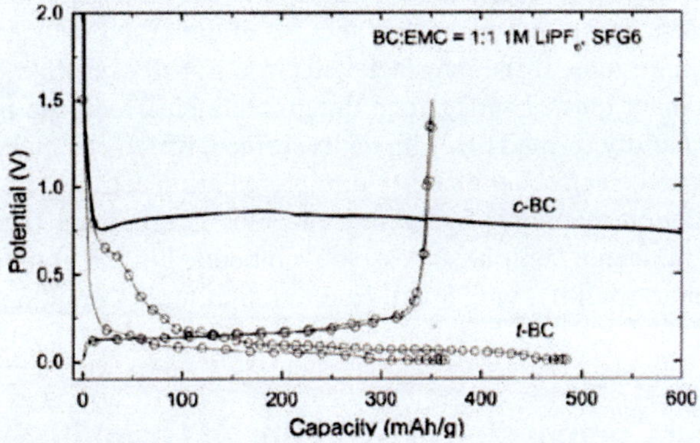

Fig. 9.1 First charge–discharge electrogram of graphite in 1 M $LiPF_6$ + t-BC + EMC and 1 M $LiPF_6$ + c-BC + EMC [7]

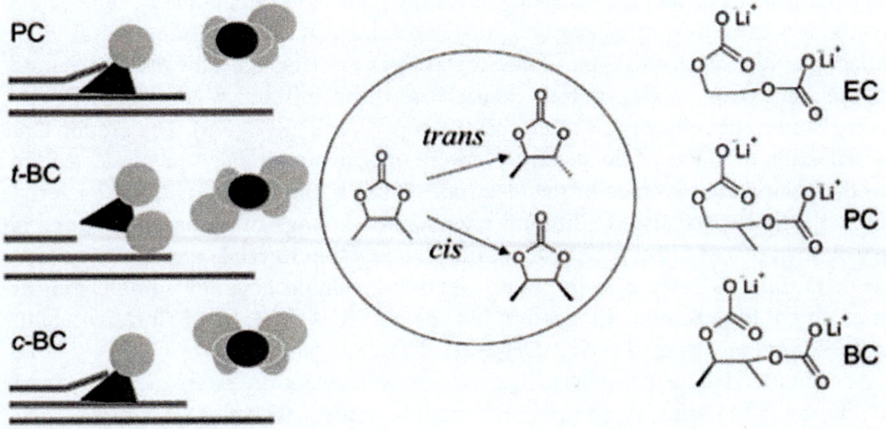

Fig. 9.2 Comparison of the relationship between PC, t-BC, and c-BC molecules and lithium-ion co-embedded graphite layers [7]

Besenhard model has been recognized by many scholars as soon as it is proposed. However, the ternary graphite-embedded compound $Li(Sol)_xC_y$ mentioned in this model cannot be fully confirmed by experiments so far.

To sum up, although both Peled model and Besenhard model can explain some experimental facts, neither of them can fully explain the formation mechanism of SEI film on graphite surface alone. For this reason, Aurbach et al. [8] considered comprehensively that the solvent molecules in the electrolyte may not only form anion free radicals by two-dimensional interfacial electron transfer reaction on the graphite electrode surface, but also form ternary graphite intercalated compounds together with lithium ions. If the interfacial electron transfer reaction of the solvent molecule is dominant (such as EC), then the graphite electrode can be completely protected by the rapidly formed SEI film on its surface, which can make the graphite electrode show good electrochemical lithium intercalation activity. On the contrary, if the co-intercalation reaction of solvent molecules is dominant (such as DME), then the solvent molecules will be excessively embedded in the graphite interlayer, resulting in graphite spalling, which will lead to the failure of graphite electrodes.

9.3 Chemical Composition of SEI Film on Graphite Surface

Although there are different views on the formation mechanism of SEI film on graphite, it is agreed that the chemical composition of SEI film comes from the reduction product of electrolyte [9]. The reduction product of the electrolyte is obviously related to the type of electrolyte used, but is it related to the material of the electrode used? For this reason, Aurbach and Cohen [9] systematically studied the distribution of reduction products of different electrolytes on lithium metal electrodes, graphite electrodes, and noble metal inert electrodes by means of accurate surface analysis methods, such as Fourier transform infrared spectroscopy, X-ray photoelectron spectroscopy, Raman spectroscopy (FTIR, XPS, RS). They found that the reduction products of the electrolyte were only related to the type of electrolyte and the polarization voltage of the electrode. It has nothing to do with the material of the electrode. Because the lithium intercalation voltage of graphite electrode is very close to that of lithium metal electrode, the SEI film formed on graphite surface and metal lithium surface in the same electrolyte should have the same chemical composition. For example, in electrolyte $LiPF_6 + EC + DEC$, the SEI film on both surfaces is composed of $(LiOCO_2CH_2)_2$, Li_2CO_3, LiF, and so on.

As mentioned in the previous article, non-aqueous electrolytes are generally used in lithium-ion batteries, more precisely proton-repellent (aprotic) electrolytes, that is, electrolytes cannot contain hydrogen ions. This is because if both lithium ion and hydrogen ion exist in the electrolyte, the intercalation reaction of lithium ion obviously cannot compete with the reduction reaction of hydrogen ion, so that the lithium-ion battery cannot work properly. Generally speaking, non-aqueous electrolytes are mainly composed of polar hydrophobic organic solvents (such as EC, PC, DMC, EMC, DEC, MF, γ-BL, THF, DME, DEE) and peracid lithium salts (such

Table 9.1 Main insoluble reduction products of organic solvents and lithium salts [9]

Organic solvents or lithium salts	Main insoluble reduction products
EC	$LiOCO_2CH_2CH_2OCO_2Li$, Li_2CO_3
PC	$LiOCO_2CH(CH_3)CH_2OCO_2Li$, Li_2CO_3
DMC	CH_3OCO_2Li, CH_3OLi
EMC	CH_3OLi, CH_3OCO_2Li, CH_3CH_2OLi, $CH_3CH_2OCO_2Li$
DEC	$CH_3CH_2OCO_2Li$, CH_3CH_2OLi
MF	HCO_2Li, CH_3Li
γ-BL	$CH_3CH_2CH_2COOLi$
THF	$CH_3CH_2CH_2CH_2OLi$
DME	CH_3OLi
DEE	CH_3CH_2OLi, $LiOCH_2CH_2OLi$
$LiClO_4$	Li_2O, LiCl, $LiClO_3$, $LiClO_2$, LiClO
$LiAsF_6$	LiF, Li_xAsF_{3-x}
$LiBF_4$	LiF, Li_xBF_{3-x}
$LiPF_6$	LiF, Li_xPF_{5-x}, $Li_xPF_{3-x}O$
$LiSO_3CF_3$	Li_2SO_3, Li_2S, Li_2O, LiF
$LiN(SO_2CF_3)_2$	Li_3N, $Li_2S_2O_4$, LiF, Li_2S, Li_2SO_3, Li_2O

as $LiClO_4$, $LiAsF_6$, $LiBF_4$, $LiPF_6$, $LiSO_3CF_3$, $LiN(SO_2CF_3)_2$). The main insoluble reduction products of the above electrolyte components are summarized in Table 9.1 [9].

It can be seen from Table 9.1 that both organic solvents and lithium salts can produce insoluble reducing substances on the electrode, so both of them will contribute to the chemical composition of SEI films more or less. However, Endo et al. found that the main chemical components in SEI films come from organic solvents, especially organic cyclic carbonate solvents. For example, the chemical composition of the SEI film formed in the electrolyte containing EC is basically contributed by EC.

In addition, because it is impossible to prepare absolutely pure substances by the actual preparation process, non-aqueous electrolytes inevitably contain trace impurities, such as H_2O, O_2, CO_2, HF. The presence of these impurities will significantly reduce the organic composition in the SEI film. For example, the presence of H_2O will transform the organic components such as alkyl lithium carbonate (such as $LiOCO_2CH_2CH_2OCO_2Li$), alkyl alcohol lithium salt (such as ROLi), and alkyl lithium salt (such as CH_3Li) in the SEI film into Li_2CO_3; HF. The existence of the SEI film will transform the alkyl lithium carbonate salt, alkyl alcohol lithium salt, alkyl lithium salt, and some inorganic lithium salt (such as Li_2CO_3) in the SEI film into LiF. The decrease of organic composition will lead to the increase of lithium-ion transfer impedance in SEI film, which leads to the decrease of battery power performance. Therefore, we must strictly control the content of impurities in the electrolyte to avoid introducing harmful components into the SEI film.

9.4 Theoretical Study on the Formation Process of SEI Film on Graphite Surface

Yan et al. [1, 10–12] carried out the following theoretical studies:

(1) The reduction reaction mechanism of the chemical composition of the electrolyte was studied by using the first-principles calculation method based on density functional theory, in order to reveal the decisive factors of the chemical composition of the SEI film.
(2) The classical phase transition theory was used to study the precipitation mechanism of the reduction products of the electrolyte in the electrolyte near the surface of the graphite electrode. The determinants of the microstructure and morphology of SEI.
(3) The theory of material transport in porous films is used to study the storage process. The growth and evolution mechanism of SEI film on the surface of graphite electrode in order to reveal the reason of performance degradation of lithium-ion battery after storage. Points (1) and (2) are introduced in this section, and Point (3) is described in Sect. 9.7.2.

Figure 9.3 schematically shows the whole formation process of SEI film on the surface of graphite electrode. As can be seen from the figure, the formation process of the SEI film can be roughly divided into two stages: in the first stage, after the graphite electrode placed in the electrolyte is first cathodic polarized to a certain degree of low potential, some chemical substance already existing in the electrolyte, such as organic solvents, lithium salt anions, impurity molecules, will undergo a series of electrochemical or chemical reactions near the surface of the graphite electrode. The result is the production of new reduced chemical substance, so-called electrolyte reduction products. In the second stage, some insoluble substances in the reduction products are then separated from the electrolyte and precipitated near the surface of the graphite electrode, resulting in the formation of a solid film covering the surface of the graphite electrode, namely SEI film.

9.4.1 Reduction Mechanism and Chemical Composition

According to the first-principles calculation method, we have studied the reduction mechanism of electrolyte $LiPF_6 + EC + DMC$ and obtained the following important results:

(1) Compared with the same free-state solvent molecules, the coordinated solvent molecules in the electrolyte are easier to obtain electrons from the outside world. For example, in the electrolyte $LiPF_6 + EC + DMC$, the coordinated EC and DMC molecules are easier to obtain electrons from the outside world than the free EC and DMC molecules.

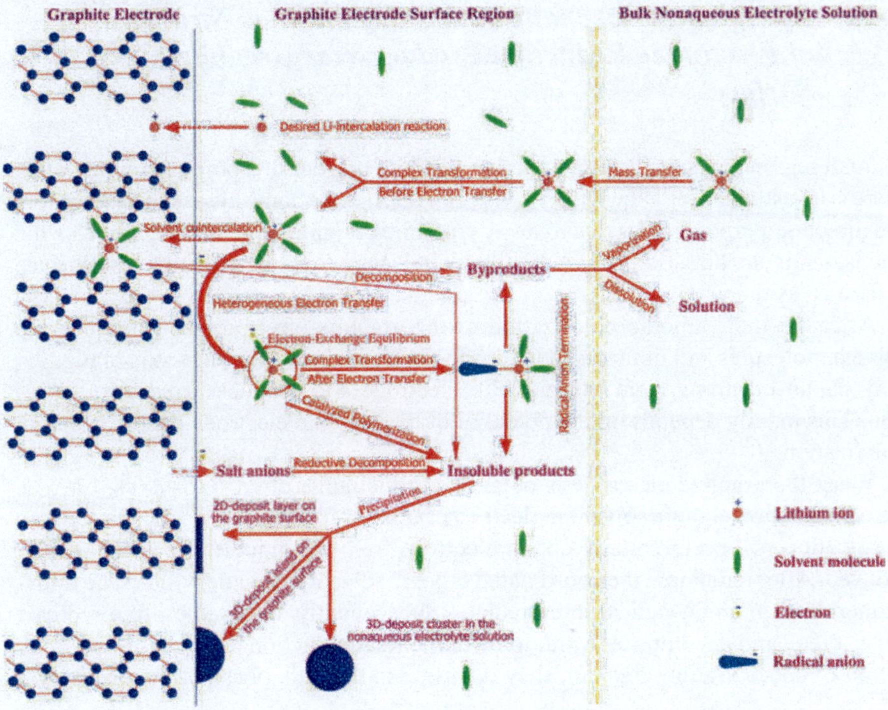

Fig. 9.3 Schematic diagram of the formation process of SEI film on the surface of graphite electrode [11]

(2) During the first cathodic polarization of the graphite electrode, a solid electrolyte interface film (SEI film) composed of the reduction products of the electrolyte will be formed on the surface of the graphite electrode. The chemical composition of SEI film mainly depends on the solvent molecules with strong solvation in the electrolyte. For example, in the electrolyte $LiPF_6 + EC + DMC$, because the solvation of EC is relatively strong, the SEI film formed on the surface of graphite electrode in the above electrolyte is mainly composed of reduction products of EC.

(3) The first-order coordination compound Li^+ is formed by lithium ion and EC molecule. (EC) can obtain an electron from the outside world. On the other hand, the anion free radicals that form the single-electron reduced state can obtain two electrons from the outside world to form the two-electron reduced anion free radicals. All these anion radicals can be formed not only by cleavage, but also by pairwise polymerization to form a series of equilibrium structures, including vinyl lithium dicarbonate $(CH_2OCO_2Li)_2$, butenyl lithium dicarbonate $(CH_2CH_2OCO_2Li)_2$, lithium carbonate Li_2CO_3, lithium carbon compound $(LiCH_2CH_2OCO_2Li)$, lithium alcohol compound $(LiOCH_2CH_2OCOCH_2CH_2OCO_2Li)$, and ethylene $(CH_2{=}CH_2)$.

9.4.2 Precipitation Mechanism and Organization Structure of Electrolyte Reduction Product Near Graphite Surface

A small region in a specific electroactive graphite powder in a porous graphite electrode is investigated, as shown in Fig. 9.4 [10–12]. For the convenience of description, the direction perpendicular to and away from the graphite end face is defined as the *x*-axis, while the direction perpendicular to and away from the graphite base surface is defined as the *y*-axis.

After the first cathodic polarization of the graphite electrode, both the organic solvent molecules and lithium salt anions in the electrolyte near the *y*-axis in Fig. 9.4 may obtain electrons from the graphite electrode and produce a reduction reaction. This mainly depends on the potential of the graphite electrode during cathodic polarization.

When the graphite electrode is polarized to the range of 1.0–0.8 V (vs. Li/Li^+), the coordination organic solvent molecules near the *y*-axis electrolyte that participate in solvation will preferentially obtain electrons from the graphite electrode and be reduced. After reduction, the coordinated organic solvent molecule will first be transformed into an anion radical intermediate. Subsequently, these anion free radicals either break and decompose in situ to form the final reduction product, or under the action of concentration gradient, they diffuse into the bulk phase of the electrolyte

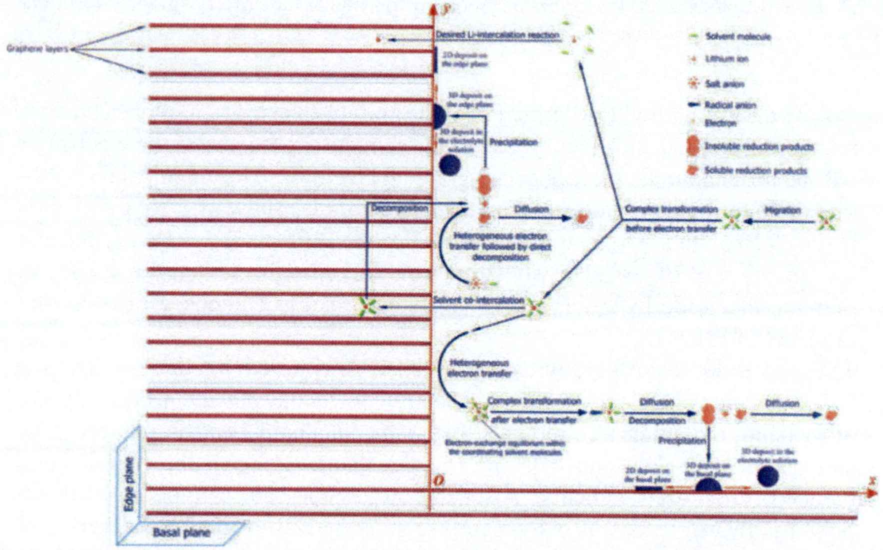

Fig. 9.4 Series of complex electrochemical or chemical reactions near the graphite surface during the first cathode and polarization

along the x-axis direction and then choose to break down to form the final reduction product. In addition, some scholars believe that the coordinated organic solvent molecules may also choose to cross the y-axis with lithium ions, thus co-embedded in the graphite layer to form a ternary graphite intercalated compound [$Li(Sol)_xC_y$]. Different from the above anion free radicals, $Li(Sol)_xC_y$ cannot move at will and can only be decomposed in situ to form the final reduction product.

When the graphite electrode is polarized to the range of 0.65–0.2 V (vs. Li/Li^+), the free organic solvent molecules and lithium salt anions that are not involved in the solvation will become like coordinated organic solvent molecules and can also be obtained from the graphite electrode. Electrons are reduced. However, it is worth noting that after obtaining electrons from the graphite electrode, lithium anions are not first transformed into an intermediate product like organic solvent molecules, but are directly reduced and decomposed into the final reduction product in the electrolyte near the y-axis.

When the graphite electrode is polarized below 0.20 V (vs. Li/Li^+), the intercalation reaction of lithium ion will become the main reaction near the surface of the graphite electrode. In addition, some impurities in the electrolyte, such as H_2O molecules, will corrode the low-reduced and unstable reduction products formed at high potential and eventually turn them all into relatively stable high reduction and stable reduction products.

To sum up, the reduction products formed by lithium salt anions can only be produced in the electrolyte close to the y-axis, while the reduction products formed by organic solvent molecules can not only be produced in the electrolyte close to the y-axis. It can also be produced in an electrolyte far away from the y-axis. In addition, the molar concentration of the reduction products formed by organic solvent molecules in the electrolyte decreases with the increase of the x-axis distance, because in the electrolyte closer to the y-axis, the number of anion free radicals that can participate in cleavage decomposition will obviously be a little more.

Based on the above results, the following important conclusions can be drawn [10–12]:

(1) The SEI film on the graphite electrode surface will inevitably show a multi-layer structure: that is, near the graphite electrode surface, the SEI film is basically composed of highly reduced and stable inorganic compounds, and its structure will be relatively tight, while near the electrolyte side, the SEI film is basically composed of low-reduced and unstable organic compounds, and its structure will be relatively loose.

(2) The SEI film on the graphite electrode substrate is mainly composed of the reduction products of organic solvent molecules, while the SEI film on the end surface of the graphite electrode is mainly composed of the reduction products of lithium salt anions, and the thickness of the SEI film on the end surface of the graphite electrode is larger than that of the SEI film on the base surface of the graphite electrode.

(3) Because the growth process of the solid cores that make up the SEI film on the graphite electrode surface is controlled by diffusion, the thickness of the SEI film increases with the increase of the square root of time.

9.5 Chemical Modification of SEI Film on Graphite Surface

Considering that the SEI film on the graphite surface is composed of insoluble reduction products of the electrolyte, many scholars have tried to add a certain chemical substance into the electrolyte in order to better modify the chemical composition of the SEI film, so that the SEI film on the graphite surface can meet the following requirements [13]:

(1) It has an electron transfer number close to zero to prevent the continuous electrochemical reduction of the electrolyte.
(2) It has high enough lithium-ion conductivity to make lithium ion quickly intercalate or de-intercalate from the graphite layer.
(3) It has smooth surface morphology and uniform chemical composition to achieve uniform current distribution on the electrode.
(4) It has good adhesion to the graphite surface to ensure the firm adhesion between the SEI film and the graphite surface.
(5) It has good resistance to mechanical strain to adapt to the volume change of graphite in the process of lithium intercalation and de-intercalate lithium.
(6) It has a chemical composition with low solubility to ensure that the SEI film is not continuously dissolved by the electrolyte.

Aurbach et al. [14] first started this tentative exploration. After adding a small amount of CO_2 to the electrolyte $LiAsF_6$ + dioxolane, they can make the graphite electrode which does not have any electrochemical lithium intercalation activity show good charge–discharge cycle performance. In addition, Ein-Eli et al. [15] found that the proper addition of SO_2 to the electrolyte $LiAsF_6$ + DMC can significantly improve the electrochemical lithium intercalation activity of the graphite electrode. This miraculous effect can be attributed to the active participation of CO_2 and SO_2 in the rapid formation of SEI films on the surface of graphite. Substances such as CO_2 and SO_2, which are used to modify the chemical composition of SEI films to significantly improve the electrochemical lithium intercalation activity of graphite electrodes, are generally referred to as SEI-forming agents in the field of lithium-ion battery research.

CO_2 and SO_2 are good film-forming agents in terms of effect, but they do not have much practical value. This is because the addition of gaseous CO_2 and SO_2 will not only increase the manufacturing cost of the battery, but also bring unnecessary safety risks to the battery. For this reason, a variety of liquid film-forming agents have been developed successively, such as catechol carbonate (CC) [16], ethylene sulfite (ES) [17], vinyl acetate (VA) [18], and vinylene carbonate (VC) [19].

For the optimization of film-forming agents, Morita et al. [20] summed up a semi-empirical rule through a large number of experiments: excellent film-forming agents should choose chemicals with high reduction voltage and fast electrochemical reaction speed, so as to ensure that before the main chemical components in the electrolyte participate in the reaction on a large scale, the film-forming agents can preferentially and quickly take place electrochemical reduction on the graphite electrode to form an effective SEI film.

The selection of CO_2, SO_2, CC, ES, and VA above obviously satisfies the semi-empirical rule of Morita. Because after the addition of the above film-forming agent, there will be an obvious high potential reduction peak on the cyclic voltammetry of the electrolyte, and this reduction peak corresponds to the preferential electrochemical reduction reaction of the film-forming agent on the graphite electrode. However, there is no obvious high potential reduction peak in the cyclic voltammogram of the electrolyte after adding VC [19]. This is because the reduction voltage of VC is almost the same as that of the main solvent (such as PC) in the electrolyte, so the reduction peaks of the two will overlap and become indistinguishable.

9.6 Organization Structure of SEI Film on Graphite Surface

Peled et al. [21] studied the transfer impedance of lithium ion in the SEI film on the graphite surface by electrochemical impedance spectroscopy (EIS). The results show that the impedance of the whole SEI film can be decomposed into four parts of RC components, some of which clearly correspond to the contact impedance between different particles. Accordingly, they believe that the SEI membrane is actually composed of solid particles of different chemical composition, thus forming a mosaic-like multi-layer solid structure, that is, the structure of the SEI membrane close to the electrode side is tight, while the structure of the SEI film close to the electrolyte side is loose. In order to further understand the chemical composition distribution of each layer of SEI film, Bar-Tow et al. [22] analyzed the chemical composition of SEI film layer by layer by X-ray photoelectron spectroscopy (XPS) and Ar^+ ion surface sputtering technique. They found that the SEI films near the electrode were mainly composed of inorganic lithium salts with high reduction degree, while the SEI films near the electrolyte side were mainly composed of organic lithium salts with low reduction degree and some organic small molecular polymers.

The formation process of SEI film on graphite surface will show obvious voltage change characteristics, that is, the structure of SEI film will change with the change of graphite cathode polarization voltage. Jeong et al. [23] used atomic force microscope (AFM) to in situ observe the change process of the surface morphology of graphite electrode during the first cathodic polarization. They found that when the graphite electrode was kept at the open circuit voltage (OCV), there was no change in its surface morphology; when the graphite electrode was polarized to the range of 1.0–0.8 V (vs. Li/Li^+), a small amount of star-shaped 1–2 nm high hill-like solid particles and 15–20 nm high water blister-like solid bulge appeared on the surface of the graphite electrode. When the graphite electrode is polarized to the range of 0.65–0.2 V (vs. Li/Li^+), more hill-like solid particles and blister solid bulges will appear on the surface of the graphite electrode. In addition, the original solid particles and solid bulges on the surface will continue to grow and eventually grow into solid particles with much higher 40 nm; when the graphite electrode is polarized below 0.20 V (vs. Li/Li^+), the solid particles on its surface will connect with each other, thus eventually forming a whole solid film, that is, SEI film. Figure 9.5 shows the real-time change of the surface morphology of the graphite electrode during the first cathodic polarization. The AFM observation results of Koltypin et al. [24] also confirm this conclusion.

9.7 Storage Stability and Evolution Mechanism of SEI Film on Graphite Surface

9.7.1 Storage Stability of SEI Film on Graphite Surface

Zane et al. [25] studied the real-time change of SEI film on the surface of charged graphite electrode after long-term storage by using SEM and executive information system (EIS). They found that during the storage process, the micro-morphology of the SEI film will change continuously, and its thickness will increase accordingly, resulting in a continuous increase in the interface impedance between the graphite electrode and the electrolyte. Broussely et al. [26] after storing the rechargeable commercial lithium-ion battery at different ambient temperatures for a long time, they found that the discharge capacity of the battery decreased gradually with the increase of storage time, and the degree of attenuation increased sharply with the increase of storage temperature. Subsequently, they dissected and analyzed the stored battery in detail and found that the attenuation of battery capacity was mainly due to the thickening of SEI film on the graphite surface. Yoshida et al. [27] measured the thickness of SEI film on the surface of graphite electrode after storage by means of focused ion beam (FIB), scanning electron microscope (SEM), and XPS. The results show that the thickness of SEI film increases with the increase of storage time and storage temperature, and the effect of temperature is more obvious.

Although it has been confirmed that the SEI film on the graphite surface will continue to thicken during long-term storage, which will lead to the decline of graphite lithium intercalation capacity, but the reason for it seems to be controversial. Broussely et al. [26] believe that although the electronic conductivity of SEI film is much smaller than its lithium-ion conductivity, its actual value is not zero. Therefore, in the process of storage, a small number of electrons may still penetrate the SEI film and reach the interface between the electrolyte and the SEI film, resulting in a new round of reduction reaction with the solvent molecules and lithium ions in the electrolyte, resulting in a new insoluble substance, resulting in the thickening and capacity attenuation of the SEI film. However, Ploehn et al. [28] do not think so.

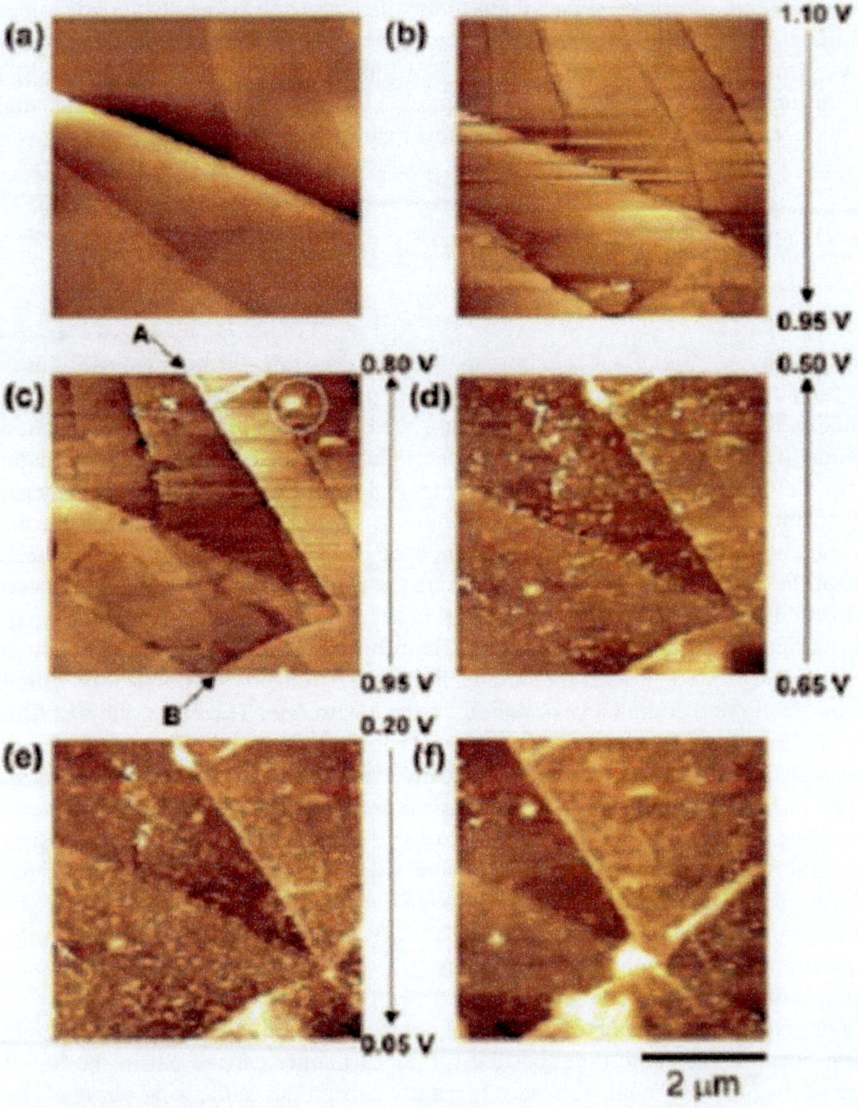

Fig. 9.5 Real-time variation of surface morphology of graphite electrode during the first cathodic polarization by atomic force microscope (AFM) [23]

After summarizing the previous understanding of the structure of SEI membrane, they think that there are many tiny pores in SEI membrane. After being stored for a long time, a few solvent molecules can still spread through tiny pores into the interface between graphite electrode and SEI film, which leads to a new round of reduction reaction with electrons and lithium ions in lithium-embedded graphite, resulting in a new insoluble substance, which leads to the thickening and capacity attenuation of SEI film. Obviously, the biggest difference between the two views is that the former believes that the flow of electrons in the SEI film causes the thickening of the SEI film during storage, while the latter believes that the flow of solvent molecules in the SEI film causes the thickening of the SEI film during storage.

9.7.2 *Storage Evolution Mechanism of Solid Electrolyte Interface Film [1]*

SEI film plays an important role in realizing the reversible electrochemical lithium intercalation activity of graphite electrode. When the lithium-ion battery is working normally, the SEI film can not only ensure the electrochemical reduction reaction without direct contact with the graphite electrode, but also has no essential effect on the normal lithium-ion intercalation and removal reaction. However, from a microscopic point of view, the SEI film is actually a layer of solid film formed by the accumulation of small solid particles composed of many insoluble reduction products on the surface of the graphite electrode, as shown in Fig. 9.6 [29]. As can be seen from the figure, there is always a part of the interstitial area between the small solid particles without solid phase filling. Therefore, when the electrolyte has enough time to fully infiltrate the SEI membrane, these interstitial areas without solid filling will be slowly and completely occupied by the electrolyte. Therefore, the SEI film fully infiltrated by the electrolyte can actually be seen as a porous thin film layer formed by the superposition of liquid electrolyte and solid particles. In the long-term storage of lithium-ion battery, some chemical components in the electrolyte (hereinafter referred to as electrolyte composition) will gradually penetrate the SEI film through slow diffusion, thus running from one side of the electrolyte to the other side of the graphite electrode. In addition, although the electronic conductivity in the SEI film is very low, it is actually not zero. Therefore, in a long enough migration time, the electrons in the graphite electrode may still slowly penetrate the SEI film, thus running from one side of the graphite electrode to the other side of the electrolyte.

In principle, the diffusion of electrolyte components and the transfer of electrons should occur at the same time, but the former is mainly carried out in the liquid phase of the SEI film, while the latter is mainly carried out in the solid phase of the SEI film. When there is little difference between the two velocities, the electrolyte composition and electrons can meet at the liquid–solid interface at any point in the SEI film, and then heterogeneous electrochemical reduction reactions occur, resulting in new insoluble reduction products. Subsequently, these newly produced insoluble reduction products precipitate near the surface of the graphite electrode, either forming new solid particles or accumulating on the old solid particles, which will eventually lead to the increase of SEI film thickness with the increase of storage time. The thickening mechanism of SEI film on the surface of graphite electrode is illustrated in Fig. 9.7.

Fig. 9.6 Typical AFM microtopography of SEI film on graphite electrode

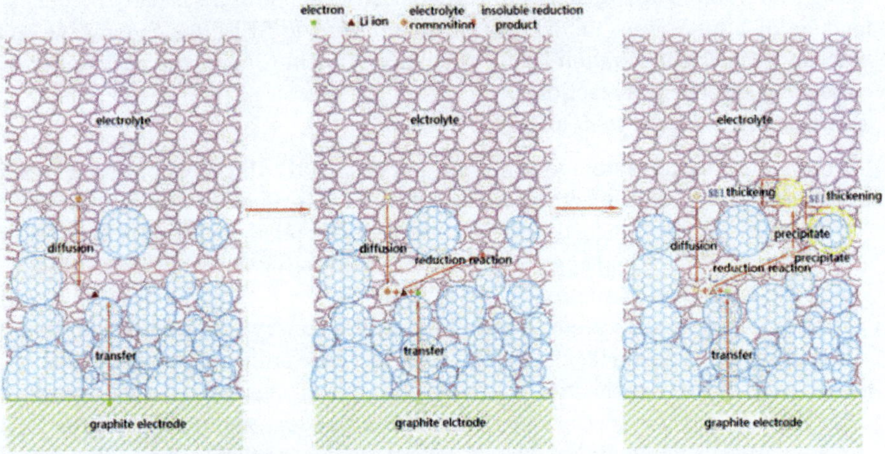

Fig. 9.7 Thickening mechanism of SEI film on the surface of graphite electrode

As can be seen from Fig. 9.7, the thickening process of SEI film actually includes the following four step-by-step steps:

(1) Diffusion steps of electrolyte components in liquid phase of SEI film.
(2) Electron transfer steps in the solid phase of SEI film.
(3) Heterophase electrochemical reduction reaction steps of electrolyte composition and electrons at the liquid–solid interface of SEI membrane.
(4) The precipitation steps of insoluble reduction products in electrolyte near the surface of graphite electrode.

It is not difficult to find that the diffusion process of electrolyte components in the liquid phase of SEI films, or the migration of electrons in the solid phase of

SEI films, can be said to be an extremely slow step compared with the electrochemical reduction reaction between electrolyte components and electrons, as well as the precipitation process of electrolyte reduction products. In addition, the composition of the electrolyte and the content of electrons in the SEI film are usually kept at a very low concentration level. Therefore, we believe that the electrolyte composition and electrons at the liquid–solid interface at any point in the SEI film (including the electrolyte/SEI membrane interface and the graphite electrode/SEI membrane interface) can react completely relatively quickly, and the resulting insoluble reduction products can also precipitate completely relatively quickly. In other words, the thickening process of SEI film can only be controlled by the diffusion of electrolyte in the liquid phase of SEI film, or by the migration of electrons in the solid phase of SEI film, while the influence of electrochemical reduction reaction or precipitation process can be ignored.

In addition, the content of lithium ion in SEI film is also an important physical quantity affecting the thickening process. Because the heterogeneous electrochemical reduction reactions involving electrolytes and electrons are generally accompanied by the consumption of lithium ions. However, because SEI films are usually excellent conductors of lithium ions, the content of lithium ions in SEI films can always be kept above a high concentration level, which is sufficient to meet the needs of any electrochemical reduction reaction.

Through our research, we found that [1]:

(1) During the long-term storage of lithium-ion battery, the thickness of SEI film on the surface of graphite electrode increases with the increase of storage time, and the increase of SEI film thickness satisfies the linear relationship with the square root of storage time and approximately conforms to the Arrhenius relationship with the storage temperature.
(2) Although the capacity attenuation of lithium-ion battery stored for a long time is mainly due to the thickening of SEI film, there are still other factors that may lead to the capacity attenuation of lithium-ion battery. Therefore, the relationship between the capacity decay rate of lithium-ion battery and the square root of storage time can only approximately satisfy the linear relationship.

References

1. Yan, Jian. 2009. Study on the formation and evolution mechanism of solid electrolyte interface film on the surface of graphite electrode for lithium ion battery. Doctoral paper of Shanghai Institute of Microsystems and Information Technology.
2. Hérold, A. 1955. Recherches Sur Les Composés D'insertion du Graphite. *Bulletin de la Societe Chimique de France* 187: 999–1012.
3. Fong, R., U. von Sacken, and J.R. Dahn. 1990. Studies of lithium intercalation into carbons using nonaqueous electrochemical cells. *Journal of the Electrochemical Society* 137 (7): 2009–2011.
4. Peled, E. 1979. The electrochemical behavior of alkali and alkaline earth metals in nonaqueous battery systems—The solid electrolyte interphase model. *Journal of the Electrochemical Society* 126 (12): 2047–2051.

5. Peled, E., and D. Golodnitsky. 2004. SEI on lithium, graphite, disordered carbons and tin-based alloys. In *Lithium-ion batteries: Solid-electrolyte interphase*, ed. P.B. Balbuena and Y. Wang, 1–69. London: Imperial College Press.
6. Besenhard, J.O., M. Winter, J. Yang, and W. Biberacher. 1995. Filming mechanism of lithium-carbon anodes in organic and inorganic electrolytes. *Journal of Power Sources* 54 (2): 228–231.
7. Chung, G.-C., H.-J. Kim, S.-I. Yu, S.-H. Jun, J.-W. Choi, and M.-H. Kim. 2000. Origin of graphite exfoliation: An investigation of the important role of solvent cointercalation. *Journal of the Electrochemical Society* 147 (12): 4391–4398.
8. Aurbach, D., H. Teller, and E. Levi. 2002. Morphology/behavior relationship in reversible electrochemical lithium insertion into graphitic materials. *Journal of the Electrochemical Society* 149 (10): A1255–A1266.
9. Aurbach, D., and Y.S. Cohen. 2004. Identification of surface films on electrodes in non-aqueous electrolyte solutions: Spectroscopic, electronic and morphological studies. In *Lithium-ion batteries: Solid-electrolyte interphase*, ed. P.B. Balbuena and Y.-X. Wang, 70–139. London: Imperial College Press.
10. Yan, Jian, Bao-Jia Xia, Jian Zhang, and Xi-Gui Zhang. 2008. Phenomenologically modeling the formation and evolution of the solid electrolyte interface on the graphite electrode for lithium-ion batteries. *Electrochimica Acta* 53: 7069–7078.
11. Yan, Jian, Yu-Chang Su, Bao-Jia Xia, and Jian Zhang. 2009. Thermodynamics in the formation of the solid electrolyte interface on the graphite electrode for lithium-ion batteries. *Electrochimica Acta* 54 (13): 3538–3542.
12. Yan, Jian, Jian Zhang, Yu-Chang Su, Xi-Gui Zhang, and Bao-Jia Xia. 2010. A novel perspective on the formation of the solid electrolyte interphase on the graphite electrode for lithium-ion batteries. *Electrochimica Acta* 55 (5): 1785–1794.
13. Peled, E., D. Golodnitsky, C. Menachem, and D. Bar-Tow. 1998. An advanced tool for the selection of electrolyte components for rechargeable lithium batteries. *Journal of the Electrochemical Society* 145 (10): 3482–3486.
14. Aurbach, D., Y. Ein-Eli, O. Chusid (Youngman), Y. Carmeli, M. Babai, and H. Yamin. 1994. The correlation between the surface chemistry and the performance of Li-carbon intercalation anodes for rechargeable 'rocking-chair' type batteries. *Journal of the Electrochemical Society* 141 (3): 603–611.
15. Ein-Eli, Y., S.R. Thomas, and V.R. Koch. 1997. The role of SO_2 as an additive to organic Li-ion battery electrolytes. *Journal of the Electrochemical Society* 144 (4): 1159–1165.
16. Wang, C., H. Nakamura, H. Komatsu, M. Yoshio, and H. Yoshitake. 1998. Electrochemical behaviour of a graphite electrode in propylene carbonate and 1,3-benzodioxol-2-one based electrolyte system. *Journal of Power Sources* 74 (1): 142–145.
17. Wrodnigg, G.H., J.O. Besenhard, and M. Winter. 1999. Ethylene sulfite as electrolyte additive for lithium-ion cells with graphitic anodes. *Journal of the Electrochemical Society* 146 (2): 470–472.
18. Abe, K., H. Yoshitake, T. Kitakura, T. Hattori, H. Wang, and M. Yoshio. 2004. Additives-containing functional electrolytes for suppressing electrolyte decomposition in lithium-ion batteries. *Electrochimica Acta* 49 (26): 4613–4622.
19. Aurbach, D., K. Gamolsky, B. Markovsky, Y. Gofer, M. Schmidt, and U. Heider. 2002. On the use of vinylene carbonate (VC) as an additive to electrolyte solutions for Li-ion batteries. *Electrochimica Acta* 47 (9): 1423–1439.
20. Morita, M., M. Ishikawa, and Y. Matsuda. 1998. Organic electrolytes for rechargeable lithium ion batteries. In *Lithium ion batteries: Fundamentals and performance*, ed. M. Wakihara and O. Yamamoto, 156–180. New York: Wiley-VCH.
21. Peled, E., D. Golodnitsky, and G. Ardel. 1997. Advanced model for solid electrolyte interphase electrodes in liquid and polymer electrolytes. *Journal of the Electrochemical Society* 144 (8): L208–L210.
22. Bar-Tow, D., E. Peled, and L. Burstein. 1999. A study of highly oriented pyrolytic graphite as a model for the graphite anode in Li-ion batteries. *Journal of the Electrochemical Society* 146 (3): 824–832.

23. Jeong, S.-K., M. Inaba, T. Abe, and Z. Ogumi. 2001. Surface film formation on graphite negative electrode in lithium-ion batteries: AFM study in an ethylene carbonate-based solution. *Journal of the Electrochemical Society* 148 (9): A989–A993.
24. Koltypin, M., Y.S. Cohen, B. Markovsky, Y. Cohen, and D. Aurbach. 2002. The study of lithium insertion-deinsertion processes into composite graphite electrodes by in situ atomic force microscopy (AFM). *Electrochemistry Communications* 4 (1): 17–23.
25. Zane, D., A. Antonini, and M. Pasquali. 2001. A morphological study of SEI film on graphite electrodes. *Journal of Power Sources* 97–98: 146–150.
26. Broussely, M., S. Herreyre, P. Biensan, P. Kasztejna, K. Nechev, and R.J. Staniewicz. 2001. Aging mechanism in Li ion cells and calendar life predictions. *Journal of Power Sources* 97–98: 13–21.
27. Yoshida, T., M. Takahashi, S. Morikawa, C. Ihara, H. Katsukawa, T. Shiratsuchi, and J.-I. Yamaki. 2006. Degradation mechanism and life prediction of lithium-ion batteries. *Journal of the Electrochemical Society* 153 (3): A576–A582.
28. Ploehn, H.J., P. Ramadass, and R.E. White. 2004. Solvent diffusion model for aging of lithium-ion battery cells. *Journal of the Electrochemical Society* 151 (3): A456–A462.
29. Hirasawa, K.A., T. Sato, H. Asahina, S. Yamaguchi, et al. 1997. In situ electrochemical atomic force microscope study on graphite electrodes. *Journal of the Electrochemical Society* 144 (4): L81–L84.

Chapter 10 Mechanism Research of Charge–Discharge Process for MH/Ni Battery

10.1 Introduction

Hydrogen-nickel (MH/Ni) battery is a typical chemical power source. Because of its high specific energy, long charge–discharge cycle life, strong ability to withstand overcharge and overdischarge, and so on, it has been widely used in many fields and become a hybrid electric vehicle power battery, which has attracted the attention of electrochemical experts, especially chemical power experts. Electrochemical chemists and chemical power experts mostly study the research and development of the battery, the performance and application of the battery, but there is almost no research on the chemical physics and conductive physical mechanism in the electrode process. For example, there is little work on the chemical and physical behavior and physical conductivity mechanism in the charge and discharge process of MH/Ni battery.

In 1966, Bode et al. [1] described the working principle of MH/Ni battery as follows:

The working principle is written as the following chemical reaction formula [2]

Positive electrode Negative electrode

$$\text{Charge } \beta - Ni(OH)_2 + OH \rightarrow \beta - NiOOH + H_2O + e^-M$$

$$+H_2O + e^- \rightarrow MH(\text{hydride}) + OH$$

$$\text{Discharge } \beta - NiOOH + H_2O + e^- \rightarrow \beta - Ni(OH)_2 + OH^-MH$$

$$+OH \rightarrow M + H_2O + e^-$$

$$\text{Total reaction } M + \beta - Ni(OH)_2 \Leftrightarrow MH(\text{hydride}) + \beta - NiOOH$$

The above description of work principle for MH/Ni batteries has been written into teaching books and academic monographs [3].

C. Yang et al., *Materials and Working Mechanisms of Secondary Batteries*,
https://doi.org/10.1007/978-981-19-5955-4_10

The above tells us that the chemical mechanism of conducting electricity is:

(1) The β-$Ni(OH)_2$ → β-NiOOH phase transition occurs during charging, and the H^+ ion is supplied by this phase transition.
(2) The transition of AB_5 → AB_5H_x (hydride) occurs during the charging process.
(3) The above two kinds of phase transitions are completely reversible in the process of charge and discharge.

The above conductive mechanism can be called phase transition theory. It can also be seen from the above that under the action of electric field, the driving force of β-$Ni(OH)_2$ → β-NiOOH is oxidation; the driving force of β-NiOOH → β-$Ni(OH)_2$ is reduction. It can be seen that oxidation–reduction occurs at the positive electrode for Ni.

In 1999, Xin et al. [4] and Wang et al. etc. [5] reported the results of charge–discharge in-situ XRD observations. The diffraction pattern of 1C charge for 3 h (300%) was identified as the coexistence of β-NiOOH and γ-NiOOH. This seems to give the experimental evidence of the conductive phase transition theory of MH/Ni battery. However, the experimental evidence of the coexistence of β-Ni $(OH)_2$ + β-NiOOH or pure β-NiOOH phase before full charge was not obtained. Therefore, it is very necessary to study the chemical and physical phenomena and physical conductivity mechanism of MH/Ni battery during charge–discharge process in detail, and it has scientific and practical significance.

10.2 Comparative Study on the Microstructure of β-$Ni(OH)_2$ Before and After Activation of MH/Ni Battery [7, 8]

The X-ray diffraction patterns of the positive active material β-$Ni(OH)_2$ of the two samples before activation (.raw) and after (.HH.raw) are shown in Fig. 10.1. The diffraction data of the samples are listed in Table 10.1. No matter from the diffraction pattern or from the data of Table 10.1, it can be seen that the effect of activation is great, and the diffraction line is obviously broadened. The results of data processing according to the methods given in Chap. 2 and Reference [8, 9] are listed in Table 10.2. From these results, we can see that:

(1) After activation, the grains are obviously refined, especially the size in the [100] direction of vertical *c*-axis is greatly reduced, so that the shape of microcrystals is transformed from chunky columnar to nearly equiaxed crystals or polyhedrons.
(2) Micro-strain was introduced into β-$Ni(OH)_2$ after activation.
(3) The structure and probability of stacking fault also change after activation.

The above three points are the reasons for the great changes in the XRD pattern of β-$Ni(OH)_2$ before and after activation. The above changes are the result of observation before and after activation. Generally, battery activation is usually carried out

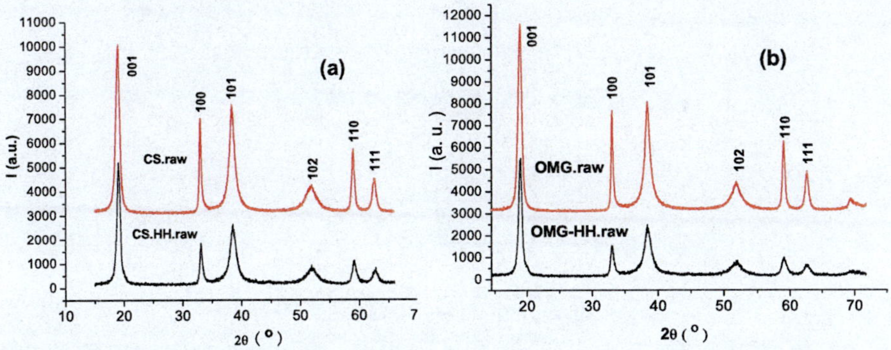

Fig. 10.1 Positive active material β-$Ni(OH)_2$ X-ray diffraction pattern of two samples before (.raw) and after (.HH.raw) activation

Table 10.1 Positive active material β-$Ni(OH)_2$ X-ray diffraction pattern of two samples before (.raw) and after (.HH.raw) activation

hkl		001	100	101	102	110	111
2θ (°)		19.16	33.20	38.67	52.25	59.16	62.55
CS-PTX	Before active	0.783	0.357	1.394	2.507	0.467	0.752
	After active	0.672	0.514	1.605	2.536	0.833	1.046
OMG	Before active	0.667	0.322	1.288	2.310	0.379	0.592
	After active	0.628	0.561	1.569	2.658	0.924	1.147
KL	Before active	0.644	0.342	1.336	2.326	0.409	0.656
	After active	0.560	0.594	1.319	2.305	1.061	1.143

according to the activation process for 2–3 cycles of charge and discharge. So how does the structure and microstructure of the electrode active material change during the first charge and discharge of the battery? Let us study this problem in more detail.

10.3 In-Situ XRD Study of Nickel Electrode β-$Ni(OH)_2$ During Charging

In the study, the charge in-situ observation was carried out by using the charge–discharge in-situ XRD device (patented product) provided by Rigaku Company in Japan [6] . The patterns of several stages of 1C charging are shown in Fig. 10.2.

We can see that in the characteristic diffraction spectrum of β-$Ni(OH)_2$ after 1C charge 50, 100, 150, and 480%, only in the case of 1C charge 480%, not only the characteristic diffraction pattern of β-$Ni(OH)_2$ is observed, but also the 003 and 006 diffraction peaks of γ-NiOOH are observed. The results show that:

Table 10.2 Analysis of three β-$Ni(OH)_2$ diffraction data

		D_{001} (nm)	nD_{100} (nm)	$\frac{D_{100}}{D_{001}}$	D_{101} (nm)	D_{102} (nm)	$\overline{D}$ (nm)	$\overline{\varepsilon}$ (10^{-3})	f_D (%)	f_T (%)	$f_D + f_T$ (%)
CS-PTX	Before active	12.4	34.6	2.792	36.8				9.35	10.09	13.44
	After active	15.0	21.8	1.420		28.6	26.2	10.284	8.31	2.37	11.04
OMG	Before active	15.1	42.7	2.828	48.9				8.79	10.73	12.52
	After active	16.3	19.0	1.165		38.1	35.3	10.703	7.11	10.27	11.38
KL	Before active	15.8	38.6	2.443	46.9				9.87	2.24	12.11
	After active	19.98	17.67	0.884		210.5	57.35	5.961	10.28	10.66	8.94

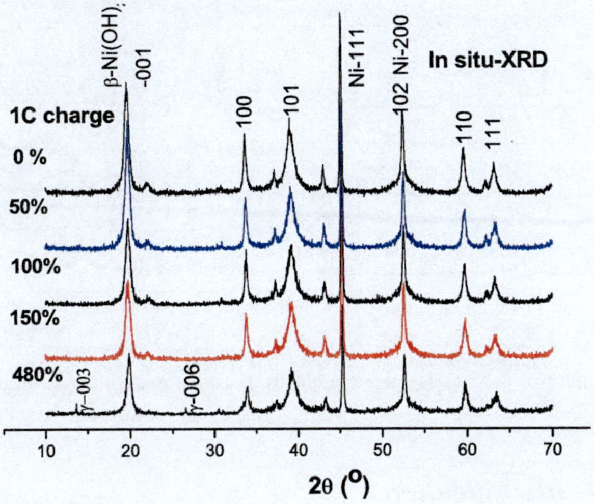

Fig. 10.2 In-situ XRD pattern of β-$Ni(OH)_2$ at different stages of 1C charge, CuK

(1) It is observed that the charging depth of the surface is much smaller than that of the interior of the battery, that is, there is a serious lag effect on the surface, so a new patent is proposed [11].
(2) In the case of overcharge, it is β-$Ni(OH)_2$ + γ-NiOOH that coexists, but is not γ-NiOOH + β-NiOOH that coexists.

10.4 Quasi-Dynamic Study of β-$Ni(OH)_2$ During the First Charge–Discharge of MH/Ni Battery [8, 9]

The so-called quasi-dynamic is sampling at different stages of charging and discharging, in other words, some stages of charging or discharging of the battery (The original, several intermediate, and final states) stopped abruptly, and then dissected the battery, positive and negative active materials were obtained as samples for XRD research and analysis. The contents of the analysis include phase identification, fine structure (lattice parameters, microstructure parameters), and so on. Figure 10.3a shows the pattern of several stages of charging. The specific analysis is as follows:

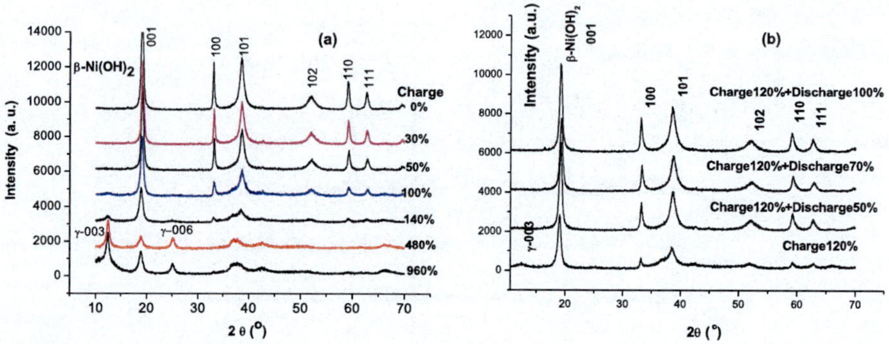

Fig. 10.3 XRD pattern of β-$Ni(OH)_2$ positive electrode active material at several stages of 0.2C charging (**a**) and discharging (**b**), CuK

10.4.1 Phase Identification

It can knows from Fig. 10.3a:

(1) When the charging depth is 0, 30, 50, 100%, the positive active material belongs to β-$Ni(OH)_2$.
(2) The γ-NiOOH is not precipitated until the charging depth reaches 140%.
(3) The content of γ-NiOOH phase increases qualitatively with the increase of charging depth.

Now let us analyze the 0.2C charge 120 and 140% XRD patterns. The result of 140% charging is shown in Fig. 10.4. It can be seen that the positive active material after 0.2C charge of 140% belongs to the mixture of β-$Ni(OH)_2$ and γ-NiOOH, rather than the mixture of γ-NiOOH and β-NiOOH. Therefore, the pattern of 1C charge for 3 h (300%) was identified by Xin et al. [4] as the mixture of γ-NiOOH and β-NiOOH. That is worth discussing.

From the relationship (Fig. 10.5) between $I_{\gamma\text{-NiOOH-003}}/I_{\beta\text{-Ni(OH)2001}}$ and overcharge percentage, it can be seen that the content of γ-NiOOH phase increases with the increase of overcharge percentage.

Therefore, we can conclude that the β-$Ni(OH)_2$ → β-NiOOH phase transition does not occur during the charging process. The experimental evidence of β-$Ni(OH)_2$ → β-NiOOH provided by Xin et al. [4] and Wang et al. [5] is worth discussing.

The XRD patterns of several stages in the discharge state are shown in Fig. 10.3b. We can see that γ-NiOOH has been decomposed, and the phase structure of β-$Ni(OH)_2$ has not changed, but their fine structure has changed.

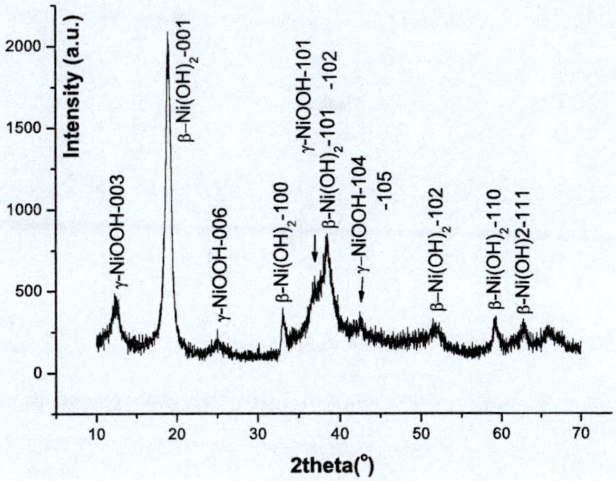

Fig. 10.4 Phase analysis of XRD pattern of cathode chemical material after 0.2 charging 140%

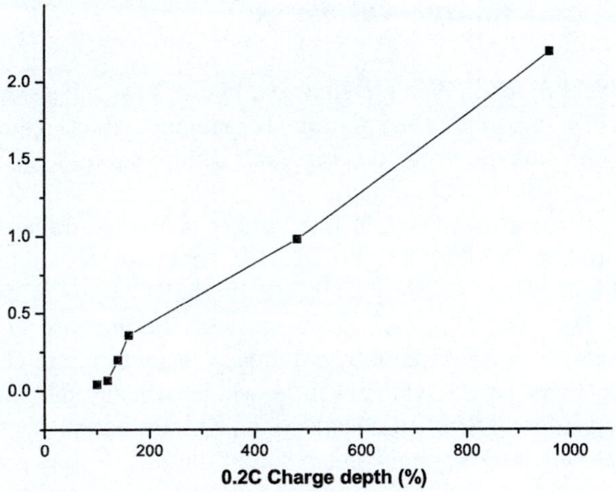

Fig. 10.5 Variation of $I_{\gamma\text{-NiOOH-003}}/I_{\beta\text{-Ni(OH)2001}}$ with overcharge percentage

10.4.2 *Lattice Parameters and Macroscopic Strain of β-Ni(OH)$_2$*

The refine in the Jade program has been used to determine the peak position and FWHM. The changes of β-Ni(OH)$_2$ phase lattice parameters a and c with charging depth are measured and by $(a-a_0)/a_0$ and $(c-c_0)/c_0$ converted into macroscopic strain, which is shown in Fig. 10.6 where a_0 and c_0 are the lattice parameters when not charged. It can be seen that the macroscopic strain of the extension along a-axis decreases with the increase of the charging depth, and the macroscopic strain of the along c-axis begins to decrease and then increases. The discharge state is roughly opposite, but not completely reversible.

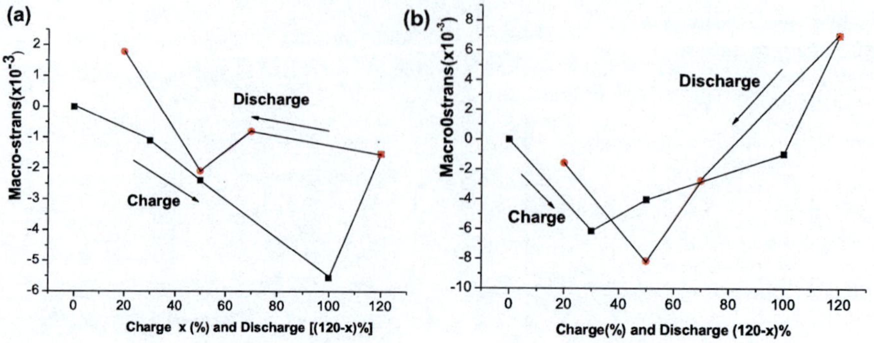

Fig. 10.6 Macro-strain changes of β-$Ni(OH)_2$ phase extending *a*-axis **a** and *c*-axis **b** during charge–discharge

10.4.3 Microstructural Changes of β-$Ni(OH)_2$ During Charge–Discharge

The diffraction data are analyzed by the least square method described in Chap. 2, which separates the multiple broadening effects caused by microcrystal-micro-strain, microcrystal-stacking fault, micro-strain-stacking fault, and microcrystal-micro-strain-stacking fault.

If there is only grain fined effect or micro-strain effect, the data of (100) and (001) about β-$Ni(OH)_2$ are listed in Table 10.3. From these data, it can be seen that if the grain is fined by charging, the fined degree of D_{100} is much faster than that of D_{001}, and D_{100}/D_{001} decreases with the increase of charging percentage, but it does not make the chubby columnar crystal turn into near-equiaxed crystal, and the discharge process changes little, which indicates that the activation before and after activation in Sect. 10.2 leads to the transformation of chubby columnar crystal into near-equiaxed crystal is the result of multiple (~3) charge and discharge.

Table 10.3 Data of β-$Ni(OH)_2$ grain size during the first charge–discharge process

Charge and discharge depth		β (°)		Grain size D (nm)			Micro-strans ε ($\times 10^{-3}$)		
		100	001	D_{100}	D_{001}	D_{001}/D_{100}	ε_{100}	ε_{001}	$\varepsilon_{001}/\varepsilon_{100}$
Charge%	0	0.447	0.158	19.3	50.6	2.768	6.59	4.11	0.624
	30	0.431	0.187	19.0	42.7	2.247	6.36	4.87	0.766
	50	0.429	0.180	19.1	44.3	2.319	6.32	4.69	0.742
	100	0.553	0.365	14.8	21.8	1.473	8.15	9.53	1.169
	120	0.569	0.383	14.6	19.9	1.363	8.26	9.99	1.209
Discharge%	50	0.455	0.293	19.0	27.2	1.511	6.70	7.63	1.139
	70	0.482	0.294	17.0	27.1	1.594	7.10	7.66	1.079
	100	0.445	0.279	19.4	28.5	1.549	6.56	7.29	1.111

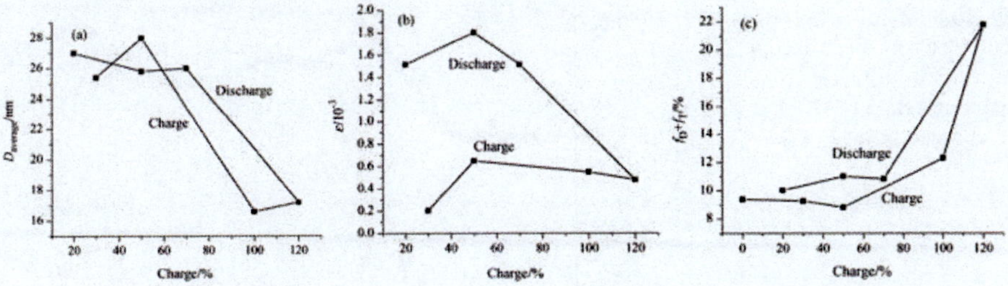

Fig. 10.7 Changes of D_{average} **a** $\varepsilon_{\text{average}}$ **b** and $(f_D + f_T)$ **c** of β-$Ni(OH)_2$ during the first charge–discharge

If charging introduces micro-strain, the strain degree in [100] and [001] direction increases with the increase of charging depth, but changes little with discharge depth, but the initial of charging ε_{001} is less than ε_{100}, indicating that there is anisotropy of micro-strain. From the data of $\varepsilon_{001}/\varepsilon_{100}$, the anisotropy of micro-strain decreases with the increase of charging depth.

In the further analysis of the mechanism caused by grain fined and microscopic strain, it is found that in the charging process, if the phase transition occurs, the new phase nucleates and grows in the parent phase, which can lead to grain fined, but in fact, there is no phase transformation, but only the hydrogen atom leaves from β-$Ni(OH)_2$ lattice. It is impossible to refine the grains, but it is possible to cause micro-strain and increase stacking faults. Therefore, we assume that there is a micro-strain-stacking fault double effect. The average micro-strain ($\varepsilon_{\text{Avearge}}$) and stacking fault probability $(f_D + f_T)$ of β-$Ni(OH)_2$ vary with the charge–discharge process, which are shown in Fig. 10.7a and b, respectively. It can be seen that the micro-strain increases when charging to 50%, and then decreases slightly, while the stacking fault first decreases and then increases, and the change trend of the discharge process is opposite to that of the charging process. But it is not completely reversible.

10.4.4 Analysis of Valence States of Ni Atom in Positive Active Materials by Photoelectron Spectroscopy

In order to further prove whether Ni^{+3} valence β-NiOOH appears in the charging process, two typical samples are selected for photoelectron spectroscopy analysis. The $2p_{3/2}$ scanning curve of the electron spectrum analysis is shown in Fig. 10.8 Generally speaking, the data cannot be obtained directly from the scan map, but the correction amount can be obtained by comparing the measured peak position of C-1s with its standard peak position (284.6 eV). It is used to calibrate the measured peak position of Ni-$2p_{2/3}$, and the results are listed in Table 10.4. It can be seen that no positive trivalent Ni^{+3} was observed in both samples. Charging 50% sample, Ni-2p peak position is 854.6 eV, which is a big difference from that of 858.2 eV.

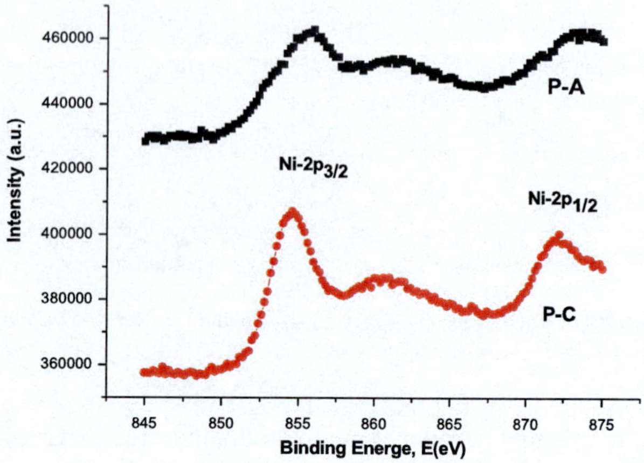

Fig. 10.8 X-ray photoelectron spectroscopy curves of β-$Ni(OH)_2$ uncharged state (*P–A*) and 0.2 charged 50% (*P–C*)

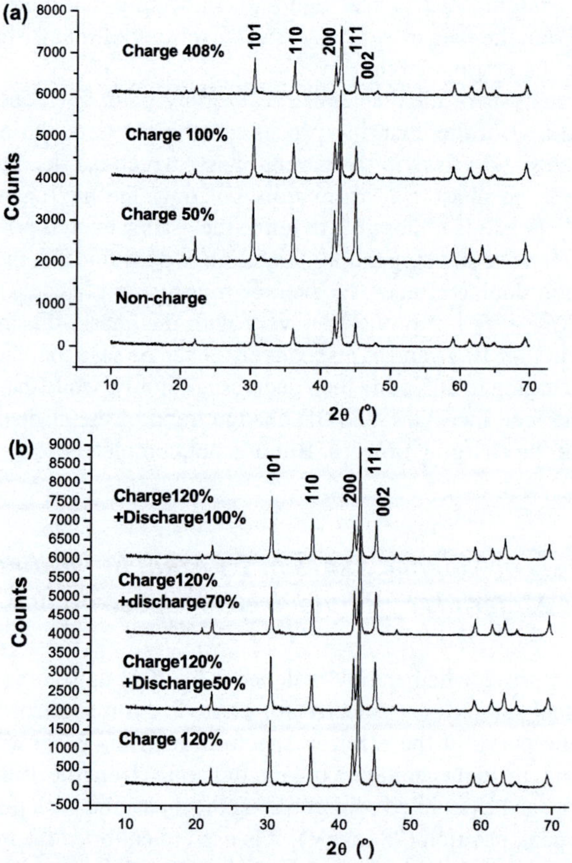

Fig. 10.9 XRD patterns of AB_5 alloy as a negative electrode active material in several stages of charging (**a**) and discharging (**b**), CuK

Table 10.4 Data analysis of X-ray photoelectron spectroscopy of typical samples

		C-1s (eV)			Ni-$2p_{3/2}$ (eV)		Standard value (eV)			
		Standard value	Experimental value	Correction value	Experimental value	Correction value	Ni	Ni(OH)	NiOOH	NiO
P–A	No charge	284.6	287.5	−2.7	858.7	856.0	852.7	856.2	858.2	854
P–C	Charge 50%		285.9	−1.3	855.9	854.6	852.0	856.0		

10.5 Quasi-Dynamic Study of Active Material AB_5 During Charge–Discharge

The XRD spectrum of the negative electrode active material AB_5 in several stages of charge and discharge is shown in Fig. 10.9. On the whole, the structure of AB_5 alloy has no obvious change during charging, but the fine structure and microstructure also change.

Figures 10.10a, b, and c show the change curve of lattice parameters *a* and *c* of AB_5 and the micro-strain ε in the charge-discharge process, respectively. It can be seen that the lattice parameters *a* and *c* increase with the increase of charging depth, and the micro-strain ε also increases with the increase of charge depth. The discharge situation is just the opposite, but it returns to the original state, which indicates that there are some irreversible factors in the charge-discharge process.

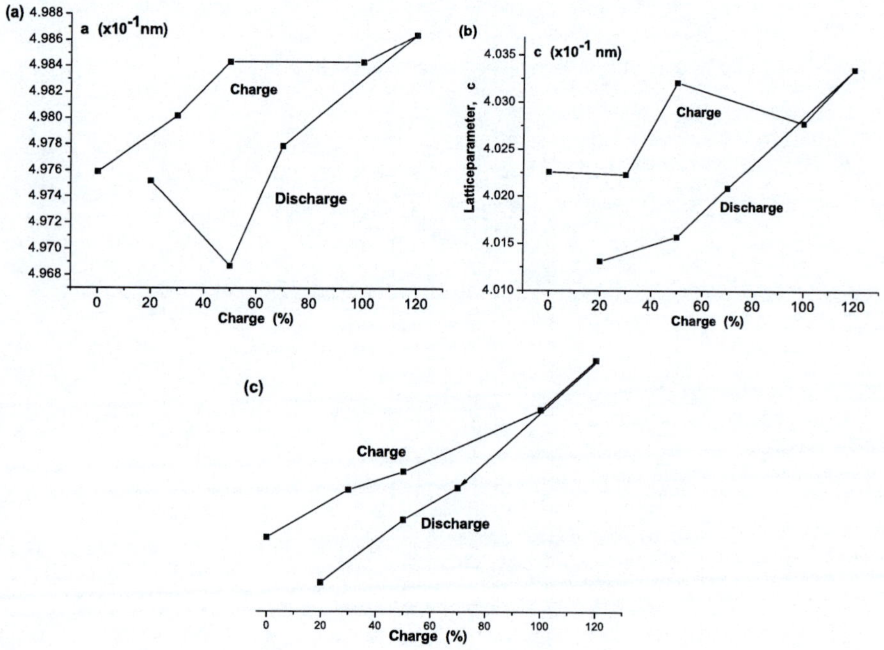

Fig. 10.10 Changes of AB_5 structure parameters during charge–discharge **a** lattice parameters *a*, **b** lattice parameters *c*, **c** micro-strain ε

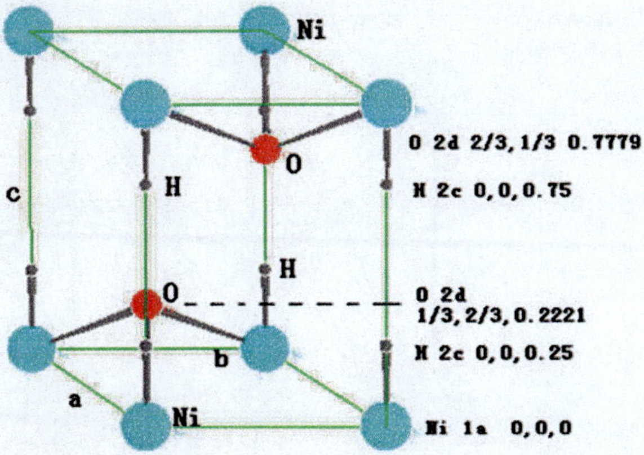

Fig. 10.11 Crystal structure model and chemical bonding of β-Ni$(OH)_2$

10.6 De-Intercalation Theory of Charge–Discharge Process for MH/Ni Battery [8–10]

10.6.1 De-Intercalation Behavior of β-Ni$(OH)_2$ During Charge–Discharge

The β-Ni$(OH)_2$ belongsβ to hexagonal structure, P-3m1 (No. 164) space group; there is one molecule in the unit cell, that is, one Ni atom, two oxygen atoms, and two hydrogen atoms, a total of five atoms. Their crystallographic positions in the unit cell are as follows:

Atomic	position	*x*	*y*	*z*.
Ni	1*a*	0	0	0
H	2*c*	0	0	±1/4
O	2*d*	1/3	2/3	±0.222

The crystal structure model and chemical bonding are shown in Fig. 10.11. As can be seen from the figure, the chemical bond between Ni and O is stronger, while that between H and O is much weaker. When there is no stacking disorder, the Ni–O layer according to ABAB… sequential stacking. The hydrogen atom is divided into two layers embedded between the Ni–O layers. These experimental results in Sects. 10.4–10.5 have fully proved that there is no phase transition from β-Ni$(OH)_2$ to β-NiOOH during the charging process of MH/Ni battery. The hydrogen ion H^+ is not provided by this phase transition. The lattice parameters *a* and *c* of β-Ni$(OH)_2$ decrease with the charging process, the macroscopic strain of the extension a-axis decreases with the increase of the charging depth, and the macro-strain of the extension *c*-axis begins to decrease and then increases. This shows that the hydrogen atom leaves the 00 ± 1/4 lattice position of β-Ni$(OH)_2$ and leaves a vacancy, which distorts the lattice of β-Ni$(OH)_2$ and the probability of stacking fault increases.

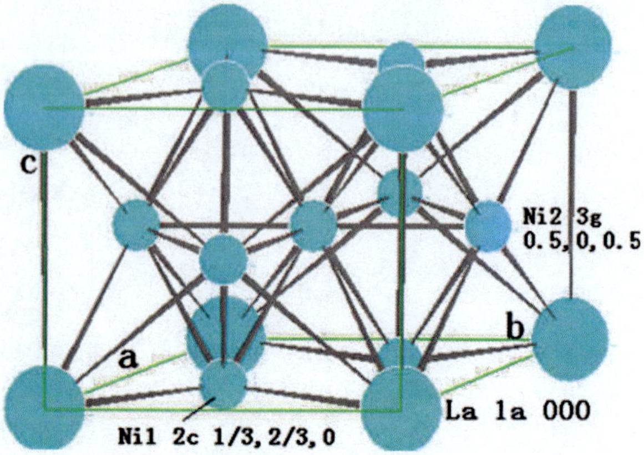

Fig. 10.12 Crystal structure model and chemical bonding of $LaNi_5$

γ-NiOOH precipitates from β-Ni $(OH)_2$ only when there are enough hydrogen atoms leaving the β-$Ni(OH)_2$ lattice from Ni:O:H to fall from 1:2:2 to 1:2:1. The + 3 valence Ni-$2p_{3/2}$ photoelectron peak was not observed in 50% of the charged samples, which further proved this point.

There is now a saying that when β-$Ni(OH)_2$ → β-NiOOH, the cell volume shrinks by 15%, and when β-$Ni(OH)_2$ → γ-NiOOH, the volume shrinks by 18%. This 18% figure seems to be wrong. When β-$Ni(OH)_2$ → γ-NiOOH, the cell volume increases to 262.9% (see Table 10.5).

More reasonably, the change in the volume of each molecule in the unit cell should be taken into account. In this way, the cell volume of each molecule is reduced by 15.0% at β-$Ni(OH)_2$ → β-NiOOH and only 9.3% at β-$Ni(OH)_2$ → γ-NOOH, as shown in the last row of Table 10.5. It can be seen that it is more reasonable to have β-$Ni(OH)_2$ transition γ-NOOH, instead of β-$Ni(OH)_2$ transition β-NiOOH phase transition.

Table 10.5 Data related to β-$Ni(OH)_2$, β-NiOOH, and γ-NiOOH

Substance	β-$Ni(OH)_2$	β-NiOOH	γ-NiOOH
Type of crystal structure	Hexagonal ($P\bar{3}m1$)	Hexagonal	Hexagonal
PDF card number	03-0177	06-0141	06-0075
Valence state of Ni	Ni^{+2}	Ni^{+2}	Ni^{+3}
a (Å)	3.126	2.81	2.828
c (Å)	4.605	4.84	20.569
V $(Å)^3$	38.97	33.10	141.45
The rate of increase of cell volume (%)	0.00	−15.07	+ 262.98
Density (g/cm^3)	3.946	4.62	3.890
Number of molecules/unit cell	1	1	4
V ($Å^3$)/molecule	38.97	33.10	35.36
The rate of change of the volume occupied by a molecule (%)	0.0	−15.9	−9.3

10.6.2 De-Intercalation Behavior of AB_5 Alloy During Charge–Discharge Process

The hydrogen storage alloy AB_5 ($LaNi_5$) has a hexagonal structure and P6/mmm (No. 191) space group. There is one molecule in the unit cell, that is, one La atom and five Ni atoms. Their crystallographic positions in the cell are as follows:

Atomic position		coordinate
La	1*a*	0 0 0
$Ni-1$	2*c*	1/3 2/3 0; 2/3 1/3 0
$Ni-2$	3*G*	1/2 01/2; 0 1/2 1/2; 1/2 1/2 1/2

The crystal structure model and chemical bonding are shown in Fig. 10.12.

From the experimental results in Sect. 10.5, it can be seen that the structure of AB_5 has no obvious change under the condition of charging and overcharging, but the lattice parameters *a* and *c* and micro-strain increase with the extend of charging process, which indicates that before a certain stage of charging, hydrogen atomic energy occupies the interstitial position of AB_5 lattice and forms AB_5–H_n solid solution. AB_5H_x compounds can be formed only when the cell volume increases to a certain percentage due to the intercalation of hydrogen atoms.

The lattice parameters *a* and *c* and the relationship between unit cell volume *V* and *x* of $LaNi_5H_x$ are listed in Table 10.6, and the experimental results of the authors are listed on the right side of Table 10.6. We can see that: (1) When $x = 0.40$ in $LaNi_5H_x$, the change of cell volume is still less than 1%, which indicates that when $x = 0.40$, the hydrogen atom is in the interstitial position. It seems that when the charging depth reaches 140%, the hydrogen atom is still in the interstitial position. (2) Because the hydrogen atom is embedded in the AB_5 lattice, the lattice parameters *a* and *c* and microscopic strain increase. The negative active material is distorted by charging or overcharging. (3) The discharge process is opposite to the charging process, but it is not completely reversible. From this point of view, even if the overcharge reaches 140%, the hydrogen atom is still in the interstitial position and has not yet formed a hydride.

The experimental results show that at least a certain percentage of hydrogen atoms are dissolved in AB_5 alloy in the form of interstitial solution to form AB_5–2xH solid solution, but not to form hydride AB_5H_x. Hong and Zu-Yan [12] in《The crystallographic analysis of the maximum hydrogen storage of $LaNi_5$》, according to the atomic radius of hydrogen is 0.045 nm (or 0.078 nm), so in theory, there are 18 interstitial positions in a $LaNi_5$ cell that can hold hydrogen atoms, and the ratio of H atom to metal atom is H/M = 1. If a unit cell has one La atom and five Ni atoms, the maximum hydrogen storage mass fraction = $(1.008 \times 6)/(138.91 + 58.693 \times 5) = 1.398\%$ can be calculated, which is basically consistent with the experimental hydrogen storage mass fraction of 1.35–1.38% [13]. So in AB_5–H_n, $n = 0$–6.

Table 10.6 Changes of lattice parameters *a* and *c* and cell volume with *x* for $LaNi_5H_x$ and AB_5H_x

$LaNi_5H_x$ data in references							In this experimental study, the data of AB_5H_x were measured		
X in $LaNi_5H_x$	PDF card No.	Space group	*a* (Å)	*c* (Å)	*V* (Å)^3	$(V-V_0)/V_0$ (%)	Charge %	*V* (Å)	$(V-V_0)/V_0$ (%)
0.00	42–1191	P6/mmm	5.013	3.984	86.70	0.00		88.148	0.00
0.13	33–0518	P6/mmm	5.025	3.988	87.21	0.60	0.2C 50	88.183	0.04
0.26	93–2139	P6/mmm	5.025	3.988	87.21	0.60	0.2C 100	88.289	0.16
0.40	84–1312	P6/mmm	5.025	3.991	87.27	0.68	0.2C 140	88.390	0.27
3.00	80–0307	P6/mmm	5.302	4.066	98.98	14.19	0.1C 480	88.614	0.53
3.12	79–2070	P6/mmm	5.302	4.066	98.98	14.19			

The AB_5H_x, is formed only when the concentration of hydrogen atom in AB_5 alloy reaches a certain value. In other words, AB_5H_x is formed when the hydrogen atom dissolved in AB_5 alloy the cell volume increases to a certain value.

10.7 Conductive Mechanism of Charge–Discharge Process for MH/Ni Battery

It is found that the phase transition of β-$Ni(OH)_2$ → β-NiOOH does not occur during the charging process, and only when fully charged and overcharged, part of β-$Ni(OH)_2$ changes into γ-NiOOH, and the two phases of β-$Ni(OH)_2$ and γ-NiOOH coexist all the time, and the hydrogen ion is not supplied by the β-$Ni(OH)_2$ → β-NiOOH phase transition during the charging process. It can be seen that in the process of charge and discharge, β-$Ni(OH)_2$ of Ni^{+2} valence is not oxidized into β-NiOOH of Ni^{+3} valence and β-NiOOH of Ni^{+3}-valent is not reduced to β-$Ni(OH)_2$ of Ni^{+2} valence; instead hydrogen atoms are oxidized and reduced at the positive and negative electrodes, respectively.

If during the charging process, β-$Ni(OH)_2$ is also oxidized, the Ni atom loses an electron. This electron does work through the external circuit together with the electron lost by the H atom. The electron does work through the external circuit together with the electron lost by the H atom and reaches the negative electrode. The H^+ ion through the internal circuit from the positive pole to the negative electrode gets an electron to become an H atom and embedded in the AB_5 lattice. In this way, there will be one more electron in the negative electrode, resulting in the accumulation of negative charge in the negative electrode and the imbalance of battery charge. This proves that there is no problem that β-$Ni(OH)_2$ is oxidized (Ni loses electrons) during the charging process.

If the phase transition occurs and the Ni atoms are oxidized and reduced during the charge–discharge process, then the oxidation–reduction occurs at the positive electrode and occurs during charging and discharging, respectively, which is one of the two necessary conditions for chemical power sources to directly convert chemical energy into electric energy: "whether charging or discharging, the process of losing electrons (oxidation) and getting electrons (reduction) in a chemical reaction must be separated on the positive and negative electrodes" to contradict each other. It can be seen that the phase transition theory of electrical conductivity can be not established.

From the above experimental study and theoretical analysis, it can be concluded that the conductive ion is hydrogen ion, which is provided by the de-intercalation of hydrogen atom in positive and negative active materials. Section 10.6 has described the de-intercalation behavior of hydrogen atoms in positive and negative active materials. The conductive mechanism is as follows.

The charging process of MH/Ni battery starts from the negative electrode–electrolyte interface. Under the action of electric field, the KOH and/or NaOH ions in the electrolyte dissociate into hydrogen ions through the negative electrode–electrolyte interface and enter the hexagonal interstitial position of AB_5 alloy. After that, the hydrogen ions migrate from the positive electrode to the negative electrode like a relay race, resulting in a decrease in the concentration of H^+ ions at the positive electrode–electrolyte interface. Then the H atom in the 2C crystallographic position of the positive active material β-$Ni(OH)_2$ leaves the lattice, loses electrons, becomes H^+ ion, and enters the electrolyte through the positive electrode–electrolyte interface. In this way, during charge the H atom on the positive side leaves the β-$Ni(OH)_2$ lattice and the H atom on the negative side is embedded in the interstitial position of the AB_5 alloy, the conductive mechanism of the charging process is H^+ ion transferred directionally from the positive electrode to the negative electrode.

Only when the Ni:O:H decreases from 1:2:2 to 1:2:1 in β-$Ni(OH)_2$ will the phase transition from β-$Ni(OH)_2$ to γ-NiOOH occur, and β-$Ni(OH)_2$ and γ-NiOOH coexist, so that the MH/Ni battery has a great overcharge. When the overcharge is large, the hydride AB_5H_x will precipitate from the negative active material.

The discharge process is contrary to the charging process. Under the action of a reverse electric field, the H^+ ions in the KOH and/or NaOH electrolyte are dissociated and reduced to hydrogen atoms through the positive electrode–electrolyte interface, which is then embedded into the β-$Ni(OH)_2$ lattice. After that, the hydrogen ions migrate from the negative electrode to the positive electrode like a relay race, resulting in a decrease in the concentration of H^+ ions at the negative electrode–electrolyte interface. Then the H atoms in the negative active material AB_5 leave the lattice, lose electrons, and become H^+ ions, which enter the electrolyte through the negative electrode–electrolyte interface. In this way, during discharge the H atom on the negative side leaves the AB_5 alloy lattice and the H atom on the positive side is embedded in β-$Ni(OH)_2$ lattice, the conductive mechanism of the discharging process is H ion transferred directionally from the negative electrode to positive the electrode.

In short, the physical conduction mechanism of MH/Ni battery is the directional migration of hydrogen ions formed by de-intercalation and re-intercalation in positive and negative electrode active materials.

Therefore, the following reaction formula can be written:

At positive At negative

Charge $\beta - Ni(OH)_2 \rightarrow \beta - Ni(OH_{1-x})_2 + 2xH^+ + 2xe^-$ AB_5
$+2xH^+ + e^- \rightarrow AB_5 - 2xH(\text{solid solution})$
$0 < X < 0.50$

Overcharge $2\beta - Ni(OH)_2 \rightarrow \beta - Ni(OH)_2 + \gamma - NiOOH$
$+H^+ + 2xe^-$ $AB_5 + H^+ + e^- \rightarrow AB_5H(\text{hydride})$

Discharge $\beta - Ni(OH_{1-x})_2 + 2xH^+ + e^- \rightarrow \beta - Ni(OH)_2 AB_5$
$-2xH \rightarrow AB_5 + 2xH^+ + 2xe^-$

Totalreaction $\beta - Ni(OH)_2 + AB_5 \leftrightarrow \beta - Ni(OH_{1-x})_2$
$+AB_5 - 2xH(\text{solid solution})$

Totalreactionofoverdischarge
$\beta - Ni(OH)_2 + AB_5 \leftrightarrow \beta - Ni(OH)_2$
$+\gamma - NiOOH + AB_5H(\text{hydride})$

References

1. Bode, H., T.K. Dehmel, and J. Witte. Zur kenntnis der nickelhydrox–idelktode—I. U ber das nickel (II)-hydroxidhydrat. *Electrochim Acta* 11: 1079–1087.
2. Shukla, A.K., S. Venugopalan, and Hariprakash, et al. 2001. Nickel-based rechargeable batteries. *Journal of Power Sources* 100: 125–148.
3. Guo Xin, L., ed. 2007. Introduction to new chemical power supply technology. Shanghai Science and Technology Press, 172–218.
4. Xin, Z.L., G.X. Li, C.Q. Wang, and J.S. Xiao. 1999. XRD in situ observation of Nickel electrode during charge and discharge. *Power Supply Technology* 23 (2): 140–142.
5. Wang, C.Q., Z.L. Xin, N. Wang, G.X. Li, and B.R. Wu. 1999. Quantitative phase analysis of-NiOOH on Nickel electrode. *Power Supply Technology* 23 (6): 328–331.
6. Ulrik, P., E. Lars, and G.G. Javier, et al. 2001. On the misuse of the structure model of the Ni electrode material. *Journal of Power Sources* 25 (1): 15–25.
7. Yu-Wan, Lou, Chuan-Zheng Yang, and Xi-Gui Zhang, et al. 2006. Comparative study on microstructure of -Ni(OH)$_2$ as cathode material for Ni-MH battery. *Science in China: Series E: Technological Sciences* 49 (3): 297–312; 36 (5): 467–482.
8. Li, Y.X., C.Z. Yang, Y.W. Lou, and B.J. Xia. 2009. Study on the physical mechanism of conductivity in the charge-discharge process of MH/Ni battery. *Acta Chimica Sinica* 67 (9): 901–909.
9. Li, Y.X. 2009. Relationship between fine structure of MH/Ni battery and electrode active material. Master paper of Shanghai Institute of Microsystems and Information Technology.

10. Yang, C.Z., Y.W. Lou, and B.J. Xia. 2009. Chemical and physical phenomena and mechanism of Ni-MH battery during charge and discharge. *Journal of Jishou University (Natural Science Edition)* 30 (6): 54–58.
11. Lou, Y.W., C.Z. Yang, J. Zhang, and B.J. Xia. An X-ray diffraction in-situ testing device for electrode charge-discharge process. Invention patent CN 100373168C.
12. Wang, Hong, and Zhu-Yan Liu. 2004. Crystallographic analysis of maximum hydrogen storage capacity of $LaNi_5$. *Rare Metal Materials and Engineering* 33 (3): 239–240.
13. Yuan Manxue, Han Jianwen, Lai Chengming 1998. Theoretical study of hydrogen storage state in $LaNi_5$. Hydrogen storage materials, *Journal of Nankai University (Natural Science)*, 31 (1): 84–89.

Chapter 11
Mechanism of Charge–Discharge Process of Graphite/$LiCoO_2$ and Graphite/$Li(Ni_{1/3}Co_{1/3}Mn_{1/3})O_2$ Batteries

In $Li(Ni_xCo_yMn_{1-x-y})O_2$, the x, y can be equal to 0–1; if $x = 0$, $y = 1$, one is $LiCoO_2$; if $x = 0.4$, $y = 0.6$, then $Li(Ni_{0.4}Co_{0.6})O_2$; if $x = 1/3$, $y = 1/3$, $Li(Ni_{1/3}Co_{1/3}Mn_{1/3})O_2$. Visible, graphite/$Li(Ni_xCo_yMn_{1-x-y})O_2$ includes many kinds of batteries. But because their positive active materials have the same crystal structure and their negative electrodes are all graphite, they are discussed together, represented by graphite/$LiCoO_2$ and graphite/$Li(Ni_{1/3}Co_{1/3}Mn_{1/3})O_2$ batteries.

Lithium-ion battery is a new generation of green high-energy battery developed in the early 1990s, which has been widely used because of its high energy density, high working voltage, long cycle life, and low self-discharge rate. $LiCoO_2$ has the characteristics of high capacity, stable discharge voltage, long life, and stable performance, so it is the preferred positive active material for lithium-ion batteries. However, in many studies, few researchers pay attention to the structural changes of electrode active materials and the de-intercalation behavior of lithium during the charge–discharge process. In 1992, Reimers and Dahn [1] used an in-situ X-ray diffraction device to study and found that when Li_xCoO_2 is in the charging process of $x = 0.5$, due to lattice distortion, O–Li–O–Co–O–O …., ABCABC… the rhombohedral structure of the stacked R-3 m is transformed into a monoclinic structure. Amatucci et al. [2] considered that the positive end-product of the battery at full charge is that CoO_2, has a hexagonal structure. MeBreen et al. [3, 4] have studied the phase transition of $Li_{1-x}CoO_2$ during charging by means of in-situ synchrotron radiation X-ray diffraction ($\lambda = 1.195$ Å). It is concluded that $Li_{1-x}CoO_2$ changes into monoclinic M2 phase at $0.75 < x < 0.85$, and from hexagonal phase H2 of $CdCl_2$-type to $O1_A$ phase of CdI_2-type at $0.77 < x < 1.00$, and finally to O1 phase. The $O1_A$ phase belongs to the P63mc space group, the $a = 4.24$ Å, the $c = 6.864$ Å, and the O1 phase is CoO_2. However, these studies only focus on the phase transition of $LiCoO_2$ materials during battery charging and do not involve the changes of their microstructure (including microcrystals and micro-strain, etc.). At the same time, they do not systematically study the structural changes of positive and negative active materials during charge and discharge of the battery as a whole.

C. Yang et al., *Materials and Working Mechanisms of Secondary Batteries*,
https://doi.org/10.1007/978-981-19-5955-4_11

At present, most of the industrialized lithium-ion batteries use $LiCoO_2$ as the positive active material. However, for the consideration of safety and comprehensive performance, $Li(Ni,Co)O_2$ and $Li(Ni,Co,Mn)O_2$ materials with layered structure are considered to be the most promising substitutes for $LiCoO_2$ [5, 6]. As the position material of lithium-ion battery, in recent years, there are many researches on $Li(Ni,Co,Mn)O_2$ [7–9], and the speed of industrialization is also gradually accelerating.

Hexagonal graphite is the most widely used negative active material for lithium-ion secondary batteries. At present, its quality (including graphitization degree *g* and stacking disorder *P*) has an important influence on the capacity and performance of the battery.

At present, the reports on the positive and negative active materials of lithium-ion battery mainly focus on the preparation of the material, the electrochemical performance, and the charge and discharge performance of the battery [10]. Few people are concerned with the changes of fine structure of positive and negative active materials during charge and discharge. In this chapter, the crystal structure and microstructure changes of positive and negative active materials of graphite/$LiCoO_2$ [11] and graphite/$Li(Ni_{1/3}Co_{1/3}Mn_{1/3})O_2$ [12] lithium-ion batteries were studied systematically and carefully by means of X-ray diffraction, in order to understand the physical process of lithium intercalation in electrode active materials and the physical mechanism of battery working.

11.1 Structural Evolution of Negative Active Materials During Charge–Discharge for 2H-Graphite/$LiCoO_2$

The XRD patterns of graphite in the typical stage of charge and discharge of 18,6502H-graphite/$LiCoO_2$ lithium-ion battery are shown in Fig. 11.1a and b, respectively. In order to make a further phase analysis of them, the local magnification map of Fig. 11.1 and the characteristic peak positions of each phase in the PDF card are shown in Fig. 11.2, respectively. It can be seen from the figure that the structure of negative graphite has changed regularly during the charge and discharge of the battery. At the initial stage of battery charging (10% charging), the negative material still shows the characteristic peak of 2H-graphite, but it shifts to a low angle and broadens at the same time; as the battery continues to charge, the negative material changed from 2H-graphite structure to LiC_{24}, LiC_{12} and LiC_6 in turn. In this series of phase transition process, there is a moment when two phases coexist in negative materials. Such as 2H-graphite and LiC_{24} coexist in the negative electrode when the battery is charged at 20%; LiC_{12} and trace LiC_6 coexist in the negative electrode when charging 40–60%; in the subsequent charging process, the intensity of the characteristic peak of LiC_{12} in the negative material decreases and the intensity of the characteristic peak of LiC_6 increases, indicating that the content of LiC_{12} and LiC_6 phase has changed until the battery is charged to 125%, the LiC_{12} phase in

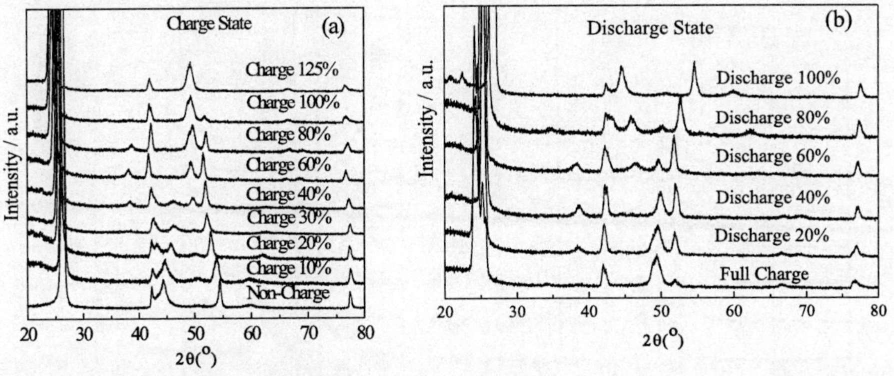

Fig. 11.1 XRD patterns of negative materials in several main stages of charge **a** and discharge **b** of 18,650 2H-graphite/$LiCoO_2$ battery

the negative electrode disappears, and the negative electrode material is completely transformed into LiC_6 phase.

The above phenomenon can be explained as follows: When the battery is charged, the Li atoms are embedded in the interstitial site between the hexagonal grids of carbon atoms in the 2H-graphite lattice, and the solid solution is first formed, which will not destroy the 2H-graphite structure, which shows that the negative electrode still shows the characteristic peak of graphite when the battery is charged at 10%, and then the solid solution gradually reaches saturation; that is, after the battery is charged for 20%, the negative electrode will precipitate LiC_{24}, LiC_{12}, and LiC_6 compound phases.

The discharge process of the battery is basically the reverse process of charging. However, by comparing the XRD patterns of several typical stages of negative materials during battery charge and discharge, it can be found that the phase transition of negative materials in the process of battery charge and discharge is not completely reversible. After fitting the XRD spectrum with Jade 6.5 software, the data obtained are calculated and analyzed as follows.

11.1.1 Change of Macroscopic Strain in Graphite During Charge–Discharge of 2H-Graphite/$LiCoO_2$ Battery

The variation curve of macroscopic strain along graphite *a*-axis and *c*-axis with the depth of battery charge and discharge is shown in Fig. 11.3. As can be seen from the figure, the macroscopic compressive strain along the *a*-axis increases with the increase of the charging percentage of the battery, then decreases gradually after 30%, and then becomes the tensile strain, while the macroscopic strain along the *c*-axis is the tensile strain and increases with the increase of the charging depth. This shows that when Li atoms are embedded in graphite, it is preferred to enter the interstitial position

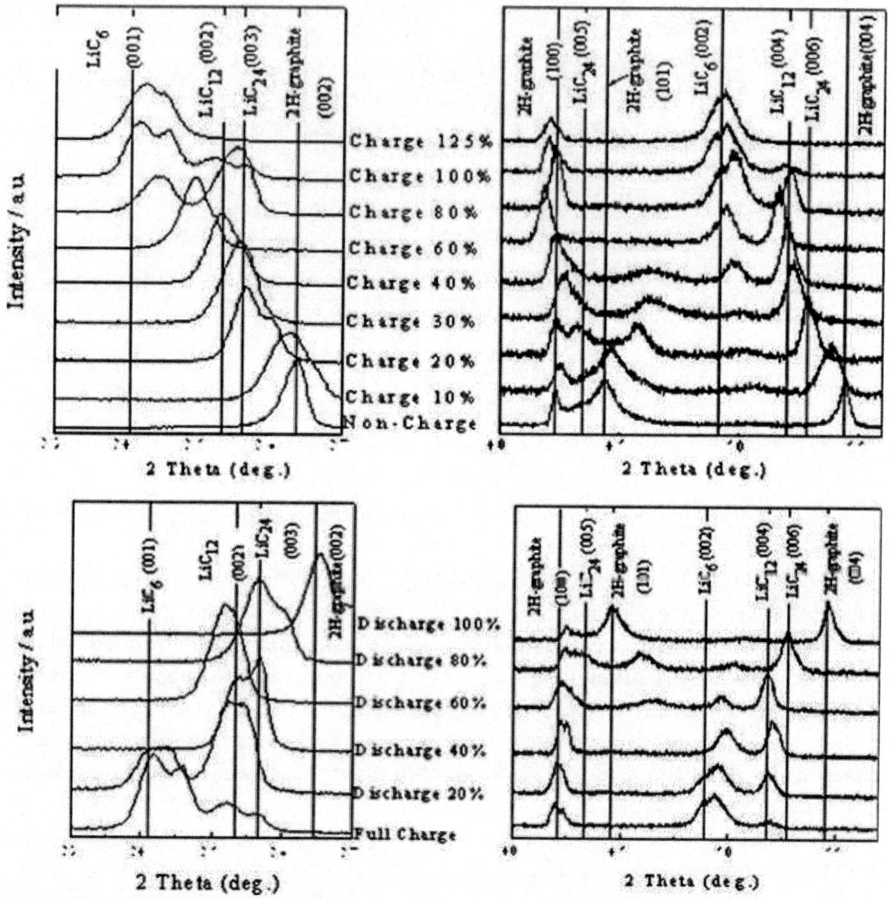

Fig. 11.2 Local magnification of **a** and **b** as well as position of characteristic peaks of each phase in PDF card

between the hexagonal grids of carbon atoms, which leads to the increase of graphite lattice parameter *c*. However, at some stage, the macroscopic strain values along the *a*-axis and *c*-axis fluctuate, which may be related to the phase transition. The change law of graphite along *a*-axis and *c*-axis in the discharge process is opposite to that in the charging process, but their change curves do not coincide with each other, and after the discharge is over, they no longer return to the uncharged value, which indicates that in the whole battery charge and discharge process, the lithium intercalation and de-lithium process of the negative electrode is not completely reversible.

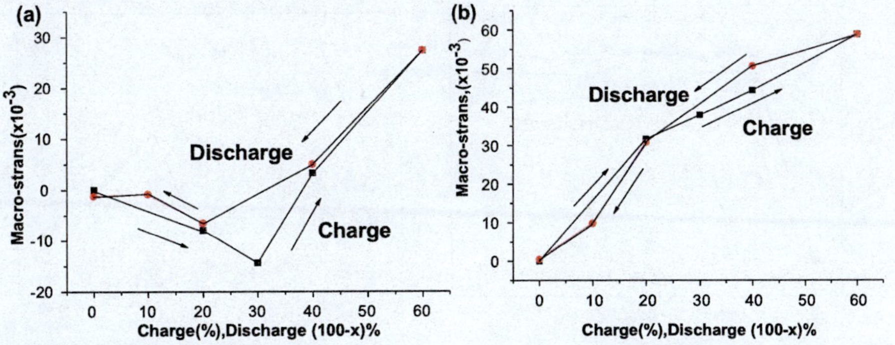

Fig. 11.3 Variation of macroscopic strain of extended graphite *a*-axis **a** and *c*-axis **b** with charge and discharge in graphite/$LiCoO_2$ battery

11.1.2 Changes of Graphite Micro-strain and Stacking Disorder in 2H-Graphite/$LiCoO_2$ Battery

The true line shape contains the information of crystal Imperfections (such as small grain size and micro-strain). The diffraction line broadening effect can be caused by microcrystal refinement, microscopic stress and crystal defects, or multiple effects exist at the same time. This method is briefly introduced in combination with the calculation process of graphite micro-strain and stacking disorder.

The change of micro-stress of graphite lattice caused by lithium-ion intercalation is reasonable, and it may also lead to the change of stacking disorder. The grain refinement effect can be ignored in the charge–discharge process, so there is only micro-strain-stacking disorder effect.

It can be seen from Fig. 11.1 that in most cases, only FWHM data of 002, 100, 101, and 004 can be obtained from the XRD pattern of negative samples, but the valid FWHM data of 102, 103, 202 cannot be obtained. In other words, it cannot satisfied the requirement of n_{even} number n_{odd} number ≥ 2 and cannot be solved by the general least square method [13–15]. However, it can be solved by the following simplified method; that is, if the shape of the microcrystal is polyhedral or nearly equiaxed, and ε_{002}, ε_{100}, and ε_{004} are roughly the same.

$$\overline{\varepsilon} = (\varepsilon_{002} + \varepsilon_{100} + \varepsilon_{004})/3 \tag{11.1}$$

Substitute the following formula

$$\frac{\beta_{101}\cos\theta_{101}}{\lambda} = \frac{\cos\varphi_{Z101}}{2c}\frac{\cos\theta_{101}}{\lambda}P + \overline{\varepsilon}\frac{4\sin\theta_{101}}{\lambda} \tag{11.2}$$

To enter the relevant data:

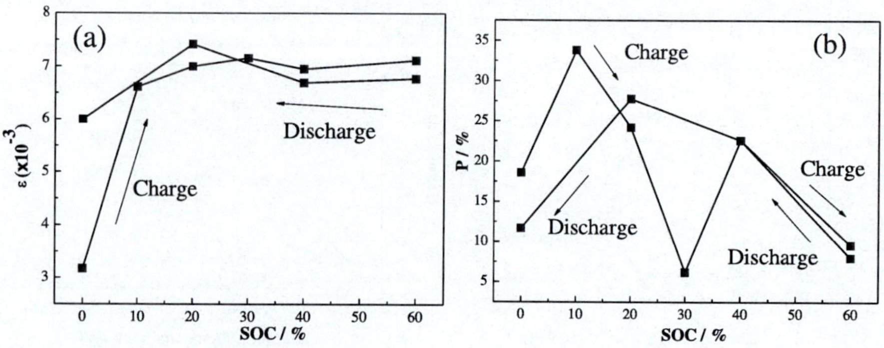

Fig. 11.4 Variation of micro-strain ε **a** and stacking disorder P **b** of graphite in graphite/$LiCoO_2$ battery with charging and discharging state (SOC)

$$0.6010\,\beta_{101} = 9.0806 \times 10^{-3} P + 0.9810\,\overline{\varepsilon} \tag{11.3}$$

P can be obtained [13]. The calculated results of ε and P are shown in Fig. 11.4a, b, respectively. As can be seen from the figure, at the initial stage of battery charging (to 10%SOC), the micro-strain of the negative material increases with the Increase of charged (SOC) increase of the charge percentage of the battery, which is caused by the intercalation of Li into the graphite layer; but the micro-strain of the negative material remains basically unchanged, which is related to the phase transition in the negative material. On the other hand, the change of stacking disorder is more complex, and it fluctuates in the process of battery charging, which is also related to the negative material phase transition. The change trend of ε and P in the discharge process of the battery is roughly opposite to that of the charging process, but the position of the inflection point does not coincide with that of charging, indicating that the microstructure change of the negative electrode material in the charging and discharging process of the battery is not completely reversible.

11.2 Structural Evolution and Microstructure of Positive Active Materials for 2H-Graphite/$LiCoO_2$ Batteries [11]

11.2.1 Phase Analysis of Positive Active Materials

The XRD patterns of the positive active material $LiCoO_2$ at different stages of charge and discharge are shown in Fig. 11.5a and b, respectively. It can be seen from the figure that in the battery charging process, the change of the (006) peak in XRD pattern is more significant. At the initial stage of the battery charging (10% of the charge), it splits at first, then moves to a low angle as the battery charge state continues

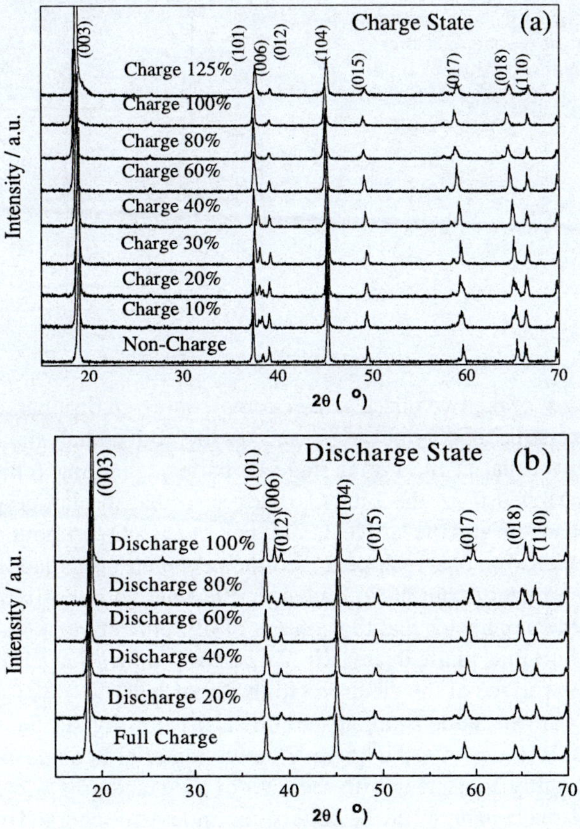

Fig. 11.5 X-ray diffraction patterns of positive materials in several main stages of charge **a** and discharge **b** of 18,650 graphite/$LiCoO_2$ battery

to increase, and finally disappears; the other diffraction peaks have no other obvious changes except a small shift.

In order to clearly investigate the structural changes of the positive material when the battery is charged by 10–30%, the three diffraction patterns are magnified in Fig. 11.6 after removing the $K\alpha_2$ component. As can be seen from the diagram, from the battery charging to 10%, the four diffraction peaks (006, 015, 107, and 018) in the spectrum are all split into two peaks. Corresponding to the crystal plane index (hkl) of each peak, it can be seen that the higher the l value or the larger the value of l, the more obvious the splitting of the diffraction lines. At the same time, it was found that at the initial stage of charging (10% charging), the intensity of the low-angle peak (left peak) was less than that of the high-angle peak (right peak). As the battery continues to charge (to 20%), the low-angle peak intensity increases and the high-angle peak intensity weakens. The intensity of the two peaks is reversed. The lattice parameters of the two peaks are calculated as two independent phases, as follows:

When the battery is charged at 20%, the lattice parameters corresponding to the low-angle peak are as follows: $a = 2.8114$ Å, $c = 14.1670$ Å, and the high-angle peaks are: $a = 2.8104$ Å, $c = 14.0983$ Å, respectively. It can be seen that the lattice

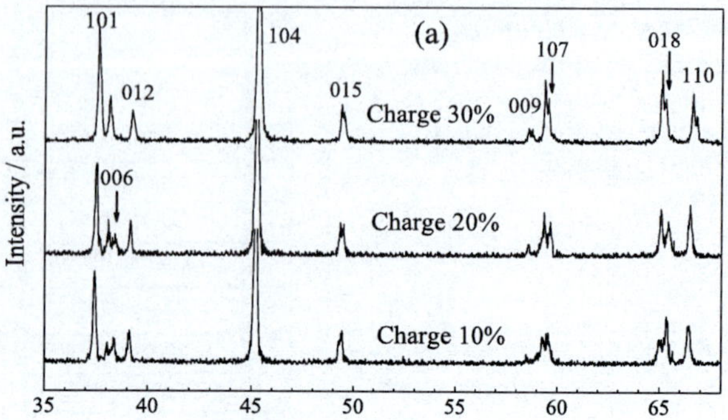

Fig. 11.6 XRD pattern of positive materials after removal of $K\alpha_2$ at 10, 20, and 30% battery charge

sizes of the two phases are different, but from the characteristic peak, the two phases are still $LiCoO_2$ phase. Based on the above analysis, the following explanation is given that at the initial stage of battery charging (charging to 10%), due to Li is detached from the $LiCoO_2$ lattice, so that the lattice parameter of the material is changed, and the lattice parameters of $LiCoO_2$ without Li are unchanged. Therefore, two R-3 m phases appear, and the new phase is called lithium deficient phase; when the battery continues to charge, the relative intensity of the two peak is reversed, which indicates that the content of lithium deficient phase increases and the content of original phase decreases. This clearly shows that $LiCoO_2$ de-Li has a process from the surface of the electrode to the inner layer.

At the same time, as can be seen from Fig. 11.5a, when the battery is charged to 30% or 40%, the positive active material becomes single-phase again, which is roughly consistent with the data of the standard $LiCoO_2$, only the lines have some displacement; when the battery continues to be charged to 60, 80, 100, 125%, the other lines are still consistent with the standard $LiCoO_2$ data, except for the disappearance of 006 lines and no new phase is generated. This is different from the reports of MeBreen, etc. [3, 4], who believe that when the battery is charged to about 60%, H2a phase appears in the positive material $Li_{1-x}CoO_2$; when the battery is charged to about 100%, the positive material $Li_{1-x}CoO_2$ is transformed into the hexagonal phase H2 of $CdCl_2$-type. Today, of the experimental data of charging, 60 and 100% are included in Table 11.1, and the relevant data that can be found are listed in Table 11.1 by comparison. It is not difficult to judge by comparing these data carefully. They still belong to R-3 m structure, and only the lattice parameters change, so that 006 and 012 overlap. For this reason, they have been indexed by R-3 m, and their lattice parameters have been calculated, which are also listed in Table 11.1.

These results further show that there is no phase transition in the charging process, as described in reference [1–4], but the diffraction pattern changes only due to the change of lattice parameters and diffraction intensity of $Li_{1-x}CoO_2$ caused by leave-Li, and there is no phase transition in the samples with full charge 125% charge.

Table 11.1 XRD data and related data of 60 and 100% positive active materials

Charge 60%	Charge 100%		$Li_{0.4}CoO_2$		Monoclinic unit cell		2H hexagonal unit cell		O1 hexagonal unit cell		CoO_2 hexagonal cell	
d (Å)	d	hkl	d	hkl	d	hkl	d	hkl	d	hkl	d	hkl
4.753	4.795	003	4.615	003	4.739	003	4.784	003	4.267	001	4.304	003
2.937	2.396	101	2.410	101	2.385	111	2.404	101			2.439	100
2.301	2.302	012	2.308	006	2.303	202	2.309	102	2.448	100	2.120	103
			2.307	012	2.288	112						
2.011	2.015	104	1.998	104	2.012	114	2.012	104	2.116	101		
1.854	1.859	015	1.834	015	1.993	204	1.853	105				
					1.846	115						
1.589	1.603	009	1.5384	009	1.571	117	1.571	107	1.602	102	1.610	106
1.566	1.572	107	1.5382	107	1.554	207						
					1.452	208						
1.441	1.448	018	1.4130	018	1.439	118	1.446	108				
1.404	1.403	110	1.4127	110	1.401	020	1.407	110	1.414	110	1.410	110
1.347	1.347	113	1.3508	113			1.350	113	1.341	111	1.341	113
R-3 m $a = 2.8117$ $c = 14.274$	R-3 m $a = 2.8105$ $c = 14.385$		R-3 m Hexagonal cell $a = 2.8254$ $c = 13.846$		Monoclinic cell $a = 4.865$ $b = 2.806$ $c = 14.420$ $\beta = 90.77°$		Hexagonal cell $a = 2.813$ $c = 14.370$		Hexagonal cell $a = 2.828$ $c = 4.237$		Hexagonal cell $a = 2.8222$ $c = 12.8787$	
Reference	PDF card 44-0146		[1]		[3]		[3]		[2]			

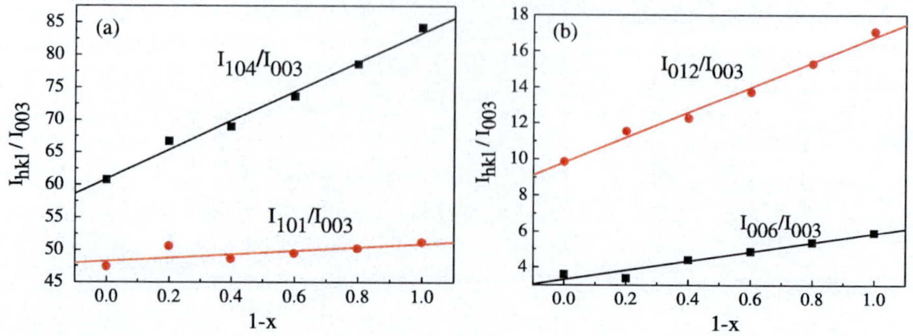

Fig. 11.7 Relationship between the intensity of the main diffraction lines of $Li_{1-x}CoO_2$ and $1-x$

There may be two reasons for the disappearance of 006 lines:

(1) The diffraction intensity of 006 decreases due to the removal of Li. For this reason, we use the power.cell program, according to the $Li_{1-x}CoO_2$ model, order $x = 0.0$ ~ to 1.0 to calculate the relative intensity of each diffraction line. The relationship between the relative intensity of the main diffraction lines and $1-x$ is shown in Fig. 11.7. It can be seen that when $1-x = 0$ the diffraction intensity of 006 decreases very low and is not visible.
(2) It is possible that 006 overlaps with 101 or 012 due to the change of lattice parameters.

Comparing Fig. 11.5a and b, it can be seen that the structural change of the positive material $LiCoO_2$ during the discharge process of the battery is basically the reverse process of the battery charging.

11.2.2 Change of Macroscopic and Microcosmic Strain of $LiCoO_2$ During Battery Charge and Discharge

(1) The change of macroscopic strain of $LiCoO_2$

Figure 11.8 shows the variation curve of macroscopic strain along *a*-axis and *c*-axis with the percentage of charge-discharge of positive active material $Li_{1-x}CoO_2$ during charge-discharge process. Thus, it can be seen that the macroscopic compressive strain along the *a*-axis increase with the increase of the charging percentage, while tension strain increases long the *c*-axis and the change rate of the charging percentage < 30% is slower; when the charging fraction is more than 30%, the change rate is much larger, especially along the *c*-axis. The discharge process is roughly reversible, but not completely reversible.

(2) The change of micro-strain of $LiCoO_2$

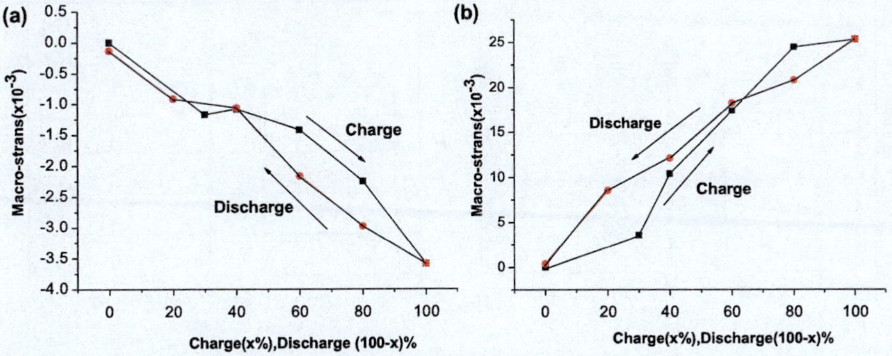

Fig. 11.8 Variation of macroscopic strain of $LiCoO_2$ extension *a* axis (**a**) and extension *c*-axis (**b**) with the charge and discharge charge state of the battery

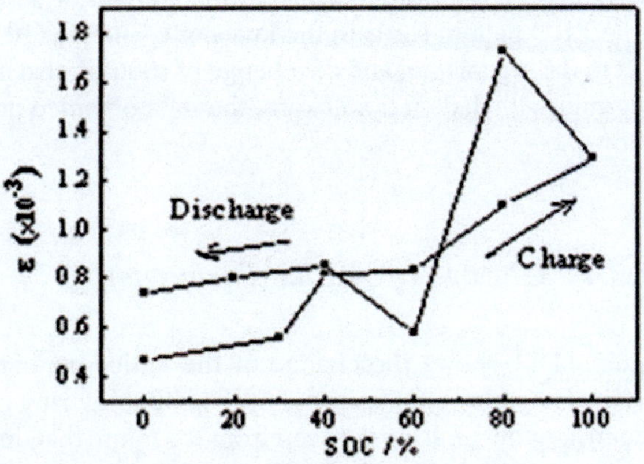

Fig. 11.9 Variation of micro-strain ε^{II} of cathode material $LiCoO_2$ **a** with battery charge and discharge percentage (SOC)

The variation of the micro-strain ε^{II} of $Li_{1-x}CoO_2$ with the percentage of charge and discharge or the amount of de-Li *X* during the charge–discharge process is shown in Fig. 11.9. Although the data are dispersed, it can still be seen that the micro-strain increases gradually with the increase of charging percentage, while the discharge is on the contrary and is not completely reversible.

11.3 Structural Evolution of Positive Active Materials During Charge–Discharge of Graphite/$Li(Ni_{1/3}Co_{1/3}Mn_{1/3})O_2$ Battery [12]

The diffraction patterns of positive active material $Li(Ni_{1/3}Co_{1/3}Mn_{1/3})O_2$ at different stages of charge and discharge are shown in Fig. 11.10a and b. It can be seen from

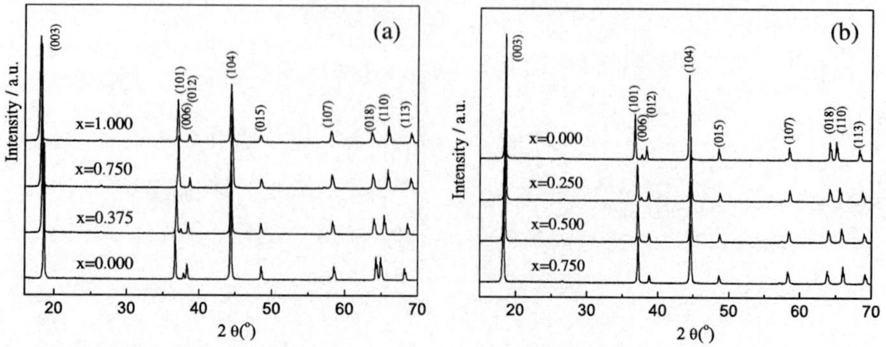

Fig. 11.10 Main stages of charge and discharge of 18,650 graphite/$Li(Ni_{1/3}Co_{1/3}Mn_{1/3})O_2$ battery X-ray diffraction pattern. **a** Charging process and **b** discharge process of positive material

the figure that with the increase of the amount of de-Li x, the 003 diffraction peak in the XRD pattern shifts to the low angle, and the (101) and (104) diffraction peaks shift to the high angle, and the change of (006) is also obvious. After fitting the XRD spectrum with Jade 6.5 software, the data obtained are calculated and analyzed as follows.

11.3.1 Change of Lattice Parameters

Figure 11.11 shows the change of the lattice parameters a and c of the positive $Li(Ni_{1/3}Co_{1/3}Mn_{1/3})O_2$ material of the battery during charge and discharge with the amount x of de-Li. It can be seen from the figure that during the charging process, the lattice parameter a decrease with the increase of de-Li quantity x, while c increases with the increase of de-Li quantity x, which is basically the same as $LiCoO_2$, while the discharge process is on the contrary, an increase with the decrease of de-Li quantity x, and c decreases accordingly. However, when x returns to 0, both a and c do not return to the original values, indicating that the lithium removal process of $Li(Ni_{1/3}Co_{1/3}Mn_{1/3})O_2$ is not completely reversible.

11.3.2 The Change of Micro-strain

Figure 11.12 shows the relationship between the micro-strain ε and the Li removal amount x of $Li(Ni_{1/3}Co_{1/3}Mn_{1/3})O_2$. The $\varepsilon_{average}$ represents the average of both ε_{003} and ε_{101}. It can be seen that the micro-strain ε obtained by using the diffraction peaks of (003) and (101) increases with the increase of the amount of de-Li of the material x.

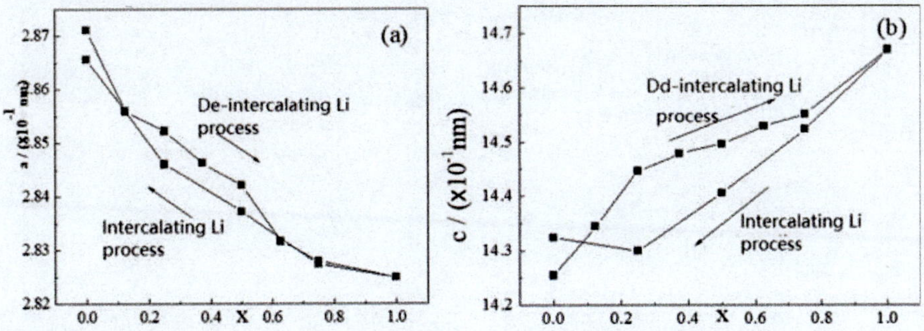

Fig. 11.11 Variation of lattice parameters *a* **(a)** and *c* **(b)** of cathode material $Li(Ni_{1/3}Co_{1/3}Mn1/3)O_2$ with de-lithium content *x*

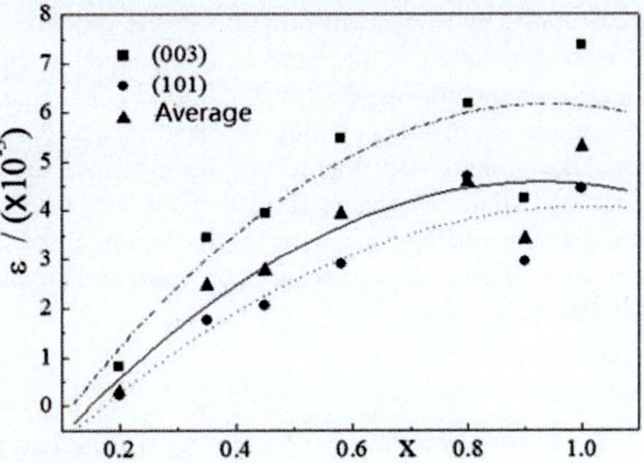

Fig. 11.12 Variation of micro-strain ε of $Li(Ni_{1/3}Co_{1/3}Mn_{1/3})O_2$ with de-lithium content *x*

11.3.3 The Change of Relative Diffraction Intensity

According to the relationship between the mixed occupation parameter *X* and the diffraction intensity ratio in Chap. 5, Table 5.5 [16, 17]. *z* occupation position of oxygen atom in 6c.

$Z = 0.24$	$Z = 0.25$
$X = 0.5269 - 0.4427(I_{003}/I_{104})^{1/2}$	$X = 0.4742 - 0.4496(I_{003}/I_{104})^{1/2}$
$X = 0.4474 - 0.2638(I_{101}/I_{012})^{1/2}$	$X = 0.4656 - 0.2635(I_{101}/I_{012})^{1/2}$
$X = 0.4505 - 0.5895(I_{101}/I_{104})^{1/2}$	$X = 0.4900 - 0.6372(I_{101}/I_{104})^{1/2}$

The relative intensity ratios I_{104}/I_{101} and I_{012}/I_{101} of three characteristic peaks (104, 012, and 101) in the $Li(Ni_{1/3}Co_{1/3}Mn_{1/3})O_2$ XRD spectrum were measured. The experimental curve of the change during the charge and discharge of the battery is shown in Fig. 11.13, and using model of $Li_{1-x}Ni_x(Li_xNi_{1/3-x}Co_{1/3}Mn_{1/3})O_2$, the variation curve of theoretical diffraction intensity ratio of $Li(Ni_{1/3}Co_{1/3}Mn_{1/3})O_2$

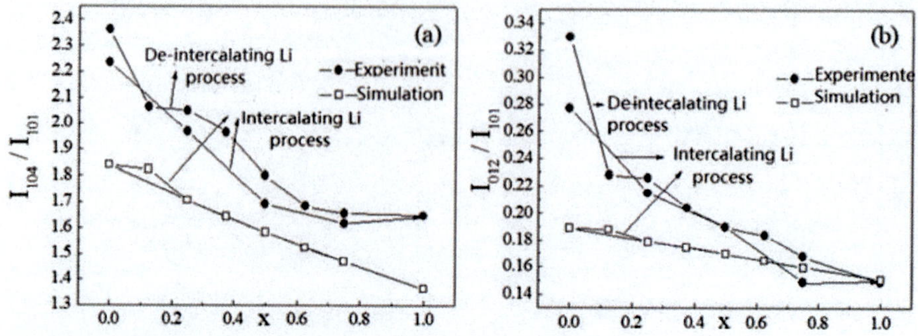

Fig. 11.13 Variation of relative intensity ratio I_{104}/I_{101} **(a)** and I_{012}/I_{101} **(b)** of $Li(Ni_{1/3}Co_{1/3}Mn_{1/3})O_2$ with de-lithium content x

is calculated by Power.cell program in the process of lithium removal [8]. It can be seen that the changing trend of the theoretical curve is consistent with that of the experimental curve. However, the theoretical charge curve and discharge curve are completely overlapped, but the experimental curve is not the case, the charge discharge process is different, and the full discharge state may not return to the uncharged state, which indicates that the actual charge–discharge process is not completely reversible, which indicates that the mixed occupation of Ni and Li may also occur in the charge–discharge process, or there may be mixed discharge of raw materials.

11.4 Comparison of Charge and Discharge Process of Graphite/$LiCoO_2$ Battery Before and After Storage [18]

11.4.1 Comparative Analysis of Negative Active Materials

Figure 11.14a shows that the XRD pattern of the negative active material in several stages of the charging process after of storage 100 days at 55 °C, (b) is partially magnified. The phase existence case of the charging process before and after storage is summarized in Table 11.2. It can be seen from Fig. 11.14 and Table 11.2 that during the charge and discharge of the battery the structure of the negative active material evolved regularly both before and after storage.

(1) The uncharged state before storage and the initial charging stage of the battery (10% charging) are 2H-graphite, and only after 20% charging does the LiC_{24}; battery continues to charge, and the negative material changes to LiC_{24}, LiC_{12}, and LiC_6 in turn. In this series of transition process, there is a moment when two different lattice parameters of the same structure coexist in the negative

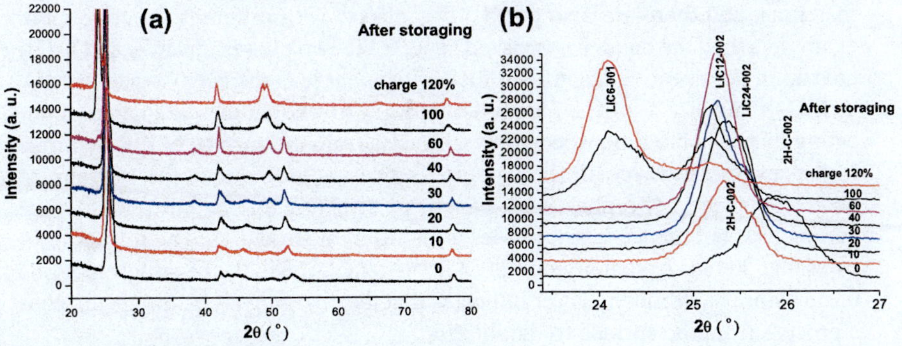

Fig. 11.14 XRD patterns of negative active materials in several stages of charging process **a** after 55 °C storage for 100 days **b** is its local magnification

Table 11.2 Results of phase analysis of negative active materials in several main stages of battery charging before and after storage

Before storage				After storage 100 days at 55 °C 100%SOC	
Charge degree	Existing phase	Discharge depth	Existing phase	Discharge depth	Existing phase
0	2H-graphite	100	2H-graphite	0	2H-graphite + little 3R-graphite
10	2H-graphite			10	2H-graphite + little LiC_{24}
20	2H-graphite + LiC_{24}	80	2H-graphite	20	LiC_{24}
30	2 kinds of LiC_{24}			30	LiC_{12} + LiC_{24}
40	LiC_{24}	60	LiC_{24}	40	LiC_{12} + little LiC_{24}
60	LiC_{12} + LiC_{24}	40	LiC_{12} + LiC_{24}	60	LiC_{12}
80	LiC_{12} + LiC_6	20	LiC_{12} + LiC_6		
100	LiC_6 + little LiC_{12}	0	LiC_6 + LiC_{12}	100	LiC_{12} + many LiC_6
				120	LiC_6 + little LiC_{12}
125	2 kinds of LiC_6				

Note The so-called two identical phases mean that they have the same structure but different lattice parameters materials, and there are also coexist two phases with different structures. For example, when the battery is charged at 60%, the two phases of LiC_{12} and LiC_{24} coexist in the negative electrode; 80% of the charge is the coexistence of LiC_6 + LiC_{12}; and 125% of the charge is the LiC_6 of two lattice parameters, indicating that the charging process has a process from the surface of the electrode to the inside. However, in the charging process after storage, there is almost no coexistence of two phases with the same structure and two different lattice parameters, but there are two phases with different structures. The former indicates that there is a phenomenon of Li homogenization in the negative electrode during storage, while the latter indicates that the charging after storage also has a process from the surface to the interior.

(2) The comparison before and after storage showed that the order of LiC_{24}, LiC_{12} and LiC_6 was also significantly different. Before storage charge, the electric depth reaches 80%, there are LiC_6; after storage, the charge depth reaches 100 and 120%. Although it is mainly LiC_6, it still contains LiC_{12} phase, which clearly indicates that the storage slows down the formation of high-carbon Li-C compounds.

(3) The typical data of FWHM of diffraction lines of the same phase before and after storage are listed in Table 11.3. It can be seen that the half width of diffraction lines increases with the increase of charging depth before storage, and on the contrary, the half width of diffraction lines decreases with the increase of charging depth after storage. If the grain refinement effect in the charging process is ignored, the micro-strain of the negative active material before storage increases with the increase of the charging depth, and the change of before and after storage is just the opposite, and decreases with the increase of the charging depth.

Table 11.3 Typical data of FWHM of diffraction lines with the same phase during charging before and after storage

Charge situation before storage				Charge situation before storage			
Depth (%)	Phase	hkl	FWHM (°)	Depth (%)	Phase	hkl	FWHM (°)
80	LiC_6	001	0.439	30	LiC_{12}	002	0.527
		002	1.139	40	LiC_{12}	002	0.406
125	LiC_6	001	0.953	60	LiC_{12}	002	0.377
		002	1.248				
				100	LiC_{12}	002	0.542
					LiC_{12}	001	0.617
						002	2.187
				120	LiC_{12}	002	0.422
					LiC_{12}	001	0.439
						002	1.925

11.4.2 Comparative Analysis of Positive Active Materials

(1) Phase analysis of positive active materials

The XRD patterns of the positive active material $LiCoO_2$ at different stages of charging after storage are shown in Fig. 11.15a. It can be seen from the figure that in the battery charging process, the change of the (006) peak in the positive material XRD pattern is more significant. At the initial stage of the battery charging (10% of the charge), it splits at first, then moves to a low angle as the battery charge state continues to increase, and finally disappears; the other diffraction peaks have no other obvious changes except a small shift.

In order to clearly investigate the structural changes of the negative material when the battery is charged from 10 to 30%, the three diffraction patterns are partially magnified. This is shown in Fig. 11.15b. Comparing Fig. 11.15b with Fig. 11.6, the battery charging process after storage is much slower than that before storage.

The experimental data of 100% charging before and after storage are included in Table 11.4, and it is not difficult to make a judgment to compare these data carefully. They still belong to R-3m structure, and only the lattice parameters change, so that 006 overlaps with 101. For this reason, they have been indexed by R-3m, and their lattice parameters have been calculated, which are also listed in Table 11.4. These results further show that there is no phase transition in the charging process, as described in reference [1–4], but the diffraction pattern changes the change of lattice parameters and diffraction intensity of $Li_{1-x}CoO_2$ caused only by de-Li, and there is no phase transition in the samples. Even 125% full charge and overcharge, there is no phase transition.

(2) Lattice parameters of positive active materials

The changes of lattice parameters *a* and *c* of positive active material $LiCoO_2$ with charging depth before and after storage are shown in Fig. 11.16; the change trend

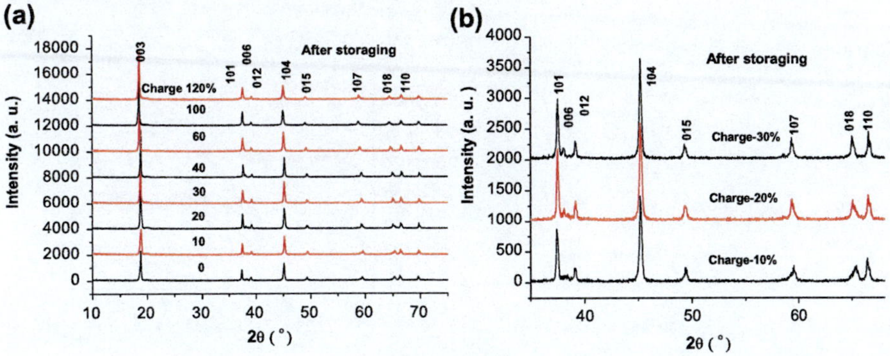

Fig. 11.15 **a** XRD pattern of positive active materials in several stages of charging process after 55 °C storage for 100 days, **b** partial magnification

Table 11.4 XRD data and related data of 60% positive active material $LiCoO_2$ before and after storage

Battery charging 100% before storage	Battery 100% charging after storage		44-0146		$Li_{0.4}CoO_2$	
d	d	hkl	d		hkl	
4.795	4.795	003	4.615		003	
2.396	2.397	101	2.410		101	
2.303	2.300	012	3.308	006	2.307	012
2.015	2.014	104	1.998		104	
1.859	1.857	015	1.834		015	
1.603						
1.572	1.571	107	1.5382	107	1.538	009
1.448	1.448	018	1.4130		018	
1.403	1.404	110	1.4127		110	
1.347	1.347	113	1.3508		113	
R-3 m, Hexagonal cell $a = 2.8117$ $c = 14.274$	R-3 m, Hexagonal cell $a = 2.8105$ $c = 14.385$		R-3 m, Hexagonal cell $a = 2.8254$ $c = 13.846$			

is consistent, but the details are different; this difference has exceeded the range of measurement error, so there is a certain difference in Li removal in the charging process of battery before and after reaction storage.

(3) Microstructure of positive active materials

Since there is no phase transition in the charging process of the positive material, and the change of lattice parameters is caused by the de-intercalation of Li atoms, then

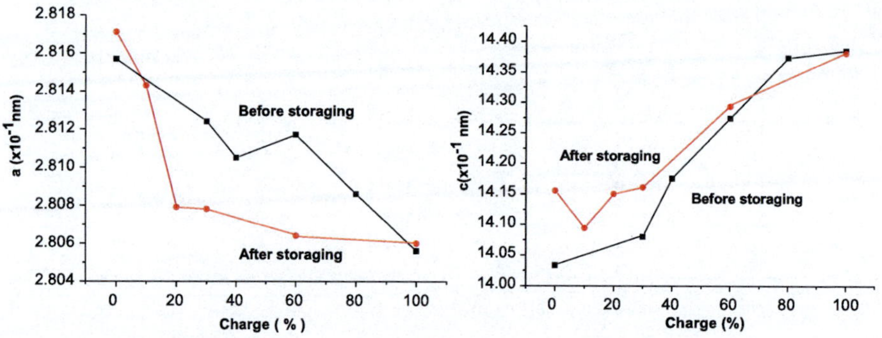

Fig. 11.16 Variation of $LiCoO_2$ lattice parameters *a* and *c* with charging depth before and after storage

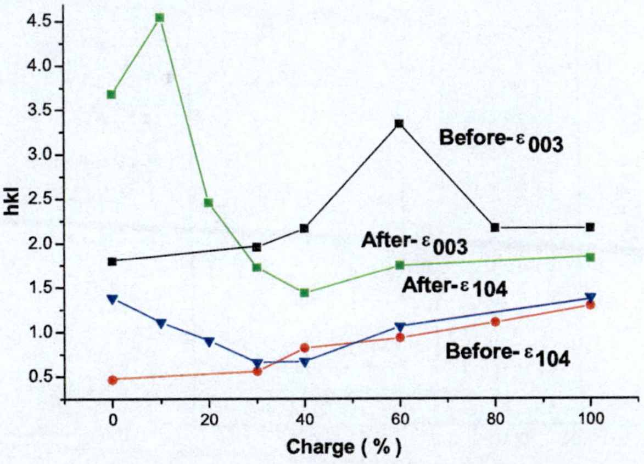

Fig. 11.17 Variation of micro-strain of battery cathode active material with charging depth before and after storage

the microstructure will change. When the grain refinement effect during charging is ignored, the variation of micro-strain with charging depth is shown in Fig. 11.17. It can be seen from Fig. 11.17 that there is little difference in the micro-strain in [104] direction before and after storage, but there is a great difference in micro-strain in [003] direction, which once again shows that there is a certain difference in de-Li during battery charging before and after storage.

11.5 Behavior of Electrode Active Materials During Charge and Discharge [11, 12]

11.5.1 Behavior of Graphite During Charge and Discharge

The structure character of graphite is the extension of the hexagonal grid plane composed of carbon atoms in the network plane, and along normal direction of the network plane, or the ABAB… stacking; or the ABCABC…stacking. The former belongs to 2H-graphite structure, P63/mmc (No. 194) space group, while the latter belongs to 3R-graphite, R3 (No.146) space group. Based on the experimental results of Sect. 11.2, some behaviors of graphite in the process of charge and discharge can be summarized.

With the beginning of the charging process, the Li atom is embedded in the graphite layer, and its priority is to enter the interstitial position between the hexagonal grid faces of graphite, so that the lattice parameters a and c of 2H-graphite, as well as the micro-strain ε and stacking disorder P, are increased, as shown in Figs. 11.3 and 11.4. Because the ABAB sequence stacking of 2H-graphite is not as dense as the dense hexagonal structure, there is a large gap space between the layers, which can hold a large number of interstitial atoms, so in general, it will not form Li-C compounds, but will produce larger strain and stacking disorder. When the stacking disorder reaches

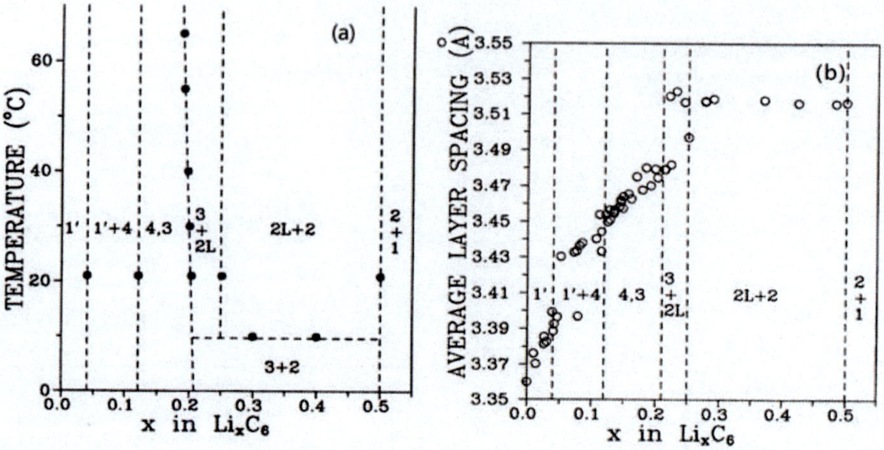

Fig. 11.18 When embedding Li in 2H-Graphite Li, Li_xC_6 phase diagram under 0–70 °C range (**a**) and the variation of average interlayer spacing of graphite grid with x measured at 21 °C (**b**) [19]

a certain value, it will be accompanied by the gradual precipitation of 3R-graphite, as shown in Fig. 11.1. Only when the battery is fully charged or overcharged will the LiC_{24}, LiC_{12}, and LiC_6 phases be formed. Due to the precipitation of the new phase, there is a process of nucleation and growth, so the coexistence of two or three phases will occur in this process.

In fact, it is not that simple. According to the results of Dahn "phase diagram of Li_xC_6" [19] and discuss of Gao Jian et al to phase diagram and phase transition [21]. it is pointed out that the Li in 2H-graphite can be divided into several stages: 1', 4, 3, 2L, 2 and 1, as shown in Fig. 11.18a. Figure 11.18b shows the average spacing of the corresponding graphite layers.

Theoretically, when $x = 0.25$, LiC_{24} precipitates, $x = 1/3$, LiC_{18}, $x = 0.5$, LiC_{12}, $x = 1$, LiC_6 precipitates. When x is not equal to 0.25, 1/3, 0.5, 1.0, the coexistence of the two phases may occur. For example, if $x < 0.25$, 2H-graphite + LiC_{24} may coexist, and for example, $0.25 < x < 0.5$ there may be $LiC_{24} + LiC_{12}$ and $0.5 < x < 1.0$, the coexistence of $LiC_{12} + LiC_6$ may occur. But the actual situation is more complicated, because there is a process from the surface of the electrode to the inside of the electrode, so there will be two phases with different lattice parameters but the same structure.

However, from the point of view of crystallography, the changes of phase lattice parameters should be investigated, as shown in Table 11.5. According to the lattice parameters in the table, compared with 2H-graphite, both a and c of LiC_{24} have little change, and Li is trapped between hexagonal grids, but the a of LiC_{18}, LiC_{12}, and LiC_6 is much larger than that of hexagonal graphite, indicating that not only Li atoms are inserted in hexagonal grids, but also Li atoms are inserted in 1/2 0 0 and 0 1/2 0 on the hexagonal base. In addition, from the point of view of their crystal structure, they all belong to the same space group P6/mmm, which seems to be a kind of continuous solid solution with different Li content.

During battery discharge, the lithium removal from the negative electrode is the opposite process; that is, under the action of the reverse electric field, the Li atoms leave the interstitial position of the carbon electrode, enter the electrolyte through the negative electrode–electrolyte interface, and then the decomposition of the new phase (Li-C compound) continues to provide Li ions. With the departure of Li atoms, the lattice parameters, micro-strain and stacking disorder of graphite decrease. As can be seen from Figs. 11.3 and 11.4, the above process is not completely reversible.

11.5.2 Behavior of $Li(Ni_xCo_yMn_{1-x-y})O_2$ During Charge and Discharge

Like $LiCoO_2$, $Li(Ni_xCo_yMn_{1-x-y})O_2$ also belongs to the R-3 m (No. 166) space group. There are 3 molecules in a unit cell, a total of 12 atoms, and their occupation in the unit cell is as follows:

$$\left.\begin{array}{ll} \text{Li} & 000 \\ \text{Ni, Co, Mn} & 001/2 \\ \text{O} & 00 \pm Z \end{array}\right\} + (000);\ (1/3\ 2/3\ 2/3);\ (2/3\ 1/3\ 1/3)\ \text{Symmetrical}$$

operation

It is more clear that:

Li (3a) 位 000; 1/3 2/3 2/3; 2/3 1/3 1/3
Ni,Co,Mn (3b) 位 001/2; 2/3 1/3 5/6; 1/3 2/3 7/6
(6c) 位 001/4; 2/3 1/3 7/12; 1/3 2/3 11/12; 00-1/4; 2/3 1/3 1/12; 1/3 2/3 5/12

The crystal structure model and bonding are shown in Fig. 11.19. As can be seen from the diagram, the chemical bond between the Li atom at position 000 and the neighboring atom is longer, and its binding force in the crystal cell is weak. When the charging starts, the Li atom at position 000 is preferentially leave from the crystal lattice, followed by the Li atoms at position 1/3 2/3 2/3 and 2/3 1/3 1/3 leave from the lattice, so when $x < 1/3$, the change rate of lattice parameters and lattice strain is larger, and then the change rate is slower, which is roughly consistent with the results of Fig. 11.9 in Fig. 11.8. In the process of discharge, it is roughly a reversible process, but it is not completely reversible.

After $Li(Ni_{1/3}Co_{1/3}Mn_{1/3})O_2$ begins to de-Li, the XRD intensity ratio of I_{104}/I_{101} can be calculated by Power.cell program according to the lithium deficiency model of $Li_{1-x}(Ni_{1/3-x}Co_{1/3}Mn_{1/3})O_2$. Compared with the experimental curve and the simulated curve, Both changing trends are same; the experimental curve is much more complex. There may be three reasons for this:

Table 11.5 Phase, crystal structure, lattice parameters, unit cell volume and relative volume change rate formed during the charging process of 2H-graphite/$LiCoO_2$ battery

Electrode materials	Phase	Structure	Space group	PDF No.	Lattice parameter ()				$(V-V_0)/V_0$
					a	*b*	*c*	*V*	
2H-graphite	C	Hexagon	P6/mmm	41-1478	2.4704		6.7244	35.54	0.0
	CLi_{24}	Hexagon	P6/mmm		2.47		6.711	35.46	−0.22
	CLi_{18}	Hexagon	P6/mmm		4.288		7.07	112.57	216.7
	CLi_{12}	Hexagon	P6/mmm	01-083-4147	4.288		7.066	112.51	216.6
	CLi_6	Hexagon	P6/mmm		4.305		7.036	112.92	217.7
$LiCoO_2$	$LiCoO_2$	Rhombic	R-3 m	44-0145	2.8237		13.8908	95.91	0.0
	$Li_{0.90}CoO_2$	Rhombic	R-3 m	01-083-5430	2.8154		14.0484	96.43	0.5
	$Li_{0.78}CoO_2$	Rhombic	R-3 m	04-015-9701	2.8150		14.1609	97.18	1.32
	$Li_{0.51}CoO_2$	Rhombic	R-3 m	01-083-6088	2.8027		14.389	97.88	2.05
	$Li_{0.45}CoO_2$	Rhombic	R-3 m		2.809		14.40	98.40	2.59
	$Li_{0.22}CoO_2$	Rhombic	R-3 m		2.810		14.00	95.73	−0.19
	CoO_2	Rhombic	R-3 m		2.822		12.897	88.94	−7.2

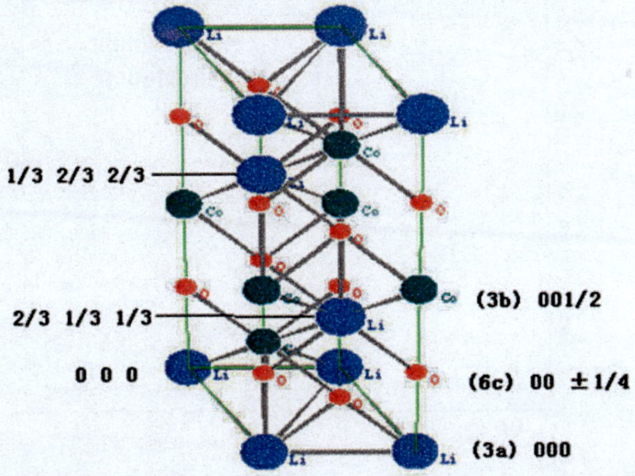

Fig. 11.19 Crystal structure model and bonding of $LiCoO_2$

First, there is a difference between the calculated de-Li amount x of the cathode material according to the charge state of the battery and the real value, which can also be seen in Figs. 11.20 and 11.21.

Second, the mixed arrangement of Li/Ni atoms at the 3*a*/3*b* position may occur in $Li_{1-x}(Ni_{1/3}Co_{1/3}Mn_{1/3})O_2$ materials during charge and discharge. In order to verify this point, we assume that the mixed arrangement parameter of Li/Ni atoms is y; then the molecular formula of positive active material can be expressed as

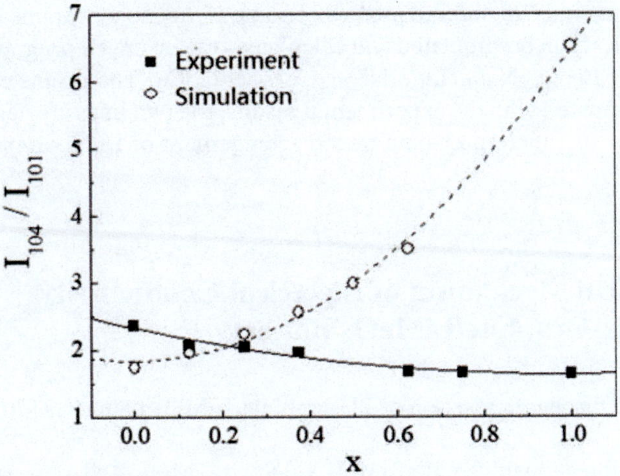

Fig. 11.20 Simulation intensity ratio curve I_{104}/I_{101} of $Li_{1-x}(Ni_{1/3}Co_{1/3}Mn_{1/3})O_2$ [model is $Li_{1-x-y}Ni_y(Li_yNi_{1/3-y}Co_{1/3}Mn_{1/3})O_2$ $y = 0.10x$)] of XRD with de-lithium x and comparison with the experimental curve

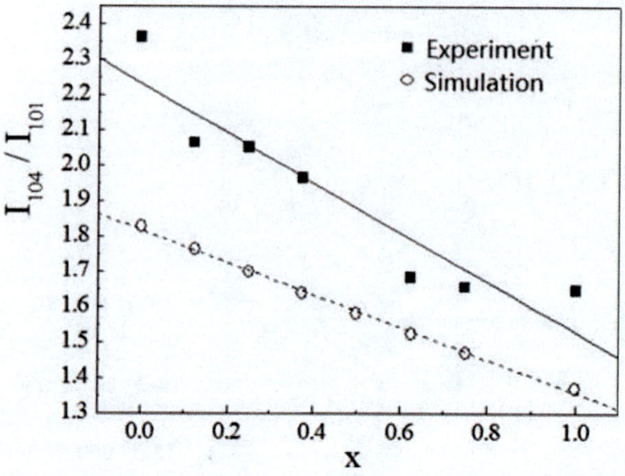

Fig. 11.21 Comparison between the simulated curve and the experimental curve of XRD intensity ratio I_{104}/I_{101} calculated according to $Li_{0.99-x}Ni_{0.01}(Li_{0.01}Ni_{0.324}Co_{1/3}Mn_{1/3})O_2$ model

$Li_{1-x-y}Ni_y(Li_yNi_{1/3-y}Co_{1/3}Mn_{1/3})O_2$, and order $y = 0.25x$ and $y = 0.10x$, the variation curve of XRD intensity ratio I_{104}/I_{101} with de-lithium content x is simulated by Power.cell program according to this molecular formula model, and the simulation curve of $y = 0.10x$ is shown in Fig. 11.20. The results show that the change trend of simulation results is just the opposite with the actual results measured by the experiment (see Fig. 11.20), which indicates that there is no mixed arrangement of Li/Ni atoms in $Li(Ni_{1/3}Co_{1/3}Mn_{1/3})O_2$ materials during battery charge and discharge.

Finally, it is possible that there is a phenomenon of mixed arrangement of Li/Ni atoms in raw materials. In order to prove this point, if the mixed parameter in the raw material is 0.01, it can be simulated and calculated by Power.cell program according to the model of $Li_{0.99-x}Ni_{0.01}(Li_{0.01}Ni_{0.324}Co_{1/3}Mn_{1/3})O_2$. The results are shown in Fig. 11.21. Compared with the experimental results, the two lines are nearly parallel, which indicates that there may be a mixed arrangement of Li/Ni atoms in the raw materials.

11.6 Physical Mechanism of Electrical Conductivity for 2H-Graphite/LiMeO$_2$ Battery

The charge and discharge reaction of 2H-graphite/LiMO$_2$ battery is [20]:

$$\text{Reaction at positive pole:} LiMO_2 \longleftrightarrow Li_{1-x}MO_2 + xLi^+ + xe^-$$

$$\text{Reaction at negative pole:} xLi^+ + 6C + xe^- \longleftrightarrow Li_xC_6$$

$$\text{Total reaction:} LiMO_2 + 6C \longleftrightarrow Li_{1-x}MO_2 + Li_xC_6$$

The chemical working mechanism of lithium-ion battery is that [20]: (1) The conductive ability of lithium-ion battery depends entirely on the migration of lithium-ion Li^+ between positive and negative electrode; (2) after lithium atom is embedded in negative electrode, lithium atom forms Li_xC_6 compound with carbon atom by covalent bond; (3) the electrode reaction in the process of battery charge and discharge is completely reversible. However, combined with the above experimental results, it can be found that the chemical working mechanism of lithium-ion battery only explains the origin of battery conductivity, but there is no clear explanation for how lithium ions come from and how they migrate. At the same time, viewpoints (2) and (3) are not completely consistent with our experimental results. According to the chemical working mechanism of the battery, combined with the behavior of the positive and negative active material described in Sect. 11.5 during the charge and discharge process, we discuss the physical mechanisms of conducting electricity of the lithium-ion battery.

The conductivity of lithium-ion batteries depends entirely on the directional migration and movement of lithium-ion Li^+ between the positive and negative electrodes. The physical mechanism of its conductivity is discussed below.

At the beginning of charging, the migration of lithium ions begins at the negative-electrolyte interface. Because the lithium ion is embedded in the negative electrode after obtaining electrons at the negative electrode–electrolyte interface, the concentration of lithium ion on the interface decreases. and Lithium ions migrate from the positive to the negative in the electrolyte like a relay race to make up for this decrease in concentration. Due to the directional migration of lithium ion to the negative electrode, the concentration of lithium ion in the positive-electrolyte interface decreases correspondingly. At this time, under the action of electric field, the Li atom in $LiMO_2$ leaves the lattice position and loses electrons to form Li^+ into the electrolyte to supplement the lithium ion in the electrolyte. When the ion flow reaches dynamic equilibrium, it corresponds to the charging platform of the battery. When the battery is fully charged, the Li atoms in the $LiMO_2$ are exhausted and continue to charge. Li^+ ions can only be supplied by the electrolyte. This is the physical mechanism of conduction in the charged state.

The discharge process starts from the positive-electrolyte interface, under the action of a reverse electric field, the Li^+ ions on the electrolyte side of the interface obtain electrons and then re-embed into $LiMO_2$. Lack of Li^+ at electrolyte-negative interface, Li atoms are separated from the negative electrode, and electrons are lost through the interface and enter the electrolyte in the state of Li^+ ions, thus forming the conductive mechanism of the battery using state (discharge) of directional movement of Li^+ ions from the negative electrode to the positive electrode.

11.7 Conclusion

The changes of crystal fine structure and microstructure of positive and negative active materials of graphite/$LiCoO_2$ and graphite/$Li(Ni_{1/3}Co_{1/3}Mn_{1/3})O_2$ 18,650 lithium-ion batteries during charge and discharge were studied in detail by XRD methods. The conclusions are as follows:

(1) When the battery begins to charge, lithium atoms are insert crystal lattice in 2H-graphite and preferentially enters the interstitial between the hexagonal grid surfaces of carbon atoms. The formation of solid solution leads to the increase of lattice parameters a and c of 2H-graphite, micro-strain ε and stacking disorder P. When the stacking disorder of the negative electrode reaches a certain value, 3R-graphite precipitates gradually. When the battery is fully charged or overcharged, LiC_{12} and LiC_6 phases are formed in the negative electrode.
(2) When the battery is charged, the Li atoms at position (000) in positive active material $LiCoO_2$ and $Li(Ni_{1/3}Co_{1/3}Mn_{1/3})O_2$ are preferentially separated from the crystal lattice, after which the Li atoms at positions (2/3 1/3 1/3) and (1/3 2/3 2/3), leave from crystal lattice respectively. With the increase of lithium removal amount of positive materials, the lattice parameter a decreases, c increases, and micro-strain ε increases. The change trend of XRD theoretical diffraction intensity ratio of Li $(Ni_{1/3}Co_{1/3}Mn_{1/3})O_2$ in the process of lithium removal is consistent with that measured by experiments.
(3) The physical mechanism of the lithium intercalation process of the electrode active material is analyzed, and it is found that the lithium-ion migration starts from the negative electrode–electrolyte interface; The lithium-ion migration starts from the positive-electrolyte interface during discharge. In charge/discharge process the lithium intercalation process of positive and negative active materials is not completely reversible.
(4) In the graphite/$LiMeO_2$ battery system, there is no structural phase transition in the positive active material of graphite/$Li(Ni,Co,Mn)O_2$ battery during the charge and discharge process, but its fine structure changes. Lithium ions leave the $LiMeO_2$ lattice and are embedded in the graphite lattice. Its oxidation–reduction is Li atoms and occurs on the positive and negative electrodes of the battery, respectively.
(5) After studying the structure of battery positive and negative active materials before and after storage, it is shown that the bulk structure of $LiCoO_2$ does not change after storage, but its microstructure changes obviously, which shows that the microcrystal size decreases, and the micro-strain changes from tensile strain before storage to compressive strain after storage, and the range of change increases with the increase of charge state during storage. After storage, the microstructure of graphite does not change obviously, and the bulk structure of $Li(Ni_{1/3}Co_{1/3}Mn_{1/3})O_2$ has no obvious change after storage, but the microstructure of $Li(Ni_{1/3}Co_{1/3}Mn_{1/3})O_2$ obviously changes from tensile strain to compressive strain, and the cationic-mixed discharge of $Li(Ni_{1/3}Co_{1/3}Mn_{1/3})O_2$ increases

after storage, which is also one of the reasons for the attenuation of battery performance.

References

1. Reimers, J.N., and J.R. Dahn. 1992. Electrochemical and in situ x-ray diffraction studies of lithium intercalation in Li_xCoO_2. *Journal of the Electrochemical Society* 139 (8): 2091–2096.
2. Amatucci, G.G., J.M. Taracin, and L.C. Klein. 1996. CoO_2, the end member of Li_xCoO_2 solid solution. *Journal of the Electrochemical Society* 143 (3): 1114–1123.
3. Yang, X.Q., X. Sun, and J. McBreen. 2000. New phases and phase transitions observed in $Li_{1-x}CoO_2$ during charge: In situ synchrotron X-ray diffraction studies. *Electrochemistry Communications* 2: 100–103.
4. Sun, X., X.Q. Yang, J. McBreen, et al. 2001. New phases and phase transitions observed in $Li_{1-x}CoO_2$-based cathode materials. *Journal of Power Sources* 97–98: 274–276.
5. Luo, Z., N.L. Di, Z.Q. Kou, Z.H. Cheng, L.J. Liu, L.Q. Chen, and X.J. Huang. 2004. Moessbauer study and magnetic properties of electrochemical material $LiFePO_4$. *Chinese Physics* 13: 2158–2161.
6. Hao, W.J., C. Li, and Y.J. Wei, et al. 2003. The change of Co^{+3} electronic state and its effect on structure evolution in $Li(Al_xCo_{1-x})O_2$ crystal. *Acta Physica Sinica* 52 (3): 1023.
7. Sun, Y.C., C.Y. Ouyang, Z.X. Wang, X.J. Huang, and L.Q. Chen. 2004. Effect of Co content on rate performance of $LiMn_{0.5-x}Co_{2x}Ni_{0.5-x}O_2$ cathode materials for lithium-ion batteries. *Journal of the Electrochemical Society* 151: A504–508.
8. Chen, Y., G.X. Wang, K. Konstantinov, H.K. Liu, and S.X. Dou. 2003. Synthesis and characterization of $LiCo_xMn_yNi_{1-x-y}O_2$ as a cathode material for secondary lithium batteries. *Journal of Power Sources* 119–121: 184–188.
9. Machheil, D.D., Z. Lu, and J.R. Dahn. 2002. Structure and electrochemistry of $Li[Ni_xCo_{1-2x}Mn_x]O_2$ ($0 \le x \le 1/2$). *Journal of the Electrochemical Society* 149: A1332–A1336.
10. Naoaki, Y., M. Yobuuchi, and O. Tsutomu. 2007. Solid-state chemistry and electrochemistry of $LiCo_{1/3}Ni_{1/3}Mn_{1/3}O_2$ for advanced lithium-ion batteries: III. Rechargeable capacity and cycleability. *Journal of the Electrochemical Society* 154 (4): A314–321.
11. Li, J., C.Z. Yang, J. Zhang, X.G. Zhang, and B.J. Xia. 2010. Study on structure evolution of electrode active material during charge and discharge of graphite/$LiCoO_2$ battery. *Acta Chemica Sinica* 68 (7): 646–652.
12. Li, J., C.Z. Yang, J. Zhang, and B.J. Xia. 2009. Study on physical mechanism of lithium intercalation and delithium removal in graphite/$Li(Ni_{1/3}Co_{1/3}Mn_{1/3})O_2$ battery. *Acta Physica Sinica* 58 (9): 6573–6581.
13. Li, H., C.Z. Yang, and F. Liu. 2008. A new X-ray diffraction method for determining the stacking disorder degree of hexagonal graphite. *Chinese Science, Series B Chemistry* 28 (9): 755–760.
14. Li, Hui, Chuanzheng Yang, and Fang Liu. 2009. Novel method of determination stacking disorder degree in hexagonal graphite by X-ray diffraction. *Science in China Series B: Chemistry* 522: 174–180.
15. Li, H., C.Z. Yang, and F. Liu. 2009. Determination of graphitization degree and disorder degree of carbon electrode materials by X-ray diffraction. *Journal of Testing Technology* 23 (2): 161–167.
16. Zhang, J., C.Z. Yang, X.G. Zhang, J. Li, and B.J. Xia. 2009. Simulation analysis and study on the structural changes of $Li(Ni,Co,Mn)O_2$ during synthesis. *Journal of Materials Science and Engineering* 27 (6): 824–828.

17. Zhang, X.G., J. Zhang, C.Z. Yang, and B.J. Xia. 2010. Simulation and experimental study on mixed arrangement of lithium and nickel in $LiMeO_2$ materials. *Journal of Inorganic Materials* 25 (1): 8–12.
18. Li, J. 2010. Study on storage performance of lithium-ion battery. Doctoral Paper of Shanghai Institute of Microsystems and Information Technology.
19. Dahn, J.R. 1991. Phase diagram of Li_xC_6. *Physical Review B* 44 (17): 9170–9177.
20. Li, G.X., ed. 2007. Introduction to new chemical power supply technology. Shanghai Science and Technology Press, 320.
21. Gao Jian, Lu Yingchun, Li Hong 2013. Basic Science of Lithium Battery (IV)-Phase Diagram and Phase transition. *Energy Storage Science and Technology*, 2 (4): 383–401.

Chapter 12 Mechanism Research of Charge–Discharge Process for Graphite/$LiFePO_4$ Battery

12.1 Introduction

The charge and discharge process of 2H-graphite/$LiFePO_4$ battery has been studied as follows:

At room temperature, the de-intercalation behavior of Li_xFePO_4 is actually a two-phase reaction process forming the two-phase interface between $FePO_4$ and $LiFePO_4$. Newman [1], Yamada [2], and Dodd [3] have systematically studied the phase transition during the charge–discharge process of Li_xFePO_4. During charging, the lithium ion migrates from the FeO_6 layers and enters the negative electrode through the electrolyte, resulting in the oxidation reaction of $Fe^{2+} \rightarrow Fe^{3+}$. In order to maintain the charge balance, the electron arrives at the negative electrode from the external circuit. The reduction reaction occurs during discharge, which is opposite to the above process. That is:

$$\text{Charge:}\quad LiFePO_4 + xLi^+ + xe^- \rightarrow xFePO_4 + (1-x)LiFePO_4$$

$$\text{Discharge:}\quad FePO_4 + xLi^+ + xe^- \rightarrow xLiFePO_4 + (1-x)FePO_4$$

Li_xFePO_4 is a typical electron–ion mixed conductor. The width of the bandgap is 0.3 eV. The electronic conductivity at room temperature is very low, about 10 ~ 9 S/cm. The room temperature ion conductivity of Li_xFePO_4 is also quite low (about 10 ~ 5 S/cm). The characteristic structure of olivine makes the bulk diffusion channel of lithium ion few (only quasi-one-dimensional diffusion can be realized). In the two-phase reaction of Li_xFePO_4 de-intercalation, the theoretical lithium-ion diffusion coefficient in $LiFePO_4$ and $FePO_4$ is about 10^{-8} cm^2/s and 10^{-7} cm^2/s, while the actual measurement shows that the "effective" diffusion coefficient of lithium ion in $LiFePO_4$ and $FePO_4$ may be 7 orders of magnitude lower than the theoretical value 1.8×10^{-14} cm^2/s and 2×10^{-16} cm^2/s, respectively. Therefore,

C. Yang et al., *Materials and Working Mechanisms of Secondary Batteries*,
https://doi.org/10.1007/978-981-19-5955-4_12

in order to use $LiFePO_4$ as positive material for lithium-ion battery, it is necessary to improve its electronic and ionic conductance and improve its electrochemical interface properties.

The corresponding reaction on the negative pole:

$$\text{Charge} \quad x\text{Li}^{+} + 6\text{C} \rightarrow \text{Li}x\ \text{C}_6$$

$$\text{Discharge} \quad \text{Li}x\ \text{C6} \rightarrow x\text{Li} + \text{C}_6$$

From the above, it can be seen that the conductive Li^+ ions are supplied by the phase transition of $LiFePO_4 \rightarrow FePO_4$ and are completely reversible. However, Shin et al. [4] based on synchrotron radiation in situ XRD method found that carbon-coated $LiFePO_4$ has asymmetric structural changes during charge/discharge.

This chapter first discusses the basis of phase identification of related materials in 2H-graphite/$LiFePO_4$ battery and then, through the experimental study on the structural evolution of positive and negative active materials during the first charge and discharge of 2H-graphite/$LiFePO_4$ battery, as well as the crystallographic analysis of $LiFePO_4$ and $FePO_4$ structure characteristics, discusses the nature, physical process, and charge–discharge conductive mechanism of $LiFePO_4$ phase transition during charge and discharge of 2H-graphite/$LiFePO_4$ battery.

For the convenience of the following research and analysis, we first understand the phase structure and powder diffraction data of $LiFePO_4$ and $FePO_4$. $LiFePO_4$ was discovered in mineral research, so its mineralogical name is lithium phosphate iron ore (triphylite), belongs to orthorhombic, Pnma (No. 62) space group, $a = 6.08$ Å, $b = 10.44$ Å, and $c = 4.75$ Å. X-ray density $D_x = 3.47$, measured density $D_{\text{m}} = 3.45$ g/cm^3, and PDF card number is 40-1499.

The substances and structures related to $LiFePO_4$ and $FePO_4$ can be found in the database of the International Centre for Diffraction Data (ICDD):

Substance	Mineralogy name	PDF number	Space group	a/Å	b/Å	c/Å
$LiFePO_4$	Lithium phosphate iron ore	40-1499	Pnma (No. 62)	6.0157	10.3915	4.7207
$FePO_4$	Iron phosphate	30-0659	Cmcm (No. 63)	5.227	7.770	6.322
(Fe,Mn)PO_4	Isophosphate iron (manganese) ore	34-0134	Pnma (No. 62)	5.824	9.821	4.786
$FePO_4$	Iron phosphate	31-0647	$P_{22}2222$ (No. 180)	5.170		11.37

Almost the same structure as $LiFePO_4$ is $(Fe(Mn)PO_4)$. The structure of $(Fe,Mn)PO_4$, is almost the same as that of $LiFePO_4$. Its mineralogical name is isophosphate iron (manganese) ore, and PDF card number is 34-0134.

12.2 Experimental Study on Charge–Discharge of Activated Graphite/$LiFePO_4$ Battery [5]

12.2.1 Structural Evolution of $LiFePO_4$ During Charge–Discharge of Activated Batteries

(1) **Phase identification**

Figure 12.1 shows X-ray diffraction patterns of positive active materials at several stages of the 0.2C charging process, which are indexed by $LiFePO_4$ with Pnma (No. 62) orthorhombic structure and $FePO_4$, represented by the letters T and H, respectively. It should be noted that 100% charging patterns can only be indexed by $FePO_4$ with orthogonal structure [Pnma (No. 62) 34-0134] and not by $FePO_4$ of other structures [Cmcm (No. 63) or $P_{22}2222$ (No. 180)]. It can be seen that there are two substances $LiFePO_4$ and $FePO_4$ in the positive active material after activation, and the content of the former is more, and with the increase of charging depth, $LiFePO_4$ decreases gradually, while $FePO_4$ increases gradually. When the charge reaches 80%, there is still a trace of $LiFePO_4$, and it is not all $FePO_4$ until the charge reaches 100%.

The diffraction patterns of several stages of the discharge process are shown in Fig. 12.2. It can be seen that the positive electrode of the fully charged battery is composed of pure $FePO_4$; When 20% discharge appears for $LiFePO_4$, and increases with the increase of discharge depth, until 80% discharge, although it is mainly $LiFePO_4$, but there is still a considerable amount of $FePO_4$. Comparing the phase

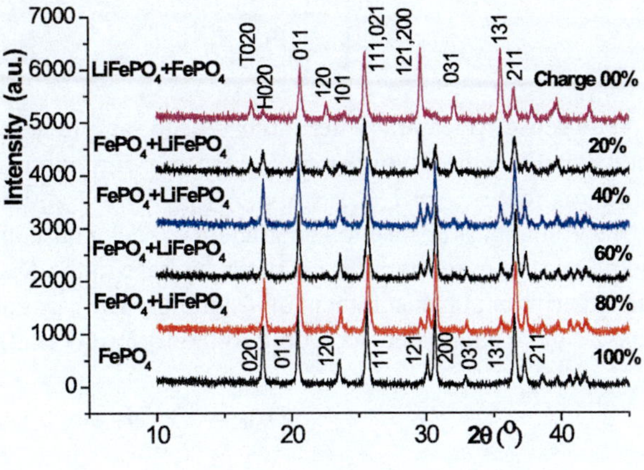

Fig. 12.1 XRD pattern of positive material after 0.2C charging process of activated battery

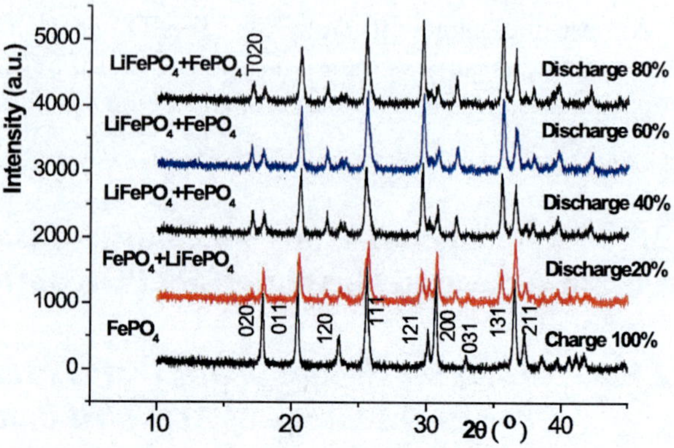

Fig. 12.2 XRD pattern of positive active material after 0.2C discharge of activated battery

composition of the positive active material of 80% discharge with that of the positive active material when charging 20%, there is an obvious difference between them, which shows that the positive active material is not completely reversible or called asymmetry in the charge–discharge process, but it indicates that there is a change of $LiFePO_4 \leftrightarrow FePO_4$ in the charge–discharge process.

(2) **Quantitative analysis of phase—experimental study on the asymmetry of charge and discharge**

In order to further analyze the hysteresis of the phase transition of the positive active material during the charge and discharge of graphite/$LiFePO_4$ battery, the relative contents of $FePO_4$ and $LiFePO_4$ in the positive active material at each stage of charge and discharge were quantitatively calculated by Zevin's non-standard sample method [6]. When the mass absorption coefficient of each phase is known

$$\begin{cases} \sum_{i=1}^{n}\left[\left(1-\frac{I_{iJ}}{I_{iK}}\right)\mu_{mi}x_{iK}\right]=0 \\ \sum_{i=1}^{n}x_{iK}=1 \end{cases} \tag{12.1}$$

The subscript capital letter indicates the sample number, and the lowercase letter indicates the phase number in the sample. Equation (12.1) includes $(n-1)+1$ equation, that is, $J=1, 2, \ldots, n, \neq K$ $(n-1)+1$ equation, and each equation has n terms. If the mass absorption coefficient of each phase in the sample to be determined is known, the mass fraction x_{iK} of each phase in the K sample can be obtained from the experimental measurement of n samples. The mass absorption coefficient of each phase can be calculated according to the formula (12.2)

$$\mu_{\mathrm{mi}}=\sum_{p}\mu_{\mathrm{m}p}\omega_{p} \tag{12.2}$$

μ_{mp} and ω_p are the mass absorption coefficient and mass fraction of p element in the phase, respectively.

Here, the phases to be measured are $FePO_4$ and $LiFePO_4$ phases. "It is found from the manual that the mass absorption coefficients of lithium, iron, phosphorus, and oxygen to CuKα radiation are, and the atomic weights are $\mu_{m_{\rm Li}} = 6.941$, $\mu_{m\rm Fe} = 308.0$, $\mu_{m\rm P} = 74.1$, and $\mu_{m\rm O} = 11.5$. The atomic weights are, $M_{Li} = 6.941$, $M_{Fe} = 55.841$, $M_P = 30.974 M_O = 15.998$, respectively. "**It can** be calculated by using the formula (12.2).

$$\mu_{\rm mLiFePO_4} = (\omega_{\rm Li}\mu_{\rm mLi} + \omega_{\rm Fe}\mu_{\rm mFe} + \omega_{\rm P}\mu_{\rm mP} + \omega_{\rm O}\mu_{\rm mO})$$

$$= \begin{pmatrix} \frac{M_{\rm Li}}{M_{\rm Li} + M_{\rm Fe} + M_{\rm P} + M_{\rm O} \times 4} \cdot \mu_{\rm mLi} + \frac{M_{\rm Fe}}{M_{\rm Li} + M_{\rm Fe} + M_{\rm P} + M_{\rm O} \times 4} \cdot \mu_{\rm mFe} \\ + \frac{M_{\rm P}}{M_{\rm Li} + M_{\rm Fe} + M_{\rm P} + M_{\rm O} \times 4} \cdot \mu_{\rm mP} + \frac{M_{\rm O} \times 4}{M_{\rm Li} + M_{\rm Fe} + M_{\rm P} + M_{\rm O} \times 4} \cdot \mu_{\rm mO} \end{pmatrix} = 128.28$$

$$\mu_{\rm mFePO_4} = (\omega_{\rm Fe}\mu_{\rm mFe} + \omega_{\rm P}\mu_{\rm mP} + \omega_{\rm O}\mu_{\rm mO})$$

$$= \begin{pmatrix} \frac{M_{\rm Fe}}{M_{\rm Li} + M_{\rm Fe} + M_{\rm P} + M_{\rm O} \times 4} \cdot \mu_{\rm mFe} + \frac{M_{\rm P}}{M_{\rm Li} + M_{\rm Fe} + M_{\rm P} + M_{\rm O} \times 4} \cdot \mu_{\rm mP} \\ + \frac{M_{\rm O} \times 4}{M_{\rm Li} + M_{\rm Fe} + M_{\rm P} + M_{\rm O} \times 4} \cdot \mu_{\rm mO} \end{pmatrix} = 128.25$$

By using the relative integral strength of the two phases in the positive active material at each stage of charge and discharge, the mass fraction of the two phases in a certain stage An is calculated, and the substitution (12.1) can be written as a set of equations.

$$\begin{cases} \left[\left(1 - \frac{I_{\rm LiFePO_4B}}{I_{\rm LiFePO_4A}}\right) 128.28 x_{\rm LiFePO_4A}\right] + \left[\left(1 - \frac{I_{\rm FePO_4B}}{I_{\rm FePO_4A}}\right) 128.25 x_{\rm FePO_4A}\right] = 0 \\ x_{\rm LiFePO_4A} + x_{\rm FePO_4A} = 1 \end{cases} \quad (12.3)$$

The relative integral intensities $I_{\rm T020}$ and $I_{\rm H020}$ of the two phases in the XRD diffraction spectra of each charge–discharge state are obtained by fitting the Jade6.5 software and solved in Eq. (12.3). The relative contents of $FePO_4$ and $LiFePO_4$ in the positive active materials in each charge–discharge state can be obtained. The results are shown in Fig. 12.3.

As can be seen from Fig. 12.3, (1) even in 0% SOC, the positive active material is not entirely $LiFePO_4$ phase, in which there is a small amount of $FePO_4$ phase (about 8%), which is due to the formation of inactive $FePO_4$ shell during the first charge; (2) under the condition of the same SOC, the content of $LiFePO_4$ in the discharge state is more than that in the charge state, on the contrary. The content of $FePO_4$ for discharge state is obviously lower than that of the charged state, which obviously reveals the hysteresis of the discharge, thus showing the asymmetry of the charge and discharge.

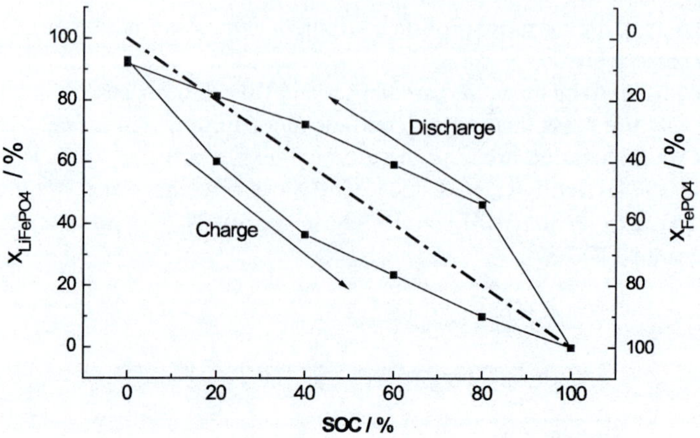

Fig. 12.3 Content change, of $LiFePO_4$ and $FePO_4$ in positive active materials during charge and discharge of graphite/$LiFePO_4$ battery. —·—·— is 45° line

12.2.2 Structural Evolution of Negative Active Materials During Charge and Discharge of Activated Batteries

Figure 12.4 shows the X-ray diffraction patterns of negative active materials at several stages of the 0.2C charging process. Its composition is listed in Table 12.1 by phase identification. It can be seen that the asymmetry of negative active materials in the process of charge and discharge is not very obvious. Obviously, the Li solid solution phenomenon in hexagonal graphite is not obvious, but the conclusion that there is no solid solution process cannot be obtained. In a word, compared with the charge–discharge process of 2H-graphite/$Li(Ni_{1/3}Co_{1/3}Mn_{1/3})O_2$ [6] and 2H-graphite/$LiCoO_2$ [7] batteries, the order of lithium intercalation in the carbon electrode and the precipitation of C–Li compounds is similar, only because of the different positive active materials. When the C–Li compound begins to precipitate, the charge–discharge depth is different.

The amount of solid solution of Li is different.

It is worth noting whether the crystal structure of 2H-graphite and LiC_{24}, LiC_{12}, LiC_6, and LiC_2 in Tables 12.1 and 11.2 is the same. After investigation, it is found that:

Name of substance	Crystal structure	*a*	*c*
2H-graphite	P63mmc	2.464	6.711
LiC_{24}	P63mmc	4.288	7.07
LiC_{12}	P63mmc	4.288	7.066
LiC_6	P63mmc	4.305	7.036

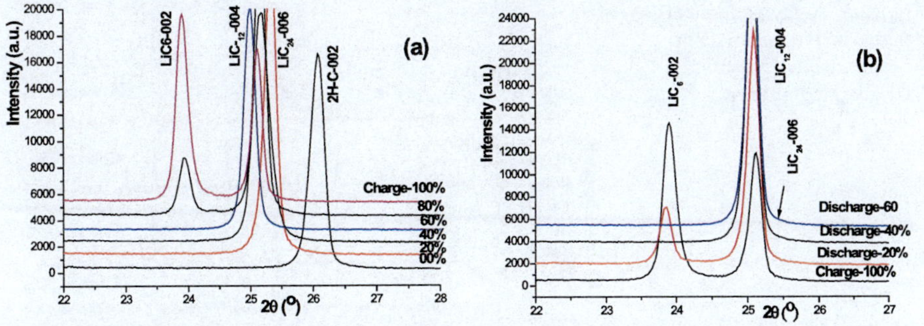

Fig. 12.4 XRD pattern of negative active material in the process of 0.2C charging (**a**) and discharging (**b**) for activated battery

Table 12.1 Phase structure of negative active materials during 0.2C charge and discharge of activated battery

Charge state		Discharge state	
Charging depth (%)	Existing phase	Discharge depth (%)	Existing phase
0	2H-石墨		
20	LiC_{24}		
40	LiC_{12}	60	LiC_{12} + LiC_{24}
60	LiC_{12}	40	LiC_{12}
80	LiC_{12} + LiC_6	20	LiC_{12} + LiC_6
100	LiC_6 + LiC_{12}	0	LiC_6 + LiC_{12}

It can be seen that their crystal structures are the same. Only the unit cell parameters *a* and *c* increase with the increase of Li content, which indicates that LiC_{24}, LiC_{12}, and LiC_6 are continuous solid solutions.

12.3 Results of the Study on the Charging Process of Unactivated Batteries [5]

In the study of Sect. 12.2.1, it has been observed that there is a phase transition of $LiFePO_4 \leftrightarrow FePO_4$ in 2H-graphite/$LiFePO_4$ battery during charge and discharge, but it is not known how the phase transition occurs. Let us carefully study the changes in the structure of positive active materials during the first charge of the battery.

The X-ray patterns of the positive active materials at several stages of the first 0.01C charging process of the unactivated 2H-graphite/$LiFePO_4$ battery are shown in Fig. 12.5. It can be observed from Fig. 12.5 that:

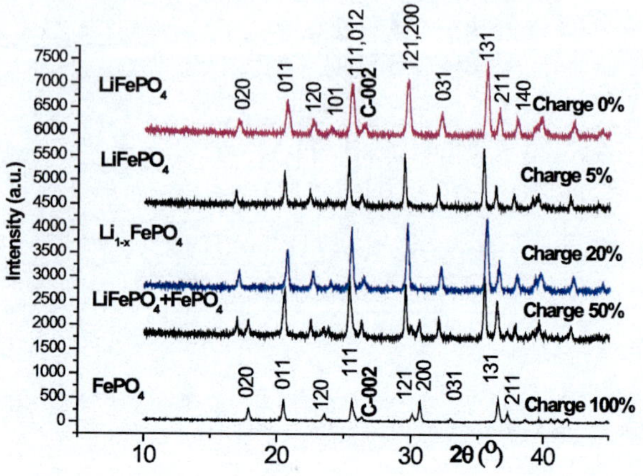

Fig. 12.5 X-ray patterns of positive active materials in several stages of the first 0.01 charging process

(1) $FePO_4$ phase does not appear until charging 50%. When charging depth is 50% and 80%, respectively, both $LiFePO_4$ and $FePO_4$ coexist, and 100% charging is pure $FePO_4$ phase, as shown in Table 12.2 and Fig. 12.5.

(2) The relevant lattice parameters are also given in Table 12.2. As far as $LiFePO_4$ is concerned, the changes of a and b are irregular, but c increases with the increase of charging depth. These two points seem to indicate that there is a process from $LiFePO_4$ to $FePO_4$; that is, there is a process of Li atoms leaving the $LiFePO_4$ lattice, and in other words, there is a process of $LiFePO_4 \rightarrow Li_{1-x}FePO_4 \rightarrow FePO_4$ ($x = 1$). If we further consider that the crystal structures of $LiFePO_4$, $Li_{1-x}FePO_4$, and $FePO_4$ are strikingly consistent, it can be considered that the change of $LiFePO_4 \rightarrow Li_{1-x}FePO_4 \rightarrow FePO_4$ ($x = 1$) is the result of the Li atoms detach from the $LiFePO_4$ lattice. It can be further inferred that the conductive Li^+ ion is not provided by the $LiFePO_4 \rightarrow FePO_4$ phase transition, but the Li atom leaves the $LiFePO_4$ lattice position under the action of the electric field and is provided by the Li^+ which loses an electron.

Table 12.2 Results of X-ray data analysis of positive active material in several stages of the first charging process

Charge depth (%)	Existing phase		a	b	c	V
0	$LiFePO_4$	ε^I (10^{-3})	6.0157	10.3915 0.00	4.7207 0.00	295.1 0.00
20	$LiFePO_4$	ε^I (10^{-3})	6.0116 −0.68	10.3654 −2.51	4.7342 −2.86	295.0 −0.34
50	$LiFePO_4$ $FePO_4$	ε^I (10^{-3})	6.0226 1.15	10.4368 4.36	4.7451 5.17	298.3 10.84
80	$LiFePO_4$ $FePO_4$	ε^I (10^{-3})	6.0182 0.42	10.4049 1.29	4.7405 4.19	296.8 5.76
100	$FePO_4$		5.8094	9.9222	4.8166	277.6

12.4 Asymmetry of Graphite/$LiFePO_4$ Battery During Charge–Discharge

12.4.1 *Asymmetry Phenomenon in the Process of Charge–Discharge*

The problem of incomplete reversibility of charge–discharge and the quantitative test results of $LiFePO_4$ and $FePO_4$ have proved the problem of incomplete reversibility of charge–discharge, that is, the asymmetry of charge and discharge process. Srinivasan and Newman [8] reported that the charge–discharge experiments at various constant current densities also show the asymmetry of $LiFePO_4$, which shows that under the current density which plays an important role in transport restrictions, the utilization rate of $LiFePO_4$ during charging is significantly higher than that during discharge and also shows the dependence of material utilization on path at high rates. They use the shrinking core model to explain this property, in order to explain the phenomenon of juxtaposed coexistence of two phase. According to their hypothesis, the transport of Li^+ and electrons occurs in the Li-rich phase during the discharge, so it has a higher activation barrier and a lower diffusion coefficient and vice versa. Therefore, Li^+ and electronic transmission are easier during charging.

Shin et al. [4] used Korean synchrotron radiation on-line XRD to study the asymmetry of Li/$LiFePO_4$ button battery in charge and discharge.

Figure 12.6 shows the charge–discharge curve of 2.5–4.5 V carbon-coated $LiFePO_4$ at 5C rate. The vertical line and ten characters on the curve indicate the beginning of XRD scanning, and the number of targets of XRD scanning is selected on the curve. The capacities given during charge and discharge are 147.5 and 143.1 mAh/g, respectively, indicating that the capacity during discharge is lower than 4.4 mAh/g.

Figure 12.7a, b shows the selection range of on-line XRD patterns collected during 5C charge and discharge. The pattern shown with thick lines is that the current used is cut off and maintained until later. The horizontal dotted line with arrows indicates the XRD pattern when the charge and discharge depth is the same. All the diffraction peaks of the pattern at the beginning of charging can be indexed by $LiFePO_4$ (Triphylite), which is shown by T in Fig. 12.7a. During the charging period, the diffraction line of the $FePO_4$ (heterosite) gradually appears with the increase of the charging depth, and the peak shape becomes sharper gradually. Figure 12.7a is represented by H. At the end of charging, only the diffraction peak of $FePO_4$ was observed, and the diffraction peak of $LiFePO_4$ disappeared completely. In the subsequent discharge process, the phase transition shown is observed, as shown in Fig. 12.7b. At the beginning of the discharge, only lines from iron isophosphate ($FePO_4$) were observed. As the discharge continues, the line from ferric isophosphate decreases gradually, and the line from lithium iron phosphate ($LiFePO_4$) gradually increases. At the end of the discharge, the diffraction pattern completely returns to

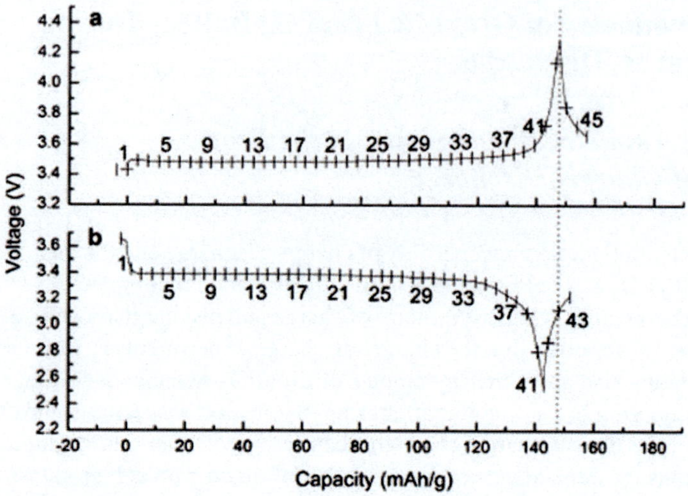

Fig. 12.6 2H-graphite battery 5C charge–discharge curve between 2.5 and 4.5 V, the number on the curve is the on-line XRD scanning point [4]

the original line of lithium phosphate iron ore. When the charge–discharge rate is increased, the deviation from the above phenomenon occurs.

Figure 12.8 shows the 2C charge–discharge curve of carbon-coated $LiFePO_4$ battery, the charge and discharge rate is 20 times of 0.2C, and its diffraction pattern is shown in Fig. 12.9. It is obvious that all the diffraction peaks have been seriously broadened, and with the increase of the charging depth, the peak position moves to

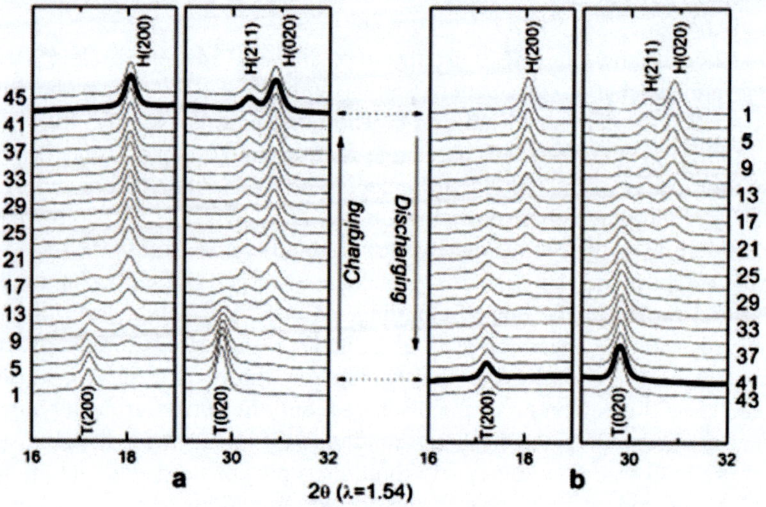

Fig. 12.7 5C on-line XRD patterns collected during charging to 4.3 V and discharge to 2.5 V [4]

a large angle, and the capacity of charging and discharge is 136.9 and 111.4 mAh/g, respectively. The difference between charge and discharge is 25.5 mAh/g, which is significantly larger than the charge–discharge difference (4.4 mAh/g) at 0.2C rate. As can be seen from Fig. 12.8a, the phase transition from lithium iron phosphate to iron isophosphate has been completed during charging. The opposite phase transition in the discharge process is far from complete, as shown in Fig. 12.9. When the 0.2C and 10C rate discharge current is cut off at the same discharge depth, the XRD scan curve of the horizontal dotted line indicates that the XRD scan curve at the same charge and discharge depth, the phase change at high rate is slower than that at low rate, in other words, the asymmetry between charge and discharge at high rate is more serious.

From the research results described in this section, it can be summarized that the asymmetry is shown as follows:

(1) Under the same charge and discharge depth (such as 80% charge and 20% discharge after 100% charge), the discharge capacity is less than charge capacity. This difference increases with the increase of charge and discharge rate, and the difference value is 4.4 mAh at 0.2C and 25.5 mAh at 10C.

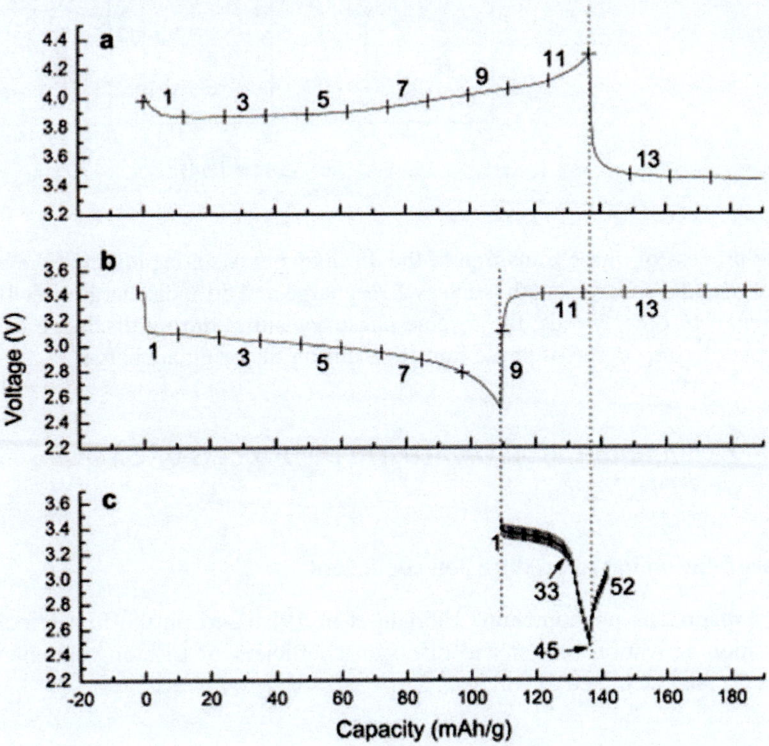

Fig. 12.8 2C charge–discharge curve between 2.5 and 4.3 V (**a**, **b**), **c** 10C further discharge to 2.5 V [4]

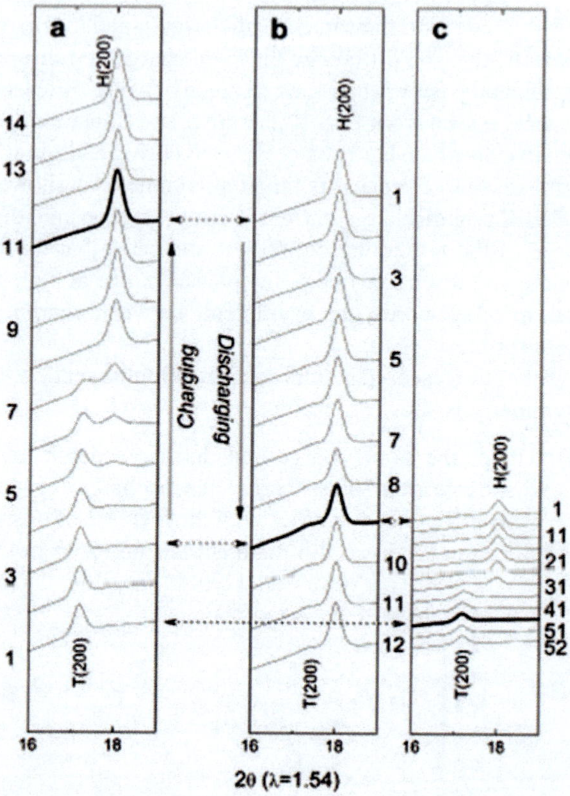

Fig. 12.9 XRD scan curve, 2C charge to 4.3 V (**a**) and discharge 2.5 V (**b**), **c** 10C further discharge curve to 2.5 V [4]

(2) The process of phase transition of the discharge exists a lag effect at the same charge and discharge depth (such as 20% charge and 80% discharge after 100% charge). In other words, the reverse phase transition during discharge cannot recover to the degree of phase transition during the original charge.

12.4.2 *Explanation of Asymmetry in the Process of Charge and Discharge*

1. **Interpretation based on diffusion coefficient**

How to explain this phenomenon? Morgan et al. [9] based on the first principle, the transition activation barrier and diffusion coefficient of Li along ID chain in $Li_{1-x}FePO_4$ are calculated as follows:

	Active barrier	Diffusion coefficient (cm^2/s)
$LiFePO_4$	270	10^{-8}
$FePO_4$	200	10^{-7}

It is concluded that the magnification problem in $Li_{1-x}FePO_4$ is caused by the limitation of electron conductivity, not by the limitation of ion conductivity. This shows an important fact that there is an asymmetry of activation barrier and diffusion coefficient in $Li_{1-x}FePO_4$ materials between charge and discharge. This asymmetry may affect the high-rate performance of $LiFePO_4$. When carbon-coated or metal-doped $LiFePO_4$, is used to improve the electronic conductivity, the relative contribution of asymmetry is more obvious.

The above data of activation barrier and diffusion coefficient can show that there is asymmetry between charge and discharge. Because the activation barrier of Li atom in $FePO_4$ is lower than that in $LiFePO_4$, the diffusion coefficient is larger, and the speed of phase transition in discharge is faster than that in charge, and the phase transition in charge should be in advance. This is consistent with the experimental facts. Therefore, the data of Morgan et al. cannot explain the hysteresis effect of discharge phase transition, let alone explain why the hysteresis of phase transition at higher rate is more serious than that at low rate at the same charge–discharge depth. The discussion is as follows:

(1) The crystal structures of $LiFePO_4$, $Li_{1-x}FePO_4$, and $FePO_4$ are the same, and they are all Pnma (No. 62) space groups, but the unit cell volume changes obviously. The actual measured $Li_{1-x}FePO_4$ cell volume is

X of in $Li_{1-x}FePO_4$	0	0.5	1.0
Cell volume (Å^3)	295.1	298.3	277.6

The diffusion velocity of the same kind of atom in $LiFePO_4$ and $Li_{1-x}FePO_4$ with little difference in atomic density should be about the same, but in $FePO_4$ with the same crystal structure and much smaller cell volume, the diffusion rate is obviously slower.

(2) The diffusion path of Li atoms is also different. The path of Li atom leaving the positive active material during charging is the change from the beginning of $LiFePO_4$ to the $LiFePO_4 + Li_{1-x}FePO_4$ and, finally, to $Li_{1-x}FePO_4 + FePO_4$, $LiFePO_4 \rightarrow Li_{1-x}FePO_4 \rightarrow FePO_4$, while the path of Li atom back-embedding positive active material during discharge is from $FePO_4$ to $Li_{1-x}FePO_4 + FePO_4$ and finally to $LiFePO_4 + Li_{1-x}FePO_4$.

(3) When charging and discharging at a low rate (such as 0.2C), the process is very slow; for example, it takes 20 h to reach full charge. It takes 20 h to complete the discharge. In this case, the actual diffusion velocity of Li atoms is lower than that corresponding to the theoretical diffusion coefficients of $LiFePO_4$, $Li_{1-x}FePO_4$, and $FePO_4$, so it cannot show the asymmetry of charge and discharge, or the asymmetry is minimal.

(4) When charging and discharging at high rate (such as 2C or 10C), the process is very fast, at the rate of 0.2C and 2C. The latter is 10 times faster; that is, it can reach full charge in 2 h, so the actual diffusion velocity of Li atoms is greater or much higher than that corresponding to the theoretical diffusion coefficients of $LiFePO_4$, $Li_{1-x}FePO_4$, and $FePO_4$. Because the unit cell volume of $FePO_4$ is 6% smaller than that of $LiFePO_4$, it can directly affect the diffusion velocity of Li atoms in $LiFePO_4$ at high rate, so that it cannot meet the requirement of diffusion speed during fast discharge, and the phase transition hysteresis effect occurs, which becomes more and more serious with the increase of charging rate.

2. **Interpretation of Andersson**

Andersson et al. [8] proposed radiation model and mosaic model to explain the hysteresis phenomenon of structural changes during charge and discharge, as shown in Fig. 12.10.

According to the radiation model, the lithium-ion de-intercalation process in $LiFePO_4$ takes place at the interface of $LiFePO_4/FePO_4$. During charging, the lithium ion in the outer layer of the $LiFePO_4$ particles is preferentially removed; thus, a $FePO_4$ shell is formed in the outer layer of the particles, and a two-phase interface appears, as B shown in Fig. 12.10a. During the subsequent charging process, the two-phase interface continues to push toward the core. The outer $LiFePO_4$ constantly are transformed into $FePO_4$, and lithium ion and electrons continue to pass through the newly formed two-phase interface to maintain the effective current, but the diffusion rate of lithium ion is constant under certain conditions. With the reduction of the two-phase

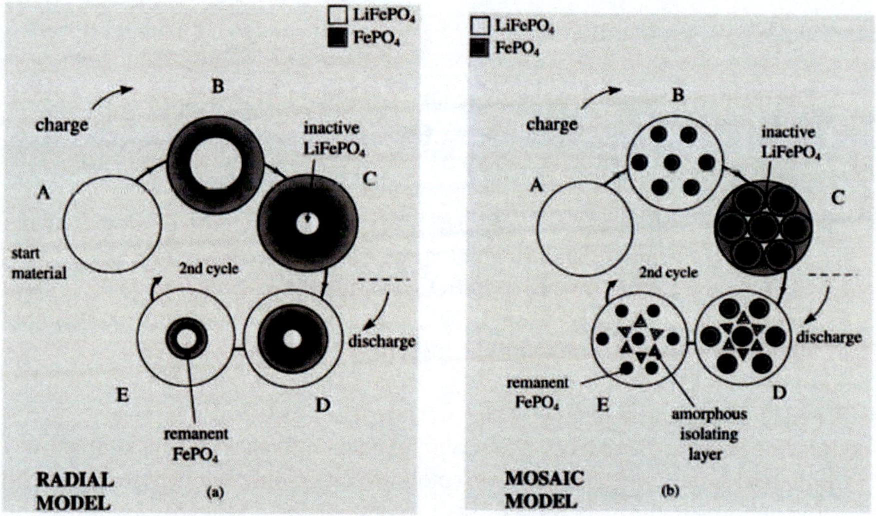

Fig. 12.10 Lithium de-intercalation model for $LiFePO_4$ charge–discharge process of Andersson [8]

interface, the diffusion amount of lithium ion will eventually be insufficient to maintain the effective current. Therefore, the $LiFePO_4$ in the core part of the particles will not be fully utilized, and an inactive $LiFePO_4$ core will be formed inside the particles after charging. During the discharge, on the contrary, the outer layer of the particles forms a $LiFePO_4$ shell, and the two-phase interface continues to push toward the core. Finally, due to the narrowing of the two-phase interface, the diffusion amount of lithium ion is not enough to maintain the effective current, and a $FePO_4$ shell is formed outside the inactive $LiFePO_4$ core inside the particles at the end of the discharge, so that the $LiFePO_4$ core inside the particles will no longer be used in the subsequent charge–discharge process, resulting in capacity loss.

The mosaic model is similar to the radiation model, except that although the lithium-ion de-intercalation process is a de-intercalation process at the two-phase interface of $LiFePO_4$/$FePO_4$, the lithium-ion de-intercalation process can occur at any position of the particles. During charging, the $FePO_4$ region increases at different points of the particles, and the edges of the region cross-contact, forming a lot of dead corners that cannot react, resulting in capacity loss. During the discharge, the reverse reaction process takes place, and the lithium ion is embedded in the $FePO_4$ phase, and the capacity loss is caused by the part where the lithium ion is not embedded in the core.

It can be seen from Fig. 12.10 that the residual $FePO_4$ is formed inside the $LiFePO_4$ particles after the first charge and discharge. Therefore, in the XRD spectrum of the formed $LiFePO_4$ positive active material, even in the discharge state, there is a small amount of $FePO_4$ peak. At the same time, during the charging process, the $FePO_4$ phase is preferentially generated in the outer layer of the particles and diffused to the inner layer, and because there is a certain incident angle of X-ray on the surface of the material particles, the X-ray diffraction of the outer $FePO_4$ phase relative to the inner $LiFePO_4$ phase has a certain blocking effect, resulting in the weakening of the X-ray diffraction intensity of the $LiFePO_4$ phase. On the contrary, the discharge process is the opposite. The outer layer of the particle preferentially forms the $LiFePO_4$ phase, and its blocking effect on the X-ray diffraction of the inner layer $FePO_4$ phase weakens the X-ray diffraction intensity of the $FePO_4$ phase. In this way, the hysteresis or asymmetry of the phase change in the charge–discharge process of $LiFePO_4$ positive active materials is shown in the XRD spectra.

Compared with graphite/$LiCoO_2$ and graphite/$Li(Ni_{1/3}Co_{1/3}Mn_{1/3})O_2$ batteries, there is asymmetry in the charge and discharge process of the three kinds of batteries, but the asymmetry of graphite/$LiFePO_4$ battery is more obvious.

12.5 De-intercalation Mechanism of Lithium in 2H-Graphite/$LiFePO_4$ Battery [5]

12.5.1 *Phase Transition Characteristics and De-intercalation Mechanism of $LiFePO_4$ During Charge–Discharge*

In the process of charge and discharge of 2H-graphite/$LiFePO_4$ battery, the positive active materials involve $LiFePO_4$, $Li_{1-x}FePO_4$, and $FePO_4$. Both $LiFePO_4$ and $Li_{1-x}FePO_4$ can be indexed by the space group, Pnma (No. 62) of (triphylite) orthorhombic system of lithium phosphate, while $FePO_4$ belongs to the isophosphite ore with orthorhombic system and can only be indexed by Pnma (No. 62) space group. The actual measured lattice parameters are shown in Table 12.3.

Each unit cell of crystalline $LiFePO_4$ contains four $LiFePO_4$ molecules, namely 4 Li atoms, 4 Fe atoms, 4 phosphorus atoms, and 16 oxygen atoms, with a total of 28 atoms. The atomic position is as follows [8].

$$
\begin{array}{llccc}
 & & x & y & z \\
\text{Li} & \text{4a} & 0 & 0 & 0 \\
\text{Fe} & \text{4c} & 0.282 & 1/4 & -0.023 \\
\text{P} & \text{4c} & 0.095 & 1/4 & 0.418 \\
\text{O1} & \text{4c} & 0.107 & 1/4 & -0.268 \\
\text{O2} & \text{4c} & 0.460 & 1/4 & 0.208 \\
\text{O3} & \text{8d} & 0.165 & 0.043 & 0.288
\end{array}
\left.\vphantom{\begin{array}{c}1\\1\\1\\1\\1\\1\end{array}}\right\} +0\ 1/2\ 0;\ \ 1/2\ 0\ 1/2;\ \ 1/2\ 1/2\ 1/2
$$

The $FePO_4$ unit cell also contains four $FePO_4$ molecules, namely 4 Fe atoms, 4 phosphorus atoms, 16 oxygen atoms, a total of 24 atoms, and their atomic positions are as follows [10]

Table 12.3 Phase, crystal structure, lattice parameters, unit cell volume, and relative volume change rate formed during clcctrochemical lithiation

Phase	Space group	Cell parameter (Å)			Cell volume (Å)^3	Change rate of cell volume (%)	Theoretical Li capacity (mAh/g)
		a	*b*	*c*			
$LiFePO_4$	Pnma	10.33	6.01	4.69	29.2	0	170
$Li_{0.5}FePO_4$	Pnma	10.07	5.90	4.73	28.10	−3.8	
$FePO_4$	Pnma	9.81	5.79	4.78	27.5	−7.2	

		x	y	z
Fe	4c	0.277	1/4	0.9449
P	4c	0.0935	1/4	0.3983
O1	4c	0.1167	1/4	0.7131
O2	4c	0.4417	1/4	0.1614
O3	8d	0.1684	0.0461	0.2513

+0 1/2 0; 1/2 0 1/2; 1/2 1/2 1/2

The crystal structure model and chemical combination of $LiFePO_4$ and $FePO_4$ are shown in Figs. 12.11 and 12.12, respectively. The structural data of the two substances (including space groups, lattice parameters, and atomic positions in the unit cell) is almost the same, and Figs. 12.11 and 12.12 show that their crystal structures are almost the same, and the change from $LiFePO_4 \rightarrow Li_{1-x}FePO_4 \rightarrow FePO_4$ ($x = 1$) is not a real structural phase transition, only because under the action of electric field, the Li atom gradually leaves the 4a position of the $LiFePO_4$ crystal lattice, which changes from $LiFePO_4$ to $Li_{1-x}FePO_4$, lack Li and then change to $FePO_4$ without Li. However, the crystal structure remains basically unchanged, but only the crystallographic position of each atom moves slightly. Therefore, the change of $LiFePO_4 \rightarrow Li_{1-x}FePO_4 \rightarrow FePO_4$ is not the phase transition of real phase structure change, but can be called pseudo-phase transition. Similarly, the phase transition during discharge, $FePO_4 \rightarrow Li_{1-x}FePO_4 \rightarrow LiFePO_4$, is also not an real phase transition caused by the gradual re-intercalation of Li atoms, and it is also a kind of pseudo-structural phase transition. It is also caused by the de-intercalation and re-intercalation of Li atoms into the crystal lattice of the positive active material.

Li atoms have four crystallographic positions in the $LiFePO_4$ lattice, namely:

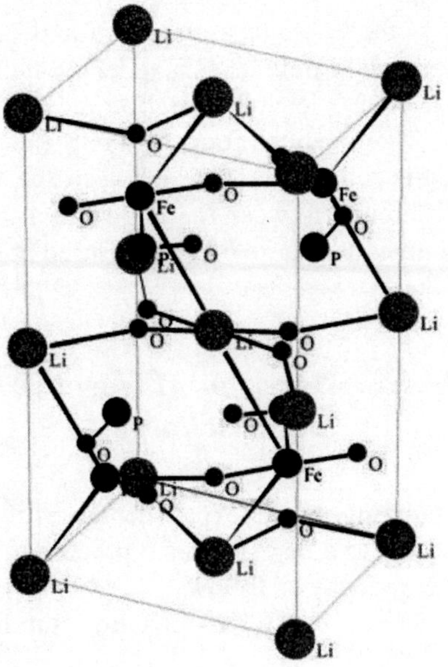

Fig. 12.11 Crystal structure model and bonding of olivine $LiFePO_4$ with Pnma (No. 62) space group structure

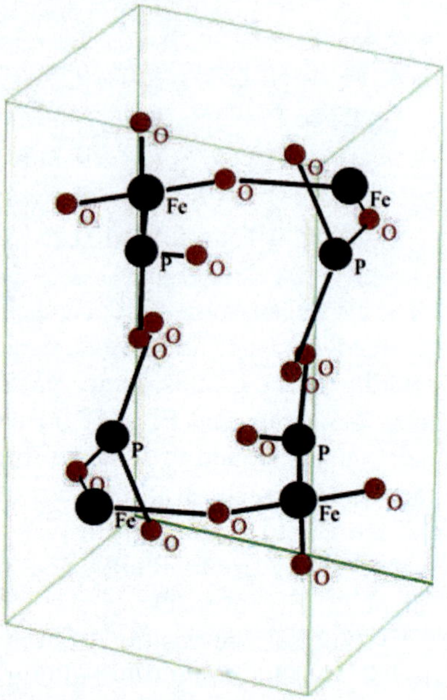

Fig. 12.12 Structural model and bonding $FePO_4$ with Pnma (No. 62) space group structure

$$0\ 0\ 0; \quad 0\ 1/2\ 0; \quad 1/2\ 0\ 1/2; \quad 1/2\ 1/2\ 1/2$$

As can be seen from Fig. 12.11, the bonding force at position 0 0 0 is the weakest, while that at position 1/2 1/2 1/2 is the strongest, and that at position 0 1/2 0 and 1/2 0 1/2 are the almost. Therefore, under the action of a small electric field, the Li atom at position 0 0 0 first leaves the $LiFePO_4$ lattice until it becomes $Li_{0.25}FePO_4$; and then the Li atoms at the position 0 1/2 0 and 1/2 0 1/2 leave successively until they become Li-deficient $Li_{0.50}FePO_4$ and $Li_{0.75}FePO_4$; until Last Li atom in the position 1/2 1/2 1/2 leave lattice, become $FePO_4$ of without Li.

There may be four steps in the charge–discharge curve of 2H-graphite/$LiFePO_4$ battery, the difference between the front and the last two steps is large, and the difference between the middle two steps is very small. This is proved by the actual charge–discharge curve measured in Fig. 12.13.

12.5.2 Behavior of Negative Active Materials During Charge and Discharge

Summing up the experimental results of Chap. 11 and Sect. 12.3 of this chapter, we can see that there is a process of C $\rightarrow$ xLi $\rightarrow$ Li_xC_6, after charging, which is theoretically as follows:

When $x < 0.25$, C–Li solid solution $\rightarrow$ C–Li solid solution + LiC_{24} is formed.

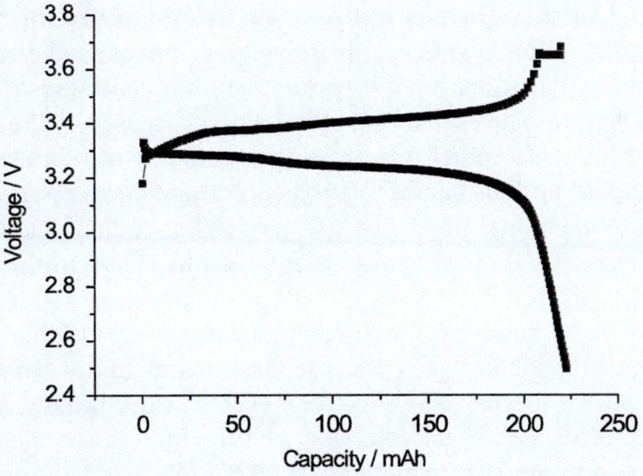

Fig. 12.13 Charge–discharge curve of 2H-graphite/$LiFePO_4$ battery

$x = 0.25$	LiC_{24}
$0.25 < x < 0.50$	$LiC_{24} + LiC_{12} \rightarrow LiC_{12} + LiC_{24}$
$x = 0.50$	LiC_{12}
$0.50 < x < 1.00$	$LiC_{12} + LiC_6 \rightarrow LiC_6 + LiC_{12}$
$x = 1.00$	LiC_6

But the actual situation may be more complicated. See Sect. 11.5.1. Such a process occurs at different stages of charging, and the discharge is a reversible process, which is not completely reversible.

12.6 Conductive Mechanism of 2H-Graphite/$LiFePO_4$ Battery [5]

In Sect. 12.1, it is mentioned that $LiFePO_4 \Leftrightarrow FePO_4$ occurs during the charge and discharge of 2H-graphite/$LiFePO_4$ battery, so that the main body of $Fe^{+3} \Leftrightarrow Fe^{+2}$ is Fe atom, and the conductive ions are provided by this kind of oxidation and reduction.

If the real structural phase transition occurs in the process of charge and discharge, and the Fe atoms are oxidized and reduced, then the oxidation–reduction occurs at the positive electrode and occurs during charging and discharging, respectively, which is one of the two necessary conditions for chemical power sources to directly convert chemical energy into electric energy: "Whether charging or discharging, the process of losing electrons (oxidation) and getting electrons (reduction) in a chemical reaction must be separated on the positive and negative electrodes" to contradict each other. It can be seen that the phase transition theory of electrical conductivity cannot be established.

Our experimental results show that the phase transition of $LiFePO_4 \Leftrightarrow FePO_4$ in 2H-graphite/$LiFePO_4$ battery during charge and discharge is not a real structural phase transition, but a pseudo-structural phase transition with no change in crystal structure. The pseudo-phase transition is caused by the de-intercalate and intercalate of Li atoms, rather than the de-intercalation of Li atoms is caused by pseudo-phase transition. This causal relationship cannot be reversed. Therefore, the conduction is the directional migration and movement of the conductive ion Li^+ formed by the de-intercalation of the Li atom under the action of the electric field. The process is as follows.

At the beginning of charging, the migration of lithium ions begins at the negative-electrolyte interface. Because the lithium ion is embedded in the negative electrode at the negative electrode–electrolyte interface, the concentration of lithium ion decreases at the interface, and the lithium ion migrates from the positive electrode to the negative electrode like a relay race to make up for this concentration gap. Due to the directional migration of lithium ion, the concentration of lithium ion produced at the positive electrode–electrolyte interface decreases correspondingly. At this time, under the action of the electric field, the Li atom at position 0 0 0 in $LiFePO_4$ leaves the lattice position and loses electrons at the positive electrode electrolyte interface to form Li^+, which enters the electrolyte to supplement the lithium ion in the electrolyte. When the ion flow reaches dynamic equilibrium, it corresponds to the charging platform of the battery. When the battery is fully charged, the Li atoms in the $LiFePO_4$ are exhausted and continue to charge. Li^+ can only be provided by the electrolyte, so there must be Li^+ ions in the original composition of the electrolyte. It can be imagined that if there is no electrolyte in the electrolyte for Li^+, charging, the migration of lithium ions must start from the positive electrode–electrolyte, transfer from the Li atoms separated from the $LiFePO_4$ lattice to the electrolyte, and then transfer to the negative electrode surface through the electrolyte and embedded graphite, which must correspond to the high charging platform.

During the discharge of the battery, the migration of Li^+ begins at the interface between the positive and the electrolyte, which is opposite to that of charging. So we obtain the conclusion:

(1) The conductive Li^+ ions during charge and discharge are not provided by the phase transition from $LiFePO_4$ to $FePO_4$, but by the de-intercalation and re-intercalation of Li atoms in $LiFePO_4$ and $FePO_4$.
(2) There is a process of pseudo-structure phase transition from $LiFePO_4$ to $FePO_4$, that is, $LiFePO_4 \rightarrow LiFePO_4 + Li_{1-x}FePO_4 \rightarrow Li_{1-x}FePO_4 + LiFePO_4 \rightarrow Li_{1-x}FePO_4 + FePO_4 \rightarrow FePO_4$.
(3) The intercalation of lithium atoms in 2H-graphite also has the process of from formation of C–Li solid solution $\rightarrow LiC_{24} \rightarrow LiC_{12} \rightarrow LiC_6$.
(4) The conductivity of 2H-graphite/$LiFePO_4$ battery cannot be explained by the mechanism of phase transition, but the directional migration and movement of Li^+ ions separated from $LiFePO_4$ lattice and embedded graphite electrode under the action of electric field.

Therefore, the reaction of the charging process in the positive electrode can be written as:

$$2LiFePO_4 \rightarrow LiFePO_4 + Li_{1-x}FePO_4 + xLi^+ \rightarrow Li_{1-x}FePO_4 + xLi^+ + FePO_4 \rightarrow FePO_4$$

The reaction in the negative electrode can be written as:

$$C + Li \rightarrow xLi - C\,(\text{solid solution}) \rightarrow xLi - C\,(\text{solid solution}) + LiC_{24} \rightarrow LiC_{24} + LiC_{12} \rightarrow LiC_{12} + LiC_6 \rightarrow LiC_6$$

(5) Whether in the process of charging or discharging, the main body of redox is Li atom, and there are two positive and negative atoms in the process of redox. The electrode is oxidized-reduced. Although the $LiFePO_4 \rightarrow FePO_4$ pseudo-structure occurs due to the de-intercalation of Li atoms during charge and discharge. At the same time, the change of $Fe^{+3} \rightarrow Fe^{+2}$ also occurs, but this so-called oxidation-reduction occurs at the positive electrode, and occurs during the process of charge and discharge respectively. According to the oxidation and reduction of chemical power supply (battery) in the process of charge and discharge, whether it is the charging process or the discharge process, which occurs on the positive and negative electrodes, respectively, so the phase transition theory of conduction can not be established. Here, it must be emphasized here that during charging, because the Li atoms leave the $LiFePO_4$ lattice the subsequently leave the $Li_{1-x}FePO_4$ lattice, $LiFePO_4 \rightarrow Li_{1-x}FePO_4 \rightarrow FePO_4$ occurs. Release lithium ions not by this phase transitio, On the contrary, the pseudo-structural transition occurs after the Li atom leave $LiFePO_4$ lattice, during discharge, because the Li atoms are re-embedding into the $FePO_4$ and the subsequent Li_{1-x}lattice, $FePO_4 \rightarrow Li_{1-x}FePO_4 \rightarrow LiFePO_4$, This causal relationship cannot be reversed.

References

1. Srinivasan, V., and J. Newman. 2004. Discharge model for the lithium iron-phosphate electrode. *Journal of the Electrochemical Society* 151 (10): A1517–A1529.
2. Yamada, A., H. Koizumi, N. Sonoyama, and R. Kanno. 2005. Phase change in Li_xFePO_4. *Electrochemical and Solid-State Letters* 8 (8): A409–A413.
3. Dodd, J.L., R. Yazami, and B. Fultz. 2006. Phase diagram of Li_xFePO_4. *Electrochemical and Solid-State Letters* 9 (3): A151–A155.
4. Shin, H.C., K.Y. Chung, W.S. Min, et al. 2008. Asymmetry between charge and discharge during high rate cycling in $LiFePO_4$, in situ X-ray diffraction study. *Electrochemistry Communications* 10: 536–540.
5. Wang, Qian, Yu-Wan Lou, Jian Zhang, et al. 2014. *Study on phase transformation character of $LiFePO_4$ during charge-discharge process of graphite/lithium iron phosphate battery.* Chemical Research in Chinese University (will be published).

6. Cheng, G.F., C.Z. Yang, and Y.H. Huang. 2010. *X-ray analysis of nanomaterials*. Chemical Industry Press. Yang, C.Z., D.C. Xie, K.Z. Chen, and F.M. Zhong. 1989. *Phase diffraction analysis*. Beijing: Metallurgical Industry Press.
7. Morgan, D., A. Van der Ven, and G. Ceder. 2004. Li conductivity in Li_xMPO_4 (M = Mn, Fe, Co, Ni) olivine materials. *Electrochemical and Solid-State Letters* 7 (2): A30–A32.
8. Wilson, A.J.C. 1950. *Structure Reports* 13: 319.
9. Andersson, A.S., and J.O. Thomas. 2001. The source of first-cycle capacity loss in $LiFePO_4$. *Journal of Power Sources* 97–98: 498–502.
10. Eventoff, W., R. Martin, and D.R. Peacor. 1972. *American Mineralogist* 57: 45–51; *Structure Reports* 38A: 314.

Chapter 13
Mechanism Research of Cycle Process for MH/Ni Battery

At present, the electrochemical research on the degradation of cycle performance of MH/Ni battery has been relatively mature, but the combination of cycle performance and the microstructure of electrode materials has not been reported. This chapter introduces the relationship between the cycle performance of MH/Ni battery and the microstructure of positive and negative active materials, in order to reveal the decay mechanism of cycle performance for MH/Ni battery in charge and discharge cycle process.

13.1 Cycle Performance of MH/Ni Battery at 20 and 60 °C [1–3]

The cycle life of the battery was tested at 1C charge and 2C discharge at room temperature (20 °C) and high temperature (60 °C), respectively. The results are shown in Fig. 13.1. It can be seen that under the two temperature conditions, the capacity attenuates slowly when the number of cycles is less than 200, and the attenuation rate is obviously accelerated after more than 200 cycles. At 60 °C, the initial capacity is 150 mAh lower than that at room temperature; the capacity attenuation rate is higher than that at room temperature, and the capacity has attenuated to 51.6% of the initial capacity when cycling 300 times, while the battery capacity retention rate is 60.8% when cycling 400 times at room temperature.

The change trend of the internal resistance is shown in Fig. 13.2. The internal resistance of the battery slowly increases to 97.2 mΩ after 300 cycles at 20 °C. Subsequently, the internal resistance increases faster; while at 60 °C, the internal resistance increase rapidly, internal resistance for 200 cycles at 60 °C exceeds the internal resistance for 400 cycles at 20 °C.

Figures 13.3 and 13.4 show the relationship between the capacity and the self-discharge rate with the cycle times under 1C charge and discharge, respectively.

C. Yang et al., *Materials and Working Mechanisms of Secondary Batteries*,
https://doi.org/10.1007/978-981-19-5955-4_13

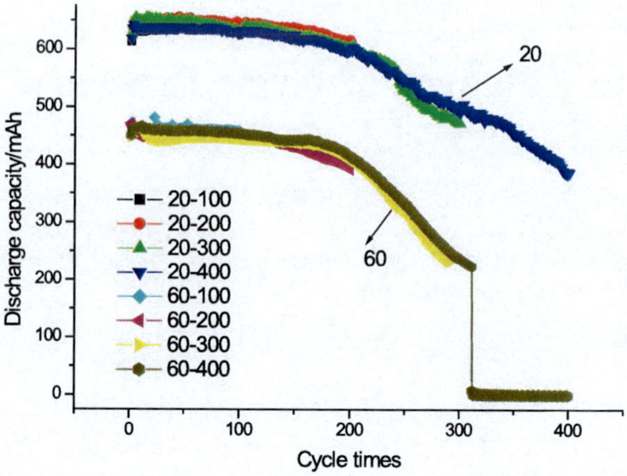

Fig. 13.1 Relationship between capacity and cycle times of MH/Ni battery during 1C charge and 2C discharge at 20 and 60 °C

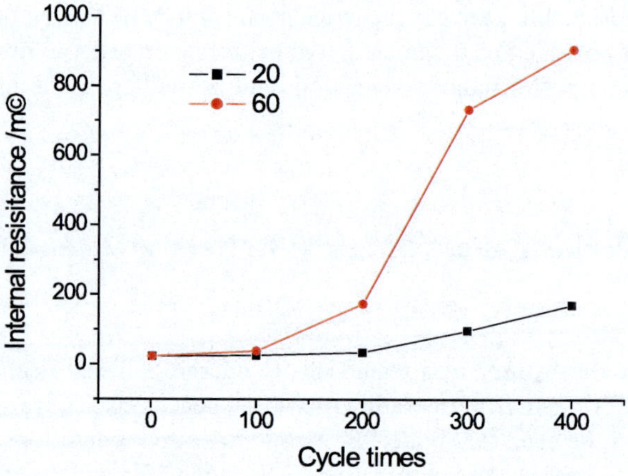

Fig. 13.2 Variation of internal resistance of battery with cycle period during 1C charge and 2C discharge at 20 and 60 °C

It can be seen that the capacity attenuation of 60 °C is much faster than that of 20 °C, and the self-discharge rate is also much higher, especially after the cycle times is more than 100.

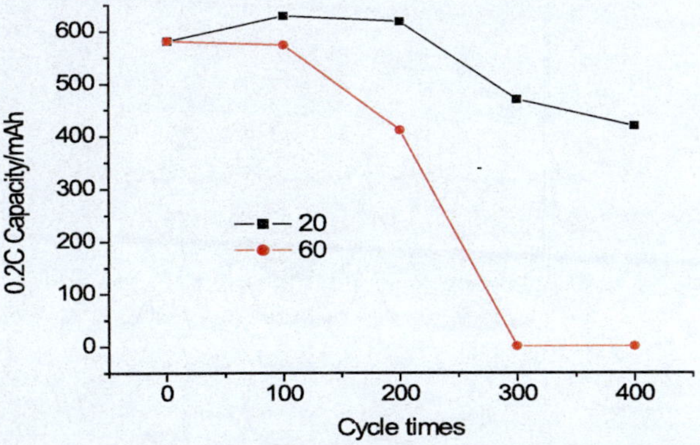

Fig. 13.3 Variation of battery capacity with cycle times under 1C charge and discharge at 20 and 60 °C

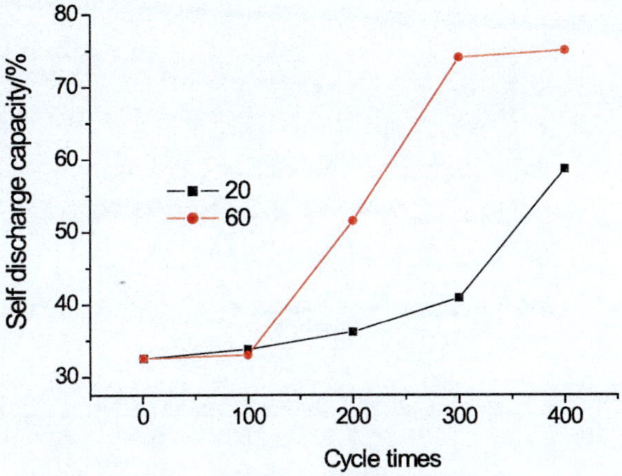

Fig. 13.4 Variation of self-discharge rate of battery with cycle times at 20 and 60 °C

13.2 Microstructure of β-Ni(OH)$_2$ During Cycling at 20 and 60 °C [2, 3]

Figure 13.5 shows the X-ray diffraction patterns of β-Ni(OH)$_2$ at different cycle stages at 20 and 60 °C. The microstructure parameters obtained by the analysis and calculation method of Chap. 2 are listed in Table 13.1, which is directly shown in Figs. 13.6, 13.7, 13.8 and 13.9.

Figure 13.6 shows that the lattice parameters of β-Ni(OH)$_2$ change little with the increase of the number of cycles, while Table 13.1 and the corresponding Figs. 13.7, 13.8 and 13.9 show that the charge–discharge cycle significantly changes the microstructure of the material: The charge–discharge cycle of AAA600 battery

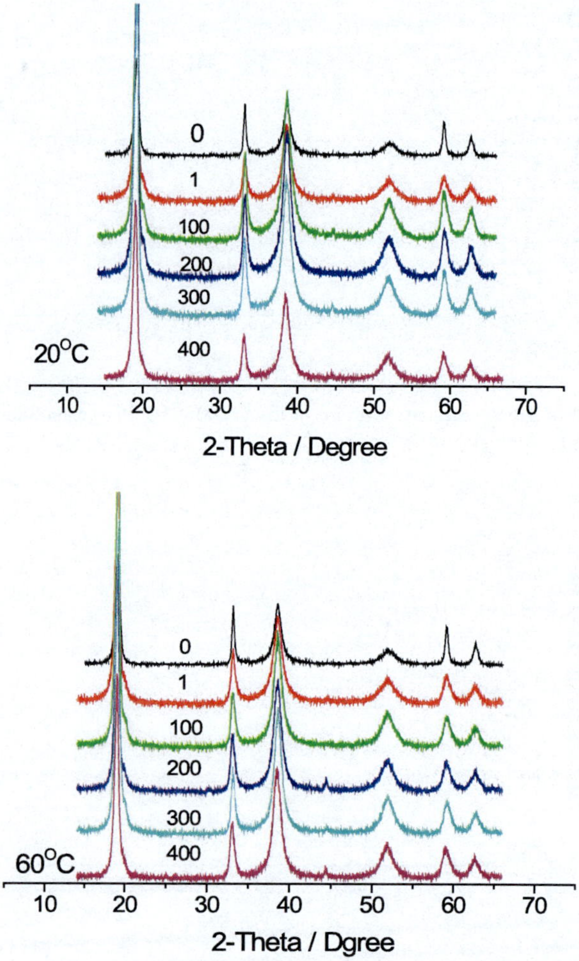

Fig. 13.5 XRD patterns of positive active material β-$Ni(OH)_2$ at different cycle stages at 20 and 60 °C

at 20 and 60 °C further refines the grains of near-equiaxed grains in β-$Ni(OH)_2$, reduces the micro-strain, and reduces the total stacking fault probability. Therefore, the change of lattice parameters of β-$Ni(OH)_2$, which is often paid attention to in the literature, cannot fully reflect its real performance.

At 20 °C, the average grain size, micro-strain, and total stacking fault probability of β-$Ni(OH)_2$ decrease with the increase of cycle times. The results show that the grains are gradually refined and the micro-strain tends to smooth after 200 cycles. This is because the micro-strain reflects the stress change of the lattice structure with the phase transformation. The stress state in β-$Ni(OH)_2$ decreases gradually with the cycle and decreases with the increase of the charge–discharge depth, which is

Table 13.1 Results of experimental data analysis of β-$Ni(OH)_2$ in several main stages before and after the cycle at 20 and 60 °C

	D_{001} (nm)	D_{100} (nm)	D_{100}/D_{001}	$D_{average}$ (nm)	ε (× 10^{-3})	f_D (%)	f_T (%)	$f_D + f_T$ (%)
20 °C								
No activate	16.50	29.40	1.782	27.2	–	9.86	1.59	11.45
Only activate	19.98	17.67	0.931	57.3	5.96	4.28	4.66	8.94
100 cycles	19.78	17.82	0.901	54.9	5.57	5.12	2.83	7.95
200 cycles	20.03	17.67	0.882	43.9	4.81	4.59	3.68	8.27
300 cycles	19.49	19.42	0.996	39.2	4.61	5.57	2.10	7.67
400 cycles	16.71	19.46	1.105	34.2	4.62	4.13	3.10	7.23
60 °C								
100 cycles	19.63	17.15	0.874	85.1	6.81	3.03	2.03	5.06
200 cycles	19.30	19.56	0.962	54.3	5.64	2.93	2.21	5.81
300 cycles	20.49	15.70	0.766	82.4	6.99	3.12	2.73	5.85
400 cycles	Because it has been short-circuited (see Fig. 13.1) and the data is not credible, it is not included							

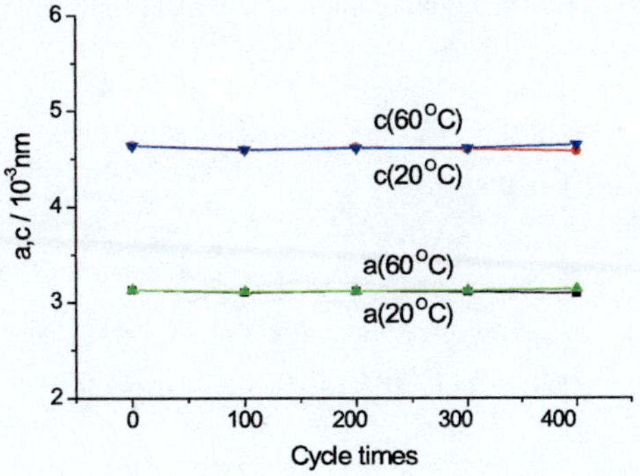

Fig. 13.6 Variation of lattice parameters *a* and *c* of β-$Ni(OH)_2$ with the cycle times

consistent with the decrease of the battery capacity. Although the variation trend of deformation stacking fault and growth stacking fault is different, the total stacking fault probability decreases with the increase of cycle times, which is consistent with

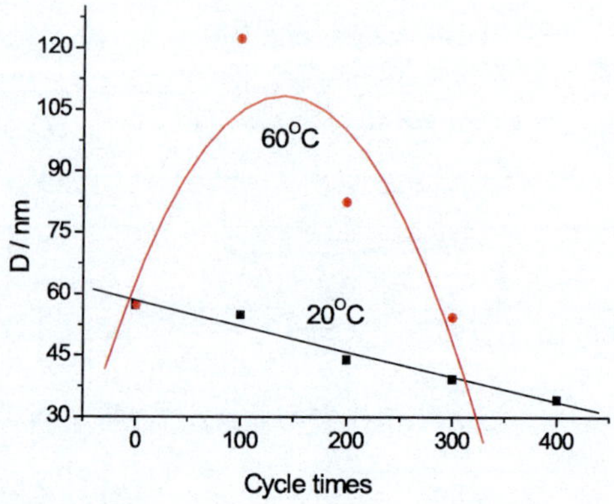

Fig. 13.7 Variation of the average grain size of β-$Ni(OH)_2$ with the cycle times

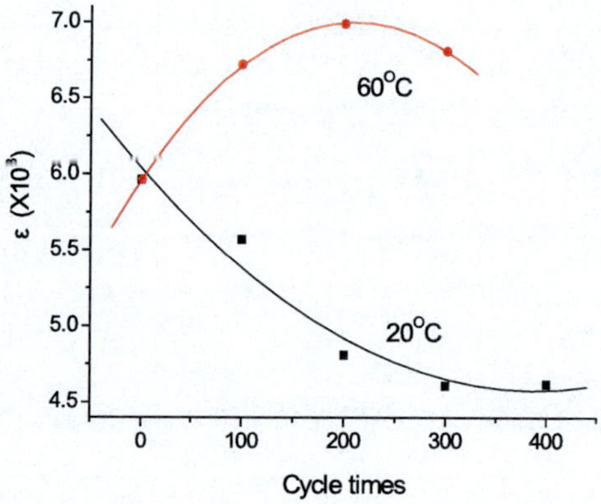

Fig. 13.8 Variation of micro-strain ε of β-$Ni(OH)_2$ with the cycle times

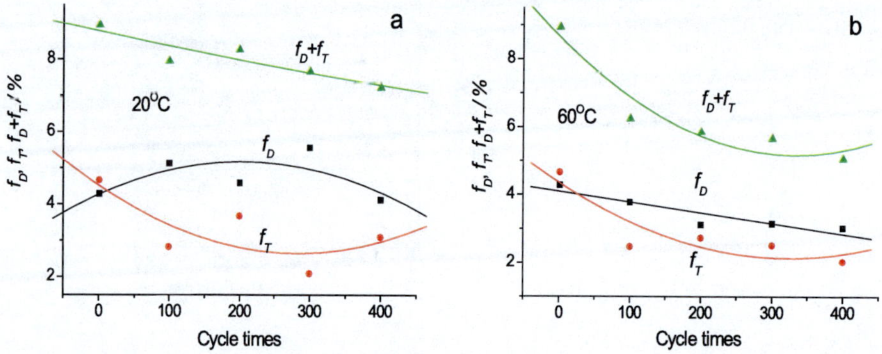

Fig. 13.9 Variation of stacking fault probability ($f_D, f_T, f_D + f_T$) in β-$Ni(OH)_2$ with the cycle times

the change of battery capacity and also consistent with the decrease of β-$Ni(OH)_2$ capacity and activity with the lower total stacking fault probability in literature.

Summing up the changing trend of the above microstructure parameters, we can know:

(1) Activation results in significant changes in microstructure parameters (grain shape, grain size, micro-strain, and stacking fault probability).
(2) After cycling at 20 °C, the performance attenuation of the battery corresponds with the continuous refinement of grains, the continuous decrease of micro-strain, and the change of stacking fault probability. When the cycle is less than 200 times at 60 °C, it is similar to that at 20 °C, but the inflection point effect is very obvious after 200 times. When the cycle is more than 200 times, the changes of average grain size, micro-strain, and stacking fault probability are more complex.

13.3 Proton Diffusion Coefficients of Different Positive Active Materials [2–4]

The cyclic voltammetry curves of positive active material in MH/Ni battery with different cycle times were measured with the scanning speeds of 1 mVs^{-1}, 2 mVs^{-1}, 4 mVs^{-1}, 6 mVs^{-1}, 8 mVs^{-1}, 12 mVs^{-1}, 16 mVs^{-1}, and 20 mVs^{-1}, respectively. The relationship between the corresponding reduction peak current and scanning speed is shown in Fig. 13.10. It can be seen that the reduction peak potential of the cyclic voltammetry curve of nickel electrode increases with the increase of scanning speed, and there is a linear relationship between the reduction peak current and the square root of scanning speed. It can be concluded that the electrode process of nickel electrode is controlled by diffusion in the range of scanning speed, but it is not completely reversible. Therefore, we choose the following formula.

$$i_L = 2.69 \times 10^5 n^{3/2} A D_0^{1/2} C_0 v^{1/2} \tag{13.1}$$

i_L—peak current of cyclic voltammetry curve;

n—number of electrons in electrode reaction;

A—surface area of electrode;

D_0—determine the diffusion coefficient of the velocity step, which in this experiment refers to the diffusion coefficient of protons;

C_0—the concentration of protons is approximately equal to the ratio of the density (3.97 g/cm^3) of nickel hydroxide to its mol mass (92.7 g/mol), that is, 0.0428 mol/cm^3.

The curve of the relationship between i_L and the square root of scanning velocity $v^{1/2}$ should be a straight line. From the slope of the straight line, the proton diffusion coefficient in the nickel electrode can be calculated according to the formula (13.1). The calculated results are shown in Table 13.2. It can be seen that the proton diffusion coefficient decreases with the increase of the times of cycles of MH/Ni battery. The

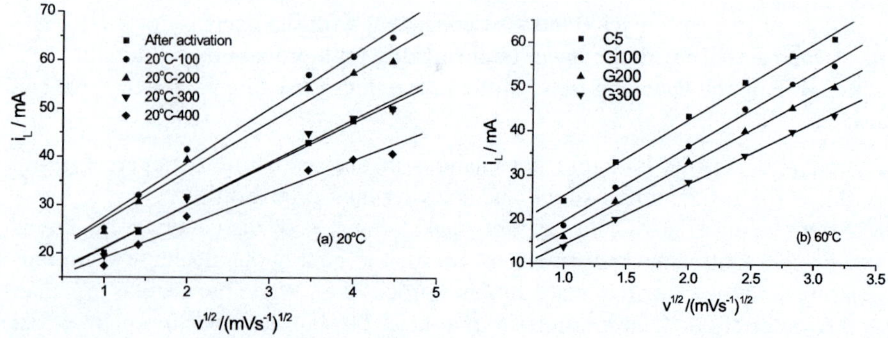

Fig. 13.10 Curve of the relationship between the peak current of cyclic voltammetry i_L and the scanning speed $v^{1/2}$ of different $Ni(OH)_2$

Table 13.2 Proton diffusion coefficients of positive active materials with different cycles

Diffusion coefficient D/cm^2 ($\times 10^{-8}$ cm^2 s^{-1})	After active, before cycle	Number of cycles under 20 °C				Number of cycles under 60 °C		
		100	200	300	400	100	200	300
	13.7	14.9	12.4	9.26	5.39	8.49	7.21	5.77

proton diffusion coefficient decreases with the increase of the number of cycles of MH/Ni battery. After cycling at 20 °C, the diffusion coefficient of β-$Ni(OH)_2$ reached the maximum at 100th cycle and then decreased gradually. While the proton diffusion coefficient of β-$Ni(OH)_2$ after cycle at 60 °C decreases greatly, the proton diffusion coefficient of β-$Ni(OH)_2$ after 100 times cycle is lower than that of 300 times cycle at 20 °C. This may be due to the fact that when the nickel electrode is charged at 60 °C, it is easier to produce γ-NiOOH with poor activity and the capacity decreases.

Combined with the microstructure change law of β-$Ni(OH)_2$ in Sect. 13.2 and the specific capacity data in Sect. 13.2, the corresponding analysis can be carried out. At 20 °C, with the increase of the number of cycles, the proton diffusion coefficient first increases and then decreases and reaches the maximum at 100th, which is consistent with the 100th maximum of the specific capacity 276 mAh/g of β-$Ni(OH)_2$ and the 0.2C capacity 674 mAh of the battery. Both the proton diffusion coefficient and the specific capacity data are for the test results of β-$Ni(OH)_2$ in a large number of electrolytes, and the 0.2C capacity of the battery is the test result of reducing the charge–discharge rate. In the microstructure parameters, the grain refinement, the decrease of micro-strain, and the decrease of stacking fault probability are consistent with the change of 1C capacity of the battery. The results show that the microstructure parameters can better reflect the real performance of β-$Ni(OH)_2$ in the battery. At 20 °C, with the increase of the number of cycles, the reason why the inflection point data do not appear on the cyclic voltammetry curve may be related to the fact that the number unit is 100 times. Therefore, the proton diffusion coefficient and microstructure of β-$Ni(OH)_2$ are related to the capacity change trend of MH/Ni cells. Because the grain size and micro-strain are related to the temperature of charge and discharge, the main microstructure parameter that leads to the change of β-$Ni(OH)_2$ performance is the total stacking fault probability, which can be used as an important content to evaluate the state of β-$Ni(OH)_2$.

13.4 Study on the Microstructure of AB_5 Alloy in Cyclic Specimen at Room Temperature and 60 °C [1–3]

Figure 13.11 shows the XRD patterns of several main samples of AB_5 alloy at 20 and 60 °C after cycling. It can see that except for the diffraction lines of AB_5 alloy, new diffraction lines appear in many places and are very broadened. According to the principle and method of "phase diffraction analysis" [5] and using Jade6.5 search/matching program [6], the unknown phase is $A(OH)_3$ (PDF card number 06-0585, $La(OH)_3$) and phase B (card number 04-0850, Ni). The lines are broadened, indicating that the grains of the new phase are very fine, which is consistent with the results of Ref. [4].

Figure 13.12 comprehensively shows the changes of microstructure parameters D and ε of AB_5 alloy in MH/Ni battery with the increase of cycle times. It can be seen that the average grain size and micro-strain increase with the increase of cycle times, especially the grain size of anode alloy in the battery cycled at 20 °C is faster than that at 60 °C.

In order to further determine the changing trend of $A(OH)_3$ and B phase with the increase of cycle period, it is expressed by the change of the relative intensity of the peak in the XRD spectrum of the hydrogen storage alloy. The observation of the related XRD pattern shows that strongest peaks 101 and 110 could not be separated at 20 °C. Therefore, the intensity ratio of $I_{A(OH)_3110+101}/I_{AB_5101}$ and $I_{A(OH)_3201}/I_{AB_5110}$ is determined. The 101 and 110 peaks of $A(OH)_3$ can be separated at 60 °C (see Fig. 13.11), so $I_{A(OH)_3101}/I_{AB_5101}$ and $I_{A(OH)_3201}/I_{AB_5110}$ are determined; so the results of the determination are shown in Fig. 13.13. It can be seen from the figure that the content of $A(OH)_3$ increases with the increase of cycle at 20 °C, but at 60 °C, when the number of cycles is greater than 200, the content of $A(OH)_3$ decreases with the increase of cycle. There may be two reasons for this result:

First, the battery after 300 cycles has been short-circuited; second, due to the precipitation of the new phase, the volume expands or contract of the electrode material, and the larger micro-strain, resulting in the cracking and loosening of the electrode surface, which makes $A(OH)_3$ fall off and dissolve in the electrolyte. The cracking and loosening of the electrode surface has been observed in the scanning electron microscope photos of Ref. [4]. Therefore, it can be considered that no matter cycle at 20 or 60 °C, the new phase $A(OH)_3$ increases with the increase of the cycle period, and the new phase B is similar.

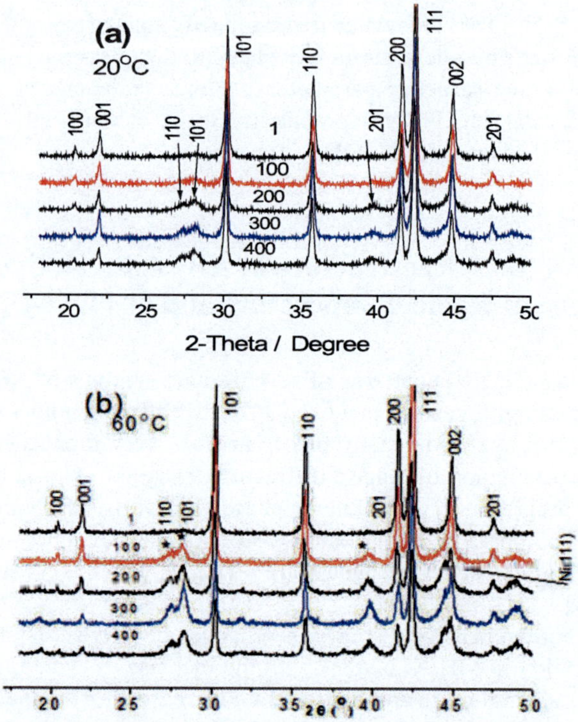

Fig. 13.11 XRD patterns of several main samples of AB_5 alloy in 20° (**a**) and 60 °C (**b**) cycle tests

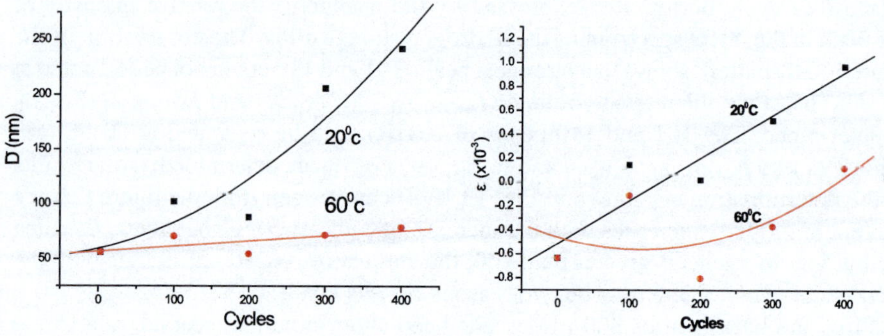

Fig. 13.12 $\overline{D}$ (left) and ε (right) of the anode material AB_5 in the cycle test of MH/Ni battery at 20 and 60 °C

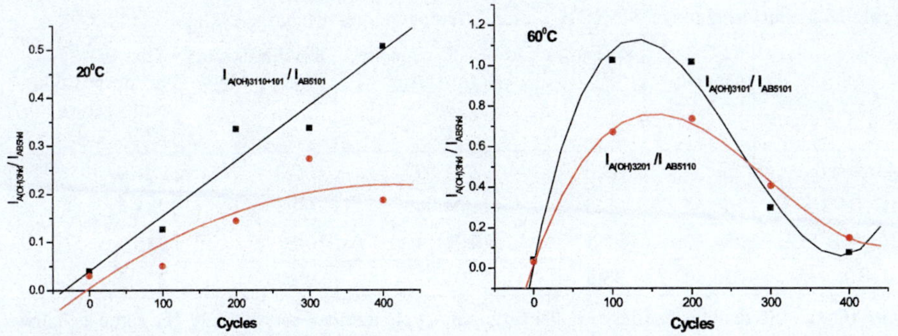

Fig. 13.13 Variation of the $I_{A(OH)_3hkl}/I_{AB_5hkl}$ ratio of anode materials with the of cycle times at 20 and 60 °C

13.5 Cycle Performance and Microstructure Under Different Charge–Discharge Rates [1–3]

The most important application fields of MH/Ni batteries are hybrid electric vehicles and power tools that meet the requirements of high-rate charge and discharge. Therefore, by studying the cycle life of high-power battery at high rate and the microstructure of positive and negative active material, the microstructure data affecting the cycle life of high-power battery are obtained to guide the preparation and use of high-power MH/Ni battery. With reference to the charge and discharge conditions of the power tool battery, the cycle life testing system is as follows:

(1) 1C current charge 1.2 h, with $-\Delta V = 10$ mV as parallel control condition, rest for 1 h, current discharge to 1.0 V, standing for 1 h, counted as a cycle, a total of 436 cycles, recorded as 1C-1C-436.
(2) 1C current charge 1.2 h, with $-\Delta V = 10$ mV as parallel control condition, rest for 1 h, then 10C current discharge to 0.8 V, standing for 1 h, counted as a cycle, a total of 117 cycles, recorded as 1C-10C-117.
(3) 3C current charging for 0.4 h, with $-\Delta V = 10$ mV as parallel control condition, rest for 1 h, and then 10C current discharge to 0.8 V, rest for 1 h, count as a cycle, recorded as 3C-10C-95.

In order to characterize the performance of the battery, in addition to measuring the charge and discharge capacity of the battery, the internal resistance of the battery and the utilization rate of β-$Ni(OH)_2$ were also tested. The performance of $Ni(OH)_2$ after different charge–discharge cycles is listed in Table 13.3. Visible from this table:

(1) Under the condition of different rate, the charge–discharge capacity is obviously different, and the capacity of low rate is obviously larger than that of high rate.
(2) The capacity after low-rate cycle is higher than that after high-rate cycle. For example, the capacity retention after 3C charge 10C discharge cycle 95 cycle is the same as that of 1C charge–discharge cycle 436 cycle, which is 98.5%.

Table 13.3 Performance of β-$Ni(OH)_2$ after different charge–discharge cycles

	0.2C capacity (mAh/g)	Retention rate (%)	10C capacity (mAh/g)	Retention rate (%)	Thermal decomposition temperature (°C)
After activate	260.50	100.0	205.89	100.0	221.9
1C-1C-436	256.64	98.5	204.80	99.5	243.4
1C-10C-117	254.10	97.5	206.44	100.3	216.9
3C-10C-95	256.60	98.5	205.67	99.9	223.2

Note 1C-1C-436 denotes charge rate-discharge rate-cycle number, respectively, the same as below

Table 13.4 Microstructure parameters of cathode material β-$Ni(OH)_2$ before and after cycle life test under different conditions

	D_{001}	D_{100}	D_{110}	D_{111}	$\overline{D}$	ε	f_D	f_T	$f_D + f_T$
	(nm)					$\times 10^{-3}$	(%)		
Inactivate	15.33	41.00	26.03	15.81	21.2	–	−5.90	15.70	9.80
Only activate	19.68	17.23	9.35	10.90	44.10	5.098	5.99	−1.12	4.87
1C-1C-436	17.11	17.23	10.69	11.75	26.7	3.411	4.70	−0.62	4.08
1C-10C-117	19.68	13.73	9.98	10.90	30.0	4.283	3.68	0.39	4.07
3C-10C-95	17.11	15.27	9.35	8.47	40.2	5.859	9.02	−5.23	3.79

(3) The thermal decomposition temperature of β-$Ni(OH)_2$ varies with the rate factor and cycle period, and the decomposition temperature after low-rate cycle is higher, indicating that it is stable.

The microstructure parameters of cathode materials before and after cycle life test under different conditions are listed in Table 13.4. From the table data, the following results can be obtained:

(1) The grain size of β-$Ni(OH)_2$ is refined by activation, especially the size perpendicular to the hexagonal *c*-axis is greatly reduced. The microcrystal shape was changed from short and fat column to near-equiaxed crystal, and the cycle life test did not change this trend.
(2) The magnitude of the micro-strain remaining in the cathode material due to repeated reversible phase transition is related to the cycling conditions, and the residual strain with more cycles is small.
(3) The stacking fault after activation and cycle life test is mainly deformation stacking fault, and the total stacking fault probability and deformation stacking fault probability vary with different cyclic condition. The reducing radian of total stacking fault probability at high-rate cycle is maximum.

Table 13.5 Microstructure data of hydrogen storage alloys after different charge–discharge cycles

	After Activation	1C-1C-436	1C-10C-117	3C-10C-95
D (nm)	88.9	23.4	34.7	32.7
ε / (10^3)	1.46	9.46	0.64	−0.02

The microstructure parameters of negative active materials under different cycle conditions are shown in Table 13.5. It can be seen that the charge–discharge cycle makes the grain of the hydrogen storage alloy further refined. In the case of 1C charge–discharge cycle, the battery has been in a deep charge–discharge state, which makes the volume change of the hydrogen storage alloy in the process of hydrogen absorption is greater, and when number of cycles is up to 436 periods, lead to alloy pulverization and serious oxidation. Under cycle condition of 3C charging and 10C discharge, the hydrogen absorption and desorption rate of hydrogen storage alloy is very fast. So the alloy has experienced rapid expansion and shrinkage, resulting in rapid change of internal stress of the alloy, thus accelerating the pulverization of the alloy. In contrast, in the case of 1C charging and 10C discharge cycle, the number of cycles is less than 1C discharge, and the expansion rate is slower than 3C charging 10C discharge, so the total pulverization is lighter. Under the condition of 1C-1C-436 with the largest depth of charge and discharge and the largest number of cycles, the grain refinement is the most obvious and the residual micro-strain is the smallest.

13.6 Physical Behavior of Positive and Negative Active Materials During Battery Cycle Process [1, 2]

The purpose of studying the cycle performance of batteries has three meanings: (1) to characterize the active materials of positive and negative electrodes and the performance of batteries, which must be carried out by electrode materials and battery developers, so there are a lot of reports in this area; (2) through the study of cycle performance and the structure (including crystal structure and microstructure) change of related electrode materials during charge–discharge and cycle process, and through the structure parameters of positive and negative electrode materials, the studying attenuation mechanism of cycle performance, which is hardly reported; and (3) on this basis, to explore the methods to improve the charge–discharge and cycle performance of the battery. About (3) will be discussed in Chap. 20.

According to the research results of Sects. 13.1–13.5, the physical behavior of positive and negative active materials during battery cycle is as follows:

(1) Activation makes the grain shape of positive active materials change obviously from chunky columnar crystals to nearly equal axial crystal and makes the material have micro-strain, and the total stacking fault probability is reduced. Cycling at room temperature further refines the grains and reduces the total stacking fault probability, which corresponds to the attenuation of cycling performance, but

at 60 °C, there is an inflection point in 200 cycles, and the total stacking fault probability does not change much. These changes also reduce the proton diffusion coefficient in β-$Ni(OH)_2$. The final grain size, micro-strain, and stacking fault probability are related to the cycle conditions under different charge and discharge rates.

(2) Whether cycling at 20 or 60 °C, the average grain size and micro-strain of negative active materials increase with the increase of cycle, while the new phases $A(OH)_3$ and B formed on the negative electrode increase with the increase of cycle.

13.7 Cycle Performance Attenuation Mechanism [1–3]

13.7.1 Relationship Between Cycle Performance and Structural Parameters

The cycle performance and the change of the structure of the electrode active material during the cycle are listed in Table 13.6 in the way of comparison. It can be seen from the table that the attenuation of cycle performance has a good corresponding relationship with the microstructure parameters and proton diffusion coefficient of positive active materials, and microstructure parameters of negative electrode as well as the amount of new phase precipitation. It can be seen that the main factor affecting the cycle life and comprehensive performance of the battery is the joint action of positive and negative materials, rather than a single factor.

Table 13.6 Relationship between cycle performance and electrode materials

Cycle performance	20 °C	60 °C
Capacity of 1C charge and 2C discharge	<200 times decreased slowly, >200 times significantly accelerated	<200 times decreased slowly, and >200 times was significantly faster
Internal resistance of 1C charge and 2C discharge	The increase of <200 times is slower, and that of >200 times is faster	The increase of <100 times is slower, and that of >100 times is faster
Capacity of 1C charge and discharge	The decrease of <200 times is slower, and that of >200 times is faster	>100 times decreased quickly

(continued)

Table 13.6 (continued)

Self-discharge rate			There is little change in <100 times, but a quick increase in >100 times	The increment rate of >100 times is faster than that of 20 °C
Parameters of positive active materials	Lattice parameter	*a* (nm)	It decreases slowly as the cycle increases	It decreases slowly as the cycle increases
		c (nm)	It decreases slowly as the cycle increases	It decreases slowly as the cycle increases
	Grain size	D_{100}	Suddenly decreased after activation and then changed little	Suddenly decreased after activation and then changed little
		D_{001}	It has not changed much	It has not changed much
		D_{Average}	Sudden increase after activation, followed by slow refinement	It suddenly increases after activation and then changes little
	Miro-strain	$\varepsilon_{\text{average}}$ ($\times 10^{-3}$)	Micro-strain occurs after activation, and the number of cycles increases and decreases slowly	Micro-strain occurs after activation, and the number of cycles does not change much with the increase of cycle times

(continued)

13.7.2 Attenuation Mechanism of Cycle Performance

The study of Zhou et al. [5] shows that the performance degradation of MH/Ni battery during cycling at low rate (less than 1C charge and discharge) is caused by positive electrode expansion, which not only squeezes the diaphragm and absorbs electrolyte, but also leads to electrolyte redistribution in the battery. In serious cases, the electrode

Table 13.6 (continued)

	Stacking fault probability	$(f_D + f_T)$ (%)	It increased slightly before 100 times and decreased significantly when it was more than 100 times	Reduce it first, it does not change much for 200 times
	Proton diffusion coefficient	D_0		All of them decreased significantly, but faster than 20 °C
Parameters of negative active materials	Grain size D (nm)		It increases faster with the number of cycles	Slow increase
	Miro-strain ε ($\times 10^{-3}$)		It increases faster with the number of cycles	Slow increase
	New phases $A(OH)_3$ and Ni		It increases rapidly with the increase of the number of cycles	It increases rapidly with the increase of the number of cycles

active material falls off, which reduces the capacity of the battery and may cause a micro-short circuit. Bernard [6] believes that the corrosion product A1 of the negative electrode pollutes the positive electrode and forms a similar water-soluble stable phase containing A1 in the positive electrode, which reduces the charging efficiency and shortens the battery life. However, Laure and Bernard [7], Leblanc et al. [8] think that the life of MH/Ni battery is directly related to the corrosion of negative alloy. Alloy corrosion consumes electrolyte and battery charging reserve capacity, which leads to the rapid increase of battery internal resistance and internal pressure, and shortens the battery life. Chen et al. [9] studied the mechanism of 1C charge–discharge cycle life attenuation of high-power D-type battery. The results show that the reason for battery failure is the pulverization, oxidation, and corrosion of negative alloy powder in alkaline medium. It can be seen that the above discussion on the attenuation mechanism of cyclic performance has not gone deep into the structure and microstructure of positive and negative materials.

Based on our experimental results, the attenuation mechanism of cycle performance is discussed as follows:

(1) After cycling at room temperature, the lattice parameters of positive active materials decrease slightly with the increase of cycle period, indicating the formation of Ni vacancy, and the average grain size, micro-strain, and total stacking fault probability decrease with the increase of cycle period. These changes gradually reduced the activity of β-$Ni(OH)_2$, and when these changes reach a certain extent, the life may end and the cycle performance decreases monotonously (see Fig. 13.1). When the cycle number is less than 200, the cycle at 60 °C is similar to that at 20 °C, but when the cycle number is greater

than 200, the average grain size, micro-strain, and total stacking fault probability change little with the increase of period. Cycle performance decays faster (see Fig. 13.1) and even short circuit.

(2) On the side of negative material, the lattice parameters of the material decrease slightly with the increase of the cycle. It indicates that there may be A vacancy (or B vacancy). The average grain size and micro-strain increase with the increase of cycle, especially the precipitated new phases $A(OH)_3$ and B increase with the increase of cycle, which is the result of corrosion of AB_5 alloy and may lead to cracking and shedding on the surface of negative plate.

(3) Due to the precipitation of Ni ions in the positive electrode and the precipitation of A and B ions in the negative electrode materials, and the transfer of corrosion products $A(OH)_3$ and B to the electrolyte, that is, the consumption and deterioration of the electrolyte, the properties of the electrolyte were also changed.

Because the combined action of the above three aspects is the main reason for the decline of cycle performance, the changes of the positive and negative electrodes and the degradation process of the electrolyte are also related to the temperature and charge–discharge rate during the cycle, and the life is also different.

References

1. Lou, Y.W., C.Z. Yang, D.N. He, and B.J. Xia. 2008. Study on cycle performance of Ni-MH battery and microdefects of electrode materials. *Acta Chemica Sinica* 66 (10): 1173–1180.
2. Yang, C.Z., Y.W. Lou, Y.X. Li, et al. 2009. Some advances in the study of the relationship between the fine structure of electrode active materials and the performance of MH-Ni batteries. *Advances in Physics* 29 (1): 108–126.
3. Li, Yu-Xia. 2009. Study on the relationship between fine structure and performance of electrode active material in MH/Ni batter. Master paper of Shanghai Institute of Microsystems and Information Technology, 7.
4. Lou, Yu Wan. 2007. XRD study on the microstructure of active material for positive and negative electrodes of metal hydride-nickel battery. Doctoral paper of Shanghai Institute of Microsystems and Information Technology.
5. Zhou, Z.Q., G.W. Lin, J.L. Zhang, et al. 1999. Degradation behavior of foamed nickel positive electrodes of Ni-MH batteries. *Journal of Alloys and Compounds* 293–295: 795–798.
6. Bernard, P. 1998. Effects on the positive electrode of the corrosion of AB_5 alloys in nickel-metal-hydride batteries. *Journal of the Electrochemical Society* 145: 456–458.
7. Laure, L.G., and P. Bernard. 2002. Life duration of Ni-MH cells for high power applications. *Journal of Power Sources* 105: 134–138.
8. Leblanc, P., C. Jorde, B. Knosp, et al. 1998. Mechanism of alloy corrosion and consequences on sealed nickel-metal hydride battery performance. *Journal of the Electrochemical Society* 145: 860–863.
9. Chen, Y.H., J.P. Wei, F. Gao, et al. 2001. Discussion on influencing factors of high rate cycle life of MH-Ni battery. *Power Supply Technology* 125: 142–145.

Chapter 14
Mechanism Research of the Cycle Process for 2H-Graphite/Li(Ni,Co,Mn)O_2 Battery

The 18650 lithium-ion battery adopts traditional lithium-ion battery cell composition and liquid electrolyte system. The shape packaging adopts cylindrical stainless-steel shell. It combines the excellent electrochemical performance of liquid lithium-ion battery and the unique advantages of cylindrical battery shape packaging. It is a widely used battery configuration in portable appliances. At present, it is used in large-scale high-power lithium-ion batteries (such as HEV power supply) because of its high safety.

In this chapter, the general law and general characteristics of the cycle performance of 18650 lithium-ion battery with and without Al_2O_3-coated graphite/Li($Ni_{0.4}Co_{0.2}Mn_{0.4}$)O_2 system are introduced, and the fine structure of positive electrode, negative electrode, and diaphragm is change law during the cycle process. The connection of the above two kinds of change laws and the battery performance attenuation mechanism is discussed, and some novel phenomena and conclusions are obtained [1–4].

14.1 Surface Morphology of Coated and Uncoated Al_2O_3 Positive Materials and Cycling Performance of Batteries [3, 4]

14.1.1 Surface Morphology of Positive Active Materials Before and After Coated with Al_2O_3

Figure 14.1a shows the SEM diagram of uncoated positive active material, and Fig. 14.1b shows the SEM diagram of after coating positive by two-step reactive under 30 + 60 and 300 °C sintering. It can be seen from the diagram that the surface of Li($Ni_{0.4}Co_{0.2}Mn_{0.4}$)O_2 before coating is smooth, while the surface of Li($Ni_{0.4}Co_{0.2}Mn_{0.4}$)O_2 after Al_2O_3 coating becomes rough, indicating that alumina

C. Yang et al., *Materials and Working Mechanisms of Secondary Batteries*,
https://doi.org/10.1007/978-981-19-5955-4_14

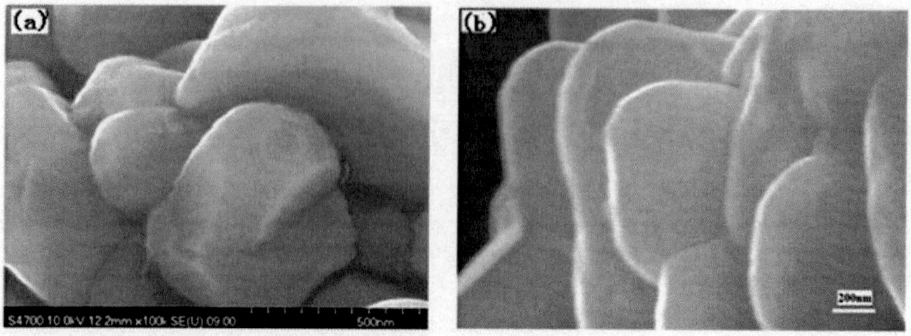

Fig. 14.1 TEM diagram of $Li(Ni_{0.4}Co_{0.2}Mn_{0.4})O_2$ before and after coating, **a** before and **b** after coating

is coated on the surface of $Li(Ni_{0.4}Co_{0.2}Mn_{0.4})O_2$. It can also be seen from the figure that the coating is uniformly distributed on the surface of $Li(Ni_{0.4}Co_{0.2}Mn_{0.4})O_2$.

14.1.2 Battery Cycle Curve

Figure 14.2 shows the first charge–discharge curve of graphite/$Li(Ni_{0.4}Co_{0.2}Mn_{0.4})O_2$ battery at 2.75–4.2 V and 0.05C rate. As can be seen from the figure, the charge–discharge curves of the two kinds of batteries almost coincide. The first discharge capacity of 0.05C battery before and after coating is 1214.2 mAh and 1218.0 mAh, and the first Coulomb efficiency is 88.0% and 88.3%, respectively.

The battery is cycled at room temperature, 2C rate and 2.75–4.2 V, and the cycle performance curve of the whole battery is shown in Fig. 14.3. It can be seen from the figure that the coating not only greatly improves the cycle stability of the battery, but

Fig. 14.2 First charge–discharge curve of graphite/$Li(Ni_{0.4}Co_{0.2}Mn_{0.4}O_2)$ battery

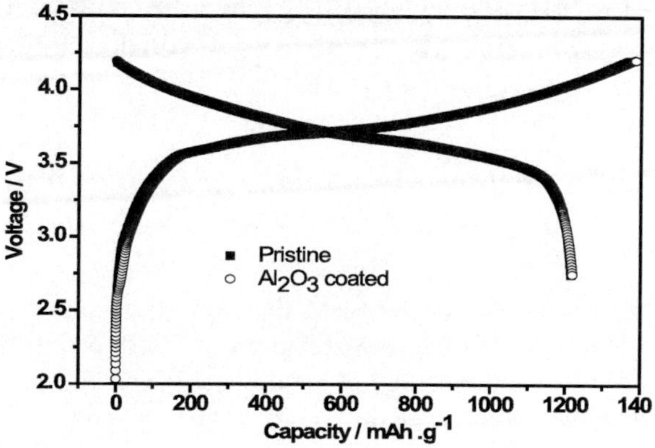

also greatly improves the cycle performance. After 200 and 400 cycles, the capacity retention of uncoated batteries decreased by 89.2% and 81.1%, respectively, while that of Al_2O_3-coated batteries was 97.4% and 94.2%, respectively.

Figure 14.4 shows the 0.2C discharge curve of coated and uncoated batteries before and after cycle. It can be seen from the figure that the initial 0.2C capacity of the battery before and after coating is 1163.3 and 1168.9 mAh, respectively, and there is little difference between them, while after 400 cycles of 2C, 2.75 V and 4.2 V, the 0.2C capacity of the battery before and after coating is 991.8 and 1101.1 mAh, respectively. The capacity retention rates are 85.3% and 94.2%, respectively. This result is different from the capacity attenuation law of the 2025 semi-battery, and the specific reasons will be analyzed in the following discussion.

Figure 14.5 shows the 2C discharge curve of the two kinds of batteries before and after coating. It can be seen that the initial 2C capacity of the battery before and after coating is 1046.7 and 1047.9 mAh, respectively, and the capacity is similar.

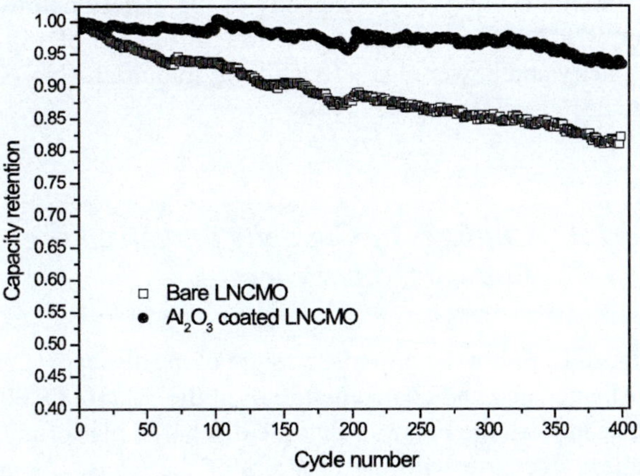

Fig. 14.3 Cycle performance of 18650 battery before and after coating at 25 °C, 2.75–4.2 V, 2C rate

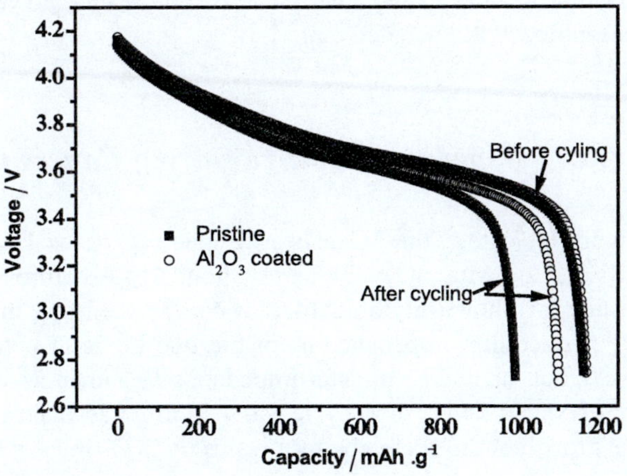

Fig. 14.4 Discharge curve of battery before and after coating at 25 °C and 0.2C rate

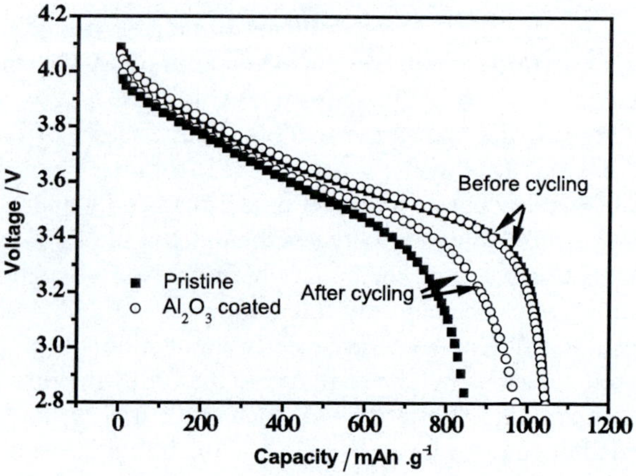

Fig. 14.5 Discharge curve of battery before and after coating at 25 °C and 2C rate

After 400 cycles, the 2C capacity of the battery before and after coating was 860.4 and 978.7 mAh, respectively, with a difference of 119.3 mAh. This shows that the capacity and power characteristics of graphite/$Li(Ni_{0.4}Co_{0.2}Mn_{0.4})O_2$ 18650 battery are decreased during the cycle.

14.1.3 Changes of Capacity and Discharge Platform During Battery Cycle Process

Figure 14.6 shows comparison figure of the discharge curves of the battery assembled with uncoated and coated materials at the 1st, 100th, 200th, and 400th cycles. It can be seen from the diagram that the discharge platform and discharge capacity of the battery decrease with the progress of the cycle. It can be seen in the comparison figure that the discharge capacity and discharge platform attenuation of the battery assembled with coated material are significantly lower than those of the battery assembled with uncoated material.

14.1.4 Impedance Change During Battery Cycle

Figure 14.7 shows the AC resistance spectrum of two kinds of batteries assembled by $Li(Ni_{0.4}Co_{0.2}Mn_{0.4})O_2$ with and without Al_2O_3 before and after cycling. The fitting results are shown in Table 14.1. It can be seen that the ohmic impedance R_e and the surface film impedance R_f of the two kinds of batteries change little after the cycle, but the charge transfer impedance R_{CT} changes obviously. The R_{CT} value of the uncoated battery after cycle was 4.53 times as much as that before the cycle, while the growth of the coated battery was only 2.27 times, which was significantly lower

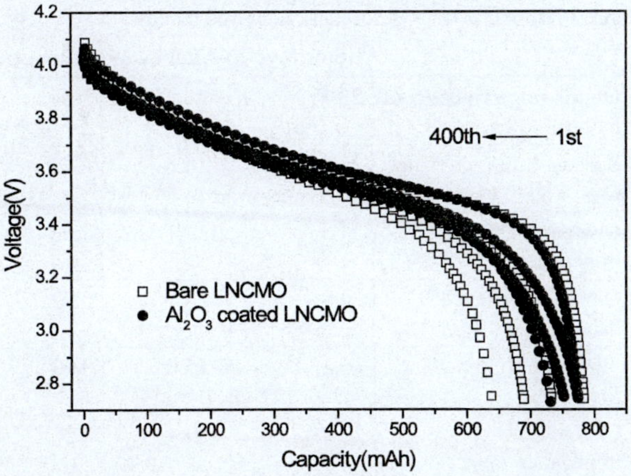

Fig. 14.6 Comparison of discharge curves of batteries assembled with uncoated and coated materials at the 1st, 100th, 200th, and 400th cycles

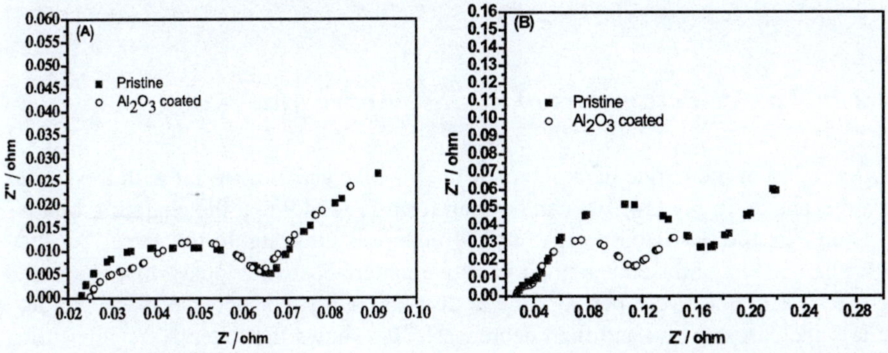

Fig. 14.7 AC impedance Nyquist curves of $Li(Ni_{0.4}Co_{0.2}Mn_{0.4})O_2$ assembled coated and uncoated Al_2O_3. **a** After activation before cycle; **b** after 400 cycles under 2C, 4.2–2.75 V

than that of the uncoated battery. It shows that the cyclic stability of the material has been obviously improved after coating.

14.2 Changes of Fine Structure and Surface Structure of Positive Active Materials During Cycling [4]

The XRD pattern of the positive active material before and after the 2C charge–discharge cycle is shown in Fig. 14.14. At first glance, all the diffraction lines belong to the R-3m structural $Li(Ni_{0.4}Co_{0.4}Mn_{0.2})O_2$, and there seems to be no change.

Table 14.1 Impedance of 18650 battery before and after 400 cycles

		Before cycle	After cycle	After cycle/before cycle
Uncoated battery	Ohmic impedance R_e of battery	23.1	24.0	1.04
	Surface film impedance R_f	21.4	14.6	0.72
	Charge transfer impedance R_{CT}	24.3	110.1	4.53
Coated battery	Ohmic impedance R_e of battery	24.9	24.1	0.97
	Surface film impedance R_f	16.7	15.0	0.90
	Charge transfer impedance R_{CT}	26.2	59.4	2.27

However, after careful analysis, it is found that there are obvious changes in its fine structure.

14.2.1 Lattice Parameters of Positive Active Materials

The variation of the lattice parameters of the positive active material with the cycle period is shown in Fig. 14.9. It can be seen from Fig. 14.9 that the changing trends of *a* and *c* of the uncoated positive active materials are roughly the same, but the lattice parameters of the coated positive active materials are obviously different with the increase of the cycle period. The *a* decrease with the increase of the cycle, while *c* increases at first and then decreases. This shows that after 100 cycles, the proportion of remaining Li atoms in the positive active material (NCM) is higher. From Fig. 14.8, it can also be seen that the (003) peak changed of NCM after 400 cycles of uncoated battery. For in-depth analysis, thc XRD of NCM before and after 400 cycles is compared, and the result is shown in Fig. 14.10. Figure 14.10a shows the full spectrum, and Fig. 14.10b shows the amplification near the main diffraction peak of (003). It can be seen that the 111 diffraction peak of $Li_{1-x}Me_2O_4$ of spinel appears in the uncoated material.

The spinel phase is the product of the side reaction of NCM with electrolyte, but the peak is not seen in the coated material. The phase change is mainly due to the oxygen evolution and the dissolution of transition metal compounds during the cycle process, and the reaction can be expressed as follows:

$$3Li_{1-x}MeO_2 - 2(1-x)e^- = MeO + Li_{1-x}Me_2O_4 + O + 2(1-x)Li^+ \quad (14.1)$$

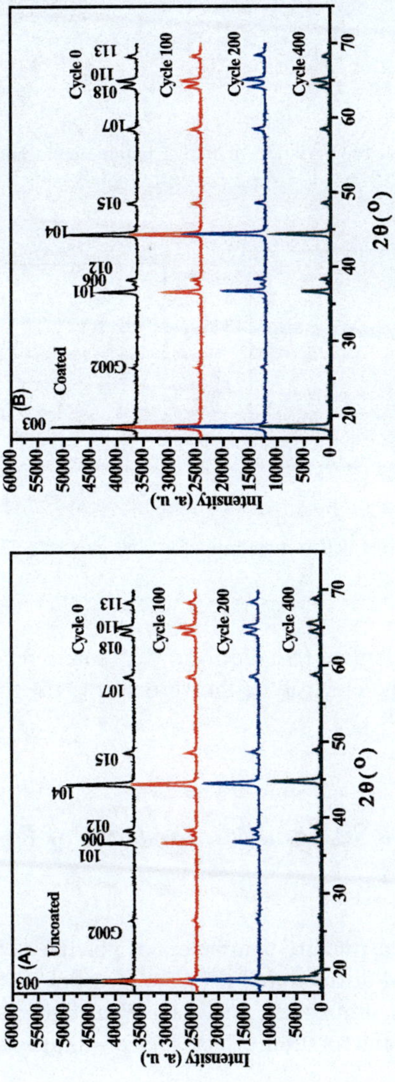

Fig. 14.8 2C XRD pattern of positive active materials before and after charge–discharge cycle **a** uncoated; **b** coated Al_2O_3

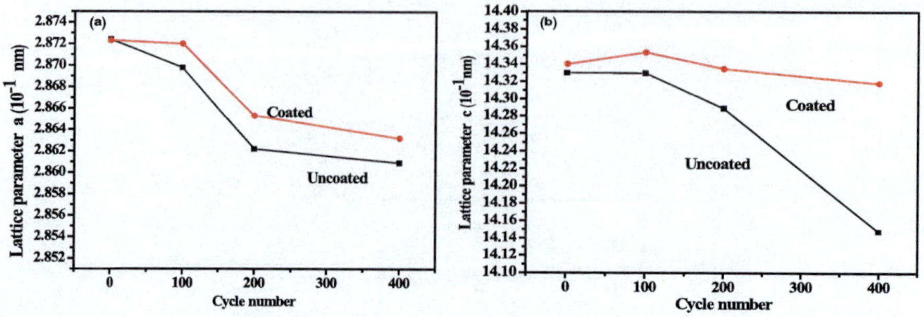

Fig. 14.9 After 2C charge–discharge, Variation of lattice parameters a and c of positive active materials with cycle periods

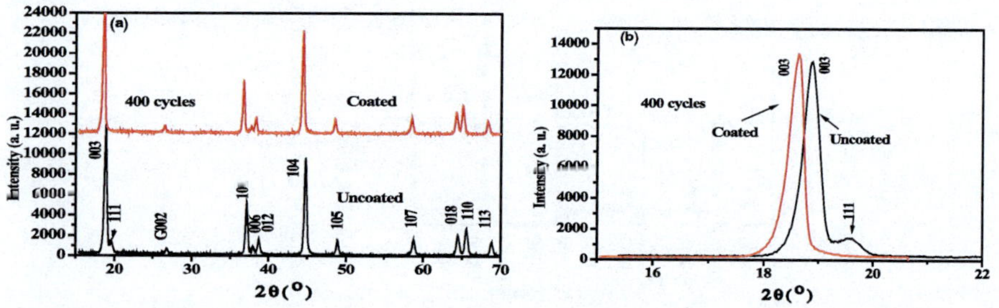

Fig. 14.10 XRD pattern; of positive active materials after 400 cycles. **a** Full spectrum; **b** local magnification of **a**

Me is a transition metal element (Ni, Co, Mn), Li^+ enters the liquid phase, and MeO is easily dissolved in the electrolyte, thus promoting the formation of spinel phase $Li_{1-x}Me_2O_4$.

14.2.2 Microcrystalline Size and Micro-strain of Positive Active Materials

In order to calculate the microstructure parameters of positive active materials, the FWHM values of 003 and 104 diffraction lines were measured. The average micro-crystal size and average micro-strain of the material are calculated by the least square method of separate crystallite-micro-strain effect. The results are shown in Fig. 14.11. It can be seen that:

(1) After charge–discharge cycle, the grain is obviously refined, the micro-strain is obviously reduced, and the grain refinement of the coated positive active material is greatly slowed down.

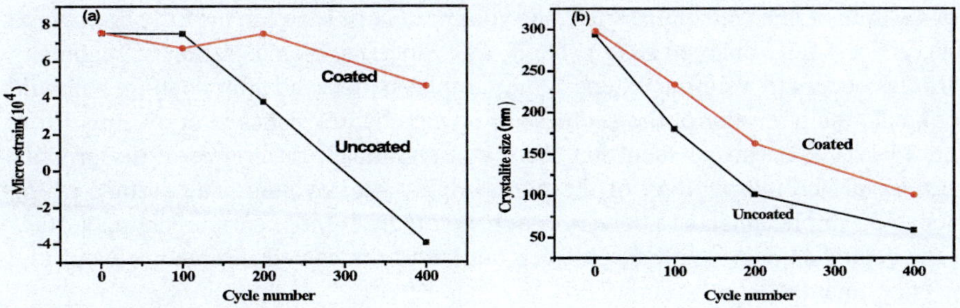

Fig. 14.11 Variation of microstructure parameters of positive active materials with cycle **a** average micro-strain; **b** average crystallite size

(2) The micro-strain of the uncoated material decreases with the increase of the cycle period and finally changes from tensile strain to compressive strain, while when the number of cycles is more than 100, the micro-strain of the coated material is larger than that of the uncoated material, especially the tensile strain is still maintained.

During the cycle process of the battery, the strain of uncoated NCM starts to be tensile strain, which is mainly due to the fact that lithium ion enters into the lattice of NCM precursor during the crystallization process. Microscopically, it can be seen as a kind of lattice expansion. With the progress of the cycle, the structure changes such as O precipitation appear in the uncoated NCM bulk phase, which leads to lattice contraction and compressive strain gradually. On the other hand, the coating can inhibit the precipitation of O, thus maintaining the state of compressive strain during the cycle. It is not difficult to see that the change law of micro-strain is consistent with the law of microcrystal refinement. This is mainly due to the accumulation and release of micro-stress caused by the change of micro-strain, which leads to the refinement of microcrystals. The coating suppresses the change of micro-strain, thus inhibiting the refinement of microcrystals.

14.2.3 Surface Morphology and Composition of Positive Electrode After Battery Cycle

Figure 14.12 is the positive SEM photograph of the two batteries after the cycle. It can be seen from Fig. 14.12a, b that the distribution of the surface conductive network of the positive plates of the two kinds of batteries after cycling is similar, indicating that the electronic path is not the main reason for the decrease of power performance. It can be seen from Fig. 14.12c, d that cracks appear in some secondary particles after cycling, which is mainly due to the volume contraction and expansion of $Li(Ni_{0.4}Co_{0.2}Mn_{0.4})O_2$ during the charge–discharge process. In fact, the coating with nanometer thickness on the surface is very difficult to restrain this volume effect.

For 424 rate of this kind of materials, the volume change between the fully discharged state and the fully charged state is about 4%. Since the disintegration of secondary particles occurs to a similar extent in the samples before and after coating, it should not be the main reason for the decrease of power characteristics after cycling. From Fig. 14.12e, f, it can be seen that there is a significant difference in the primary particle surface morphology of the two samples after cycling: The surface of the uncoated sample appears to be very rough, and there is more enrichment on surface of the coated sample, while the surface outline of the coated sample is clearer and the enrichment is less.

In order to study the composition of the surface enrichment, the energy dispersion spectrum (EDS) test was carried out. The content of the test results is shown in Table 14.2. It can be seen that after cycle 100 period, the F/(Mn + Co + Ni) values of uncoated and coated cathode are 1.51 and 0.80, respectively. The results show that the deposition amount of fluoride on the surface of the coated $Li(Ni_{0.4}Co_{0.2}Mn_{0.4})O_2$ electrode is obviously reduced, which is consistent with the results shown in the SEM diagram of Fig. 14.12e, f.

Figure 14.13 shows the X-ray photoelectron spectrum (XPS) of positive surface for 2025 battery charged to 4.5 V versus Li^{+}/Li after cycle 100 period. The peak position in the picture is corrected by C_{14} and separated by XPS peak program. The spectrum peak of 857.1 eV is Ni spectrum corresponding to Ni^{4+} in $Li(Ni_{0.4}Co_{0.2}Mn_{0.4})O_2$, and the spectrum peak of 854.2 eV is Ni spectrum of NiO. It can be seen from the figure that the composition of NiO on the surface of uncoated $Li(Ni_{0.4}Co_{0.2}Mn_{0.4})O_2$ cycle increases, while the coating has an obvious inhibitory effect on it. During the cycle, the dissolution of transition metal oxides and the precipitation of oxygen from the surface are the main reasons for the formation of NiO in the surface layer, as shown in reaction formula (14.1). Me is a transition metal ion, and MeO is mainly composed of NiO.

14.3 Fine Structure and Composition of Negative Active Materials [3, 4]

14.3.1 Fine Structure of Negative Active Materials

The XRD pattern of the negative plate is shown in Fig. 14.14, which is partially magnified as shown in Fig. 14.15. It can be seen that the peak position and half width of 002 diffraction peak of 2H-graphite are obviously different. Because the negative electrode plate is directly used as the sample, the diffraction spectrum includes three diffraction peaks of 111, 200, and 220 of the current collector Cu. We also measured the grain size of the extend *c*-axis direction [002] (see Fig. 14.16). Using Cu 111 as the internal standard, the position of the 002 diffraction peak of graphite was accurately determined. According to the method [5–7], the $2\theta_{\text{exp.}}^{\text{grap.}}$ value was determined by experiment, and the following formula was used.

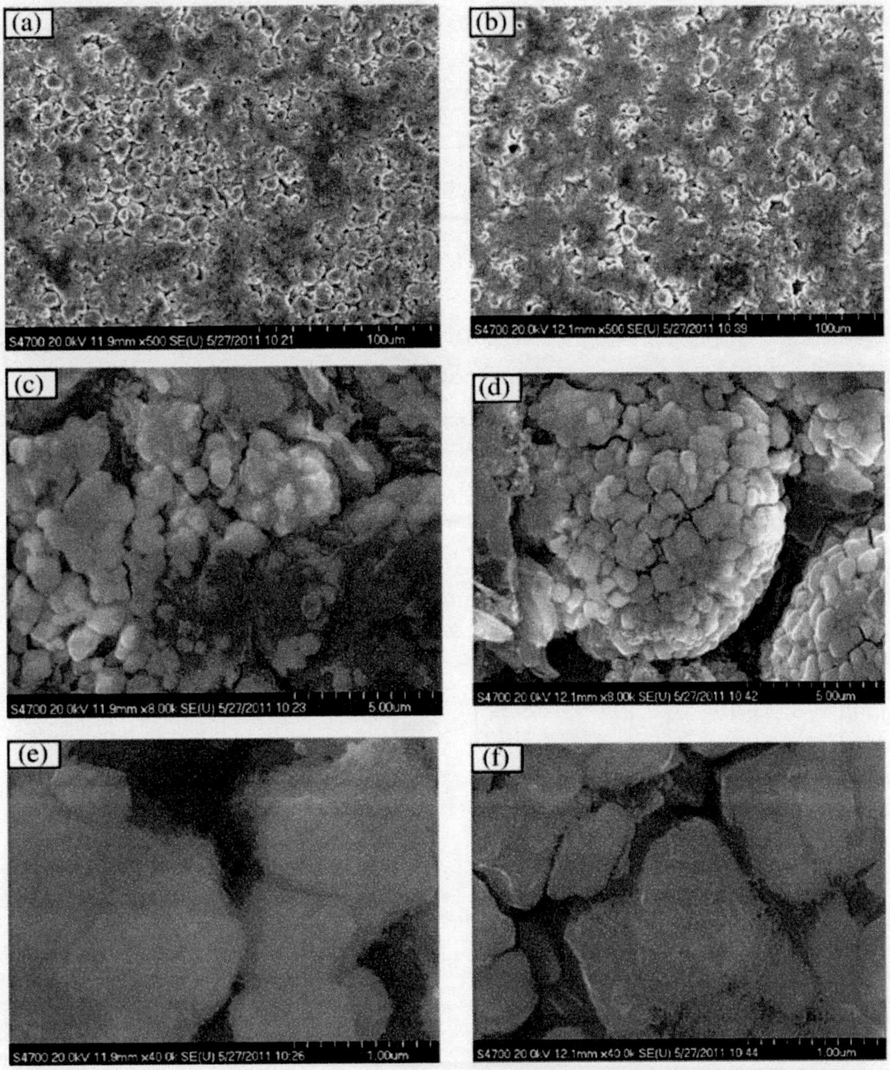

Fig. 14.12 SEM diagram of the positive plate after the cycle. **a** Uncoated, 500×; **b** coated, 500× ; **c** uncoated 8000 times; **d** coating, 8000 times; **e** uncoated, 40,000 times; and **f** coating, 40,000 times

$$2\theta_{\text{corr.}}^{\text{grap.}} = 2\theta_{\text{exp.}}^{\text{grap.}} + \left(2\theta_{\text{stand.}}^{\text{Cu}} - 2\theta_{\text{exp.}}^{\text{Cu}}\right) \tag{14.2}$$

Obtain the exact $2\theta_{\text{corr.}}^{\text{grap.}}$ value of graphite. Then, according to Bragg formula $2d_{002}\sin\theta_{\text{corr.}}^{\text{grap.}} = \lambda_{\text{CuK}\alpha 1} = 1.54056$ Å to calculate the accurate crystal plane spacing d_{002} along c-axis direction of 2H-graphite, and finally using the following formula:

Table 14.2 Distribution of elements on the surface of positive plates after 100 cycles

Elements	Atomic ratio		
	Uncoated	Coated	Coated/uncoated
C	19.40	20.37	1.11
O	42.62	44.01	1.03
F	23.46	14.80	0.67
Mn + Co + Ni	14.53	19.82	1.28
F/(Mn + Co + Ni)	1.51	0.80	0.53

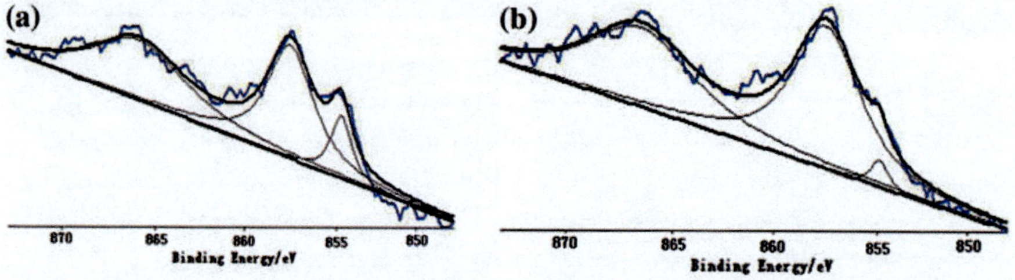

Fig. 14.13 XPS spectra of Ni in 2025 positive plates after 100 cycles **a** uncoated; **b** coated

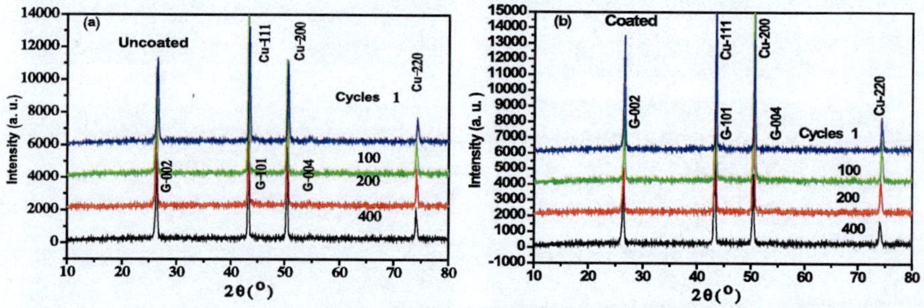

Fig. 14.14 XRD pattern of graphite before and after 2C charge–discharge cycle **a** is not coated; **b** coated with Al_2O_3

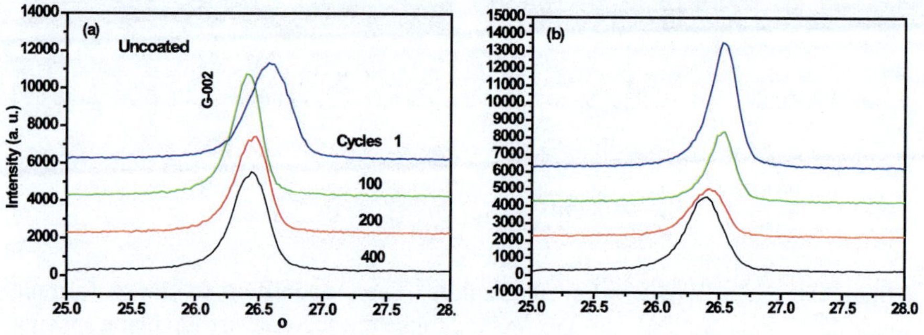

Fig. 14.15 Partial magnification of Fig. 14.14

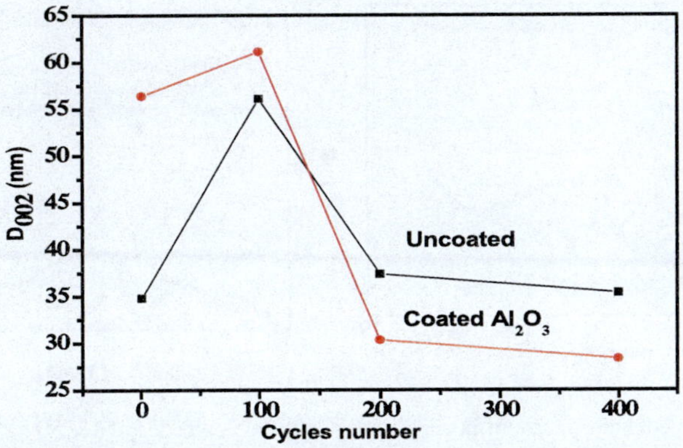

Fig. 14.16 Variation of grain size d_{002} of graphite along c-axis with cycle

$$P_{002} = \frac{d_{002} - 3.354}{3.440 - 3.354} \tag{14.3}$$

the stacking disorder degree P_{002} of graphite is calculated, and the results are shown in Fig. 14.17a, b, respectively. It can be seen that: (1) When the number of cycles is more than 200, d_{002} decreases with the increase of cycles, indicating that the grains are obviously refined during the cycle, and (2) the lattice parameters c and stacking disorder P_{002} of graphite change obviously with the number of cycles, but there is no obvious difference between coated and uncoated.

In order to study the cause of the rate decay in the graphite/Li($Ni_{0.4}Co_{0.2}Mn_{0.4}$)O_2 system, after dissecting the 18650 battery after 400 cycles, the positive and negative plates were, respectively, combined with lithium to form a 2025 battery, and the AC impedance was tested. The results are shown in Fig. 14.18. It can be seen that after 400 cycles, the increase of the impedance of the positive electrode is obviously larger than that of the negative electrode, which is the main reason for the decrease of the rate performance of the battery after 400 cycles.

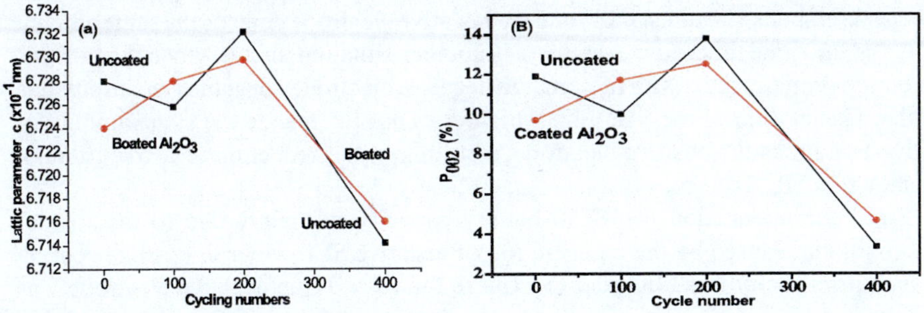

Fig. 14.17 Variation of lattice parameter c (**a**) and stacking disorder P_{002} (**b**) of negative electrode active material graphite with cycle period

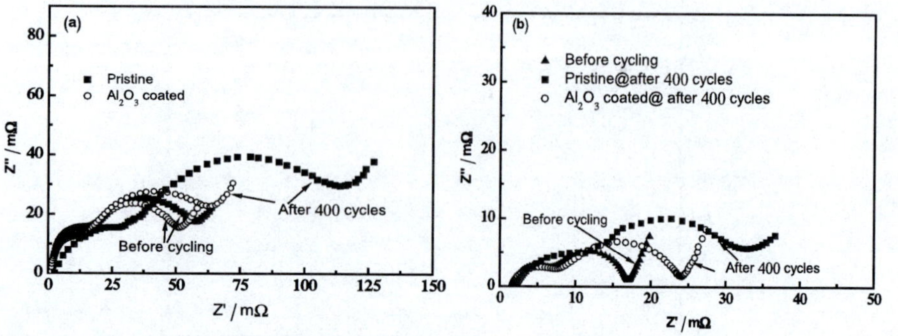

Fig. 14.18 AC impedance spectra of positive and negative electrodes of 18650 battery before and after cycling **a** positive and **b** negative electrodes

It can also be seen from the figure that the increase of the impedance of the positive and negative electrodes of the coated battery is obviously smaller than that of the uncoated battery. The reason lies in the inhibition of the deposition of LiF and the formation of NiO on the surface, and the improvement of the ability to maintain the negative pole ratio is due to the reduction of the deposition of Ni, Co, Mn, etc., and the inhibition of the reformation of SEI film.

14.3.2 Analysis of Related Components on the Surface and Body Interior of Negative Electrode

Under the premise that there is no electrochemical side reaction on the positive and negative electrodes, the Faraday charge is de-embedded on the positive and negative electrodes, and the amount of active lithium remains constant. Fig. 14.19a is a schematic diagram of charge and discharge without loss of activity. It can be seen that when charging, the positive electrode detaches 100 units of Li^+, while the negative electrode embeds the same amount of Li^+; when discharge, the positive electrode embeds 100 units of Li^+ and the negative electrode embeds the same amount of Li^+ at the same time. However, there is another situation, that is, when the positive electrode is intercalated with lithium, the negative electrode consumes electricity due to the regeneration of the SEI film, which does not intercalate the same amount of active lithium as the positive electrode, resulting in the reduction of active lithium, as shown in Fig. 14.19b.

Since the attenuation of 18650 battery capacity is mainly due to the loss of active lithium caused by the regeneration of negative SEI, why can positive coating reduce its capacity loss? In order to explore the reason, the negative electrode was sampled and tested by EDS. The semi-quantitative results of C, F, and P are shown in Table 14.3. It can be seen that the deposition of F and P in the negative electrode of coated battery is obviously less than that of uncoated battery.

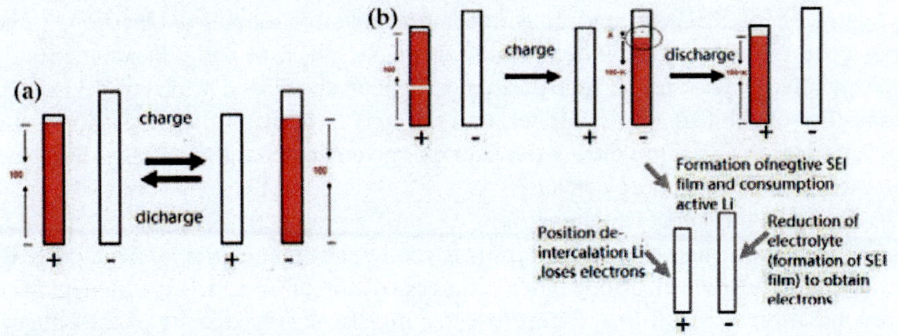

Fig. 14.19 Charge–discharge diagram. **a** Theoretical amount of de-intercalation Li^+; **b** amount of de-intercalation Li^+ during actual regeneration of SEI film

Table 14.3 Distribution of elements on the negative sheet surface of button battery and 18650 battery after cycle

Distribution of elements on the surface of negative electrode of button battery after 100 cycles			The amount of Ni, Co, and Mn in the negative electrode of button battery (μg)		
	Atomic ratio				
	Uncoated	After coated	Elements	Uncoated	After coated Al_2O_3
C	19.40		Ni	8	3
O	42.62		Co	24	6
P	23.46		Mn	32	7
Ni + Co + Mn	14.53				
F/(Ni + Co + Mn)					
Distribution of elements on the surface of negative electrode after 400 cycles of full battery			The amount of Ni, Co, and Mn in the negative electrode after 400 cycles of the whole battery (μg)		
	Atomic ratio				
	Uncoated	After coated	Elements	Uncoated	After coated
C	94.12	96.31	Ni	17	11
F	2.88	0.67	Co	54	21
P	0.17	0.05	Mn	61	19

In order to further quantitatively analyze the amount of Ni, Co, and Mn, the negative electrode was dissolved by the above-mentioned method and then determined by plasma excitation-atomic emission spectrometry (ICP-AES). The results are shown in Table 14.3. It can be seen that the contents of Ni, Co, and Mn in the negative electrode of coated battery are obviously lower than those of uncoated battery.

The deposition of Ni, Co, and Mn destroyed the integrity of the formed SEI film, and the electrolyte continued to be reduced on its surface during charging, thus thickening the formation of SEI film. At the same time, Ni, Co, and Mn can catalyze the formation of SEI film and consume more negative electrode electricity. The Al_2O_3 coating layer reduces the dissolution of Ni, Co, Mn, and other compounds in the positive electrode, so that the deposition on the negative electrode is less, the stability of the SEI film is higher, and the loss of active lithium caused by the reformation of the SEI film during the later charge and discharge process is less, and the battery can release more capacity.

To sum up, the capacity attenuation of 18650 graphite/$Li(Ni_{0.4}Co_{0.2}Mn_{0.4})O_2$ battery is due to the loss of active lithium caused by the regeneration of negative SEI. The coating of positive electrode inhibits the deposition of Co and Mn, which inhibits the regeneration of SEI film and improves the dischargeable capacity of the battery. The attenuation of the magnification is mainly caused by the rapid increase of the electrochemical impedance of the positive electrode, but it is not the action of LiF. Combined with the analysis of the reasons for the attenuation of the magnification performance, it can be considered that the formation of NiO on the surface of the cathode is a more important factor.

14.4 Changes of the Fine Structure of the Diaphragm with the Cycle [3, 4]

The diaphragm of this kind of battery is composed of three layers of PP-PE-PP; PP is polypropylene film, belongs to monoclinic structure, $P2_1/c$, No. 14 space group; and PE is polyethylene, belongs to orthorhombic, Pnma (No. 62) space group. The XRD pattern of the diaphragm before and after 2C charge–discharge cycle is shown in Fig. 14.20, in which the diffraction peaks have been indexed; it can be seen that:

(1) The diffraction peaks of PP-131 and PE-110, PP-111 and PE-200 overlap.
(2) The positions of amorphous scattering peaks of PP and PE are at $2\theta = 16.30°$ and 19.50°, respectively, and it can be seen that the crystallinity of all samples is very high, and the amorphous part is very few.

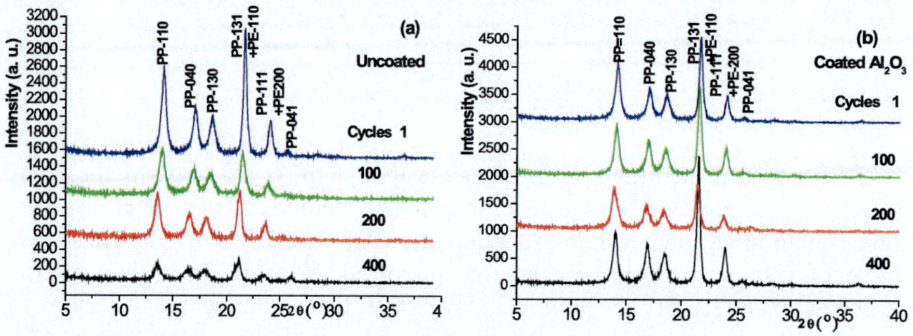

Fig. 14.20 XRD pattern of diaphragm before and after charge and discharge cycle at 2C **a** uncoated, **b** coated Al_2O_3

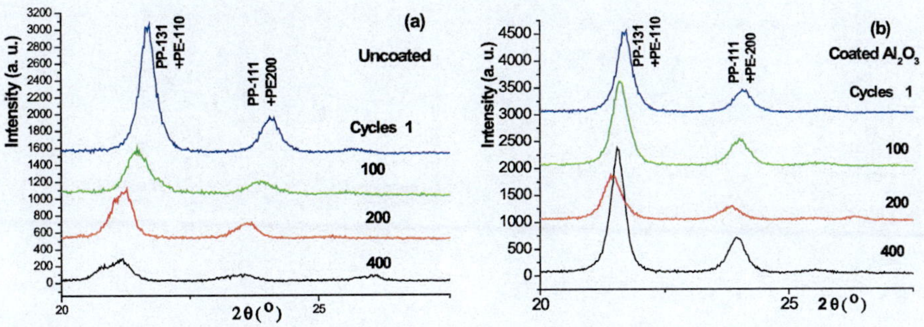

Fig. 14.21 Partial magnification of XRD pattern of diaphragm before and after charge–discharge 2C cycle, **a** uncoated; **b** coated Al_2O_3

(3) The peak position shifts before and after the cycle, and the lines are obviously broadened.
(4) Compared with the uncoated, the broadening phenomenon of the coated is obviously reduced.

In order to investigate the situation of two pairs of overlapping peaks, the enlarged pattern is shown in Fig. 14.21. It can be seen that the coated and uncoated peaks are obviously different, and the latter peak displacement and broadening are much more serious. After 200, 400 cycles, the PP-131 and PE-110 of the uncoated samples have obviously split, indicating that the surface (PP) of the diaphragm is seriously damaged and thinned during the cycle.

14.4.1 Lattice Parameters of Polypropylene Diaphragm

Since the diaphragm consists of two layers of polypropylene and one layer of polyethylene, we use the peak position data of six diffraction peaks of PP to calculate the lattice parameters of polypropylene which are shown in Fig. 14.22. It can be seen that the three lattice parameters all change with the increase of cycle period, but the changes of coated samples are slower than those of uncoated samples, a and b are smaller than uncoated ones, and c has a similar situation. This shows that the coating of positive active materials has a certain protective action on the diaphragm.

14.4.2 Microstructure of Polypropylene

There is an obvious difference in the broadening phenomenon of the observed diffraction peaks, especially between coated and uncoated, which indicates that there is an obvious difference in the microstructure (microcrystal size and/or micro-strain) of

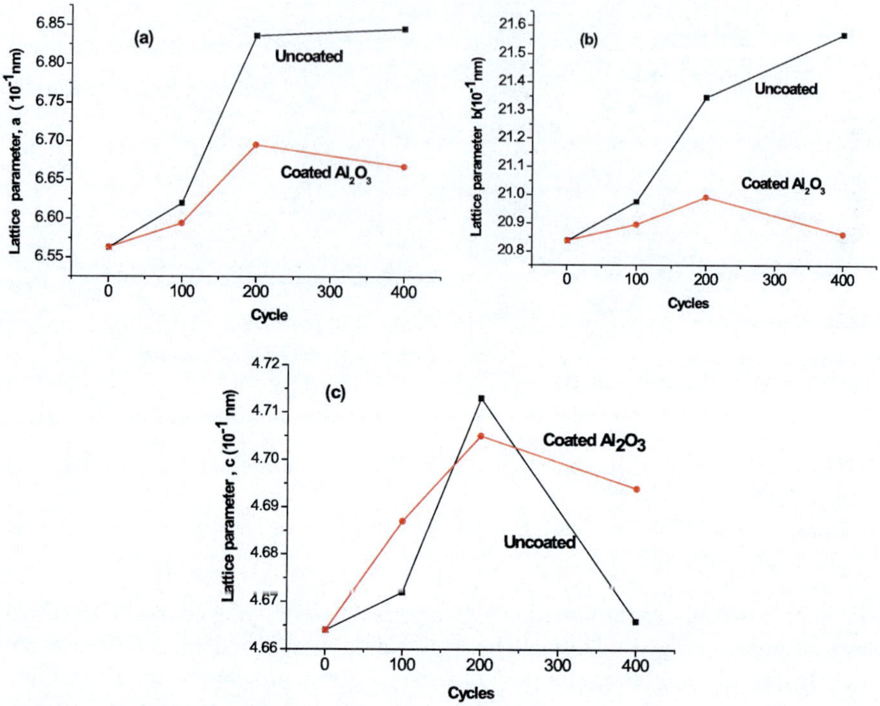

Fig. 14.22 Variation of diaphragm polypropylene lattice parameters before and after 2C charge–discharge cycle

the diaphragm. The data are processed according to the double effect of microcrystal-micro-strain and the existence of both separately. The results show that the treatment of double effect is obviously unreasonable and the microcrystal scale is negative. The calculation results that the two exist separately are shown in Fig. 14.23. It can be seen from Fig. 14.23 that the change of uncoated data is irregular, but the effect of coating is obvious: The coating of positive active material slows down the grain refinement of the diaphragm, or the change of micro-strain is also reduced.

14.4.3 Surface Morphology of the Diaphragm

Figure 14.24 shows the SEM diagram of the diaphragm before and after the cycle, and Fig. 14.24 (a) shows the morphology before cycle, Fig. 14.24 (b) and (c) shows respectively the morphology of diaphragm on the positive side for uncoated and Al_2O_3-coated NCM battery after 400 cycles. It can be seen that after 400 cycles, the morphology of the diaphragm of the uncoated battery changed greatly, and the micro-pores of the diaphragm were largely closed. However, the diaphragm morphology

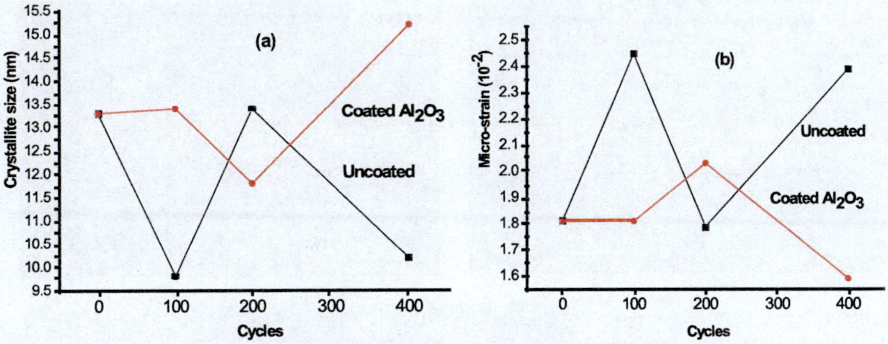

Fig. 14.23 Variation of microcrystalline size (**a**) and micro-strain (**b**) of polypropylene with cycle

of the coated battery changed little. The micropore of the diaphragm is the place to hold the electrolyte and the channel to conduct lithium ion in the diaphragm. The closure of the micropore will decrease the power performance of the battery, and the reason for the closure is related to the temperature of the diaphragm. The higher the temperature, the easier the softening of PP and PE (PP, PE melting point is 165C and 140C, respectively), and the easier the micropore is closed. From the charge–discharge curve of the battery under 2C after 400 cycles (Fig. 14.25), it can also be seen that the charge voltage of the coated battery is low and the discharge voltage is high, so the heat release of the battery in the charge–discharge cycle is smaller and the temperature is lower. Therefore, NCM coating can obviously inhibit the micropore closure of the diaphragm during circulation.

14.5 Mechanism of Cycle Performance Degradation and Coating Role Mechanism [8]

14.5.1 The Relationship Between the Attenuation of Cycle Performance and the Change of Positive and Negative Electrode and Diaphragm Structure

The changes of cycle performance and fine structure of positive and negative active materials and diaphragm are summarized in Table 14.4. Through careful comparison and analysis of these data, it is clear that whether the positive and negative active materials are coated or not, there is a good corresponding relationship between the attenuation of the cycle performance and the changes of the positive and negative active materials and the fine structure of the diaphragm during the cycle. It also shows that the coating plays an important role in improving the cycle performance

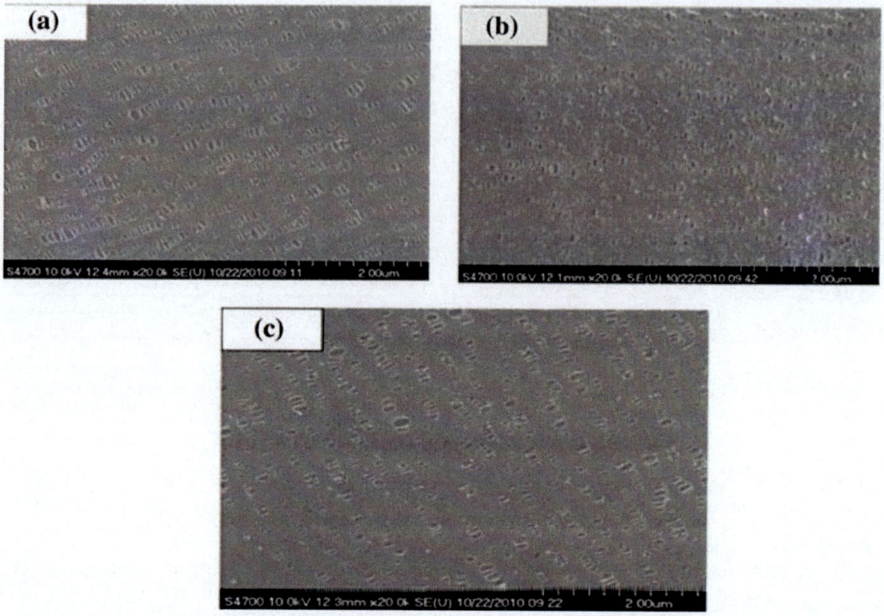

Fig. 14.24 SEM diagram of the diaphragm **a** before, **b**, **c** after 400 cycles uncoated and coated batteries

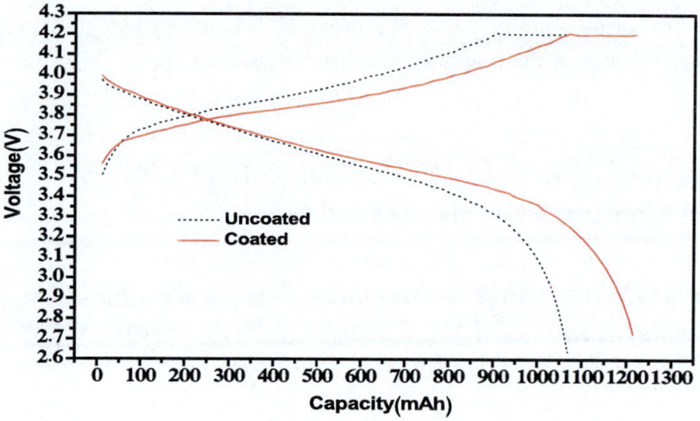

Fig. 14.25 Charge–discharge curve after 400 cycles at 25 °C, 4.2–2.75 V, 2C

and slowing down the change of the positive and negative active materials and the fine structure of the diaphragm during the cycle.

Table 14.4 Changes of cycle performance and fine structure of positive and negative active materials as well as diaphragm

Battery performance	Initial capacity	The initial capacity of the two kinds of batteries with and without coat is the same, and the charge–discharge curves almost coincide
	Cycle performance	It attenuates with the increase of cycle period, and the decay of the positive uncoated battery is much faster than that of the coated battery The capacity of the whole battery before and after coating was 0.2C after 2C, 2.75–4.2 V, 400 cycles, 991.8 and 1101.1 mAh, respectively. The capacity retention rates are 85.3% and 94.2%, respectively Both the capacity and the discharge platform decrease with the increase of the cycle period, but the decline rate of the coated one is much lower than that of the uncoated one
	Impedance	There is no difference between coated and uncoated impedance before cycle, but the increase of impedance after 400 cycles coated battery was significantly lower than that of uncoated battery In particular, the uncoated charge transfer impedance is increased by 4.53 times, while the coated impedance is only 2.27 times
Positive active material	Fine structure	The change trend of uncoated lattice parameter *a* and *c* is the same, but the general trend of coated lattice parameter *a* is decreasing, while *c* is increasing The microcrystal size and micro-strain decrease with the increase of cycle period, and the coated ones are much larger than the uncovered ones
	Surface morphology	The uncoated particles before the cycle show that the particles are smooth, the surface is rough after the coating, and the alumina is covered uniform on the surface
	Surface composition	After 100 cycles, the deposition of F on the surface without coating is only 1.5 times of that of coating, F/(Ni + Co + Mn). The uncoated one is only 1.98 times of the coated one
	Related performance	There is little difference in specific surface area, vibrating density, and diffusion coefficient of $Li(Ni_{0.4}Co_{0.2}Mn_{0.4})O_2$ materials before and after coating

(continued)

Table 14.4 (continued)

Diaphragm material	Fine structure	The lattice parameters *a*, *b*, and *c* increase slightly with the increase of cycle period. But the change of coated samples is slower than that of uncoated samples Both *a* and *b* are smaller than uncoated ones, and *c* has a similar situation. This shows that the coating of positive active materials has a certain protective effect on the diaphragm The change of the uncoated microstructure data is not regular, but the effect of the coating is obvious: the coating of the positive active material slows down the grain refinement of the diaphragm, or the change of micro-strain is also reduced
	Surface morphology	The shape of the diaphragm of the uncoated battery changes greater, and a large number of micropores of the diaphragm are closed, and the closure of micropores will reduce the power performance of the battery
Negative active material	Surface morphology	The lattice parameter *c* and stacking disorder increase slightly with the increase of cycle period when the cycle is less than 200 cycles. Large than 200 cycles all decrease obviously with the increase of cycle period When the cycle is less than 100, D_{002} increases slightly, and when the cycle is more than 100, the grain is obviously refined. Coating Al_2O_3 can obviously slow down the process of the above changes
	Fine structure	After 100 cycles, Ni, Co, and Mn were deposited on the surface. The coated ones are much lower than the uncoated ones, and it is the same after 400 cycles After 400 cycles, the deposition of F and P, uncoated was 4.3 and 3.4 times higher than that of coated, respectively

14.5.2 The Mechanism of Cycle Performance Attenuation

In the discussion of the attenuation mechanism of cycle performance of lithium-ion secondary battery, many researchers attribute it to the effect of electrode surface, such as the chemical corrosion of electrolyte on electrode surface and diaphragm surface and so on. Of course, this kind of corrosion should not be ignored. First, let us focus on the changes of the internal structure of the positive and negative active materials and of the diaphragm during the cycle, that is the bulk effect, and first discuss the uncoated situation and combine the results of surface analysis.

(1) The lattice parameters *a* and *c* of the positive electrode decrease with the cycle times increase, indicating that there are more and more inactive Li atoms. The grains and micro-strain of the positive active material become smaller with the increase of the cycle, and these two points are corresponding to each other. As a result, the activity of positive active materials will be reduced.
(2) The negative lattice parameter *c* and stacking disorder increase slightly at first and then decrease with the increase of cycle period, and the cycle refines the grains, which is not beneficial to the intercalation of Li atoms. In other words, the activity of graphite decreases with the increase of cycle period. Coating Al_2O_3 can obviously slow down the process of all kinds of changes mentioned above.
(3) The three lattice parameters of the diaphragm change with the increase of the cycle period, especially with the increase of the cycle, the damage of the surface layer of the diaphragm is more serious. The combined action of the above three aspects is the main reason for the decline of cycle performance.
(4) The performance attenuation mechanism of 2025 $Li/Li(Ni_xCo_yMn_{1-x-y})O_2$ battery was studied, which was caused by the increase of electrochemical impedance caused by the deposition of fluoride and NiO on the surface of $Li(Ni_xCo_yMn_{1-x-y})O_2$.
(5) The capacity attenuation mechanism of 18650 graphite/$Li(Ni_xCo_yMn_{1-x-y})O_2$ battery was studied. The deposition of Ni, Co, and Mn in the negative electrode during charge and discharge destroyed the integrity of SEI, and the regrowth of SEI film promoted the loss of active lithium during the cycle.
(6) The attenuation mechanism of the rate performance of 18650 MCMB/$Li(Ni_xCo_yMn_{1-x-y})O_2$ battery was studied, which was mainly caused by the deposition of fluoride and NiO on the surface of the positive electrode, and the electrochemical reaction impedance of the positive electrode increased faster than that of the negative electrode.
(7) The mechanism of coating to improve the cycle performance of $Li(Ni_xCo_yMn_{1-x-y})O_2$ material and graphite/$Li(Ni_xCo_yMn_{1-x-y})O_2$ battery was studied. Coating reduces the deposition of fluoride and NiO on the surface, and reduce the deposition of Ni, Co, and Mn on the negative surface.
(8) The rule of coating to improve the storage performance of 18650 graphite/$Li(Ni_xCo_yMn_{1-x-y})O_2$ battery at high temperature was studied. The coating inhibited the attenuation of battery capacity and rate performance and inhibited the deposition of Ni, Co, and Mn on the negative electrode.

It can be seen that in addition to the bulk effect of positive and negative active materials and diaphragm materials, the surface effect also plays an important role. The coating not only alleviates the surface effect, but also affects the bulk effect. It can be seen that the coating of positive active materials is very effective to raise and improve the performance of lithium-ion batteries.

14.6 Conclusion

Through the comparative study of the cycle performance of the positive active material $Li(Ni_xCo_yMn_{1-x-y})O_2$ coated and uncoated Al_2O_3 and through the comparative study of the changes of fine structure of positive and negative active materials and the diaphragm during the cycle, the conclusions are as follows:

(1) There is a good corresponding relationship between the attenuation of cycle performance and the change of positive active material and diaphragm fine structure.
(2) Although the decline of cycle performance is related to the change of the surface membrane structure of the negative electrode, it is also closely related to the fine bulk structure change of the positive and negative diaphragm.
(3) The surface coating of the positive active material obviously improves the cycle performance of the battery, reduces the oxidable ability of the positive material itself, reduces this side reaction of the positive $Li(Ni_{0.4}Co_{0.2}Mn_{0.4})O_2$ with the electrolyte and the diaphragm, slows down the process of fine structure change of the positive and negative electrode and diaphragm, and improves the surface film performance. It is the comprehensive result.

References

1. Liu, Hao-Han, Jian Zhang, Yu-Wan Lou, Chuan-Zheng Yang, Xiao-Hua Xie, and Bao-Jia Xia. 2012. Structure evolution and electrochemical performance of Al_2O_3-coated $Li(Ni_{0.4}Co_{0.2}Mn_{0.4})O_2$ during charge-discharge cycling. *Chemical Research in Chinese Universities* 28 (4): 686–690.
2. Liu, H.H., J. Zhang, Y.W. Lou, and B.J. Xia. 2012. XANES study on the stability of Al_2O_3 coated $Li(Ni_{0.4}Co_{0.2}Mn_{0.4})O_2$ material. *Acta Chemica Sinica* 70 (9): 1055–1059e.
3. Liu, H.H., J. Zhang, Y.W. Lou, and B.J. Xia. 2012. Effect of $Li(Ni_{0.4}Co_{0.2}Mn_{0.4})O_2$ coated Al_2O_3 on the morphology and structure of separator, battery. *Battery* 42 (2): 59–61.
4. Liu, Hao Huan. 2011. Study on failure mechanism and coating modification of $Li(Ni_xCo_yMn_{1-x-y})O_2$ material. Doctoral paper of Shanghai Institute of Microsystems and Information Technology, 6.
5. Li, H., C.Z. Yang, and F. Liu. 2008. A new X-ray diffraction method for determining the stacking disorder of hexagonal graphite. *Chinese Science, Series B Chemistry* 28 (9): 755–760.
6. Li, Hui, Chuanzheng Yang, and Fang Liu. 2009. Novel method of determination stacking disorder degree in hexagonal graphite by X-ray diffraction. *Science in China Series B: Chemistry* 52 (2): 174–180.
7. Li, H., C.Z. Yang, and F. Liu. 2009. Determination of graphitization degree and disorder degree of carbon electrode materials by X-ray diffraction. *Journal of Testing Technology* 23 (2): 161–167.
8. Liu, H.H., J. Zhang, C.Z. Yang, and B.J. Xia. 2015. Comparative study on the cycle process of 2H-graphite/$Li(Ni_{0.4}Co_{0.2}Mn_{0.4})O_2$. *Battery*.

Chapter 15
Cycle Mechanism of Graphite/[$Li(Ni_{0.4}Co_{0.2}Mn_{0.4})O_2$ + $LiMn_2O_4$] Battery

In Chap. 14, our research results on the cycle process of graphite/$Li(Ni_{0.4}Co_{0.2}Mn_{0.4})O_2$ battery have been introduced. The results show that the attenuation of the cycle performance has a good corresponding relationship with the change of the fine structure of the positive and negative active materials and the diaphragm, and the cycle performance can be greatly improved by coating the positive electrode or adding positive additive [1]. The method of Li $(Ni_{0.4}Co_{0.2}Mn_{0.4})O_2$ + $LiMn_2O_4$ composite positive active material is used to reduce the battery cost. In this chapter, the cycle performance and the attenuation mechanism of cycle performance of this kind of battery are studied systematically, in order to understand the feasibility of this kind of battery, in order to mass-produce lithium-ion batteries.

15.1 Cycle Performance of Graphite/[$Li(Ni_{0.4}Co_{0.2}Mn_{0.4})O_2$ + $LiMn_2O_4$] Battery [2]

15.1.1 Battery Preparation and Charge–Discharge and Cycling Process

In this study, $Li(Ni_{0.4}Co_{0.2}Mn_{0.4})O_2$:$LiMn_2O_4$=1:1 mass ratio was used as positive active material. 18,650 lithium-ion battery with artificial hexagonal graphite as negative electrode. The positive and negative electrodes are separated by PP-PE-PP diaphragm. After winding and forming according to the cylindrical lithium-ion battery manufacturing process, the positive and negative electrodes are assembled into type 18,650 lithium-ion battery. After drying, the battery is injected into the glove box, and the electrolyte component is 1 mol/L $LiPF_6$/ (EC: DMC: DEC = 1:1:1).

C. Yang et al., *Materials and Working Mechanisms of Secondary Batteries*,
https://doi.org/10.1007/978-981-19-5955-4_15

Table 15.1 Sample number of cyclic experiment

Sample number	Cycle temperature (°C)	Charge and discharge rate	Cycle number
18	Room temperature 25		1
18–16	Room temperature 25	1C charge, 1C discharge	250
18–10	Room temperature 25	1C charge, 1C discharge	300
18–17	Room temperature 25	1C charge, 1C discharge	1000
18–14	55	1C charge, 1C discharge	500
18–24	Room temperature 25	3C charge, 10C discharge	500

The battery 1C was charged to 4.2 V, discharged to 2.5 V, and then discharged to 0.2 V for a cycle. The cycle temperature is divided into room temperature and 55 °C. In addition, the experiment of 3C charge and 10C discharge has been carried out. The sample number is listed in Table 15.1.

15.1.2 Test Results of Cycle Performance

The cycle curves of the three batteries are shown in Fig. 15.1. It can be seen from the figure that the capacity retention rate decreases with the increase of the cycle period. The rate of charge and discharge and the temperature of the cycle have a great influence on the cycle performance. Also at 500 cycles, their capacity retention is shown in Table 15.2. It can be seen that the effect of temperature is more significant; that is, the higher the cycle temperature, the faster the cycle performance decreases, and the lower the capacity retention rate.

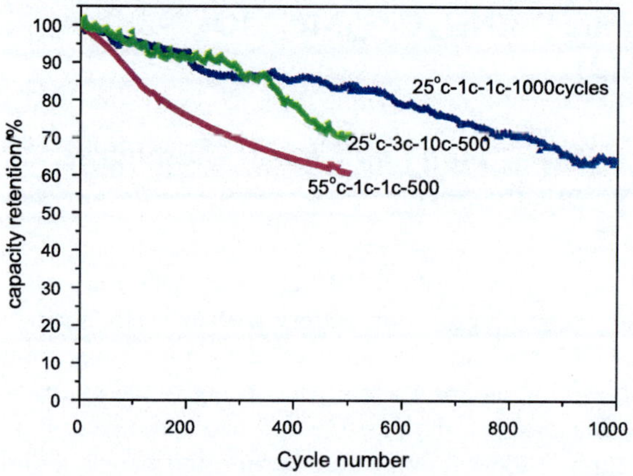

Fig. 15.1 Cycle curves of three samples under different cycle conditions

Table 15.2 Capacity retention for 500 cycles of three batteries

Sample number	Temperature (°C)	Charge and discharge rate	Capacity retention at 500 cycles (%)
18–17	25	1C charge 1C discharge	83.4
18–14	25	1C charge 1C discharge	61.1
18–24	25	3C charge 10C discharge	70.3

15.2 Fine Structure of Positive Active Materials in Graphite/$Li(Ni_{0.4}Co_{0.2}Mn_{0.4})O_2$ + $LiMn_2O_4$ Battery During Cycle

The X-ray pattern of the composite cathode active material is shown in Fig. 15.2, which shows that it is composed of $Li(Ni_{0.4}Co_{0.2}Mn_{0.4})O_2$ and $LiMn_2O_4$. The former is trigonal, 3-Rm (No. 166) space group, and the latter is cubic structure, Fd3m (No. 277) space group. Figure 15.3 shows the X-ray diffraction patterns of the positive active materials of the cyclic samples listed in Table 15.1. It is relatively knowable.

(1) R-003 with rhombic structure and C-111 with cubic structure cannot be separated in some samples, but obviously separated in some samples, indicating that the lattice parameters of the two phases change during the cycle.
(2) The diffraction lines of R-101 and C-311 are obviously separated in all samples, and there is an obvious change in the separated angle.
(3) The FWHM and relative intensity of R-101 and C-311 are broadened obviously with the cycle.

Let us analyze it carefully.

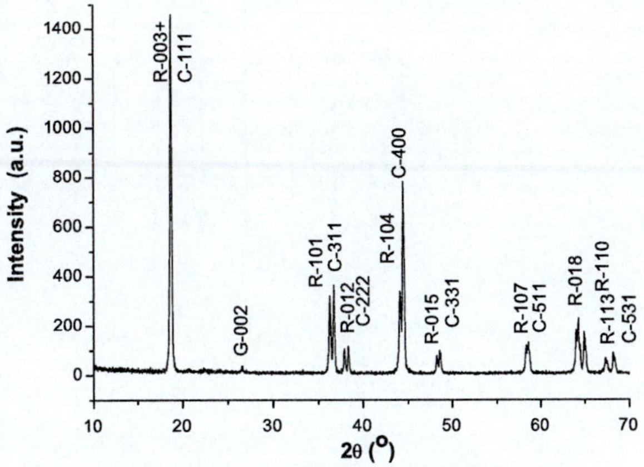

Fig. 15.2 X-ray patterns of pre-cycle composite positive active materials

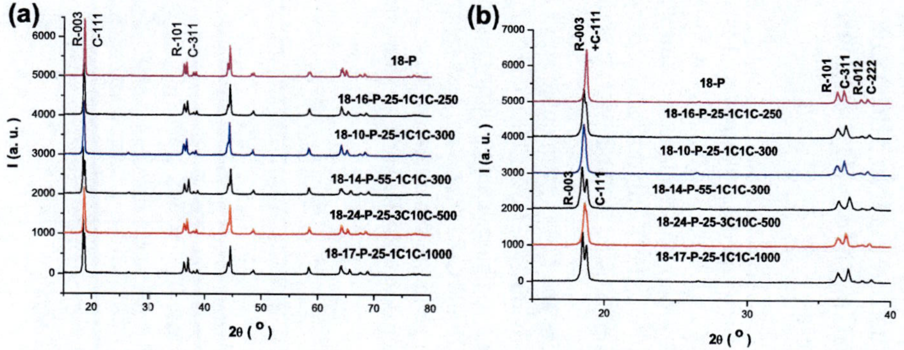

Fig. 15.3 X-ray diffraction patterns of composite positive active materials cycled in different cycles under different cycle conditions (**a**), (**b**) is the partial magnification of (**a**)

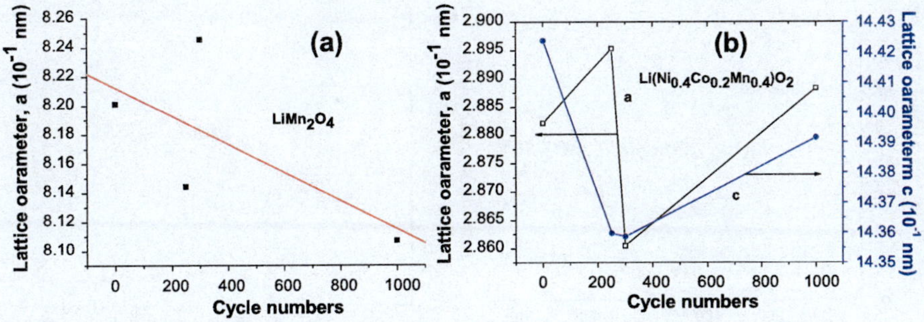

Fig. 15.4 Variation of lattice parameters of cubic (**a**) and $Li(Ni_{0.4}Co_{0.2}Mn_{0.4})O_2$ (**b**) with cycle numbers

15.2.1 Variation of Two-Phase Lattice Parameters

The lattice parameter variation of the cubic phase and Li $(Ni_{0.4}Co_{0.2}Mn_{0.4})O_2$ with the cycle period of 1C charge 1C at room temperature (25 °C) is shown in Fig. 15.4. It can be seen that the data are scattered, which may be caused by different cycle conditions of each sample, but the general trend is that the lattice parameters of cubic phase decrease with the increase of period. Therefore, the diffraction lines move in the direction of high angle; the lattice parameters a and c of $Li(Ni_{0.4}Co_{0.2}Mn_{0.4})O_2$ decrease in low period and increase in high period.

15.2.2 Variation of Microstructure and Relative Quantity of Two Phases

The variation of the microcrystal size of the two phases with the cycle times shown in Fig. 15.5a, b. The general trend is to refine the grains by cycle.

The change of the relative quantity of the two phases is qualitatively characterized by the integral intensity ratio $I_{3R\text{-}101}/I_{C\text{-}311}$. The results are shown in Fig. 15.6. It can be seen that the ratio increases with the increase of the cycle period, and the rate of increase is faster before the cycle period is less than 300. After that, it does not change much. The rate of increase is 2%.

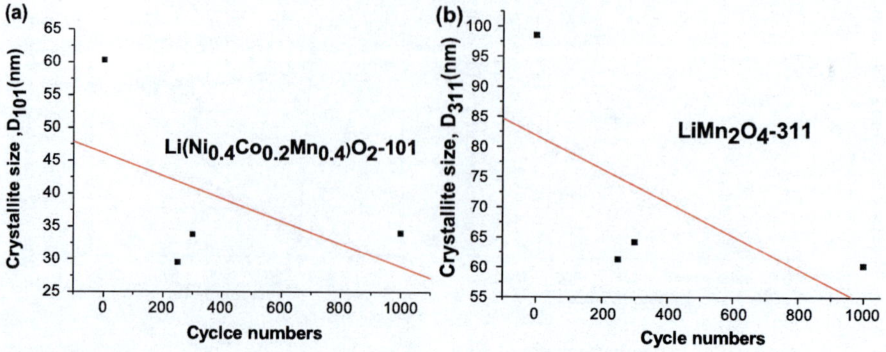

Fig. 15.5 Variation of microcrystal size with cycle numbers for Li ($Ni_{0.4}Co_{0.2}Mn_{0.4}$)O_2 (**a**) and $LiMn_2O_4$ (**b**)

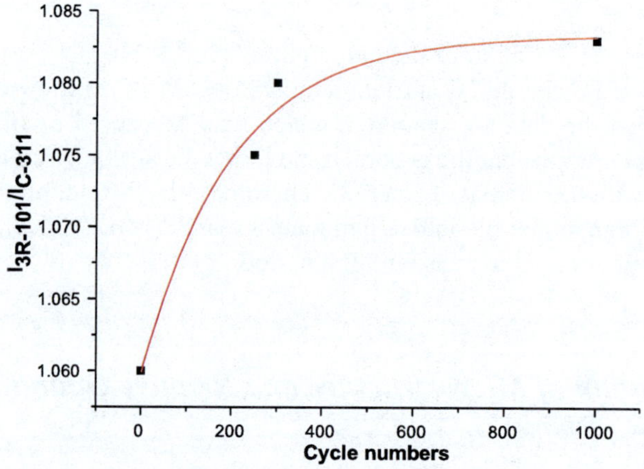

Fig. 15.6 Variation of the relative quantity with the cycle numbers of the two phases, characterization with the variation of the integral intensity ratio $I_{3R\text{-}101}/I_{C\text{-}311}$

15.2.3 Cyclic Temperature Effect and Magnification Effect of Positive Active Materials

The temperature effect and magnification effect are shown in Table 15.3 respectively. It can be seen that the lattice parameters of cubic phase cycling at high temperature or high magnification are smaller, and others seem to be irregular.

Table 15.3 Temperature effect and rate effect of charge–discharge cycle

Microstructure parameter	Cubic phase	Trigonal phase	$D_{3R\text{-}101}$	$D_{C\text{-}311}$
Cycle condition and period	*a* (Å)	*a* (Å) *c* (Å)	(nm)	
1C charge25°C cycle 250 period	8.2456	2.8605 14.3587	33.8	64.2
1C charge 55 °C cycle 300 period	8.1264	2.8637 14.3953	36.7	51.4
3C charge 10C discharge 25 °C cycle 500 period	8.1311	2.8938 14.2983	30.3	78.1

15.3 Fine Structure of Negative Active Materials [1]

The X-ray patterns and partial magnification of the negative active materials are shown in Fig. 15.7a and b, respectively. It can be seen that no carbon-lithium compounds are found except 2H-graphite, but the 002 diffraction lines of 2H-graphite have obvious shift and broadening effect. The further analysis is as follows.

15.3.1 Change of Lattice Parameters

The changes of hexagonal graphite lattice parameters a and c with the cycle are shown in Fig. 15.8. It can be seen that a decreases almost monotonously, while c increases rapidly in the low cycle and only increases slowly after the 300th cycle, which indicates that there are more Li atoms in the hexagonal grid of carbon atoms in the low cycle time gap, and the interstitial Li atoms no longer increase after the 300th cycle.

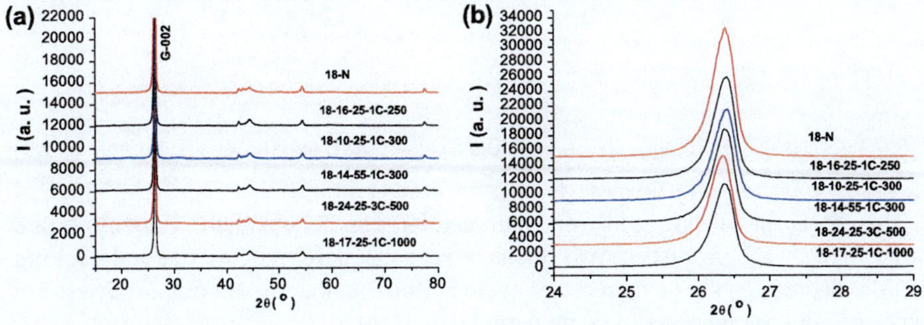

Fig. 15.7 X-ray pattern (**a**) and partially magnified (**b**) of negative active materials after cycling under different conditions

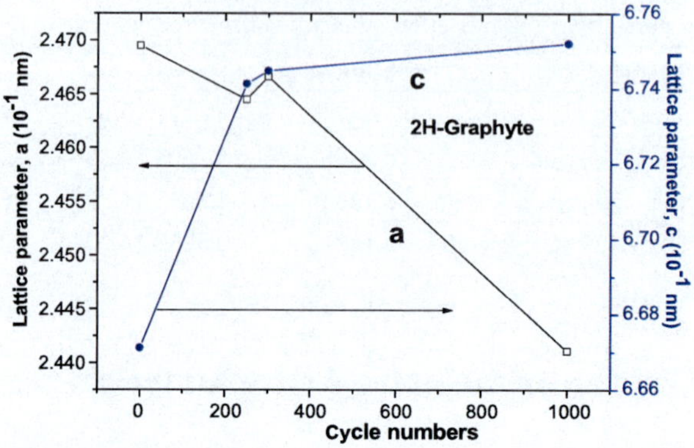

Fig. 15.8 Variation of lattice parameters a and c of hexagonal graphite with cycle numbers

15.3.2 Microstructure and Stacking Disorder of Negative Active Materials

In the battery cycle experiment, the microstructure of the negative active material should be the triple effect of microcrystal-strain-stacking disorder. According to the introduction of reference [2, 4] the solution to the triple broadening effect requires data equal to or greater than three diffraction lines, where the intensity of 102,103 is very weak and the broadening is so serious that the effective FWHM data not be obtained and not be solved. However, the FWHM data of 002, 100, 004, and 110 can be used to solve the microcrystal-micro-strain double effect and be replaced by the following formula.

$$\frac{\beta_{101}\cos\theta_{101}}{\lambda} = \frac{0.89}{\overline{D}} + \overline{\varepsilon}\frac{4\mathrm{Sin}\theta_{101}}{\lambda} + \frac{\cos\varphi_{Z101}}{2c}\frac{\cos\theta_{101}}{\lambda}P \tag{15.1}$$

To enter the relevant data:

$$0.601\beta_{101} = \frac{0.89}{\overline{D}} + 0.981\overline{\varepsilon} + 9.806 \times 10^{-3}P \tag{15.2}$$

Therefore, the stacking disorder parameter P can be obtained. The results are shown in Fig. 15.9a–c, respectively. It can be seen that the crystallite size and stacking disorder decrease with the increase of cycle period, but the micro-compressive strain increases with the increase of cycle period.

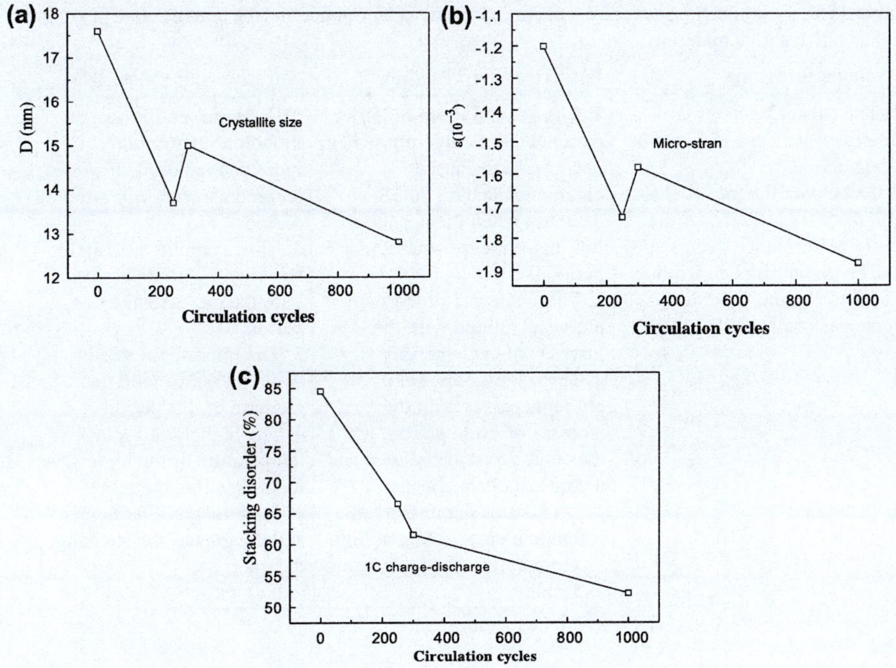

Fig. 15.9 Variation of microcrystal size D (**a**), micro-strain ε (**b**) and stacking disorder (**c**) of negative graphite with circulation cycles

15.3.3 Temperature Effect and Multiple Ratio Effect of Negative Electrode Cycle

The temperature effect and multiple ratio effect of the negative electrode cycle are listed in Table 15.4, which shows that the higher the cycle temperature or the higher the multiple ratio, the more serious the grain refinement and the greater the stacking disorder.

Table 15.4 Temperature effect and multiple ratio effect of charge–discharge cycle negative electrode

Microstructure parameter	$\overline{D}$	ε	Stacking disorder
Cycle condition and period	(nm)	(10^{-3})	(%)
1C charge 25 °C cycle 300 period	26.3	– 0.583	61.5
1C charge 55 °C cycle 300 period	22.2	– 0.818	64.8
3Ccharge 10C discharge 25 °C cycle 500 period	20.4	– 1.078	82.0

Table 15.5 Relationship between the decline of cyclic performance and the fine structure of positive and negative active materials

Cycle performance	Positive active material	Negative active material
1. The capacity retention decreases with the increase of cycle period 2. The higher the rate of charge and discharge, the faster the cycle performance decreases 3. The higher the temperature of the cycle, the faster the cycle performance decreases	1. The general trend of lattice parameters of cubic phase is to decrease. The lattice parameters a and c of 3R decrease when the period is low, but increase when the period is high 2. The grain size of the two phases is refined with the increase of cycle period 3. The relative amount of 3R phase increases with the increase of cycle period, while the cubic phase decreases, and the rate of change is about 2% 4. The lattice parameters of cubic phase cycling at high temperature or high rate are small	1. A decreases almost monotonously, while c increases rapidly at low cycles and increases slowly after 300 period 2. The crystallite size and stacking disorder decrease with the increase of cycle period 3. The micro-compressive strain increases with the increase of this period 4. The higher the cycle temperature or the higher the magnification, the more serious the grain refinement and the greater the stacking disorder

15.4 The Mechanism of Cycle Performance Degradation and the Role of $LiMn_2O_4$ [1]

15.4.1 The Relationship Between the Decline of Cyclic Performance and the Fine Structure of Positive and Negative Active Materials

The previous research results are summarized in Table 15.5. It can be seen that there is a good corresponding relationship between the decline of cycle performance and the fine structure change of positive and negative active materials.

15.4.2 Discussion on the Attenuation Mechanism of Cycle Performance

From the corresponding relationship in Table 15.5, it can be seen that the attenuation of cycle performance is due to the good corresponding changes in the fine structure of the positive and negative active materials that make up the battery during the cycle process, and in other words, the main results are as follows:

(1) Due to the change of lattice parameters of the positive active material during the cycle, it is shown that in addition to the de-intercalation and re-embedding of Li, positive ions enter the electrolyte and change some properties of the electrolyte under the action of electric field, and the grain refinement of the positive active material also reduces its properties.
(2) Due to the change of the lattice parameters of the negative active material during the cycle, it is shown that there are more Li atoms in the hexagonal grid of carbon atoms in the low cycle gap, and the interstitial Li atoms no longer change after 300 cycles, and the microcrystal size and stacking disorder decrease with the increase of the cycle period, which reduces the activity of the negative active materials, especially at high temperature or high rate.

To sum up, the positive and negative active materials and electrolytes change during the cycle, and this change will also affect the quality of the electrode surface film, resulting in the attenuation of the cycle performance of the battery.

15.4.3 *The Role of $LiMn_2O_4$ in Mixed Cathode Active Materials*

In terms of electrochemical performance, $Li(Ni_{0.4}Co_{0.2}Mn_{0.4})O_2$ is superior to $LiMn_2O_4$, so there is no doubt that the comprehensive performance of 2H-graphite/$Li(Ni_{0.4}Co_{0.2}Mn_{0.4})O_2$ battery is better than that of 2H-graphite/$LiMn_2O_4$. However, in the cycle process of lithium-ion battery made from this composite cathode active material.

(1) The lattice parameter of cubic phase decreases preferentially;
(2) The relative content of cubic phase decreases with the cycle process, while

3-Rm increases relatively. These two points indicate that $LiMn_2O_4$ gives priority to consumption during the cycle; in other words, $LiMn_2O_4$ can protect $Li(Ni_{0.4}Co_{0.2}Mn_{0.4})O_2$ during the cycle.

During the cycle of graphite/$Li(Ni_{0.4}Co_{0.2}Mn_{0.4})O_2$ + $LiMn_2O_4$) battery, the capacity decreases with the increase of cycle, and the fine structure of positive and negative active materials also changes with the increase of cycle, and there is a good corresponding relationship between them. The attenuation of cycle performance is the result of comprehensive action of the change of positive and negative electrode active material in the cycle process, which reduces its performance, and then affects the quality of electrolyte and electrode surface film.

The addition of $LiMn_2O$ has a certain protective effect on Li $(Ni_{0.4}Co_{0.2}Mn_{0.4})O_2$.

References

1. Zhang, J., H.H. Liu, C.Z. Yang, and B.J. Xia. 2003. Study on cycle process of Graphite/[Li($Ni_{0.4}Co_{0.2}Mn_{0.4}$) O_2+$LiMn_2O_4$]. *Power Supply Technology* (2): 60–64.
2. Li, H., C.Z. Yang, and F. Liu. 2008. A new X-ray diffraction method for determining the stacking disorder of hexagonal graphite). *Chinese Science, Series B Chemistry* 28(9): 755–760.
3. Li Hui, Yang Chuanzheng , Liu Fang 2009. Novel method of Determination Stacking Disorder Degree in Hexagonal Graphite by X-ray Diffraction, *Science China Chemistry B, Chemistry,* 5(22): 174–180.
4. Li Hui, Yang Chuanzheng , Liu Fang 2009. Determination of graphitization degree and disorder degree of carbon electrode materials by X-ray diffraction). *Journal of testing Technology* 23(2): 161–167.

Chapter 16
Mechanism of Cycle Process for Graphite/$LiFePO_4$ Battery

If lithium-ion battery wants to become the power supply of electric vehicle, it must have reliable safety and good performance stability. Among many positive active materials, lithium iron phosphate ($LiFePO_4$) is considered as the most promising candidate material for lithium-ion power battery because of its high safety, good cycle stability and low price. However, at present, in a large number of studies on lithium iron phosphate, people mainly focus on the improvement of its performance. However, little attention has been paid to the performance change and change mechanism of the battery as a positive active material in the process of long-term cycle use. This chapter focuses on the research results in this area [1–4].

16.1 The Change Law of Cycle Performance [1]

Figure 16.1 shows the discharge capacity curve (a) and capacity retention rate (b) of 452340 type 2H-graphite/$LiFePO_4$ battery under 25 °C of 1C charge–discharge cycle 3–1000 periods. It can be seen from the figure that the capacity increases with the increase of the cycle period. And, its peak is at 186 periods, and then decreases slowly; until 1000 cycles, the capacity retention rate is still nearly 90%.

The charge–discharge curve of the battery in several stages of the cycle process is shown in Fig. 16.2. It is obvious that the charging platform decreases with the increase of the cycle period, while the discharge platform increases at first and then decreases with the increase of the cycle period. The specific performance data are listed in Table 16.1. It can be seen that both the charging capacity and the discharge capacity increase, and then decrease gradually, and the maximum value is near the 200 periods; the charging platform voltage decreases at first and then increases slightly,

C. Yang et al., *Materials and Working Mechanisms of Secondary Batteries*,
https://doi.org/10.1007/978-981-19-5955-4_16

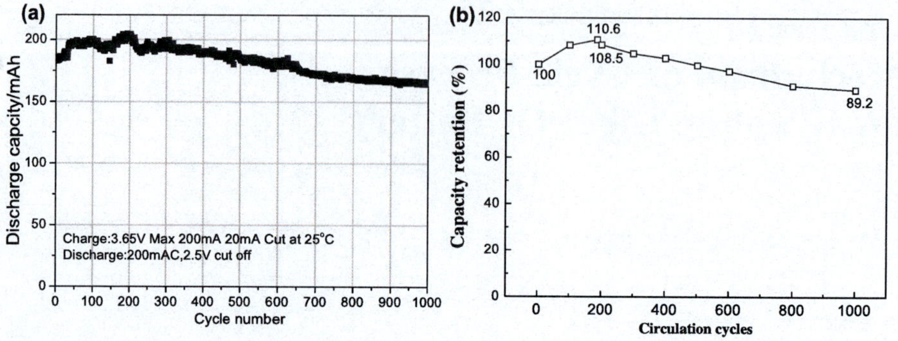

Fig. 16.1 (**a**) 25 °C cycle curve (**b**) capacity retention curve of 2H-graphite/$LiFePO_4$ battery

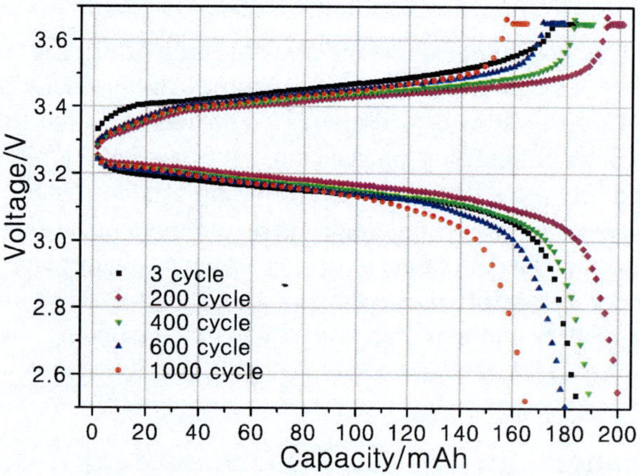

Fig. 16.2 Battery charge–discharge curves at several stages of the cycle

and the minimum value is around the 200 periods, which corresponds to the change of the charging capacity; and the discharge platform voltage also increases at first and then decreases, which is consistent with the discharge capacity rising at first and then decreasing.

16.2 The Fine Structure Change of Positive Active Material During Cycling

X-ray diffraction patterns of positive active materials at several stages of the battery cycle are shown in Fig. 16.3, where (b) is a partial amplification of (a). It can be seen that in the whole cycle process, no matter how much the cycle is, the positive active

Table 16.1 Charge–discharge performance of 2H-graphite/$LiFePO_4$ battery after cycle

Number of cycles	Charging capacity (mAh)	Charging platform (mV)	Discharge capacity (mAh)	Discharge platform (V)
3	183.5	3.462	185.2	3.143
100	198.8	3.428	200.5	3.165
200	201.4	3.430	201.0	3.168
300	195.3	3.435	193.9	3.161
400	189.6	3.436	190.2	3.154
500	185.4	3.438	184.6	3.158
600	178.8	3.446	180.1	3.146
800	168.9	3.440	168.4	3.147
1000	164.8	3.435	165.2	3.150

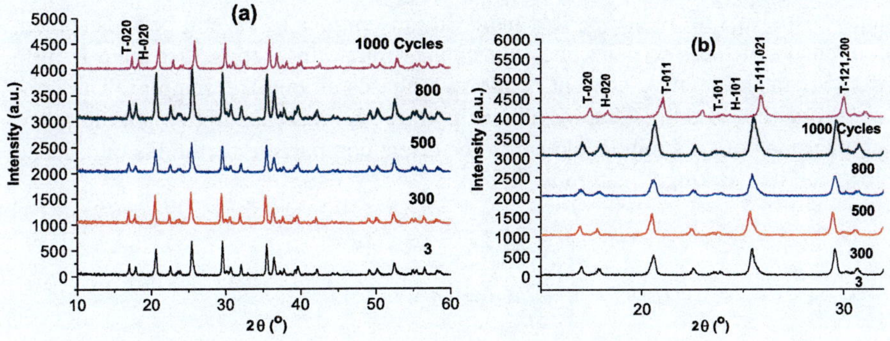

Fig. 16.3 **a** X-ray diffraction patterns of positive active materials at several stages during battery cycle. **b** Is partial magnification of (**a**)

materials are composed of $LiFePO_4$ and $FePO_4$. The main original diffraction data are listed in Table 16.2, and the analysis is as follows:

16.2.1 The Change of the Relative Content of Two Phases with the Cycle

It can be seen from Fig. 16.3 that the diffraction lines of $LiFePO_4$ (T) and $FePO_4$ (H) are almost overlapping except that T020 and H020, T101, and H101 can be separated, so the relative content of the two phases can only be determined by the

Table 16.2 Original diffraction data of positive active materials

Cycle period	$\frac{I_{LiFePO4-020}}{I_{FePO4-020}}$	2θ (°)					FWHM/(°)			
		020	011	111	121	131	011	111	121	131
3	1.398	16.014	20.604	25.386	29.564	35.442	0.260	0.302	0.192	0.186
100	1.365	16.030	20.652	25.490	29.623	35.532	0.333	0.362	0.240	0.248
200	1.396	16.014	20.608	25.387	29.562	35.441	0.231	0.185	0.167	0.157
300	1.874	16.935	20.519	25.296	29.462	35.344	0.247	0.334	0.176	0.186
400	2.087	13.980	20.605	25.404	29.559	35.450	0.266	0.273	0.233	0.236
500	1.317	16.990	20.605	25.380	29.567	35.451	0.309	0.378	0.242	0.246
600	1.226	16.227	20.827	25.639	29.777	35.669	0.284	0.325	0.246	0.240
800	1.109	16.059	20.656	25.486	29.599	35.513	0.315	0.400	0.259	0.293
1000	1.320	16.451	21.058	25.838	30.023	35.891	0.250	0.198	0.183	0.197

intensity analysis of T_{020}-H_{020} line pairs. In the samples of several periods shown, the intensity of the line pairs does not change much, indicating that there is little difference in the relative content of the two phases among the samples. The non-standard method used in Chap. 12 above to know the mass absorption coefficient of each phase is not very applicable here. The direct comparison method is introduced below, and its working equation is [2]:

$$\begin{cases} \frac{I_m}{I_i} = \frac{K_m}{K_i} \cdot \frac{\rho_i}{\rho_m} \cdot \frac{X_m}{X_i} \\ \sum_{i=1}^{n} X_i = 1 \end{cases} \tag{16.1}$$

$$K = \left(N^2 \cdot P_{hkl} \cdot F_{hkl}^2 \cdot \frac{1 + \cos^2 2\theta_{hkl}}{\sin^2 \theta_{hkl} \cdot \cos \theta_{hkl}} \cdot \mathrm{e}^{-2M} \right) \tag{16.2}$$

Among them, N is the number of unit cells in unit volume (1 cm^3), P_{hkl} is a multiplicity factor, and F_{hkl} is a structural factor. The two-phase related data are shown in Table 16.3.

When the temperature factor ($e^{-2M} = 1$) is ignored, the relevant data are substituted into:

$$\begin{cases} \frac{I_{LiFePO4-020}}{I_{FePO4-020}} = 0.8664 \times \frac{X_{LiFePO4}}{X_{FePO4}} \\ X_{LiFePO4} + X_{FePO4} = 1 \end{cases} \tag{16.3}$$

Table 16.3 Data related to $LiFePO_4$ and $FePO_4$

	hkl	N^2	P	F^2	$\frac{1+\cos^2 2\theta_{020}}{\sin^2 \theta_{020} \cos \theta_{020}}$	K	ρ(g/cm³)
$LiFePO_4$	020	11.485×10^{42}	2	21,688.453	88.60	4.4138×10^{49}	2.138
$FePO_4$	020	12.974×10^{42}	2	24,348.482	81.95	5.1775×10^{49}	2.173

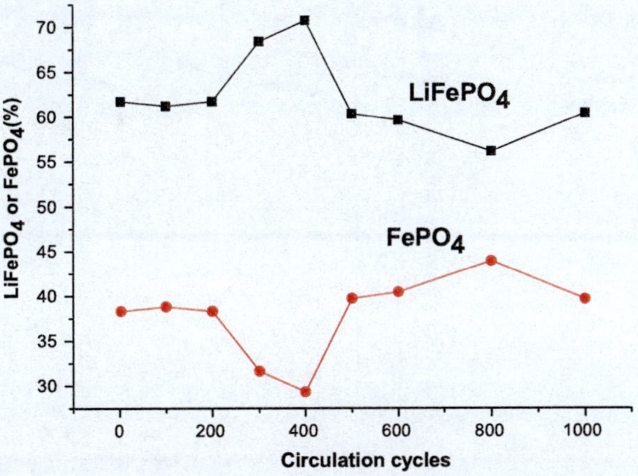

Fig. 16.4 Changes of relative contents of $LiFePO_4$ and $FePO_4$ during cycling

The average integral intensity ratio of the whole pattern and only 020 line pairs is obtained by using the Jade program (see the second column in Table 16.2). It can be seen in Fig. 16.4 that the content of $LiFePO_4$ changes little with the increase of the cycle period and then increases with the increase of the cycle, with a maximum value around 400 cycles, which roughly corresponds to the change of the cycle capacity.

16.2.2 Change of Lattice Parameters of $LiFePO_4$ with Cycle

The lattice parameters measured by the 2θ values of 020,011,111,121,131 of $LiFePO_4$ vary with the cycle. The variation curve of lattice parameters with cycle is shown in Fig. 16.5. It can be seen that: 1) the fine structure inside the diaphragm is not conducive to the passage of ions and becomes more and more serious with the extension of the cycle, especially after 800 cycles; 2) the degree of damage to the surface of the diaphragm and the thinning of the surface PP layer become more and more serious with the increase of the cycle.

16.2.3 Variation of Microstructure Parameters with Cycle in $LiFePO_4$

The double broadening effect of microcrystal and micro-strain is carried out by using the FWHM values of 011, 111, 121, 131 of $LiFePO_4$. The results of the solution are shown in Fig. 16.6. It can be seen that the overall change trend increases with the cycle, but the micro-strain becomes compressive strain.

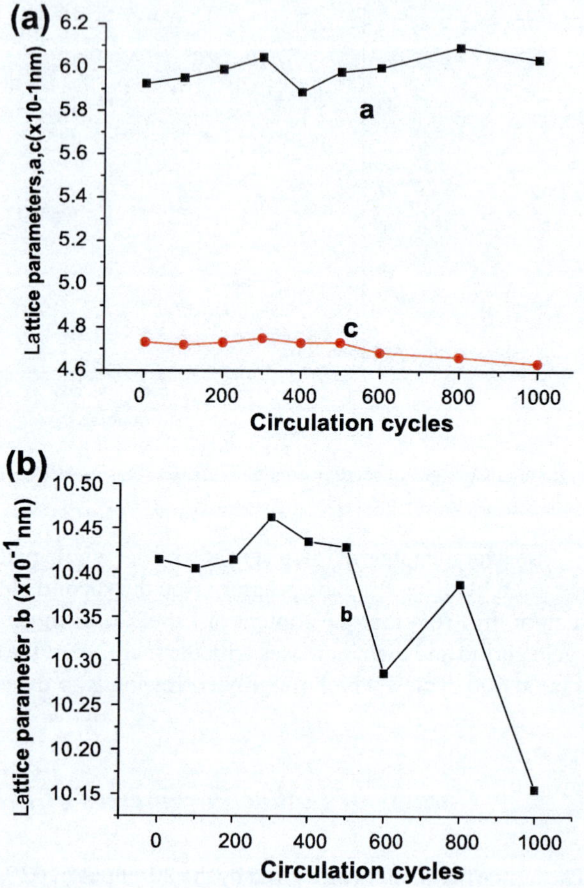

Fig. 16.5 Variation of lattice parameters of $LiFePO_4$ with cycle, **a** display a and c, **b** display b

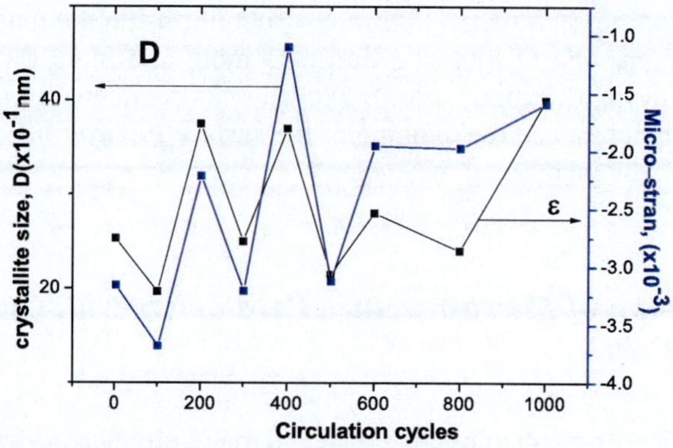

Fig. 16.6 Changes in microstructure D and ε of $LiFePO_4$

16.3 Fine Structure Change of Negative Active Materials During Cycling

The X-ray diffraction patterns of the negative active material graphite electrode at several stages of the battery cycle are shown in Fig. 16.7. It can be seen that the phase structure does not change, and no undecomposed Li-C compounds are observed, but the 002 peak shifts obviously, and the main original diffraction data are listed in Table 16.4. The lattice parameters are calculated by using the 2 θ values of the five original lines 002, 100, 101, 004, and 116. The FWHM of 002, 100, 004, and 110 are used to solve their D and ε, and then, the stacking disorder degree P is obtained by using the FWHM of 101 and the previous D and ε. The variation of the fine structure parameters with the cycle period is shown in Figs. 16.8 and 16.9.

The lattice parameters a and c do not change obviously, and the microstructure parameters D and ε do not seem to have much regularity, but the general change trend of stacking disorder increases with the increase of cycle period.

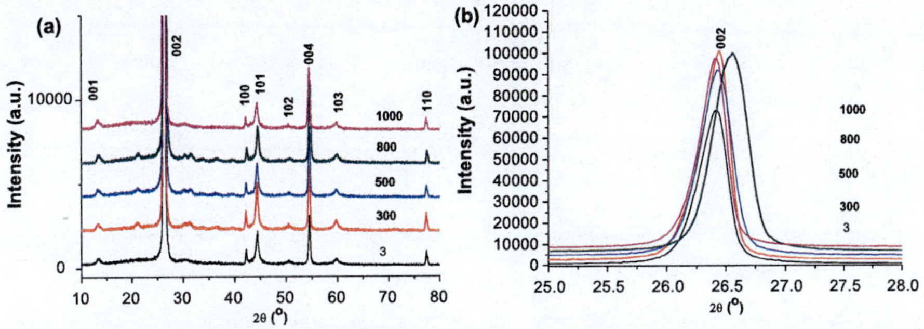

Fig. 16.7 X-ray diffraction patterns of graphite electrodes at several stages of cell cycle show that (**a**), **b** is a local amplification of (**a**)

Table 16.4 Original diffraction data of negative electrode active material graphite

Cycle period	2θ (degree)					FWHM (degree)				
	002	100	101	004	110	002	100	101	004	110
3	26.431	42.237	44.394	54.539	77.354	0.280	0.255	0.461	0.341	0.269
100	26.518	42.342	44.503	54.632	77.452	0.271	0.333	0.629	0.297	0.230
200	26.433	42.206	44.410	54.541	77.396	0.268	0.278	0.536	0.313	0.234
300	26.419	42.170	44.393	54.516	77.355	0.306	0.272	0.517	0.315	0.274
400	26.478	42.297	44.476	54.596	77.419	0.271	0.290	0.571	0.282	0.234
500	26.538	42.411	44.537	54.652	77.476	0.314	0.361	0.548	0.335	0.258
600	26.357	42.198	44.370	54.475	77.311	0.253	0.235	0.487	0.302	0.248
800	26.397	42.245	44.393	54.497	77.358	0.254	0.270	0.610	0.321	0.273
1000	26.402	42.194	44.399	54.501	77.375	0.276	0.214	0.599	0.304	0.220

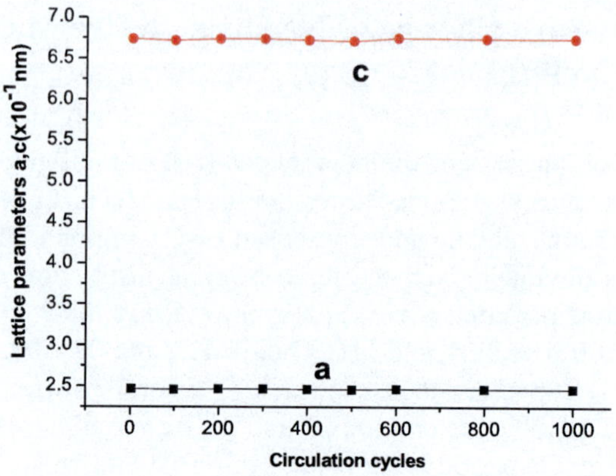

Fig. 16.8 Variation of lattice parameters of graphite electrode with circulation cycles

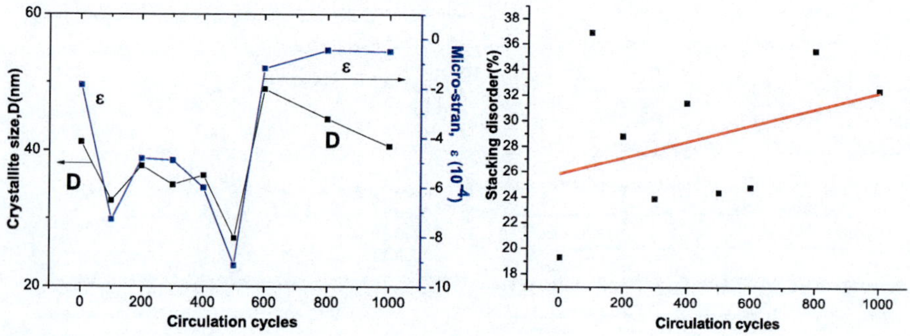

Fig. 16.9 Variation of grain size D, micro-strain ε (**a**) and stacking disorder P of graphite with circulation cycles

16.4 The Fine Structure Changes of Diaphragm Materials

The diaphragm is composed of three layers of PP-PE-PP, PP is polypropylene film, belongs to monoclinic structure, $P2_1/c$ (No.14) space group; PE is polyethylene, belongs to orthorhombic, Pnam (No.62) space group. X-ray diffraction patterns of the diaphragm material at several stages of the battery cycle are shown in Fig. 16.10, where (b) is a partial amplification of (a). The diffraction peaks have been indexed in the figure, it can be seen that (1) the diffraction peaks of PP-131 and PE-110, PP-111 and PE-200 overlap, and (2) the positions of amorphous scattering peaks of PP and PE are at = 16.30C and 19.50C, respectively, indicating that the crystallinity of all samples is very high, and the amorphous part is very few. (3) several diffraction peaks of PP shift to low angle with the increase of cycle period, and (4) the half width of the peak also changes obviously. The main raw data are listed in Table 16.5. The analysis is as follows.

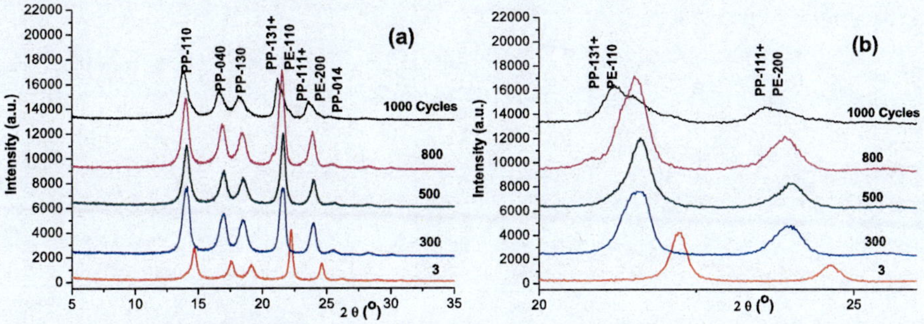

Fig. 16.10 **a** X-ray diffraction patterns of diaphragm materials at several stages during battery cycle, **b** is partial magnification of (**a**)

Table 16.5 Original diffraction data of diaphragm

Cycle period	2θ (degree)						FWHM (degree)		
	110	040	130	131	111	014	110	040	130
3	14.716	16.566	19.138	22.226	24.605	26.192	0.649	0.470	0.619
100	13.964	16.837	19.397	21.463	23.857	25.532	0.676	0.689	0.753
200	14.195	16.145	19.657	21.816	24.196	25.511	0.806	0.832	0.976
300	14.024	16.961	19.487	21.594	23.964	25.471	0.640	0.651	0.719
400	14.119	16.037	19.565	21.657	24.039	25.430	0.518	0.574	0.576
500	14.045	16.963	19.519	21.624	24.003	25.637	0.650	0.607	0.768
600	14.312	16.254	19.781	21.910	24.284	25.877	0.657	0.651	0.742
800	13.997	16.896	19.445	21.553	23.913	25.442	0.602	0.625	0.652
1000	13.755	16.564	19.181	21.176	23.564	25.329	0.739	0.808	0.882

16.4.1 Variation of Lattice Parameters of Polypropylene with Cycle

According to the 2θ of polypropylene 110, 040, 131, 131, 111, and 014 diffraction lines, the lattice parameters of polypropylene (PP) with the cycle are given in Fig. 16.11, in which the total variation trend of a and b increases with the increase of the cycle, while the change of c is more complex.

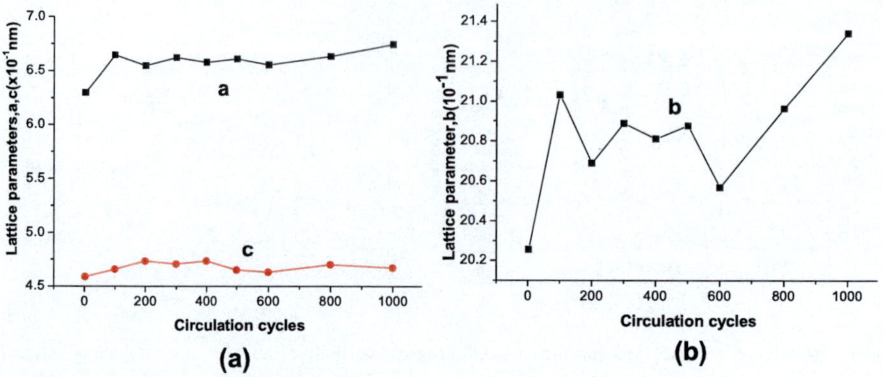

Fig. 16.11 Variation of lattice parameters of polypropylene (PP) with circulation cycle

16.4.2 Variation of Microstructure Parameters of Polypropylene with Cycle

According to the FWHM of 110, 040, 130 diffraction lines of polypropylene, the microstructure parameters D and ε obtained by solving the microcrystal-micro-strain double broadening effect are shown in Fig. 16.12. It can be seen that the general change trend of ε increases with the increase of the cycle, and changes from compressive stress to tensile stress at the beginning of the cycle.

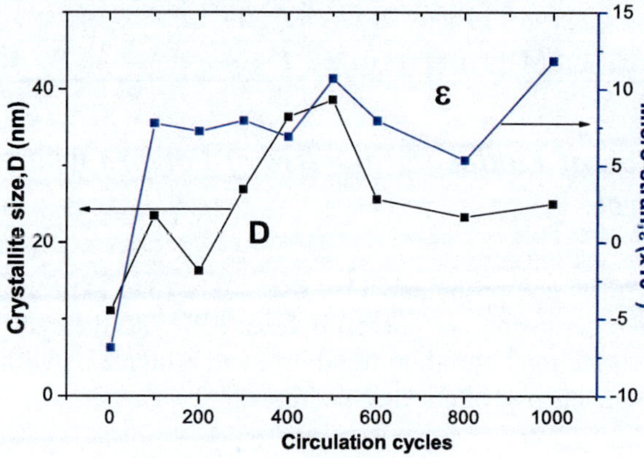

Fig. 16.12 Variation of polypropylene microcrystal D and micro-strain ε with circulation cycles

16.5 Cycle Performance Change Law and Mechanism [1]

16.5.1 The Relationship Between the Change Law of Cycle Performance and the Fine Structure of Positive and Negative Electrode and Diaphragm Fine Material

In order to explore the relationship between cycle performance and the fine structure of positive and negative electrodes and diaphragm materials, the experimental results of Sect. 16.1–16.4 are summarized in Table 16.6. It can be seen that the change law of cycle performance has a good corresponding relationship with the change of the relative content of $LiFePO_4$ in the positive electrode, the change of lattice parameters b and c, the change of disorder in negative graphite, and the change of lattice parameters a, b and micro-strain of PP in the diaphragm.

Table 16.6 Relationship between cycle performance and fine structure changes of positive and negative electrodes and diaphragm

Cyclical energy	Positive active material	Negative active material	Diaphragm
1. Change of cycle performance and capacity retention rate with cycle periods, the increased at first, and then decreased slowly, The maximum values is about 200 cycles 2. The charge capacity and discharge capacity obtained from the charge–discharge curve increase slightly with the increase of the cycle period, and then decrease gradually, and their maximum values are both around 200 cycles 3. The charging platform decreases at first, then increases slowly, while the discharge platform increases at first and then decreases	1. The relative content of $LiFePO_4$ decreases at first and then increases with the cycle period, and the maximum value is about 400 cycles 2. The lattice parameters c and b of $LiFePO_4$ increase slightly at first and then decrease with the cycle period, and the maximum value is about 300 cycles 3. The general trend of D and ε of $LiFePO_4$ is to increase	1. The general change trend of stacking disorder is that it increases with the increase of cycle period 2. The lattice parameters of graphite change little with the cycle 3. The microstructure parameters D and ε do not change much	1. The lattice parameters a and b of polypropylene increase with the increase of cycle period, but the change of c is irregular 2. The micro-strain ε increases with the increase of cycle period 3. After a long cycle, PP-113 and PE-110 split obviously, indicating that the surface of the diaphragm was seriously damaged and thinned during the cycle

16.5.2 *The Mechanism of Change Law of Cycle Performance*

From the chemical and physical study of charge and discharge process of 2H-graphite/$LiFePO_4$ battery in Chap. 12, it is known that there is hysteresis effect in the discharge process, or charge–discharge asymmetry; from the study of storage process of 2H-graphite/$LiFePO_4$ battery in Chap. 19, it is known that the hysteresis effect of discharge process is greatly reduced by storage energy [3].

The so-called cycle is that the a time of charge–discharge is one cycle period, so the tested samples are in the discharge state. Because of the hysteresis effect of discharge, how to change the hysteresis effect in the cycle is the first problem that should be paid attention to. It seems that this problem can be revealed from the change of the relative content of $LiFePO_4$ and $FePO_4$ with the cycle and the change of $LiFePO_4$ lattice parameters with the cycle. If the $LiFePO_4$ content after the subsequent cycle is greater than that in the previous cycle, the discharge lag effect is weakened, and on the contrary, the discharge hysteresis effect is enhanced.

The main results are as follows:

(1) After the beginning of the cycle, the relative content of $LiFePO_4$ increases and the content of $FePO_4$ decreases, indicating that there is more Li in the positive active materials, which increases the lattice parameters b and c of $LiFePO_4$ and can provide more Li ions, so the capacity after the cycle is increased, up to about 10%, resulting in a certain weakening of the discharge hysteresis effect. However, when the cycle exceeds 200 to 300 cycles, the amount of Li embedded into $LiFePO_4$ after discharge decreases, and the lattice parameters b and c of $LiFePO_4$ begin to decrease, so the capacity decreases, and the discharge hysteresis effect increases until 1000 cycles, and some fluctuations may occur in the cycle.

(2) The diaphragm is subjected to compressive strain at the initial stage of the cycle and then becomes tensile strain, and increases with the increase of the cycle, and the lattice parameters a and b increase with the positive pole of the cycle. In the later stage, the two pairs of overlapping lines of PP and PE are obviously broadened and split. All these indicate that there are obvious changes in the diaphragm during the cycle and reveal that: 1) the fine structure inside the diaphragm is not conducive to the passage of ions and becomes more and more serious with the extension of the cycle, especially after 800 cycles; 2) the degree of damage to the surface of the diaphragm and the thinning of the surface PP layer become more and more serious with the increase of the cycle.

The combined effect of the above two reasons make the cycle performance of graphite/$LiFePO_4$ battery improve at the beginning and then gradually decline with the change of cycle period. Although the negative active material does not play a great role, it reflects the result of the first effect. Due to the decrease of the atomic weight of Li embedded in $LiFePO_4$ after discharge, in other words, there are more Li atoms dissolved in graphite, which increases the disorder degree in graphite.

References

1. Zhang, J., H.H. Liu, C.Z. Yang, and B.J. Xia. 2012. Study on cycle process of Graphite/$LiFePO_4$ battery. *Power Supply Technology*) 36(2): 165–168.
2. Yang, C.Z., D.C. Xie, G.Z. Chen, and F.M. Zhong. 1989. Metallurgical Industry Press, Chap. 5. Cheng, G.F., C.Z. Yang, and Y. H. Huang. 2016, June. *X-ray analysis of nanomaterials*. Chemical Industry Press.
3. Li, Jia. 2016, June. Study on storage performance of lithium-ion battery. Doctoral paper of Shanghai Institute of Microsystems and Information Technology.
4. Wu, X.F., and Z.B. Wang. 2010. Study on cycle performance and safety performance of $LiFePO_4$ battery. *Battery industry* 15(3): 156–159.

Chapter 17
Mechanism Research of Storage Process for MH-Ni Battery

For practical batteries, such as MH-Ni batteries, as the power supply of power tools and hybrid electric vehicles, it is of practical significance to study its storage performance and attenuation law, so it has attracted wide attention of battery researchers and developers, and there have been many reports. To sum up, there are the following results:

(1) Lichtenbery et al. [1], Pralong et al. [2], Hideo Haigu et al. [3] considered that the reduction of CoOOH or the partial reduction of β-$Ni(OH)_2$ to metal [4] in the positive electrode during storage led to the decline of the performance of the positive electrode.
(2) Bernard [5] thinks that the deposition of corrosion products of negative alloy in the positive active material will reduce the capacity of the positive electrode and then affect the performance of the battery.
(3) The passivation [6] of negative alloy during storage and the deterioration [7, 8] of diaphragm performance may also have an effect on the battery.

The research of Li Xiao-Feng, Xia Bao-Jia et al. [9, 10] shows that the CoOOH that makes up the positive conductive network will be reduced. And it cannot be completely demobilized during the subsequent charging process, which leads to the decline of battery storage performance. If nickel powder is added to the positive electrode as a conductive agent, this effect can be alleviated to a certain extent, and the storage performance of the battery can be improved [11]. It is also found that the corrosion and deposition of negative alloy elements, especially Al and Mn, will not cause the irreversible decay of MH-Ni battery capacity during storage, which is different from the result of Bernard [5].

To sum up, due to the differences in experimental conditions and the perspective of the analysis of the problem, although the understanding of storage performance degradation of MH-Ni battery is somewhat the same, but there is no consistent view yet. In this chapter, through the comparative study of the performance before and after storage in the states of charge and discharge, and the comparative study of the structure of active materials of electrodes (positive and negative electrodes) by

C. Yang et al., *Materials and Working Mechanisms of Secondary Batteries*,
https://doi.org/10.1007/978-981-19-5955-4_17

X-ray diffraction (XRD), and the relationship between performance and structure, the reason and mechanism of battery storage performance degradation are discussed, which has not been reported.

17.1 Storage Process and Calculation of Capacity Decay Rate

17.1.1 Battery Manufacturing, Activation and Storage Process

The positive material is co-deposited spherical β-$Ni(OH)_2$ containing 2.4 wt% $Co(OH)_2$ and 8.4 wt% $Zn(OH)_2$, and mechanically mixed according to the weight ratio of β-$Ni(OH)_2$:CoO = 100:8. The positive electrode is made of plaster and sintered with nickel foam respectively. The composition of the negative electrode AB_5 alloy is $RENi_{3.35}Co_{0.75}Al_{0.2}Mn_{0.5}$ (RE stands for mixed rare earth), which is also coated on nickel foam to make a negative electrode, and then assembled into an AAA sealed battery.

The battery was activated by 0.2C charge–discharge cycle for 4 periods.

Storage system in discharge state: The discharge current is 0.1C to 1.0 V, then it is stored in an open-circuit environment of (60$\pm$ 2 °C) for 20 days, and the open-circuit voltage is measured every two days by a digital multimeter.

Storage system in charge state: Charge the battery at 0.2 current for 7 h, and then store it in the environment of (60$\pm$2°C) for 20 days in the open state.

For 20 days, 0.2C current was used to replenish electricity for 3 h every 7 days during storage.

17.1.2 Calculation of Capacity Decay Rate

The last activated 0.2C discharge capacity was used as the initial capacity of the battery. After 20 days of storage in charge state and discharge state, the last discharge capacity of 4 times of 0.2C cycle is taken as the storage capacity, so the capacity decay rate is calculated according to the following formula:

$$\text{Capacity decay rate} = \left[(\text{Before storage capacity} - \text{After storage capacity})/\text{Before storage capacity}\right] \times 100\% \quad (17.1)$$

17.2 Storage Performance of MH-Ni Battery in Discharge and Charge States

Figure 17.1 shows the curve of the relationship between open-circuit voltage and storage days of self-made paste-coated and sintered batteries. It can be seen that there is a large self-discharge rate under the condition of 60°C. When the time reaches 10 days, the two kinds of batteries decrease from 1.2V to 1.0V and 0.7V, respectively, and the whole decline is not the same.

Figure 17.2 charge–discharge curves of 0.2C before and after 20 days storage of (60 ± 2 °C) in plaster (a, left) and sintered (b, right) batteries respectively. It is relatively knowable that the charging voltage of the paste battery increases after storage, while the sintered battery is almost unchanged. Figure 17.3 shows the 0.2C charge–discharge curve of plaster battery after storage 20 days at (60 ± 2 °C).

Compared with Fig. 17.2 (a, left), the increase of charge and discharge voltage of paste battery after discharge state storage is smaller than that of charge state storage. Now, the relevant data are listed in Table 17.1. Comparing these data, we can see:

(1) Also in the discharge state, the performance of the sintered battery is almost the same as that before storage, while the capacity of the plaster battery is obvious decrease, internal resistance also obviously increases;

(2) For the same plaster battery, after 20 days of charge state storage at (60 ± 2 °C) almost maintained the before storage capacity, and the internal resistance increased by about 11.6%, while after 20 days of storage in the discharge state, the capacity decreased by about 12%, and the internal resistance increased by 42%.

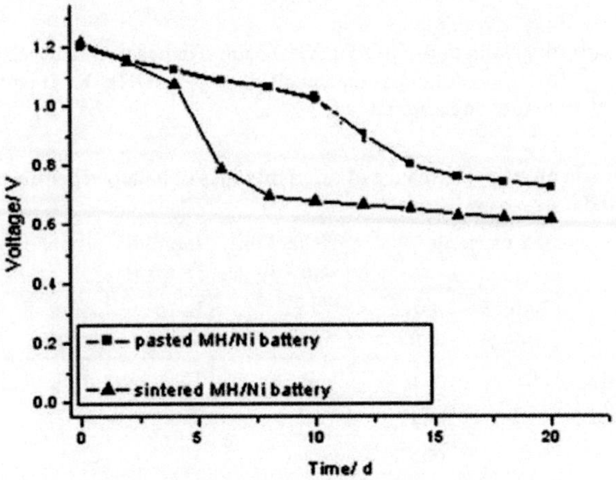

Fig. 17.1 Open circuit of paste (pasted) and sintered (sintered) batteries stored in 60 °C discharge state. The curve of the relationship between voltage and storage days

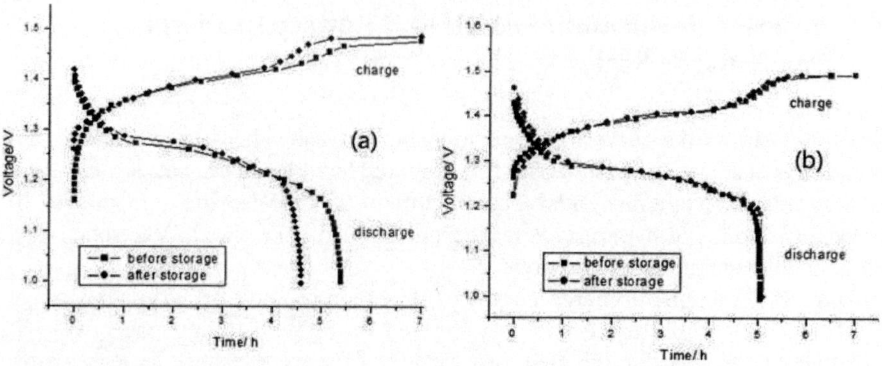

Fig. 17.2 Charge–discharge curve of 0.2C of plaster (**a**) sintered (**b**) battery before and after 20 days of discharge storage at (60 $\pm$2 °C)

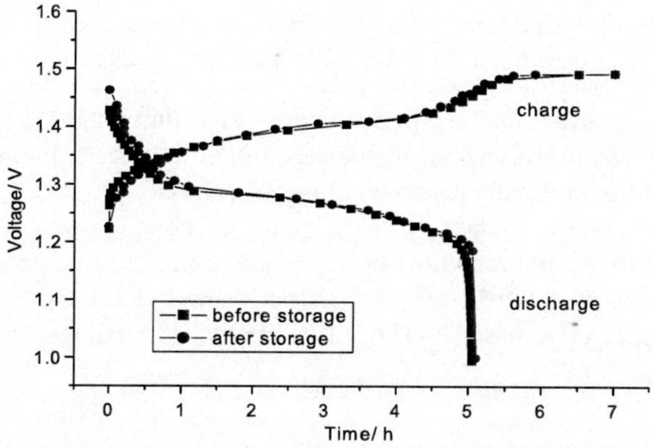

Fig. 17.3 (**a**) Phase identify result of β-$Ni(OH)_2$ XRD pattern in positive materials of paste battery before and after storage (**b**) Microstructural data analysis of β-$Ni(OH)_2$ XRD patterns in positive materials of paste battery before and after storage

Table 17.1 Comparison of the performance of different types of batteries before and after storage at (60 ± 2 °C) for 20 days

Battery type	Storage state	0.2 capacity		Reduction rate% of capacity decay	Internal resistance (mΩ)		Increase rate of internal resistance%
		Before storage	After storage		Before storage	After storage	
Sintered	Discharge	508	513	– 0.98	15.3	15.9	3.92
Pasted	Discharge	650	571	12.2	20.2	28.7	42.10
Pasted	Charge	648	655	-1.1	19.8	22.1	17.62

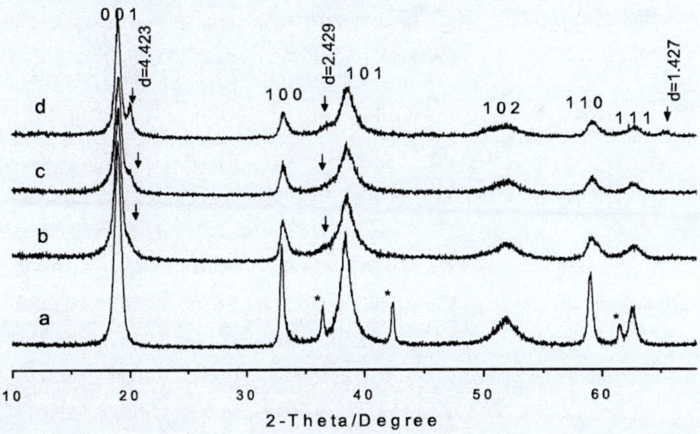

Fig. 17.4 XRD pattern of battery positive materials, **a** activation before, **b** after activation, **c** after charging storage 20 days, **d** after discharge storage 20 days

17.3 XRD Comparative Analysis of the Structure of Positive Active Materials Before and After Storage

17.3.1 *Structure of Positive Active Material β-Ni(OH)$_2$*

After dissecting un-activated, only activated and two kinds of paste-coated batteries stored in charge state and discharge state for 20 days, the XRD patterns of battery positive materials are shown in Fig. 17.4, the original data are listed in Table 17.2, and the analysis results are listed in Table 17.3. The results of phase identification are listed in Table 17.3a. The CoOOH reference PDF card number is 7–0169, which belongs to the R-3m (No. 166) space group, $a = 2.855$ Å, $c = 9.156$ Å. The lattice parameters of β-Ni(OH)$_2$ and the microstructure parameters obtained by processing the data according to the method introduced in reference [12, 13] are listed in Table 17.3a and b, respectively.

17.3.2 *Phase Structure Analysis of Insoluble Substance of Positive Materials in Acetic Acid Before and After Discharge Storage*

After separating the positive material from the matrix before and after discharge storage, the positive material was repeatedly impregnated with 50% volume ratio acetic acid in 60 °C until the impregnation solution changed from green to colorless, and the black insoluble substance was filtered and dried and tested by XRD. The

Table 17.2 Original data of β-Ni (OH) 2 XRD in positive materials before and after paste battery storage

hkl		001	100	101	102	110	111
$\beta^0_{1/2}$ (0)		0.140	0.130	0.133	0.124	0.125	0.126
Before activation	$B_{1/2}$ (^{0})	0.652	0.329	1.037	2.653	0.395	0.649
	2θ (^{0})	19.988	32.993	38.344	51.857	58.935	62.513
After activation	$B_{1/2}$ (^{0})	0.664	0.595	1.682	2.479	1.400	
	2θ (^{0})	19.013	33.091	38.419	51.966	58.940	
After storage 20 days at charge state	$B_{1/2}$ (^{0})	0.782	0.631	1.686	2.742	1.044	1.166
	2θ (^{0})	19.037	33.014	38.508	51.977	59.059	62.737
After storage 20 days at discharge	$B_{1/2}$ (^{0})	0.640	0.602	1.631	2.650	1.220	1.300
	2θ (^{0})	19.027	33.029	38.577	51.902	59.014	62.610

Table 17.3 (**a**) Phase identify result of β-Ni(OH)$_2$ XRD pattern in positive materials of paste battery before and after storage. (**b**) Microstructural data analysis of β-Ni(OH)$_2$ XRD patterns in positive materials of paste battery before and after storage

(a)

Sample state	Existing phase	Lattice parameter of β-Ni(OH)$_2$(Å) a c	
Before activation	β-Ni(OH)$_2$ + CoO	3.1124	4.6039
Before storage after activation	β-Ni(OH)$_2$ + Trace amount CoOOH	3.1353	4.6305
After storage at charge state	β-Ni(OH)$_2$ + Small amount CoOOH	3.1059	4.5928
After storage at discharge state	β-Ni(OH)$_2$ + Large amount CoOOH	3.1506	4.4669

(b)

Sample state	001	100	110	111	$\overline{D}$	$\overline{\varepsilon}$	f_D	f_T	$f_D + f_T$
	D_{hkl} (nm)				(nm)	(10^{-3})	(%)		
Before activation	15.56	41.16	33.42	17.57			2.90	10.95	13.85
Before storage after activation	15.20	17.62	7.08		108.2	8.59	7.87	0.74	8.61
After storage at charge state	12.41	16.35	9.82	8.85	20.0	3.67	7.26	4.39	17.65
After storage at discharge state	15.93	17.36	8.24	7.83	59.1	7.07	7.07	3.36	10.43

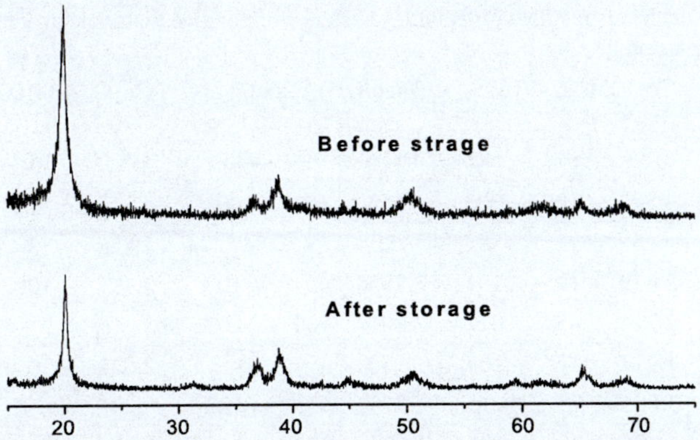

Fig. 17.5 XRD pattern of insoluble substance of positive powder in acetic acid before and after storage of MH/Ni battery

diffraction pattern is shown in Fig. 17.5. The results of diffraction data analysis are given in Table 17.4. From the analysis of the data (including d value and relative strength I/I_1), it is known that the main phase is CoOOH, rather than β-$Co(OH)_2$. Compared with the PDF card number of 7–0169, it belongs to the R-3m (No.166) space group, which is completely consistent with the analysis results of Sect. 17.3.1, and there are few Co_3O_4, and the PDF card number is 43–1003. In addition, the lines before storage are wider than those after storage, indicating that the grains of CoOOH are growing during storage.

17.4 XRD Comparative Analysis of AB_5 Hydrogen Storage Alloy Before and After Storage

The XRD patterns of negative materials AB_5 alloy before and after storage are shown in Fig. 17.6. There is no change in the phase structure and no negative corrosion products $A(OH)_3$ and B (i.e., $La(OH)_2$ and Ni), whose original data are 2θ values of 12 lines from 101 to 220 are self-compiled to solve the hexagonal lattice parameters [14], and the parameter of microstructure obtained from FMHM data of six diffraction lines from 001 to 002 is listed in Table 17.5. It can be seen that state the lattice parameter c becomes smaller, the grains are refined after storage, and there is compressive stress. The grain refinement and compressive stress of discharge storage is smaller than that of charge storage.

Table 17.4 XRD analysis results of insoluble substance of positive materials in acetic acid before and after discharge storage

Data to be analyzed		CoOOH, 07–0169			β-$Co(OH)_2$, 30–0443			Co_3O_4 43–1003		
d(Å)	I/I_1	d(Å)	I/I_1	hkl	d(Å)	I/I_1	hkl	d(Å)	I/I_1	hkl
4.419	100	4.384	100	003	4.653	60	001			
2.855	2							2.858	33	220
2.434	23	2.429	12	101	2.755	45	100	2.437	100	311
2.317	33	2.314	85	012	2.371	100	101			
2.019	7	1.977	12	104				2.021	20	400
1.802	13	1.803	75	015	1.778	45	102			
1.502	7	1.497	20	107	1.591 1.506	35 25	110 111	1.556	32	511
1.427	17	1.427	30	110				1.429	38	440
1.360	9	1.371 1.357	10 25	018 113	1.379 1.351	5 14	200 103			
1.228	5	1.214	10	202						

Note The no. 01-0357 PDF card, which was earned in 1948, has long been deleted and has been replaced by 30-0443

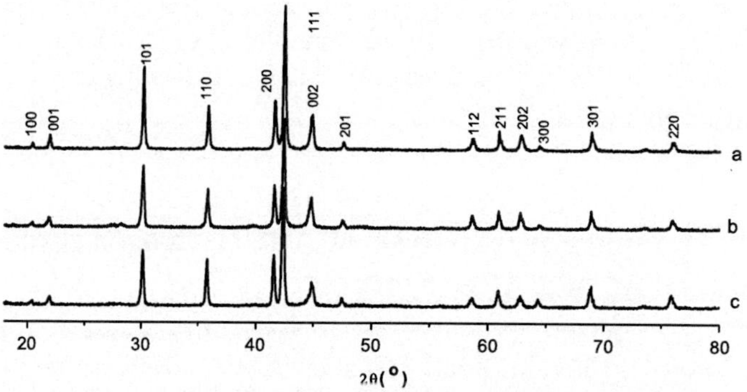

Fig. 17.6 XRD patterns of AB_5 alloy before and after storage, **a** before storage. **b** after storage 20 days at charged state, **c** after storage 20 days, at discharge

Table 17.5 Results of XRD data analysis of negative electrode material AB_5 alloy before and after storage

Sample state	Lattice parameters			Micro-structure parameters	
	a (Å)	c (Å)	V $(Å)^3$	$\overline{D}$ (nm)	$\overline{\varepsilon}$ ($\times 10^{-3}$)
Before storage after activation	5.0080	4.0421	87.79	71.4	-0.46
After storage at charge state	5.0056	4.0391	87.64	31.6	-1.88
After storage at discharge state	5.0082	4.0342	87.63	42.3	-0.77

17.5 Chemical Physics Behavior of Electrode Active Materials During Battery Storage

17.5.1 Phase Structure Changes of Positive and Negative Active Materials During Storage

From the data of Fig. 17.4 and Table 17.3, it can be seen that the positive material before activation is β-Ni $(OH)_2$ and CoO. After activation, the diffraction lines of CoO disappear; according to the law of immortality of matter, it is not that CoO disappears, but it is transformed into other substances. There are two possibilities:

(1) CoO is dissolved in alkaline solution and reprecipitated to form β-$Co(OH)_2$. That is

$$CoO + H_2O \rightarrow \beta - Co(OH)_2$$

It is either coated around β-$Ni(OH)_2$ particles or reacts with β-$Ni(OH)_2$ to form a solid solution β-$(Ni,Co)(OH)_2$. Front, the satellite peak should appear in the low angle side of the 100 and 101 diffraction lines of β-$Ni(OH)_2$, because

$2\theta^{(0)}$	β - $Ni(OH)_2$	β - $Co(OH)_2$
001	19.013	19.058
100	33.262	32.473
101	38.419	37.917

It was not observed in the experiment.

(2) The second possibility is β-$Co(OH)_2$ formed by CoO + H_2O, which occurs during activation as follows:

$$Co(OH)_2 + OH \rightarrow CoOOH + H_2O$$

Generate the CoOOH of the Co^{+3} valence. Careful observation of Fig. 17.4 shows that the high angle side of β-$Ni(OH)_2$ (001) and the low angle side of (101) in (b) and (c) are obviously asymmetrical, especially the three characteristic peaks of $d = 4.423$, 2.429, and 1.427 appear in Fig. 17.4d (see arrow), which are carefully identified to belong to CoOOH (PDF card number 07-0169). The XRD analysis of cathode materials in acetic acid (see Sect. 17.3.2) further proves this point, and it can be Judge from the intensity change of dang 4.423 peak in Fig. 17.4b–d that there is a very small amount of CoOOH after activation and increases in the subsequent storage process, especially the increase of discharge storage is more obvious.

It can be seen from Fig. 17.6 and Table 17.5 that there is no significant change in phase structure, lattice parameters and cell volume of the negative electrode before and after storage.

17.5.2 Microstructural Changes of Positive and Negative Materials Before and After Battery Storage

From the comparison of Fig. 17.4 and Table 17.3b, it can be seen that great changes have taken place in the microstructure of positive electrode before and after activation:

1. The change of D_{001} from 15.6 to 15.2 (nm), D_{100} to 41.2 to 17.6 (nm), shows that the grain shape changes from chubby columnar grain before activation to nearly equiaxed grain.
2. The total stacking fault probability is from 13.9% to 8.6%. The deformation stacking fault increases, while the twin stacking fault decreases greatly, which is consistent with the results of reference [15].

It can also be seen from Table 17.3b that the total stacking fault probability of both charged storage and discharge storage is smaller than that before activation, but increased compared with that before storage, and there is little change before and after storage deformation stacking fault, but twin stacking faults increase obviously.

It can also be seen from Table 17.3b that activation reduces the grain size and storage further refines the grains, especially in the charged state storage, which is related to supplementary electricity every 7 h during storage.

It can be seen from Table 17.4 that the grains of the negative materials are further refined and the compressive strain increases during storage, especially the grain refinement and compressive strain of the charged storage are larger than those of the discharged. This may also be related to supplementary electricity every 7 h during the charged storage.

17.5.3 Relationship Between Battery Storage Performance and Electrode Material Structure and Microstructure

The storage performance of the battery and the microstructure parameters of the electrode active material are listed in Table 17.6 by comparison. It can be seen that the discharge storage with larger capacity decay rate and larger internal resistance change rate corresponds to the higher CoOOH content and grain fine degree in the positive material, and has no corresponding relationship with the stacking fault rate of the positive material, and has no corresponding relationship with the phase structure and microstructure parameters of the negative material.

Table 17.6 Comparison of storage performance of paste-coated battery with microstructure parameters of electrode material

Plaster battery	Storage performance			Position electrode material				Negative electrode material	
	Capacity decay rate	Internal resistance increase rate	Content of CoOOH	Grain fine degree	f_D	f_T	$f_D + f_T$	Grain fine degree	Strain increase
Charge storage	– 1%	17.6%	Low	20%	Mostly un-change	Increase lager	Increase lager	56%	4 times
Discharge storage	12%	42.1%	Higher	50%	Mostly un-change	Increase littler	Increase littler	41%	1.8 times

17.6 Mechanism of Performance Degradation During Battery Storage

From the introduction, it can be seen that the corrosion of the negative alloy is one of the factors affecting the storage performance of the battery. From the results of this experimental study, this is not an important factor, so that it is less than 20 days of storage time; the reduction of CoOOH in cathode materials leads to the activity decline of cathode materials and affects the storage performance. The previous work of Li Xiao-feng and Xia Bao-jia [9, 10] supports this view, which is based on the fact that the identification results of insoluble substances in acetic acid after storage are $Co(OH)_2$ + small amount of Co_3O_4,. However, the identification results of the main phase need to be further studied. The d and I/I_1 spectrum data of CoOOH, 07–0169 and β-$Co(OH)_2$, 30–0443 listed in Table 17.4 are carefully compared with the data to will be analyzed. Obviously, the data of CoOOH are in better agreement with the data to will be analyzed, so the judge that main phase is Co^{+3}OOH rather is not β-$Co(OH)_2$, which is completely consistent with the new phase CoOOH of cathode materials before and after storage. So, it is based on the following reaction.

$$4\text{CoOOH} + 2\text{H}_2\text{O} \rightarrow 4\beta - \text{Co(OH)}_2 + \text{O}_2.$$

That is, the view that CoOOH is reduced and destroys the conductive network provided by CoOOH in positive charge, which leads to the irreversible attenuation of battery capacity is worthy of discussion and research. If the conductive network provided by CoOOH plays the main control role, because the content of CoOOH is after discharge storage > after charging storage > before storage, then the capacity should be after discharge storage > after charging storage > before storage, while the internal resistance is after discharge storage < after charging storage < before storage, which is contrary to the experimental results.

According to the research results of Chap. 10, the following reaction formulas for the charge and discharge process of Ni-MH battery can be written.

At the positive pole At the negative pole

Charge $\beta\text{-Ni(OH)}_2 \rightarrow \beta\text{-Ni(OH}_{1-x})_2 + 2 \times \text{H}^+ + \text{e}^-$ $\text{AB}_5 + 2 \times \text{H}^+ + \text{e}^- \rightarrow \text{AB}_5 - 2 \times \text{H}$ (solid solution)

$0 < X < 0.50$

Over - charge $2\beta\text{-Ni(OH)}_2 \rightarrow \beta\text{-Ni(OH)}_2 + \gamma\text{-NiOOH} + \text{H}^+ + \text{e}^-$

$\text{AB}_5 + \text{H}^+ + \text{e}^- \rightarrow \text{AB}_5\text{H}$ (hydride)

Discharge $\beta\text{-Ni(OH}_{1-x})_2 + 2 \times \text{H}^+ + \text{e}^- \rightarrow \beta\text{-Ni(OH)}_2$ $\text{AB}_5 - 2 \times \text{H} \rightarrow \text{AB}_5 + 2 \times \text{H}^+ + \text{e}^-$

Total reaction of charge $\beta\text{-Ni(OH)}_2 + \text{AB}_5 \leftrightarrow -\text{Ni(OH}_{1-x})_2 + \text{AB}_5 - 2 \times \text{H}$. (solid solution)

Total reaction over - charge $\beta\text{-Ni(OH)}_2 + \text{AB}_5 \leftrightarrow \beta\text{-Ni(OH)}_2 + \gamma\text{-NiOOHAB}_5\text{H}$ (hydride)

What measures can promote the above reversible reaction smoothly? That is, it can increase the capacity of the battery, improve the performance of the battery, and vice versa. From the previous experimental research and analysis results, it is known that there is a very small amount of CoOOH, in the cathode material before storage, which is not enough to affect the dominant reaction of battery charge and discharge, so this effect is not observed in normal charge–discharge experiments and cycle life experiments; while in the process of storage, with the continuous increase of CoOOH, especially in discharge storage, the amount of CoOOH increases faster and more, CoOOH may be partially encapsulated on the surface of β-$Ni(OH)_2$ particles, which hinders the binding of β-$Ni(OH)_2$ to ions. This makes

$$\text{Charge } \beta\text{ - Ni(OH)}_2 + \text{AB}_5 \rightarrow -\text{Ni(OH}_{1-x})_2 + \text{AB}_5 - 2 \times \text{H. (solid solution)}$$
$$\text{Discharge } \beta\text{ - Ni(OH}_{1-x})_2 + 2 \times \text{H}^+ + \text{e}^- \rightarrow \beta\text{ - Ni(OH)}_2$$

It is hindered to a certain extent. In other words, the effective utilization rate of β-$Ni(OH)_2$ decreases in the process of charge and discharge, thus the capacity decreases and the internal resistance increases; and after the battery is stored, the higher the CoOOH content, the greater the effect, the greater the reduction of capacity, the greater the increase of internal resistance.

In addition, another effect in the process of storage is grain fining, the fining degree of discharge state storage is larger than that of charge state storage, and the fining degree of discharge state is 2.5 times higher than that of charge state. Grain refining is accompanied by more precipitation of CoOOH, and the new surface of the refined grain may also be partially wrapped by CoOOH. The dominant reversible reaction that hinders the battery charging and discharging process also increases, resulting in a decrease in capacity and an increase in internal resistance.

If nickel powder is added to the cathode as conductive agent, the grain fining, the precipitation of CoOOH can be reduced to a certain extent, and the activity of β-$Ni(OH)_2$ can be increased; thus, the storage performance of the battery can be improved [11].

The conclusions of this chapter are as follows:

(1) The attenuation of the storage performance of Ni-MH battery is the result of the double action of CoOOH precipitation and grain fining on the positive material during storage, and its effect is that the CoOOH partially coated on the surface of β-$Ni(OH)_2$, hinders the dominant reaction of battery charge and discharge. The more the content of CoOOH is, the finer the grain is; the greater the hindrance is, and the greater the decrease of capacity is, the greater the increase of battery internal resistance is.
(2) That attenuation of storage performance and the change of stacking fault probability of cathode materials during storage, and whether the negative electrodes produce corrosion and microstructure parameter. There is no corresponding relationship; it does not play an important role, at least within 20 days of storage.

(3) Among the two roles of the conductive network provided by CoOOH and the dominant reaction of CoOOH hindering the charge and discharge of the battery, the latter plays the main role, while the former plays a smaller role.

References

1. Lichtenbery, F., and K. Kleisorgen. 1996. Stability enhancement of the CoOOH Conductive network of Nickel hydroxide electroxide electrodes. *Journal of Power Sources* 62: 207–211.
2. Pralong, V., A. Delahaye-Vidal, B. Beandion, et al. 2000. Binucth-enhanad electrochemical stabilityof Cobalt hydroxide word ar an additive in Ni/Cd and Ni/Metal hydric batteries. *Journal of the Electrochemical Society* 147: 2096–2103.
3. Kaitani, Hideo, Michiyo, Akimoto, and Taishi Tani. 1995. ペースト-style water acidification of Nipsel electrodes for low-potential retention characteristics. *Electrochemical Chemistry* 63: 952–959.
4. Ji, Jin. 2000. Study on storage performance of MH/Ni battery. *Battery* 30(5): 219–220.
5. Bernard, P. 1998. Effects on the positive electrode of the corrosion of AB_5 alloys in nickel-metal-hydride batteries. *Journal of the Electrochemical Society* 145: 456–458.
6. Meli, F., T. Sakai, A. Zutlel, et al. 1995. Passivation behavior of AB_5 type hgdrogen storage alloy for battery electrode application. *Journal Alloys and Comp* 221: 284–290.
7. Wang, Z.X. 2003. Evaluation of long-term storage performance of MH/Ni battery. *Battery* 33: 368–370.
8. Ji, D.F., Y. Yang, and Z.G. Lin. 1998. Study on the decline of storage performance of metal hydride-nickel battery. *Current Technology* 22: 236–239.
9. Li, X.F., L.P. Ma, Y.W. Lou, et al. 2004. Storage performance of MH-Ni battery under discharge state condition. *Power Supply Technology* 28(6): 364–368.
10. Li, X.F., and B.J. Xia. 2004. Effect of charged state storage on MH-Ni battery. *Power Supply Technology* 28(12): 737–739.
11. Li, X.F., L.P. Ma, and B.J. Xia. 2004. Study on the improvement of storage performance of MH/Ni battery and its mechanism. *Electrochemistry* (10): 425–429.
12. Qin, P., Y.W. Lou, C.Z. Yang, and B.J. Xia. 2006. A new method and calculation program for separating multiple broadening effects of X-ray diffraction. *Acta Physica Sinica* 55(3): 1325–1335.
13. Yang, C.Z., and J. Zhang. 2008. Some progress in the study of microstructure of nanomaterials by X-ray diffraction. *Acta Physical Sinica* 28(3): 280–313.
14. Lou, Y.W., L.P. Ma, X.F. Li, B.J. Xia, and C.Z. Yang. 2006. XRD study on the microstructure of rare earth hydrogen storage alloy in MH/Ni battery. *Rare Metal Materials and Engineering* 35(3): 412–417.
15. Li, X.F. 2005. Study on storage performance of metal hydride-nickel battery. Doctoral paper of Shanghai Institute of Microsystems and Information Technology.

Chapter 18
Mechanism of Storage Process for Graphite/$LiCoO_2$ and Graphite/$Li(Ni_{1/3}Co_{1/3}Mn_{1/3})O_2$ Batteries

In order to study the general rules and general characteristics of the storage performance of lithium-ion battery, this chapter firstly uses 2H-graphite/$LiCoO_2$ and 2H-graphite/$Li(Ni_{1/3}Co_{1/3}Mn_{1/3})O_2$ system aluminum-plastic film packaging lithium-ion battery as the research object to study the effects rule of storage conditions on the storage performance, structures of positive and negative active materials, and diaphragm of lithium-ion battery. The selected storage conditions include: (1) different charged states of (SOC) (0, 25, 50, 75, and 100%) and (2) different temperatures (−20, 25, 55 °C).

18.1 Macro-observation of Graphite/$LiCoO_2$ Battery Before and After Storage [1, 2]

Figure 18.1 shows the picture of A-type battery before and after storage at 55 °C for 100 days in different charged states of graphite/$LiCoO_2$ battery. It can be seen that serious gas expansion occurs after 100% SOC storage, but gas expansion not obvious after 50% SOC and 0% SOC storage. This may be due to the serious side reaction in the battery after storage at high temperature and high charge, resulting in a large amount of gas. However, we observed that the apparent morphology of the battery did not change significantly after different charged states were stored at −20 and 25 °C for a longer time (65 weeks). This shows that the storage temperature also has a great influence on the storage of the battery.

C. Yang et al., *Materials and Working Mechanisms of Secondary Batteries*,
https://doi.org/10.1007/978-981-19-5955-4_18

Fig. 18.1 Photography of A-type graphite/LiCoO2 battery before and after storage at 55 °C for 100 days. **a** before storage; **b** after 0% SOC; C: 50% SOC; D: 100% SOC storage

18.2 Changes in Battery Performance Before and After Storage [1–4]

18.2.1 Change of 0.2C Capacity of Battery Before and After Storage

Figure 18.2 shows the variation of 0.2 C (240 mA) capacity of 2H-graphite/$LiCoO_2$ and 2H-graphite/$Li(Ni_{1/3}Co_{1/3}Mn_{1/3})O_2$ batteries stored at 55 °C in different charged states with storage days. It can be found that:

(1) The capacity of all batteries stored at high temperature (55 °C) decays with the increase of storage days.
(2) The capacity attenuation of the battery increases with the increase of the charge state during storage. The capacity decay rate of fully charged state (100% SOC) storage battery is significantly faster than that of other charged state batteries,

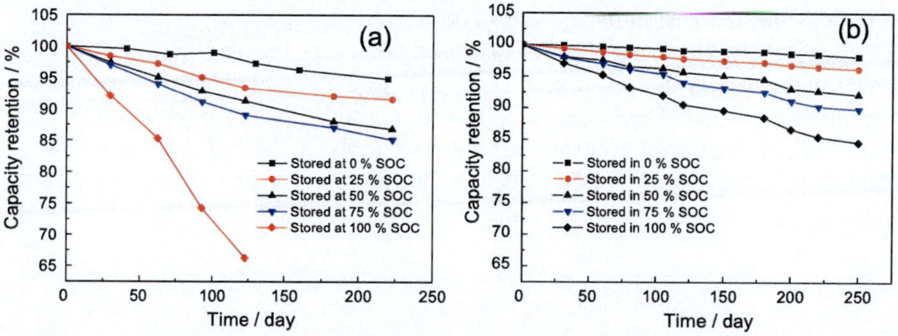

Fig. 18.2 0.2C (240 mA) capacity attenuation of A-type battery during storage at 55 °C with different charge states and time (days). **a** 2H-graphite/$LiCoO_2$ battery, **b** 2H-graphite/Li ($Ni_{1/3}Co_{1/3}Mn_{1/3}$)O_2 battery

and the attenuation is the most severe. Combination to Fig. 18.1, compared with other charged states, the internal side reaction of the battery stored in the fully charged state is more severe. At the same time, it is found that capacity attenuation of the battery gradually decreases with the extension of storage days, which is consistent with the report in the literature [1]. The reason is that the rate of side reaction causing capacity decay decreases with extension of storage time, but the fully charged storage battery does not follow this law.

Compared with the two kinds of batteries, the storage performance of graphite/Li($Ni_{1/3}Co_{1/3}Mn_{1/3}$)O_2 battery in full state is greatly improved. Figure 18.3 shows the relationship between the storage capacity decay rate and the charge degree (SOC). It can be seen that the graphite/Li($Ni_{1/3}Co_{1/3}Mn_{1/3}$)O_2 battery can greatly improve the storage performance with high charge degree. These changes can be explained by the thickening of the passive film on the electrode surface after storage. The passivation film on the electrode surface is composed of the reaction products of lithium and electrolyte, and the increase of its thickness will consume the effective lithium removal amount of the positive electrode, resulting in the attenuation of battery capacity [3]. At the same time, when Li^+ diffuses between the electrode and the electrolyte, it must pass through the passivation film on the electrode surface. It is proved that the diffusion rate of Li^+ in this film is lower than that in the electrode and electrolyte. Therefore, the diffusion of Li^+ in this film becomes the speed control step of the whole process. With the increase of the thickness of the film, the diffusion resistance of Li^+ increases, which leads to the increase of the impedance of the electrode and the polarization of the battery in the process of charge and discharge.

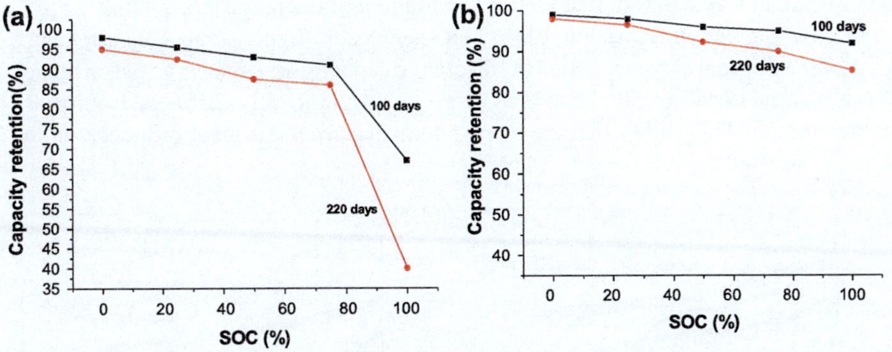

Fig. 18.3 Relationship between storage capacity decay rate and charge degree (SOC). **a** 2H-graphite/$LiCoO_2$ battery, **b** 2H-graphite/Li($Ni_{1/3}Co_{1/3}Mn_{1/3}$)O_2 battery

18.2.2 Changes in Battery Power Performance After Storage [1, 2, 4]

(1) **2H-graphite/$LiCoO_2$ battery** [2]

Figure 18.4 shows the 1C and 10C discharge curves of B-type 2H-graphite/$LiCoO_2$ battery stored at 55 °C for 100 days and at 25 and − 20 °C for 450 days. From Fig. 18.4a, it can be found that the discharge capacity of 1C and 10C of the battery decreased by 37.5% and 91.5%, respectively, after being stored at 55 °C in 100% SOC for 100 days. After storage in other charged states, the attenuation of 10C capacity of the battery was also significantly greater than that of its 1C capacity. At the same time, it can also be seen that the 10C discharge performance attenuation of the battery increases with the increase of the charge state after storage at 55 °C. After 100% SOC storage, the 10C discharge platform disappears, but after 0% SOC storage, the battery can still maintain a better 10C discharge performance, indicating that high temperature storage has a greater effect on the high current (10C) discharge capacity than low current (1C) discharge capacity. And the influence degree increases with the increase of the charged state during storage.

From Fig. 18.4b, it can be found that results are similar to those of the high temperature storage experiment mentioned above. Compared with 1C discharge performance, room temperature storage has a greater effect on the 10C discharge capacity of the battery, and with the increase of the charged state during storage, the attenuation of high current discharge capacity increases gradually. However, it is obvious that compared with high temperature storage, even if stored at room temperature for a longer time, the performance degradation of the battery is much smaller; at the same time, it is also found that after low temperature storage, except that the 10C capacity of the battery stored in 100% SOC decreases slightly (and only attenuates 2.5% of the original capacity), the 10C discharge capacity and discharge performance of the battery stored in 50 and 0% SOC are not significantly different from those before storage. This shows that the storage temperature has a great influence on the

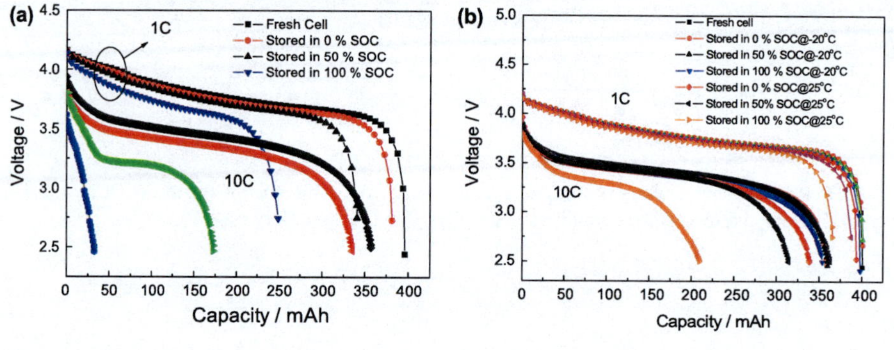

Fig. 18.4 1C and 10C discharge curves of 2H-graphite/$LiCoO_2$ type B battery before and after storage of **a** at 55 °C for 100d and **b** at – 20 and 25 °C for 450d at different charge states

high current discharge capacity of the battery after storage, which means that the storage at low temperature and low charge state is beneficial to the maintenance of the high current discharge capacity of the battery.

The direct current internal resistance (DCR) of the battery directly reflects the advantages and disadvantages of its power characteristics. The larger the DC internal resistance is, the worse the power performance of the battery is. The change of DC internal resistance of 2H-graphite/$LiCoO_2$ battery that has been detected during storage is shown in Fig. 18.5. From Fig. 18.5a, it can be found that the DC internal resistance of the battery shows an upward trend during 55 °C storage. Among them, the DC internal resistance of the battery after 100% SOC storage increases the fastest and the range is the largest. The DC internal resistance of the battery stored at 55 °C for 100 days increases by 5.2 times. The battery stored in 75% SOC takes the second place. The DC internal resistance of the battery stored at 55 °C for 100 days also increases nearly 3 times. The DC internal resistance of the battery stored below 50% SOC does not increase significantly. These rules are similar to the attenuation of 10C discharge capacity of the battery. Based on the law of 10C capacity attenuation and DC internal resistance increase during storage, it can be shown that the power performance attenuation of lithium-ion battery increases with the increase of charge state during storage.

At the same time, it can be found from Fig. 18.5b that similar to high temperature storage, the DC internal resistance of the battery storage at room temperature shows an upward trend, in which the DC internal resistance of the battery increases the largest after 100% SOC storage, and the DC internal resistance increases by 0.78 times after 450 days storage. However, compared with the battery stored at 55 °C with 100% SOC, the change of DC internal resistance of the battery is much slower during storage at room temperature. During the storage at – 20 °C, the DC internal resistance of the battery did not change obviously. Combined with the previous test results of high current discharge capacity, it can be shown that the storage temperature has a great influence on the power performance of the battery after storage, and low

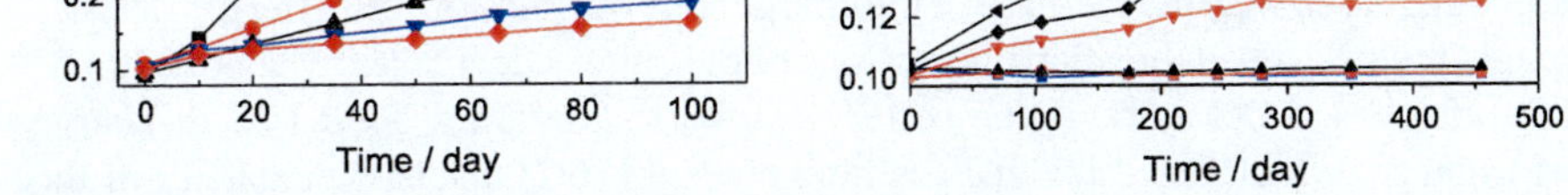

Fig. 18.5 During **a** storage at 55 °C and **b** storage at – 20 and 25 °C in different charged states. Change of DC internal resistance of type B battery of 2H-graphite/$LiCoO_2$ battery

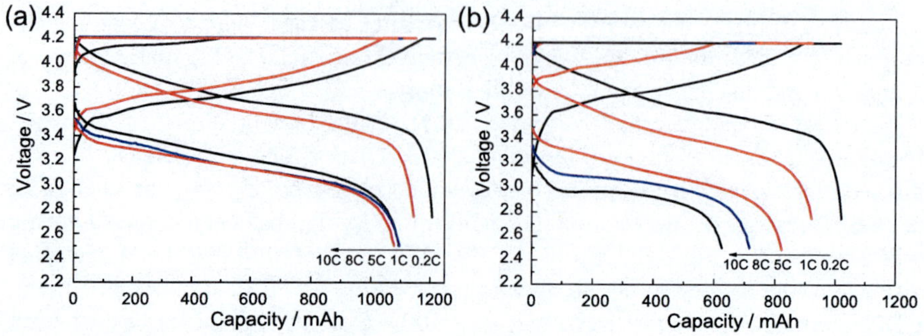

Fig. 18.6 0.2C–10C charge–discharge curve of graphite/$Li(Ni_{1/3}Co_{1/3}Mn_{1/3})O_2$ battery before (**a**) and after (**b**) 100% SOC storage at 55 °C 250 days

temperature and low SOC storage are beneficial to the maintenance of battery power performance.

(2) **2H-graphite/$Li(Ni_{1/3}Co_{1/3}Mn_{1/3})O_2$ battery** [4]
Figure 18.6 shows the 0.2C–10C charge–discharge curve of 2H-graphite/Li ($Ni_{1/3}Co_{1/3}Mn_{1/3}$)O_2 battery stored in 100% SOC for 250 days under 55 °C. It can be found from Fig. 18.6 that before storage, with the increase of the charge–discharge rate, the capacity of the battery gradually decreases, and the charge–discharge platform voltage difference increases, indicating that the battery polarization increases with the increase of the charge–discharge current. At the same time, it is found that when the current increases to 5C, the discharge curve has little difference, indicating that the rate performance of the battery is good.

After storage, the charge–discharge capacity of the battery attenuates, and the charge–discharge platform voltage difference increases at the same rate. At the same time, with the increase of the charge–discharge rate, the lag phenomenon is more obvious; after storage, the charge–discharge curves of 5C, 8C, and 10C were distinguished obviously, among which the 10C capacity and charge–discharge performance of the battery decreased the most. This confirms the conclusion of the previous chapter that storage has a greater adverse effect on the high current charge–discharge capacity of lithium-ion batteries.

Figure 18.7 shows the 10C discharge curve of graphite/Li ($Ni_{1/3}Co_{1/3}Mn_{1/3}$) O_2 battery stored for 250 days in different charge states at 55 °C. It can be seen that after storage 250 days at 55 °C, the 10C discharge performance attenuation of the battery increases with the increase of the charge state during storage. After 100% SOC storage, the 10C discharge platform voltage of the battery decreases by about 210mV, while after 0% SOC storage, the battery can still maintain a good 10C discharge performance, which indicates that the high current (10C) discharge capacity of the battery after storage increases with the increase of the charge state.

Figure 18.8 shows the comparison of DC internal resistance of 2H-graphite/Li ($Ni_{1/3}Co_{1/3}Mn_{1/3}$)O_2 battery and 2H-graphite/$LiCoO_2$ battery during storage at 55 °C

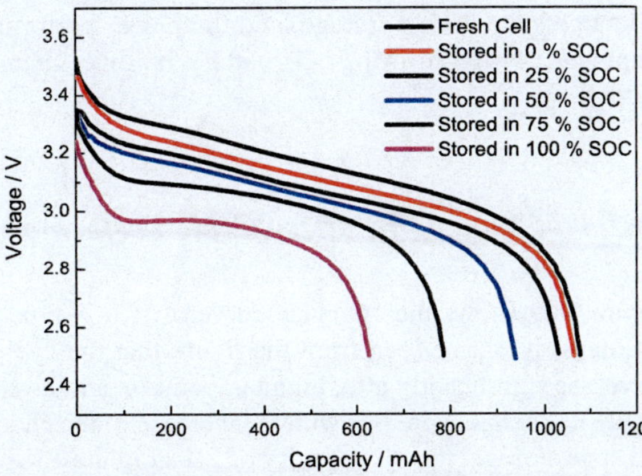

Fig. 18.7 Graphite/$Li(Ni_{1/3}Co_{1/3}Mn_{1/3})O_2$ stored in different charged states at 55 °C for 250 days. 10C discharge curve of battery

in different charged states. It can be found that the DC internal resistance of the battery showed an upward trend during storage, in which the DC internal resistance of the battery after 100% SOC storage increased the highest; after stored at 55 °C for 250 days, the DC internal resistance increased by 0.9 times. The battery stored in 75% SOC takes the second place; after storage under the same conditions, the DC internal resistance of the battery also increased by nearly 0.6 times, and the DC internal resistance of the battery stored lower than that of 50% SOC did not increase significantly. This rule is similar to the change of DC internal resistance of graphite/$LiCoO_2$ battery after storage, but by comparison, we can see as shown in Fig. 18.8, under the same storage conditions, but the increase rate of DC internal resistance of graphite/$Li(Ni_{1/3}Co_{1/3}Mn_{1/3})O_2$ is obviously slowed down.

Combining the law of 10C capacity attenuation and DC internal resistance increase of the two kinds of batteries during storage, it can be shown that the power performance attenuation of lithium-ion battery increases with the increase

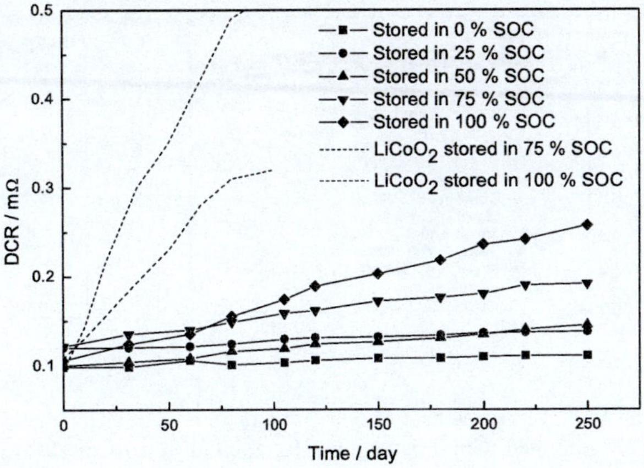

Fig. 18.8 Comparison of DC internal resistance of 2H-graphite/$Li(Ni_{1/3}Co_{1/3}Mn_{1/3})O_2$ and 2H-graphite/$LiCoO_2$ batteries stored in different charged states at 55 °C

of charge state during storage, and the power performance attenuation rate of 2H-graphite/$Li(Ni_{1/3}Co_{1/3}Mn_{1/3})O_2$ battery is lower than that of 2H-graphite/$LiCoO_2$ battery.

18.2.3 Change of Battery Cycle Performance After Storage

Figure 18.9 shows the 1C cycle curve of two A-type batteries stored at 55 °C for 100 days. It can be seen from the figure that the cycle performance of the battery decreases significantly after high temperature and high charge state storage, and the degree of decline increases with the increase of the charge state during storage. After full state storage, the 1C capacity of 2H-graphite/$LiCoO_2$ battery decays to 75% of its original capacity after 50 charge-discharge cycles, and the capacity of the battery stored in 50% SOC decreases to 89% of its original capacity after 50 cycles.

After the discharge state was stored at high temperature, the cycle performance of the battery did not decrease, on the contrary, it was slightly better than the new battery, and its capacity decreased by only about 1% after 50 cycles.

The relationship between the cycle performance retention rate and the charge degree (SOC) of the two batteries stored in 55 °C for 100 days is shown in Figs. 18.10a and b, respectively. By comparison, it is known that decreases ratio of the cycle performance of 2H-graphite/$Li(Ni_{1/3}Co_{1/3}Mn_{1/3})O_2$ battery. The decline rate is much smaller than that of 2H-graphite/$LiCoO_2$ battery. The slopes of the three lines in Fig. 18.10 are − 0.212, − 0.0312 (50 cycles), and − 0.0732 (150 cycles); it can be seen that the capacity decreasing rate of graphite/$Li(Ni_{1/3}Co_{1/3}Mn_{1/3})O_2$ battery cycle 500 period is much smaller than that of 2H-graphite/$LiCoO_2$ battery cycle 50 period.

The study believes that the cycle life of lithium-ion battery is closely related to the quality of solid electrolyte interface (SEI) film on the surface of the electrode

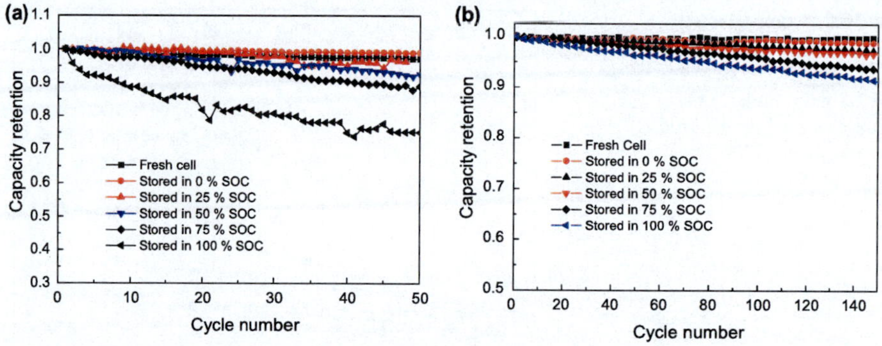

Fig. 18.9 1C cycle curve of type A battery stored in different charge states at 55 °C for 100 days. **a** 2H-graphite/$LiCoO_2$ battery, **b** 2H-graphite/$Li(Ni_{1/3}Co_{1/3}Mn_{1/3})O_2$ battery

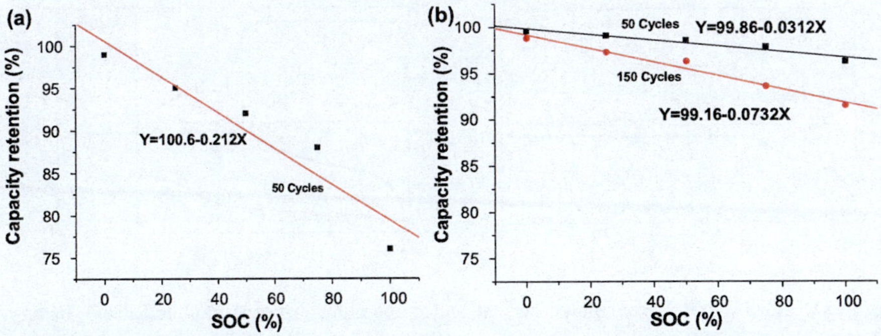

Fig. 18.10 Relationship between cycle performance retention and degree of charge (SOC) after 55 °C storage for 100 days. **a** 2H-graphite/$LiCoO_2$ battery, **b** 2H-graphite/$Li(Ni_{1/3}Co_{1/3}Mn_{1/3})O_2$ battery

(especially the negative electrode). The poor stability of SEI film will lead to its structural re-organization in the process of battery cycle, resulting in the consumption of reversible de-intercalated lithium ion, resulting in the capacity attenuation of the battery during the cycle. Therefore, after high temperature and high charge state storage, the stability of the passivation film on the electrode surface not only increases, but the thickness of the film also decreases. At the same time, the surface properties of the electrode are more stable in low charge state storage.

18.2.4 Resistance to Overcharge and Thermal Stability

Figure 18.11 shows the 3C–5 V overcharge voltage, current, and temperature curves of graphite/$LiCoO_2$ battery before and after storage at 100% SOC 55 °C for 100 days. Figure 18.12 shows 3C–5 V overcharge voltage, current, and temperature curve of graphite/$Li(Ni_{1/3}Co_{1/3}Mn_{1/3})O_2$ battery before and after 100% SOC 55 °C storage 250 days. The experimental results of other storage conditions are summarized and Tables 18.1a and b, respectively. There are four stages in Fig. 18.11.

Stage A: Lithium atoms are irreversibly detached from the $LixCoO_2$ positive electrode and deposited on the surface of the carbon negative electrode to form metal lithium. The battery voltage increases slowly with the increase of the amount of lithium removal from the positive electrode, and the battery temperature increases slowly.

Stage B: The decomposition exothermic reaction of the electrolyte occurs on the positive and negative electrodes, and the temperature of the battery increases rapidly. At this time, the fluctuation of the charging current results from the decomposition reaction of the electrolyte, because after the battery enters the constant voltage charging stage, the charging current will gradually decrease due to the decrease of

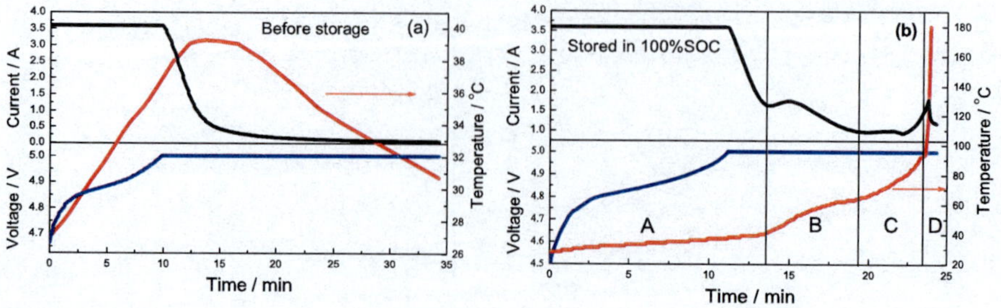

Fig. 18.11 3C–5 V overcharge voltage, current, and temperature curves of graphite/$LiCoO_2$ battery before and after storage 100 days at 55 °C

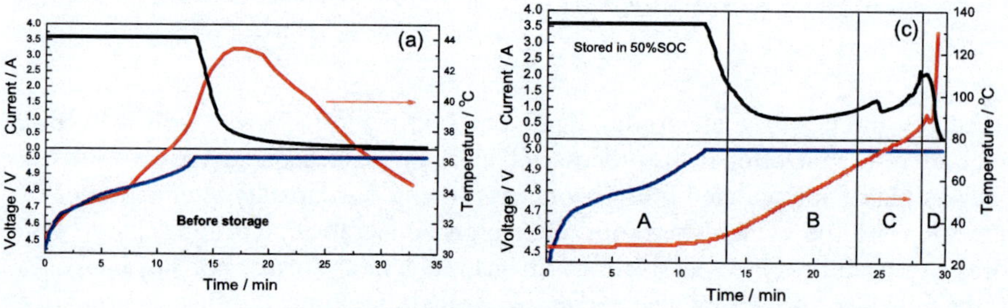

Fig. 18.12 3C–5 V overcharge voltage, current, and temperature curve of graphite/$Li(Ni_{1/3}Co_{1/3}Mn_{1/3})O_2$ battery before and after storage 250 days at 55 °C

Table 18.1 **a** 3C–5 V overcharge test results of graphite/$LiCoO_2$ battery stored at 55 °C for 100d. **b** 3C–5 V overcharge test results of graphite/$Li(Ni_{1/3}Co_{1/3}Mn_{1/3})O_2$ battery stored at 55 °C for 250d

(a)						
SOC/%	**Before store**	**0**	**25**	**50**	**75**	**100**
Maximum temperature/°C	32	42	58	180	195	213
Results	Pass	Pass	Pass	Burning	Burning	Burning
(b)						
Maximum temperature/°C	35	45	52	54	185	205
Results	Pass	Pass	Pass	Pass	Burning	Burning

the amount of lithium removed from the positive electrode, but at this time, the electrolyte is decomposed. The reaction will provide extra electrons to the electrode, so that the current stops falling and fluctuates.

Stage C: The impedance of the battery will increase with the increase of the amount of lithium removal from the positive electrode and the consumption of electrolyte, and the Joule heat generated by the current through the battery will increase, resulting in an increase in battery temperature.

Stage D: When the battery temperature rises to the decomposition temperature of the negative SEI film (about 100C), the negative SEI film disappears. The structure of highly de-lithium $Li_{1-x}CoO_2$ collapses and releases active oxygen. The highly active lithium-intercalated carbon negative electrode and the metal lithium deposited on the surface of the negative electrode react violently with the electrolyte in the presence of reactive oxygen species. A severe exothermic reaction occurs, cause the battery thermal to get out of control.

The comparison can know that there is no difference in the overcharge resistance of the two batteries before storage, but after storage, the overcharge resistance of graphite/$Li(Ni_{1/3}Co_{1/3}Mn_{1/3})O_2$ battery is better than that of graphite/$LiCoO_2$ battery, no matter from the highest temperature after overcharge or the final result.

The curve of the relationship between the self-heating rate (SHR) and temperature of the graphite/$LiCoO_2$ battery and the graphite/Li $(Ni_{1/3}Co_{1/3}Mn_{1/3})O_2$ battery is shown in Fig. 18.13. It can be seen that the results of the two batteries are similar and the thermal stability decreases, mainly due to the decrease of the initial temperature of the self-exothermic reaction in the battery, the increase of the self-heating rate of the battery, and the decrease of thermal stability increases with the increase of the charge state during storage. Among them, the variation curve of self-heating rate with temperature after 0% SOC storage is close to that of before storage battery, indicating that the thermal stability of battery after 0% SOC storage is reduced, but the decrease is not significant; after 50% and 100% SOC storage, the thermal stability of the battery decreased greatly, the initial temperature of the self-exothermic reaction decreased by 6 and 19 °C, respectively, and the self-heating rate increased obviously with the increase of temperature. At the same time, compared with the self-heating curve of graphite/$LiCoO_2$ battery stored at 55 °C for 100 days, it can be seen that although the storage time is longer, the decrease of thermal stability of graphite/Li $(Ni_{1/3}Co_{1/3}Mn_{1/3})O_2$ battery is much smaller after storage, which indicates that the effect of storage on the thermal stability of graphite/$Li(Ni_{1/3}Co_{1/3}Mn_{1/3})O_2$ battery is less than that of graphite/$LiCoO_2$ battery.

Combined with the previous battery overcharge experimental results, it can be shown that the safety of graphite/$Li(Ni_{1/3}Co_{1/3}Mn_{1/3})O_2$ battery decreases after storage, and it is more obvious with the increase of charge state during storage, but the safety retention ability of graphite/$Li(Ni_{1/3}Co_{1/3}Mn_{1/3})O_2$ battery is stronger than that of graphite/$LiCoO_2$ battery [5].

18.3 Fine Structure Changes of Battery Positive and Negative Active Materials Before and After Storage [1]

In order to study the changes of the microstructure and surface morphology of the electrode active material during storage, the new battery and the battery stored under different conditions were discharged to 2.75 V at 0.2C, making the battery at 0%

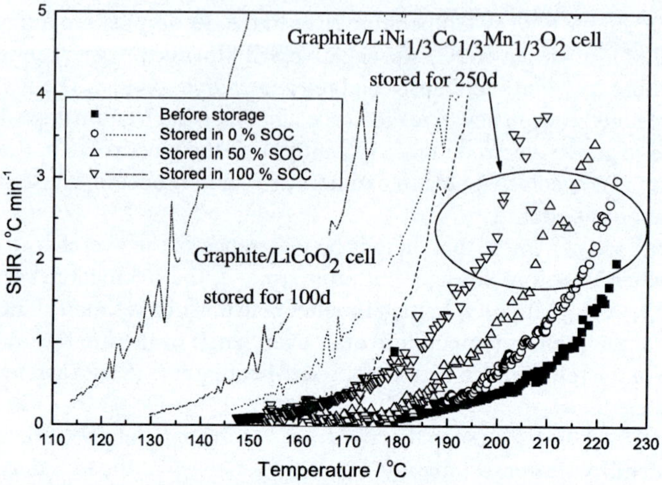

Fig. 18.13 Relationship between self-heating rate and temperature of graphite/$LiCoO_2$ battery and graphite/Li $(Ni_{1/3}Co_{1/3}Mn_{1/3})O_2$ battery before and after storage at 55 °C

SOC. In the glove box filled with argon, the battery was dissected, several positive and negative plates were removed, washed repeatedly with DMC solvent, and dried naturally. The positive and negative active materials were scraped off respectively, and the X-ray diffraction analysis was carried out on the Rigaku D/max-2200PC X-ray diffractometer. The contents of the analysis included phase structure identification, lattice parameters, microstructure, and so on. The surface morphology of the residual electrode was observed on the field emission scanning electron microscope (FE-SEM) of Shimadza Company, Japan.

18.3.1 Microstructural Changes of Negative Materials Before and After Storage

Figures 18.14a, b are the XRD patterns of negative active materials before and after storage of different charge (SOC) at 55 °C for 2H-graphite/$LiCoO_2$ battery and 2H-graphite/$Li(Ni_{1/3}Co_{1/3}Mn_{1/3})O_2$, respectively.

The lattice parameters, cell volume, microcrystal, micro-strain and stacking fault of 2H-graphite, the negative electrode active material of graphite/$LiCoO_2$ and graphite/$Li(Ni_{1/3}Co_{1/3}Mn_{1/3})O_2$ batteries are calculated according to the method described in Chap. 8. The results are listed in Tables 18.2 and 18.3, respectively. It can be found that the test results of the two kinds of batteries are similar, and the lattice parameters of the negative active material 2H-graphite have no obvious change after storage. At the same time, except for the slight increase of micro-strain after storage, there is no obvious change in microcrystals and stacking faults, indicating

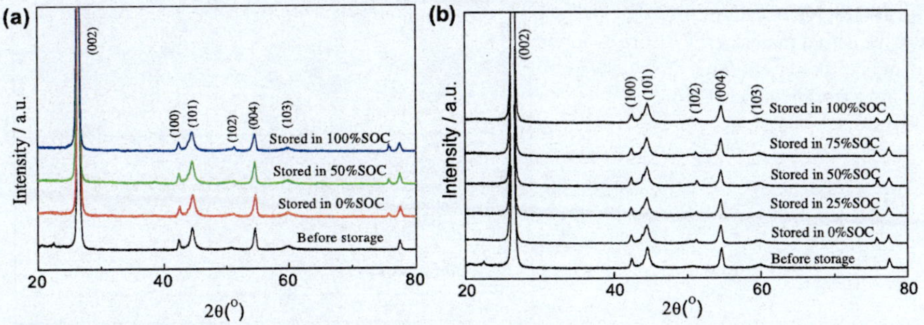

Fig. 18.14 XRD patterns of negative active materials before and after storage at 55 °C. **a** graphite/$LiCoO_2$ battery before and after storage for 100 days. **b** graphite/$Li(Ni_{1/3}Co_{1/3}Mn_{1/3})O_2$ before and after storage for 250 days

that storage has no significant effect on the bulk phase structure and microstructure of negative active material graphite. This once again shows that the decline in the storage performance of lithium-ion batteries is due to the positive electrode.

Table 18.2 Structure parameters of 2H-graphite as anode material before and after storage of graphite/$LiCoO_2$ battery in different charged states at 55 °C

		Before storage	0% SOC after storage	50% SOC after storage	100% SOC after storage
Lattice parameter/Å	a	2.4627	2.4538	2.4527	2.4517
	c	6.7207	6.6960	6.7196	6.7219
Microstructure parameter	D (nm)	52.9	75.7	111.8	65.8
	ε ($\times 10^{-3}$)	0.983	2.350	2.404	2.400
	P (%)	40.0	45.6	52.2	43.4

Table 18.3 Structure parameters of negative electrode active material 2H-graphite before and after storage of graphite/$Li(Ni_{1/3}Co_{1/3}Mn_{1/3})O_2$ battery with different charged states at 55 °C

	Before storage	0% SOC after storage	25% SOC after storage	50% SOC after storage	75% SOC after storage	100% SOC after storage	
Lattice parameter Å	a	2.4496	2.4486	2.4534	2.4520	2.4650	2.4528
	c	6.7115	6.7233	6.7338	6.7256	6.7213	6.7331
Microstructure parameter	D (nm)	84.5	73.8	68.4	88.7	79.5	95.8
	ε ($\times 10^{-3}$)	1.087	2.051	2.115	2.312	2.089	1.554
	P (%)	45.1	54.2	45.9	53.8	51.2	48.7

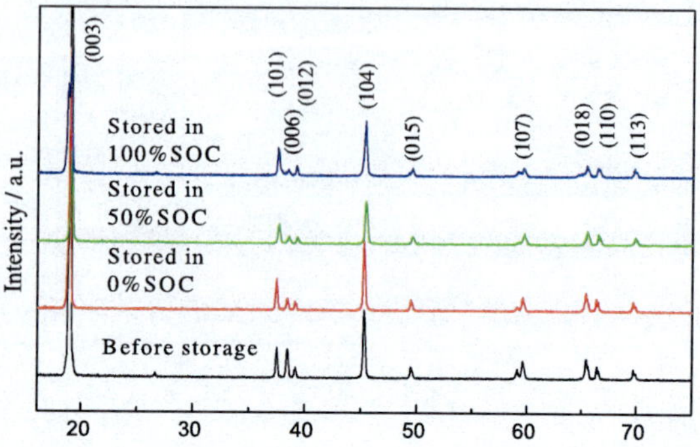

Fig. 18.15 XRD pattern of positive active material $LiCoO_2$ of battery before and after storage at 55 °C for 100 days

18.3.2 Microstructural Changes of Positive Material $LiCoO_2$ Before and After Storage [2]

Figure 18.15 shows the XRD pattern of the positive active material $LiCoO_2$ before and after storage at 55 °C for 100 days. It can be seen from the figure that there is no obvious change in the diffraction peak position of the positive active material before and after storage, but it still clearly shows the characteristic peak of $LiCoO_2$, but the relative intensity of some peaks has changed. For example, the intensity of the (006) peak is very high before storage, but its intensity decreases greatly after high temperature storage, which indicates that the phase structure of the material does not change after storage. At the same time, it is found that some of the XRD peaks (such as 003,101) of the positive active material $LiCoO_2$ are broadened after high temperature and high charge storage. At the same time, it can be found that the X-ray diffraction peak shape and position of the positive active material are almost the same before and after storage, indicating that the bulk phase structure does not change after storage, but its fine structure changes. The half-height widths and calculated results of (003), (101), and (104) are given in Tables 18.4 and 18.5, respectively. It can be seen that the crystallite size decreases greatly with the increase of SOC, while the micro-strain changes from tensile strain to compressive strain after storage, and the compressive strain increases with the increase of SOC.

18.3.3 Microstructural Changes of Positive Material $Li(Ni_{1/3}Co_{1/3}Mn_{1/3})O_2$ Before and After Storage [4]

Figure 18.16 is the XRD pattern of the positive active material $Li(Ni_{1/3}Co_{1/3}Mn_{1/3})O_2$ before and after storage at 55 °C for 250 days with different charged (SOC). It can be seen from the figure that there is no obvious change in the diffraction peak position of the positive active material before and after storage, but the characteristic peak of $Li(Ni_{1/3}Co_{1/3}Mn_{1/3})O_2$ is still clearly shown, which indicates that there is no new

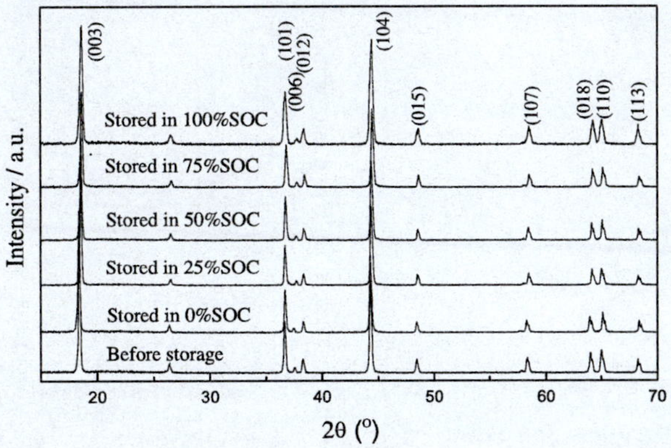

Fig. 18.16 XRD pattern of positive active material $Li(Ni_{1/3}Co_{1/3}Mn_{1/3})O_2$ of battery before and after storage at 55 °C for 250 days

Table 18.4 Half peak width (o) of positive active material $LiCoO_2$ (003), (101), and (104) before and after storage

hkl		Before storage	0% SOC after storage	50% SOC after storage	100% SOC after storage
003	Half peak width (°)	0.134	0.161	0.206	0.364
101		0.146	0.148	0.196	0.173
104		0.153	0.154	0.195	0.204

Table 18.5 Lattice parameters, microcrystals, and micro-strain of positive active material $LiCoO_2$ before and after storage in different charged states

		Before storage	0% SOC after storage	50% SOC after storage	100% SOC after storage
Lattice parameters(Å)	a	2.8156	2.8126	2.8119	2.8123
	c	14.0159	13.9975	13.9604	14.0072
Microstructure parameter	D (nm)	336.1	116.9	67.2	18.4
	ε ($\times 10^{-3}$)	0.269	– 0.187	– 0.352	– 3.447

phase in the positive active material after storage, and its matrix phase structure does not change. However, after careful observation, it is found that the relative intensity of the diffraction peak of the sample seems to have changed slightly before storage, and some of the diffraction peaks have been broadened, which indicates that the fine structure of the sample may have changed after storage. The calculation results of the microstructure parameters are given in Table 18.6. It can be seen that the law of the microstructure parameters is similar to that of $LiCoO_2$, but the change rate slows down obviously.

It is found that for the positive active materials of lithium-ion battery containing Ni (such as $LiNiO_2$, $Li(Ni_{1-x}Co_x)O_2$ and $Li(Ni_{1-x-y}Co_xMn_y)O_2$, etc.), the cationic mixed arrangement phenomenon is easy to occur in these materials, because of

Table 18.6 Lattice parameters of the $Li(Ni_{1/3}Co_{1/3}Mn_{1/3})O_2$ battery before and after 250 days storage at different charge states at 55 °C and microstructure parameters

Graphite/$LiNi_{1/3}Co_{1/3}Mn_{1/3}O_2$ battery		Before storage	0% SOC after storage	25% SOC after storage	50% SOC after storage	75% SOC after storage	100% SOC after storage
Lattice parameters(Å)	a	2.8610	2.8665	2.8661	2.8685	2.8643	2.8689
	c	14.2736	14.2720	14.3151	14.3127	14.4502	14.4196
Microstructure parameter	D (nm)	239.2	208.7	182.6	165.5	122.3	95.4
	ε ($\times 10^{-3}$)	0.2367	− 0.1367	− 0.1439	− 0.2984	− 0.468	− 0.9641

the small difference between the atomic radii of Li and Ni, that is, some Li at 3a position and Ni at 3b position occur site conversion, once Li atom enters 3b site, it loses its de-intercalation activity. Therefore, the mixed arrangement of cations directly disturbs the detachment and intercalation of Li atom in the process of charge and discharge, resulting in the decline of the electrochemical performance of the materials [6]. It is considered that the phenomenon of cationic mixed discharge may occur in the synthesis stage and charge–discharge cycle stage of the material, and this phenomenon may also occur in the storage stage of the battery due to the thermal vibration of lithium-nickel and the spontaneous re-intercalation of lithium atoms. Therefore, it is necessary to determine the crystallographic occupation of Li and Ni in $Li(Ni_{1/3}Co_{1/3}Mn_{1/3})O_2$ before and after storage.

At present, based on the comprehensive study and analysis of the main diffraction characteristics of this kind of Ni-containing materials, we propose a new simulation method, that is, according to the addition and subtraction relationship of the contribution of 3a and 3b atoms to the diffraction intensity of each hkl crystal plane, we propose to use the diffraction integral intensity ratio of additive and subtractive lines to study the crystallographic occupation of Li and Ni in this kind of materials. For detailed derivation and explanation, see Refs. [6–8] and Chap. 5. The method for calculating the mixed occupancy parameter x of $Li(Ni_{1/3}Co_{1/3}Mn_{1/3})O_2$ is briefly introduced below.

According to the mixed space occupying model of $Li_{1-x}Ni_x$ $(Li_xNi_{1/3-x}Co_{1/3}Mn_{1/3})$ O_2, the curves of the relationship between the strength ratio I_{104}/I_{003}, I_{012}/I_{101}, I_{104}/I_{101} or $(I_{003}/I_{104})^{1/2}$, $(I_{101}/I_{012})^{1/2}$, $(I_{101}/I_{104})^{1/2}$ and the mixed occupying parameter x are simulated with the help of Powder Cell1.0 program, as shown in Fig. 18.17.

It can be seen from Fig. 18.17 that the relationship between $(I_1/I_2)^{1/2}$ and the mixed occupying parameter x is linear, and the linear relationship is analyzed as a linear equation listed in Table 18.7. From the diffraction integral intensity ratio measured by the experiment, the corresponding mixed occupying parameter x can be calculated from the linear relations of $(I_{003}/I_{104})^{1/2}$, $(I_{101}/I_{012})^{1/2}$, and $(I_{101}/I_{104})^{1/2}$.

It should be pointed out here that the value of I_{104}/I_{003} is relatively sensitive to the z value, so it is required to confirm the z value first. According to our experimental results, z=0.24 is more practical for $Li(Ni_{1/3}Co_{1/3}Mn_{1/3})O_2$; the ratio of I_{104}/I_{101} is

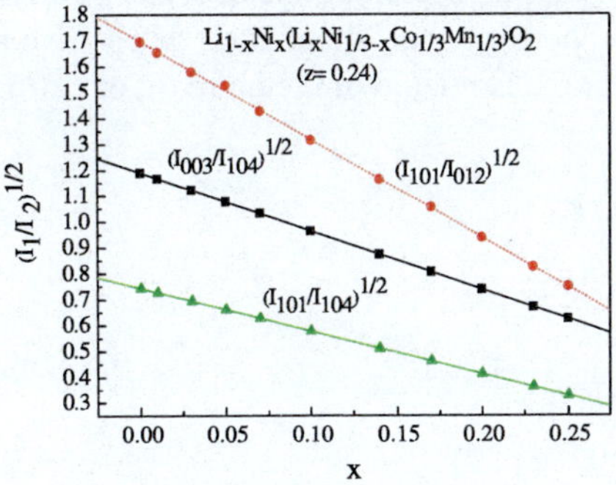

Fig. 18.17 Curve of the relationship between the $(I_1/I_2)^{1/2}$ of $Li(Ni_{1/3}Co_{1/3}Mn_{1/3})O_2$ and the mixing occupying parameter x

Table 18.7 Linear equations of $(I_1/I_2)^{1/2}$ for $Li(Ni_{1/3}Co_{1/3}Mn_{1/3})O_2$ materials and x

$z = 0.24$	$z = 0.25$
$(I_{003}/I_{104})^{1/2} =$ 1.2093–2.2062x	$(I_{003}/I_{104})^{1/2} =$ 1.0630–2.1693x
$(I_{101}/I_{012})^{1/2} =$ 1.6435–3.6632x	$(I_{101}/I_{012})^{1/2} =$ 1.7109–3.4944x
$(I_{101}/I_{104})^{1/2} =$ 0.7586–1.6055x	$(I_{101}/I_{104})^{1/2} =$ 0.7840–1.6016x

not sensitive to z value, that is, when the z value of 6c seat cannot be determined, the measured data of I_{104}/I_{101} can be used to solve x; the value of I_{012}/I_{101} is not also insensitive to z value, but because I_{012} is relatively too weak, the reliability of result is poor. It can not be solved because the measured I_{012}/I_{101} < critical value (the critical value of z=0.24 is 0.351, and the critical value of z=0.25 is 0.321). Therefore, when making relativity measurement, the I_{101}/I_{104} or $(I_{101}/I_{104})^{1/2}$ data of 101–104 line pairs are more convenient and reliable and avoid the influence of the abnormal enhancement of 003 diffraction intensity due to the abnormal preferred orientation and the uncertainty of Z value in 6c position, so the following formula is used to calculate the mixed space occupying parameters.

$$(I_{101}/I_{104})^{1/2} = 0.7586 - 1.6055x \tag{18.1}$$

Therefore, the $K\alpha_2$ component was removed by Jade program, and then the XRD spectra were fitted by Fitting all line and Refine, and the I_{101} and I_{104} of each sample were obtained, which are listed in Table 18.8. It can be found from the table that the mixed occupation parameter of $Li(Ni_{1/3}Co_{1/3}Mn_{1/3})O_2$ before storage is very small, indicating that the synthesis process of this material is excellent, so the performance of this material is also superior from the previous experiments. After storage, the mixed occupying parameters of the materials increase with the increase of SOC. This shows that the mixed arrangement of cations occurs in the positive active material $Li(Ni_{1/3}Co_{1/3}Mn_{1/3})O_2$ after storage, and the amount of lithium-nickel atoms mixed arrangement increases with the increase of the charge state of the battery during storage. This phenomenon can correspond to the decline of the performance of the positive electrode and the whole battery after storage. The results show that the performance attenuation of positive active material $Li(Ni_{1/3}Co_{1/3}Mn_{1/3})O_2$ after storage is related to the increase of cationic mixed occupying.

Table 18.8 (I_{101}/I_{104}) of positive active material $Li(Ni_{1/3}Co_{1/3}Mn_{1/3})O_2$ before and after storage and calculated mixed space occupying factor x

	Before storage	0% SOC after storage	25% SOC after storage	50% SOC after storage	75% SOC after storage	100% SOC after storage
I_{101}	4964	4967	5247	5129	5367	13,277
I_{104}	8910	9151	10,086	10,529	11,549	31,423
I_{101}/I_{104}	0.5571	0.5428	0.5202	0.4871	0.4647	0.4225
X	0.0076	0.0136	0.0233	0.0378	0.0479	0.0676

18.4 Surface Morphology and Composition of Positive and Negative Electrodes Before and After Storage [1, 2, 4]

18.4.1 Surface Morphology of Positive and Negative Electrodes of Graphite/$LiCoO_2$ Battery Before and After Storage [1, 2]

Figure 18.18 shows the SEM photographs of the positive and negative active materials of each graphite/$LiCoO_2$ battery before and after storage100 days at 55 °C. It is considered that [4] after the activation of lithium-ion battery, a layer of electronically insulated and ion-conductive solid electrolyte interface (or SEI) film will be formed on the negative surface of lithium-ion battery. It is mainly formed by the reduction and decomposition of the electrolyte on the negative electrode surface, including inorganic substances (such as Li_2CO_3, LiF), organic compounds (such as $ROCO_2Li$, ROLi), and some polymers. The properties and quality of the film directly affect the charge-discharge performance and safety of the electrode. It is also considered that there is a similar passivation film on the surface of the cathode after long-term immersion in the electrolyte. The formation mechanism is mainly the nucleophilic reaction of the positive active material with the organic solvent of the electrolyte and the reaction of the positive active material with trace water or acid in the solution. Compared with the negative SEI film, the thickness of the passivation film on the positive surface is two orders of magnitude smaller. Compared with these photos, it can be found that the passivation film on the surface of the positive and negative electrodes of the battery has changed obviously after being stored in different charged states. Before storage, the surface of positive $LiCoO_2$ particles is relatively smooth, and the edge of the particles is clearly visible, but after storage, a thicker passivation film is formed on the surface of the particles. The passivation film thickeness increases with the increase of the charge state of the battery during storage. After discharge storage, the film is discontinuous speckled; after 50% SOC storage, it is denser granular; after 100% SOC storage, it becomes a thick layer covering the entire positive active material particles. There is a similar phenomenon of passive film thickening in the negative electrode. At the same time, it can be found that the passivation film thickness of the negative electrode increases with the increase of the storage charge state of the battery. This phenomenon is consistent with the results of Edstrom et al. Due to the structural defects on the electrode surface and the deviation of activation conditions, the SEI film on the electrode surface is usually not dense and there are microcracks, which cannot completely protect the electrode. The electrolyte will pass through these microcracks and continue to react with the electrode active materials. Therefore, with the extension of the reaction time, the thickness of the passive film on the electrode surface will continue to increase. The stronger the oxidizability of the positive electrode and the reducibility of the negative electrode, the stronger the reaction activity with the electrolyte, which is characterized by serious gas expansion of the battery (as shown in Fig. 18.1). Therefore, after the battery is stored in a high charge state, the thickness increase of passive film on the electrode surface is more obviously.

In order to further study the changes of the surface composition of the positive active material before and after storage, the EDS analysis of the electrode was carried out. The results of composition analysis are listed in Table 18.9. It can be found that there is no element F on the positive surface of the battery before storage, but with the increase of the storage charge state, the content of F on the surface increases gradually, especially after 100% SOC storage, the atomic ratio of F on the positive surface reaches 13.23%. It is considered that the F element on the cathode surface is likely to come from the LiF generated on the cathode surface during storage process. Because the electrolyte salt of the battery is $LiPF_6$, solution of $LiPF_6$ will occur during the long-term storage of the battery. It will decompose during the long-term storage of the battery as shown in Formula (18.2)

$$LiPF_6 \rightarrow LiF + PF_5 \tag{18.2}$$

$$PF_5 + H_2O \rightarrow PF_3O + 2HF \tag{18.3}$$

produces LiF and PF_5, later in turn reacts with trace water in the electrolyte to form HF, as a cathode material for alkaline oxides, which is easily eroded by HF to form LiF. It can be seen that the amount of LiF on the positive surface increases with the increase of the charge state during storage. Because LiF is the product of side reactions on the electrode surface, these side reactions lead to the performance degradation of the battery during storage, which also explains that the performance degradation of the battery increases with the increase of the charge state of the battery during storage.

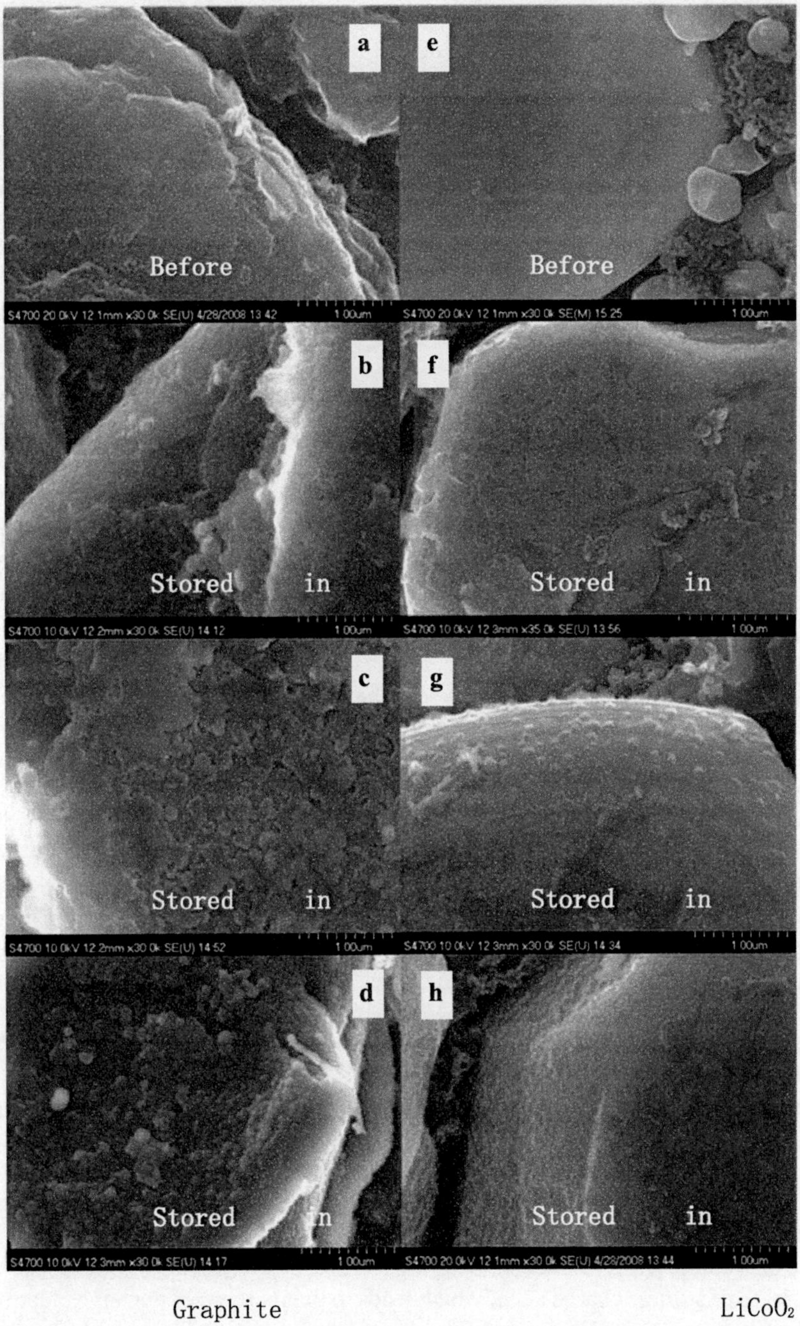

Fig. 18.18 SEM photographs of positive and negative active materials of graphite/$LiCoO_2$ battery before and after storage 100 days at 55 °C. **a** negative electrode before storage; **b** negative electrode after 0% SOC storage; **c** negative electrode after 50% SOC storage; **d** negative electrode after 100% SOC storage; positive electrode before **e** storage; **f** positive electrode after 0% SOC storage; **g** positive electrode after 50% SOC storage; **h** 100% SOC

Table 18.9 Surface composition of positive active materials for graphite/$LiCoO_2$ batteries obtained from EDS before and after storage

	Composition/(at %)			
	Co	O	C	F
Fresh battery	26.45	50.97	22.58	0
0% SOC storage 100 days	20.1	49.55	27.88	2.48
50% SOC storage 100 days	20.12	43.85	29.18	6.79
100% SOC storage 100 days	18.72	39.03	28.02	13.23

18.4.2 Surface Morphology and Composition of Positive and Negative Electrodes for Graphite/$Li(Ni_{1/3}Co_{1/3}Mn_{1/3})O_2$ Battery Before and After Storage

Figure 18.19 shows the SEM photographs of the positive and negative active materials of each battery before and after storage 250 days at 55 °C. EDS analysis was carried out with positive electrodes. The results of the analysis are listed in Table 18.10. As can be seen from Table 18.10. In the before storage, the surface of the positive electrode does not contain any other elements; F is detected on the surface of the positive electrode after storage. It can be seen that the content of F increases obviously with the increase of SOC. This is very similar to the results of the above-mentioned $LiCoO_2$. This kind of substance should be the decomposition product of $LiPF_6$ during battery storage, and the product LiF formed after HF erodes the positive active material. The electron and lithium-ion conduction capacity of LiF is very poor, which leads to the increase of AC impedance of the positive electrode after storage. At the same time, it can be found that compared with $LiCoO_2$, the amount of F formed on the surface of, $Li(Ni_{1/3}Co_{1/3}Mn_{1/3})O_2$ is much less than that of $LiCoO_2$ even after longer storage, which indicates that, $Li(Ni_{1/3}Co_{1/3}Mn_{1/3})O_2$ is more stable than $LiCoO_2$ in electrolyte, under the same storage conditions, which is one of the reasons for the better performance of graphite/$Li(Ni_{1/3}Co_{1/3}Mn_{1/3})O_2$ batteries after storage.

It can also be found from Fig. 18.19 that the surface texture of negative graphite particles is clearly visible before storage; after 0% SOC storage, the surface morphology of graphite particles changed little, but with the increase of storage charge state, the surface of graphite particles became blurred after storage, and granular substances appeared on the surface of graphite particles after stored in 100% SOC. This phenomenon is similar to the change of negative electrode surface of graphite/$LiCoO_2$ battery after storage, indicating that the SEI film thickness on the surface of graphite particles increases during storage, and its thickness increases with the increase of storage charge state, which leads to the increase of negative electrode impedance after storage. However, by comparing the impedance changes before and after positive and negative electrode storage, it can be found that the increase of negative electrode impedance after storage is much smaller than that of positive

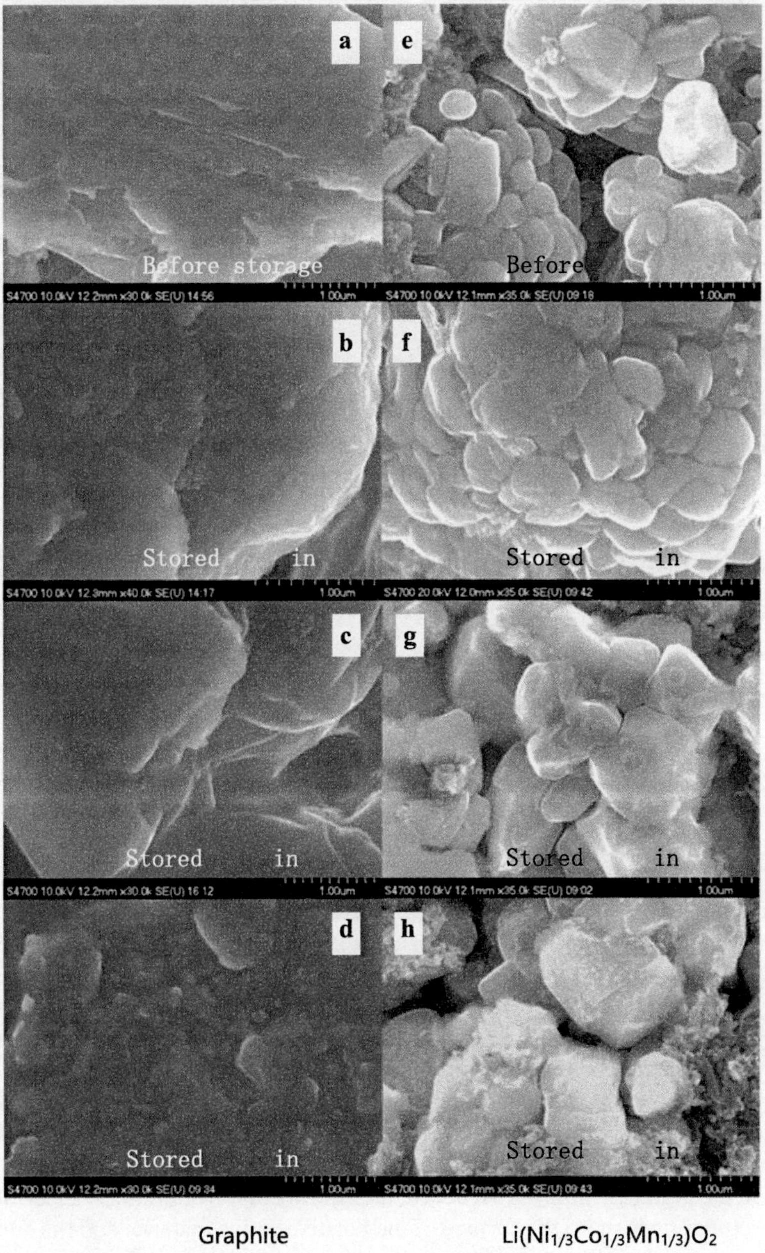

Fig. 18.19 SEM photographs of electrode active materials of graphite/$Li(Ni_{1/3}Co_{1/3}Mn_{1/3})O_2$ battery before and after storage at 55 °C for 250 days. **a** Negative electrode before storage; **b** Negative electrode 0% SOC after storage; **c** Negative electrode after 50% SOC storage; **d** Negative electrode after 100% SOC storage; **e** Positive electrode before storage; **f** Positive electrode after 0% SOC storage; **g** Positive electrode after 50% SOC storage; **h** Positive electrode after 100% SOC storage

Table 18.10 Surface composition of positive active material graphite/$Li(Ni_{1/3}Co_{1/3}Mn_{1/3})O_2$ batteries obtained from EDS before and after storage

	Compositions/(%, atom fraction)					
	Ni	Co	Mn	O	C	F
Fresh cell	8.85	9.36	8.41	58.77	14.61	0
Stored in 0% SOC	7.92	9.46	8.83	55.81	16.86	1.12
Stored in 50% SOC	9.21	8.34	9.11	54.34	16.97	2.03
Stored in 100% SOC	7.92	7.11	9.41	52.89	16.15	6.52

electrode, so it can be explained that the proliferation of negative electrode surface film during storage effects on performance of the whole battery is little effect, and the performance attenuation of the battery after storage is mainly due to the positive electrode.

18.5 Observation of the Morphology of the Diaphragm [1, 2, 4]

18.5.1 Diaphragm of Graphite/$LiCoO_2$ Battery [1, 2]

The morphology of the battery diaphragm before and after storage was studied. Figure 18.20 shows the SEM photographs of the battery diaphragm opposite to the positive and negative electrodes before and after 55°C storage for 100 days. It can be found that the morphology of the side of the diaphragm opposite to the negative electrode changes little before and after storage, and the diaphragm is still a porous structure, but the porosity of the diaphragm decreases with the increase of the charged state during storage. However, the morphology of the side of the diaphragm opposite to the positive electrode changed greatly after storage, and the porosity of the diaphragm decreased obviously with the increase of the charged state during storage. At the same time, it was found that a granular substance was formed on the surface of the positive side diaphragm after 100% SOC storage, which was found by EDS analysis. The substance mainly contains F and P elements, indicating that it is the decomposition product of electrolyte $LiPF_6$. Combined with the positive EDS results, it can be found that when the battery is stored in 100% SOC, the $LiCoO_2$ in the high de-lithium state accelerates the decomposition of the electrolyte, and its products deposit on the surface of the positive active material and the surface of the diaphragm, resulting in an increase in the positive electrode impedance and blocking the micropores of the diaphragm, thus hindering the conduction of lithium ion between the positive and negative electrodes, which is an important reason for the severe decline of battery performance after 100% SOC storage.

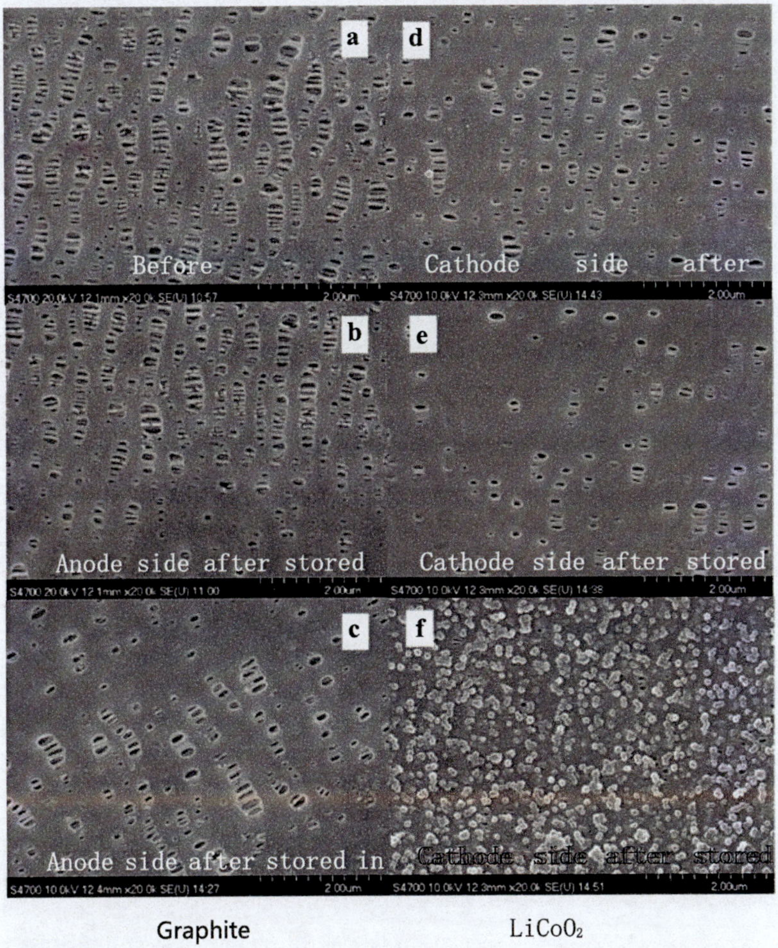

Fig. 18.20 SEM photographs of graphite/$LiCoO_2$ battery separator before and after storage100 days at 55 °C. **a** before storage; **b** negative side after 0% SOC storage; **c** negative side after 100% SOC storage; **d** positive side after 0% SOC storage; **e** positive side after 50% SOC storage; **f** positive side after 100% SOC storage

18.5.2 Diaphragm of Graphite/Li($Ni_{1/3}Co_{1/3}Mn_{1/3}$)O_2 Battery [1, 4]

Figure 18.21 shows the SEM photographs of the diaphragm relative to the positive and negative electrodes in the graphite/Li ($Ni_{1/3}Co_{1/3}Mn_{1/3}$) O_2 battery before and after storage at 55 °C for 250 d. It can be found that, similar to the graphite/$LiCoO_2$ battery, the morphology of the diaphragm opposite to the negative electrode changes little before and after storage, and the diaphragm is still porous structure, but the porosity of the diaphragm decreases with the increase of the charge state during storage. Here the photographs of the negative side of the separator after 100% SOC storage are given for comparison. In contrast, the morphology of the diaphragm opposite to the positive electrode changes more greatly after storage, and the porosity of the diaphragm decreases obviously with the increase of the charge state during storage. Here is a picture of the diaphragm on the negative side of the 100% SOC storage for comparison. In contrast, the morphology of the diaphragm opposite to the positive electrode changes more greatly after storage, and the porosity of the diaphragm decreases obviously with the increase of the charge state during storage, which is due to the decomposition of the electrolyte and the product blocking the micropores during storage. However, compared with the graphite/$LiCoO_2$ battery, it was found that the positive side diaphragm of $Li(Ni_{1/3}Co_{1/3}Mn_{1/3})O_2$ only showed a decrease of micropores after storage, and there was no obvious product on its surface, which indicated that the decomposition degree of electrolyte on the surface of $Li(Ni_{1/3}Co_{1/3}Mn_{1/3})O_2$ positive electrode was significantly lower than that of $LiCoO_2$ electrode during storage. This once again shows that the $Li(Ni_{1/3}Co_{1/3}Mn_{1/3})O_2$ electrode is more stable than the $LiCoO_2$ electrode during storage in the electrolyte, especially in the highly charged state.

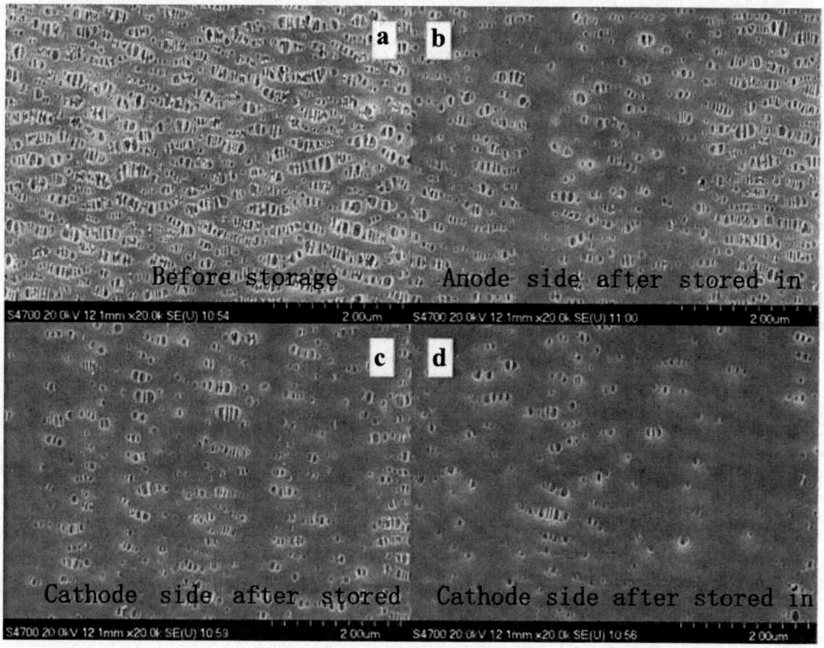

Fig. 18.21 SEM photographs of graphite/$Li(Ni_{1/3}Co_{1/3}Mn_{1/3})O_2$ battery diaphragm before and after storage 250 days at 55 °C. **a** before storage; **b** negative side after 100% SOC storage; **c** positive side after 0% SOC storage; **d** positive side after 100% SOC storage

18.6 Comparison of Two Kinds of Batteries and Analysis of the Mechanism Effecting the Storage Performance of Li-Ion Batteries

18.6.1 Temperature and Charge Effects of Storage Performance Decay of Battery

From the above experimental results, it can be seen that the storage temperature and the charge state of the battery have a great influence on the performance change of the battery during storage, and the performance attenuation of the battery increases with the increase of the storage temperature and the charge state of the battery during storage. In order to further analyze the influence of the two conditional parameters on the storage performance of the battery, firstly, the typical performance parameters (0.2C capacity retention rate and DC internal resistance increase rate) of the battery at different temperatures and different charge states stored for the same time (100 days) were selected, and the storage temperature was plotted as abscissa, and the Fit option of Origin 7.0 was used to fit the data. The fitting curve of graphite/$LiCoO_2$ battery is also shown in Fig. 18.22, and the results are shown in Table 18.11. As can be seen from the figure, when the charge state of the battery is the same during storage, the performance attenuation of the battery (0.2C capacity retention rate decreases and DC internal resistance increase rate rises) does not increase linearly with the increase of storage temperature. After fitting, it is known that the performance attenuation of the battery and the storage temperature are in good agreement with Formula (18.4)

$$y = A \cdot \exp Bx + C \tag{18.4}$$

the exponential relationship described where x is the storage temperature and y represents the performance parameters of the battery after storage. At the same time,

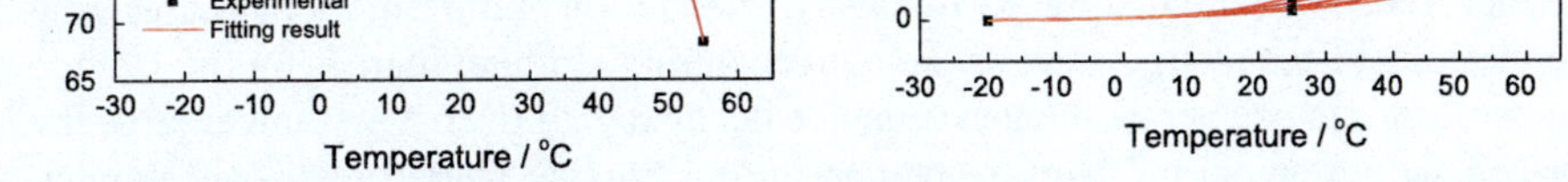

Fig. 18.22 0.2C capacity retention (**a**) of graphite/$LiCoO_2$ battery stored in different charged states for 100 days. The relationship between the storage temperature and the increase rate of DC internal resistance (**b**)

Table 18.11 0.2C capacity retention rate (C_r) of graphite/$LiCoO_2$ batteries stored in different charged states for 100 days. The fitting results of the relationship between the increase rate of DC internal resistance (D_i) and storage temperature

SOC / %	The fitting results of capacity retention rate (C_r)	The fitting results of increase rate of DC internal resistance (D_i)
100	$Cr = -0.4 \exp 0.081T + 100$	$Di = 0.028 \exp 0.093T - 0.0044$
75	$Cr = -0.3 \exp 0.065T + 100$	$Di = 0.034 \exp 0.074T - 0.0078$
50	$Cr = -0.1 \exp 0.093T + 100$	$Di = 0.029 \exp 0.067T - 0.0073$
25	$Cr = -0.1 \exp 0.072T + 100$	$Di = 0.032 \exp 0.059T - 0.0098$
0	$Cr = -0.05 \exp 0.050T + 100$	$Di = 0.034 \exp 0.052T - 0.0092$

Note Cr is the capacity retention rate, unit%; Di is the DC internal resistance increase rate; T is the storage temperature, unit ° C; S is the storage charge state, unit%

it can be seen from Table 18.8 that with the increase of the storage charge state, the absolute value of A or B value in the fitting formula increases, indicating that when the battery is in a slightly higher charge state, the temperature has a greater effect on its storage performance. In order to further analyze the influence of the charge state of the battery during storage, the typical performance parameters (0.2C capacity retention rate and DC internal resistance increase rate) of the battery stored at different temperatures for 100 days were plotted with the storage charge state as the abscissa, as shown in Fig. 18.22.

The Fit option of Origin 7.0 is also used to fit the data, and the fitting curve is also shown in Fig. 18.23, and the results are listed in Table 18.12. As can be seen from the figure and table, under the condition of – 20 °C, the state of charge has almost no effect on the storage performance of the battery, and the decline of the battery performance after storage is not obvious. When the temperature exceeds 25 °C, the storage performance of the battery becomes worse with the increase of the storage charge state. When the storage temperature is the same, with the increase of the storage charge state, the performance attenuation of the battery after storage (the decrease of 0.2C capacity retention and the increase of D_C internal resistance) does not increase linearly. After fitting, it is found that the relationship is also consistent with the exponential relationship shown in Formula (18.4), where x is the storage charge state and y represents the performance parameters of the battery after storage. At the same time, it can be seen from Table 18.12 that with the increase of storage temperature, the absolute value of A (for C_r and D_i) or the value of B (for D_i) in the fitting formula increases. It shows that when the battery is at a slightly higher temperature, the effect of the charged state on the storage performance is more obvious. It can be seen that the performance attenuation of the battery increases with the increase of the charge state of the battery during storage. Increasing the storage temperature can accelerate the performance decay rate of the battery and enlarge the performance attenuation law of the battery during storage. Therefore, the accelerated storage experiment can be carried out by increasing the storage temperature, so as to shorten the experimental period.

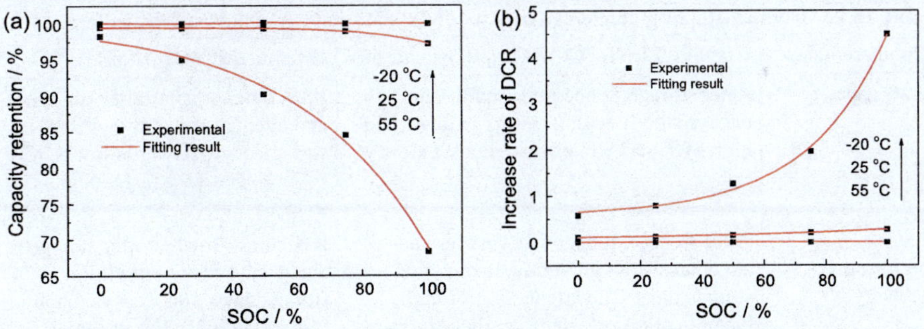

Fig. 18.23 Relationship between the storage charge state and the 0.2C capacity retention rate (**a**) and DC internal resistance increase rate (**b**) of graphite/$LiCoO_2$ battery stored at different temperatures for 100 days

Table 18.12 0.2C capacity retention rate (C_r) of graphite/$LiCoO_2$ battery stored at different temperatures for 100 d. The fitting results of the relationship between the increase rate of DC internal resistance (Di) and the stored charge state

T / °C	The fitting results of capacity retention rate (C_r)	The fitting results of increase rate of DC internal resistance (D_i)
55	$Cr = -2.5 \exp 0.025S + 100$	$Di = 0.079 \exp 0.039S + 0.57$
25	$Cr = -0.01 \exp 0.054S + 100$	$Di = 0.029 \exp 0.019S + 0.082$
− 20	$Cr = -0.01S + 100$	$Di = 0.002S$

Note Cr is the capacity retention rate, unit%; Di is the DC internal resistance increase rate; T is the storage temperature, unit ° C; S is the storage charge state, unit%

At the same time, low temperature can effectively reduce the performance degradation of the battery during storage. If the battery needs long-term storage, storage at low temperature is beneficial to maintain a good initial performance. Similarly, similar results can be obtained for the treatment of graphite/$Li(Ni_{1/3}Co_{1/3}Mn_{1/3})O_2$ battery.

18.6.2 Comparison of Graphite/$LiCoO_2$ and Graphite/$Li(Ni_{1/3}Co_{1/3}Mn_{1/3})O_2$ Batteries

From the previous Sects. (18.1–18.5) and the results, it can be seen that there are obvious differences between graphite/$LiCoO_2$ and graphite/Li $(Ni_{1/3}Co_{1/3}Mn_{1/3})O_2$ batteries, it is listed in Table 18.13 by comparison of performance and structure.

Although all the above are satisfied $y = A \cdot \exp Bx + C$, the difference is revealed by the values of A and B. When B is 0, it is linear relationship; when B $\neq$ 0, the larger the B value, the faster the change; A < 0 means attenuation, the greater the

Table 18.13 Comparison of graphite/$Li(Ni_{1/3}Co_{1/3}Mn_{1/3})O_2$ and graphite/$LiCoO_2$ batteries

Compare items	Graphite/$Li(Ni_{1/3}Co_{1/3}Mn_{1/3})O_2$ batteries	Graphite/$LiCoO_2$ batteries
0.2C capacity	After storage, it becomes smaller and decays slowly with the storage days, and high SOC (>75%) also attenuates slowly	It becomes smaller after storage and attenuates rapidly with the storage days, slower at low SOC (≤75%) and very fast at high (>75%)
Rate performance capacity	After storage, it becomes smaller and attenuates slowly with the increase of storage days. The attenuation of high magnification is faster than that of low magnification; slow attenuation of 0.2C and 5C; accelerated attenuation of 8C and 10C	It becomes smaller after storage and decays faster with the storage days, 10C decays much faster than 0.2C. For example, after 55 °C 100% SOC storage for 100 days, the decay rates of the two rates are 37.5% and 91.5%, respectively
DC resistance	It increases slowly with the increase of storage days and increases with the increase of SOC; when 55 °C is stored for 250 days, 100% SOC increases by 0.9 times, 75% SOC increases by 0.6 times, and 50% does not change much	It increases rapidly with the increase of storage days, and the increase rate increases with the increase of SOC; especially when the SOC is high, it increases very fast (see Fig. 18.8)
Temperature effect of capacity attenuation	The capacity attenuation increases with the increase of storage temperature and satisfies the following formula $y = A \cdot \exp Bx + C$	The capacity attenuation increases with the increase of storage temperature and satisfies the following formula $y = A \cdot \exp Bx + C$
Charge effect of capacity attenuation	The capacity attenuation increases with the increase of charge capacity and satisfies the following formula $y = A \cdot \exp Bx + C$	The capacity attenuation increases with the increase of electric charge and satisfies the following formula $y = A \cdot \exp Bx + C$
The relationship between the increase of DC resistance and temperature	With the increase of storage temperature, the DC resistance increases faster and satisfies the following formula $y = A \cdot \exp Bx + C$	With the increase of storage temperature, the DC resistance increases faster and satisfies the following formula $y = A \cdot \exp Bx + C$
The relationship between DC resistance and charge	With the increase of storage charge, the DC resistance increases faster and satisfies the following formula $y = A \cdot \exp Bx + C$	With the increase of storage charge, the DC resistance increases faster and satisfies the following formula $y = A \cdot \exp Bx + C$

absolute value of A, the faster the attenuation; $A > 0$ means increasing, the larger the A value, the faster it increases.

Compare items	2H-graphite/Li($Ni_{1/3}Co_{1/3}Mn_{1/3}$)O_2 batteries	2H-graphite/$LiCoO_2$ batteries
Cycle performance	It decays slowly with the increase of storage time and accelerates with the increase of SOC, but it is much slower than that of graphite/$LiCoO_2$ battery	It decays quickly with the increase of storage days and accelerates with the increase of SOC, but it is much faster than that of graphite/Li ($Ni_{1/3}Co_{1/3}Mn_{1/3}$) O_2 battery
Resistance to overcharge and thermal stability	There is no difference in overcharge resistance between low SOC storage and before storage, but the overcharge resistance becomes worse with the increase of SOC until combustion. The self-heating rate increases obviously with the increase of temperature. Both of its properties are superior to those of graphite/$LiCoO_2$ battery	There is no difference in overcharge resistance between low SOC storage and before storage, but the overcharge resistance becomes worse with the increase of SOC until combustion. The self-heating rate increases obviously with the increase of temperature
Lattice parameters of negative material	A changes little with the increase of SOC, while C increases after storage and changes little with the increase of SOC	A decreases after storage, but does not change much with the increase of SOC; C does not change much
Microstructure parameters of negative material	D decreases after storage and increases after 100% SOC storage; the ε tensile stress increases after storage, first increases and then decreases with the increase of SOC; P increased after storage, but changed irregularly with SOC	D decreased after storage, but changed little with the increase of SOC; the ε tensile stress increases after storage, first increases and then decreases with the increase of SOC; P increased after storage, first increased and then decreased with the increase of SOC
Lattice parameters of positive material	A does not change much after storage C decreases after storage, first decreases and then increases with SOC	Both A and C have not changed much
Microstructure parameters of positive material	D becomes smaller after storage and refines with the increase of storage SOC, but the speed is very slow after storage, the ε stress changes from tensile stress to compressive stress and increases with the increase of storage SOC, but the size and speed of increase are less than that of graphite/$LiCoO_2$ battery; the mixed parameters of x (Ni and Li) become larger after storage and increase with the increase of SOC	D becomes smaller after storage and refines with the increase of storage SOC, and the refinement speed is very fast After storage, the ε changes from tensile stress to compressive stress and increases with the increase of storage SOC, and the increasing speed is fast
Surface morphology of negative electrode	A passivation film was formed on the surface after storage and thickened with the increase of SOC. After 250 days of 100% SOC storage, a granular SEI film was formed	After storage, a passivation film is formed on the surface and thickens with the increase of SOC

(continued)

(continued)

Compare items	2H-graphite/$Li(Ni_{1/3}Co_{1/3}Mn_{1/3})O_2$ batteries	2H-graphite/$LiCoO_2$ batteries
surface morphology of positive electrode	The surface of the particles is smooth before storage, and the aggregated secondary particles are formed on the surface after storage and grow with the increase of SOC	The surface of the particles is smooth before storage, forms a passivation film after storage, and thickens with the increase of SOC
Surface chemical composition of positive electrode	The content of F increases with the increase of SOC, and the amount of LiF on the positive surface yet increase. After 100% SOC storage, the F content is 3.5%	The content of F increases with the increase of SOC, and the amount of LiF on the positive surface yet increase. After 100% SOC storage, the content of F is 13.2%
Diaphragm morphology	The morphology of the negative side changed little before and after storage, and it was still a porous structure, while the positive side changed greatly, and the porosity decreased with the increase of SOC, and the micropores decreased after 100% SOC storage	The morphology of the negative side changes little before and after storage, and it is still a porous structure, while the positive side changes greatly, resulting in granular electrolyte decomposition

18.6.3 Corresponding Relationship Between Storage Performance and Fine Structure of Battery Materials

Summarizing the previous research results and comparing Table 18.13, the change in law of performance for 2H-graphite/$LiCoO_2$ lithium-ion battery and 2H-graphite/$Li(Ni_{1/3}Co_{1/3}Mn_{1/3})O_2$ battery before and after storage is obtained.

The main results are as follows:

(1) It is found that the capacity, charge–discharge performance, cycle performance, power characteristics, and safety of the battery are attenuated after storage at high temperature (55 °C). The attenuation degree decreases with the increase of storage days, and the attenuation rate increases with the increase of charge state.
(2) After storage at room temperature (25 °C), the performance decay law of the battery is the same as that of 55 °C storage, but the attenuation rate decreases obviously slow.
(3) After storage at low temperature (−20 °C), the performance of the battery did not decrease obviously.
(4) The storage temperature and the charge state of the battery have great influence on the storage performance of the battery. The relationship of between performance retention of battery and storage temperature, DC internal resistance increase rate and temperature is consistent with the $y = A \cdot \exp Bx + C$ described

exponential relationship; the relationship between battery performance retention rate and charge (SOC), DC internal resistance increase rate and storage charge state is also consistent with this exponential relationship.

(5) It can be seen that graphite/$Li(Ni_{1/3}Co_{1/3}Mn_{1/3})O_2$ battery is better than graphite/$LiCoO_2$ battery in all aspects.

Summarize the previous results and compare Table 18.13, it can obtain law of the changes of positive and negative active materials and diaphragm structure of graphite/$LiCoO_2$ lithium-ion battery and graphite/$Li(Ni_{1/3}Co_{1/3}Mn_{1/3})O_2$ battery before and after storage.

(1) Change of the microcrystal size and micro-strain of negative active materials is irregular with the increase of storage charge after storage.
(2) In the positive active material, D becomes smaller after storage and refines with the increase of storage SOC; tensile stress in the positive electrode after storage changes into compressive stress and increases with the increase of storage SOC, and the increase rate of $LiCoO_2$ is much faster than that of $Li(Ni_{1/3}Co_{1/3}Mn_{1/3})O_2$. The mixing parameter x of Ni and Li in $Li(Ni_{1/3}Co_{1/3}Mn_{1/3})O_2$ becomes larger after storage and increases with the increase of SOC.
(3) SEI film is formed on the surface of negative electrode and thickness increase with the increase of SOC, but the SEI film of graphite/$Li(Ni_{1/3}Co_{1/3}Mn_{1/3})O_2$ battery has obvious concavity and convexity.
(4) Passivation film is also formed on the surface of positive electrode, and the surface of $LiCoO_2$ is relatively dense. The surface of $Li(Ni_{1/3}Co_{1/3}Mn_{1/3})O_2$ is not dense, and the content of F in the film of the former battery is much larger than that of the latter battery.
(5) The effect of the diaphragm on the negative side of the storage is not obvious, but the storage has a greater influence on the diaphragm on the positive side. In particular, the diaphragm on the positive side of the graphite/$LiCoO_2$ battery is seriously damaged, showing a multi-hole shape.

By connecting the above two five points, it can be found that there is a good corresponding relationship between the storage performance attenuation and the fine structure of the electrode active material. In particular, it has a better corresponding relationship with the fine structure change of the positive active material and the damage of the diaphragm on one side of the positive.

18.6.4 Discussion on the Decay Mechanism of Storage Performance of Battery

To sum up, the mechanism of storage performance degradation is discussed as follows: There are two functions of the battery during storage: (1) self-discharge of the battery and (2) chemical/electrochemical interaction between the electrolyte and the positive and negative electrode surfaces and the two surfaces of the diaphragm. Let us first discuss the self-discharge effect.

(1) In the process of storage, the battery will self-discharge phenomenon. This leads to obvious changes in the fine structure of the battery materials (including positive and negative active materials and diaphragms) and makes the microcrystal fining and micro-strain of positive active materials change from tensile strain to compressive strain, which obviously affects the de-intercalation behavior of Li ions in the process of battery charge and discharge after storage and causes a larger potential barrier for Li-ion de-intercalation, which hinders the de-intercalation of Li ions, which shows the increase of DC internal resistance and the decrease of capacity.
(2) This self-discharge phenomenon becomes more and more serious with the increase of the percentage of charge during battery storage. With the increase of the degree of battery charge, the grain fining rate of the positive active material is accelerated, and the compressive strain increases rapidly, so the capacity decreases rapidly. The internal resistance also increases rapidly, so the performance change faster with the increase of the degree of charge, especially performance delay greatly speed up in the case of high charge.
(3) The higher the storage temperature, the more serious the self-discharge. At low temperature (- 20 °C), the performance attenuation is linear relationship with a small slope, and at high temperature (55 °C), the performance attenuates is exponentially relationship.
(4) Graphite/$Li(Ni_{1/3}Co_{1/3}Mn_{1/3})O_2$ and graphite/$LiCoO_2$ batteries have the same negative electrode, different positive active materials, and different self-discharge effect during storage. The self-discharge phenomenon of the former battery is smaller than that of the latter battery, so the storage performance of the former battery is better than that of the latter. This is due to the fact that the fine structure of the negative active material does not change much, but there are obvious changes in the fine structure of the positive electrode and the organization structure of the surface film. The grain fining rate of $LiCoO_2$ in 2H-graphite/$LiCoO_2$ battery is faster than that of $Li(Ni_{1/3}Co_{1/3}Mn_{1/3})O_2$ in 2H-graphite/$Li(Ni_{1/3}Co_{1/3}Mn_{1/3})O_2$, and the compressive strain of $LiCoO_2$ in 2H-graphite/$LiCoO_2$ battery is much larger than that of $Li(Ni_{1/3}Co_{1/3}Mn_{1/3})O_2$ in 2H-graphite/Li $(Ni_{1/3}Co_{1/3}Mn_{1/3})O_2$.

Now let us study the chemical/electrochemical interaction between electrolyte and positive and negative electrode surface and diaphragm surface.

The experimental results show that compared with the two kinds of batteries, the surface morphology of the negative electrode and the diaphragm on one side of the negative electrode do not change much, and there is no significant difference between the two kinds of batteries; however, the surface morphology and composition of the positive electrode are obviously different, the morphology of the diaphragm on one side of the cathode is also different, the content of F on the surface of $LiCoO_2$ is also higher, and the damage of the diaphragm on the positive side is also greater, so the graphite/$Li(Ni_{1/3}Co_{1/3}Mn_{1/3})O_2$ battery is superior to the graphite/$LiCoO_2$ battery in

Table 18.14 Capacity change rate of graphite/Li($Ni_{1/3}Co_{1/3}Mn_{1/3}$)O_2 batteries and positive and negative electrodes 0.20 C after storage

SOC/%	Capacity change rate/%		
	Full battery	Li($Ni_{1/3}Co_{1/3}Mn_{1/3}$)O_2	Graphite
0	98.10	98.44	99.21
25	96.05	96.02	98.06
50	92.03	92.03	94.46
75	89.71	89.71	91.41
100	84.49	84.49	87.32

terms of capacity, cycle performance, storage performance, and rate performance. This advantage is shown in the effect of storage charge and storage temperature.

In order to understand more clearly the effect of the change of the specific capacity of positive and negative electrodes on the capacity of the battery after storage, the capacity change rate of 0.20C of graphite/Li($Ni_{1/3}Co_{1/3}Mn_{1/3}$)O_2 battery and positive and negative electrodes after storage was calculated. The results are shown in Table 18.14.

It can be seen from Table 18.14 that the capacity decay rate of full battery, positive and negative electrodes after storage decreases with the increase of SOC during storage. The capacity decay rate of the negative electrode is slower than that of the positive electrode, and the change of the positive capacity is very close to that of the full battery capacity. It can be seen that the capacity attenuation of Li($Ni_{1/3}Co_{1/3}Mn_{1/3}$)O_2 positive electrode stored in 55 °C is larger, which is close to that of the full battery; at the same time, the impedance increases, the dynamic performance decreases; the performance attenuation of graphite negative electrode after storage is small. This shows that the performance attenuation of the battery after storage is mainly due to the positive electrode. The reason for the performance degradation of the battery after storage is that the $LiPF_6$ in the electrolyte is decomposed during storage, and the resulting HF erodes Li($Ni_{1/3}Co_{1/3}Mn_{1/3}$)O_2, resulting in capacity attenuation, and high impedance deposits such as LiF are formed on the surface of the positive electrode, which increases the impedance and decreases the dynamic performance.

It can be seen that in the lithium-ion secondary battery with graphite as the negative active material, the positive active material is the core and key of the problem in terms of increasing the capacity, improving the cycle performance and improving the storage performance. Therefore, the positive coating is advantageous [9].

In order to analyze the impedance changes of positive and negative electrodes, the impedance spectra are fitted, in which R_e represents ohmic resistance, R_f and R_{SEI} represent positive and negative surface film resistance, respectively, and R_{CT} represents charge transfer impedance. The fitting results are shown in Table 18.15.

It can be seen from Table 18.15 that before and after storage, the R_e of the positive electrode changed little, but the R_f increased, indicating that during the storage of the battery, there was a side reaction between Li($Ni_{1/3}Co_{1/3}Mn_{1/3}$)O_2 and the electrolyte, and the product deposited on the surface of the positive electrode, resulting in the increase of R_f, and the charge of the battery accelerated the process. In contrast, the

Table 18.15 EIS simulation results of positive and negative electrodes of battery before and after storage. Unit: $(m\Omega)$

Sample	Positive electrodes $Li(Ni_{1/3}Co_{1/3}Mn_{1/3})O_2$			Negative electrodes graphite		
	Re	R_f	R_{CT}	Re	R_{SEI}	R_{CT}
Before storage	1.87	35.20	54.70	2.37	6.40	15.20
0%SOC after storage	2.05	38.50	66.50	2.35	6.20	18.70
75% SOC after storage	2.12	39.60	87.30	2.28	9.10	23.50
100% SOC after storage	2.98	52.40	115.20	2.45	13.40	29.40

change of R_{CT} of positive electrode after storage is more obvious, which is consistent with the results of the literature. R_{CT} increases with the increase of storage SOC, in which the storage of 100% SOC is more than doubled, indicating that the change of positive R_{CT} after storage leads to the increase of overall impedance. Compared with the positive electrode, the R_{SEI} and R_{CT} of graphite negative electrode are much smaller. After storage, the R_{SE}I and R_{CT} of the negative electrode increase with the increase of the storage SOC, indicating that the more lithium embedded during storage, the more intense the reaction between the negative electrode and the electrolyte. The deposition of reaction products not only thickens the surface film and increases the impedance, but also increases the R_{CT} of the negative electrode. By comparing the impedance results of positive and negative electrodes, it can be seen that both positive and negative electrodes contribute to the increase of the impedance of the whole battery after storage, but the positive electrodes play a leading role.

References

1. Li, Jia. 2010, June. Study on storage performance of lithium-ion battery. Doctoral thesis of Shanghai Institute of Microsystems and Information Technology.
2. Li, J., J. Zhang, X.G. Zhang, C.Z. Yang, and B.J. Xia. 2009. Study on the performance of 2H-graphite/$LiCoO_2$ lithium-ion battery after storage). *Power Supply Technology*, 33(7): 552–572
3. Li, J., J. Zhang, X.G. Zhang, et al. 2010. Study of the storage performance of a Li-ion cell at elevated temperature. *Electrochimica Acta* 55: 927–934.
4. Li, J., X.H. Xie, B.J. Xia, X.F. Gao. 2011. Performance degradation mechanism of graphite/Li $(Ni_{1/3}Co_{1/3}Mn_{1/3})$ O_2 battery after storage). Battery, 41(6): 293–296
5. Li, J., L. He, and Y. Du. Safety performance of lithium-ion battery after high temperature storage. Battery 40(3): 158–160.
6. Li, J., X.G. Zhang, J. Zhang, C.Z. Yang, and B.J. Xia. 2011. A new method for studying the mixed occupation of Li/Ni atoms in $Li(Ni,Me)O_2$ and its application. *Rare Metal Materials and Engineering* 40(8): 1348–1354.
7. Zhang, J., C.Z. Yang, X.G. Zhang, J. Li, and B.J. Xia. 2009. Simulation analysis and study on the structural changes of $Li(Ni,Co,Mn)O_2$ during synthesis. *Journal of Materials Science and Engineering* 27(6): 824–828.

8. Zhang, X.G., J. Zhang, C.Z. Yang, B.J. Xia. A new method for studying and analyzing the mixed occupation of Lithium and Nickel in Li(Ni,Co,Mn) O_2). *Journal of Inorganic Materials* 25 (1): 8–12
9. Liu, H.H. 2012 Study on coating Modification and Application of positive material Li ($Ni_xCo_yMn_{1-x-y}$) O_2 for Li-ion Battery. Doctoral paper of Shanghai Institute of Microsystems and Information Technology.

Chapter 19
Mechanism Research of Storage Process for Graphite/$LiFePO_4$ Battery

With the continuous improvement of the performance of lithium-ion battery and the gradual rise of performance-to-price ratio, the application field of lithium-ion battery is constantly expanding, and it is developing toward the direction of large-scale power supply. In recent years, in order to deal with social problems such as environmental pollution and shortage of petroleum resources, all countries are actively carrying out the research of clean energy electric vehicle EV and hybrid electric vehicle HEV. If lithium-ion battery wants to become the power supply of electric vehicle, it must have reliable safety and performance stability. Among many positive active materials, lithium iron phosphate ($LiFePO_4$) is considered as the most promising candidate material for lithium-ion power battery because of its high safety, good cycle stability, and low price. However, at present, in a large number of studies on lithium iron phosphate, people mainly focus on the improvement of its performance, but pay little attention to the performance change and change mechanism of lithium iron phosphate as an electrode after long-term storage in electrolyte. According to the research results of Chap. 18, the performance of lithium-ion batteries with lithium cobalt and ternary materials as positive electrodes attenuates after storage, which is mainly due to the difference of positive active materials. Then, what will be the storage performance of lithium-ion batteries with lithium iron phosphate as positive active material is indeed a problem worth studying. Therefore, this chapter uses the commercial aluminum-plastic film packaging lithium-ion battery (rated capacity 200 mAh) with $LiFePO_4$ as the positive active material and graphite as the negative material, carries on the accelerated storage experiment under 55 °C, systematically studies the change law of the performance of 2H-graphite/$LiFePO_4$ battery after storage, and reveals the mechanism of the change of battery performance through the study of positive and negative active materials before and after storage.

C. Yang et al., *Materials and Working Mechanisms of Secondary Batteries*,
https://doi.org/10.1007/978-981-19-5955-4_19

19.1 Performance Change of 2H-Graphite/LiFePO$_4$ Battery After Storage [1]

19.1.1 The Change of Capacity and Charge–Discharge Performance of 2H-Graphite/LiFePO$_4$ Battery After Storage

Figure 19.1 shows the change curve of 0.2C capacity of 2H-graphite/LiFePO$_4$ battery during storage at 55 °C. It can be found that, similar to the lithium cobalt and ternary material batteries described in Chap. 18, the charge state of the battery also has a great influence on the storage performance of the lithium-ion battery with lithium iron phosphate as the positive electrode. However, it can be found that, different from the other two positive active materials, the capacity of 2H-graphite/LiFePO$_4$ battery does not decay monotonously with the prolongation of storage time. As can be seen in Fig. 19.1, the capacity of the battery first increases and then decays during 25–75%SOC storage, especially in 50%SOC storage, which is most obvious, when the storage time reaches 135 days. Its capacity reached the highest value (increased to 101.4% of the original capacity). The capacity of the battery stored at 0% and 100%SOC decreases with the increase of storage time, but it can be found that compared with the other two positive active materials, the capacity decay rate of the battery is also greatly reduced. And the capacity of the battery stored for 260 days only attenuates 1.9 and 2.0% of the original capacity.

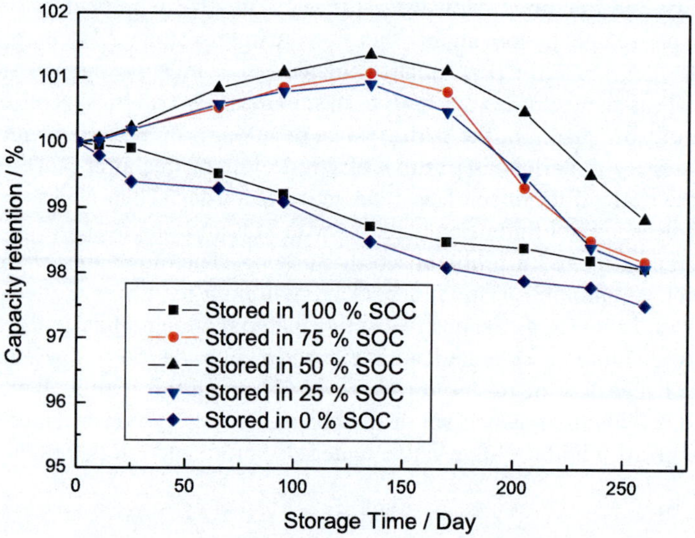

Fig. 19.1 0.2C (40 mA) capacity change of 2H-graphite/LiFePO$_4$ battery stored at different charged states at 55 °C

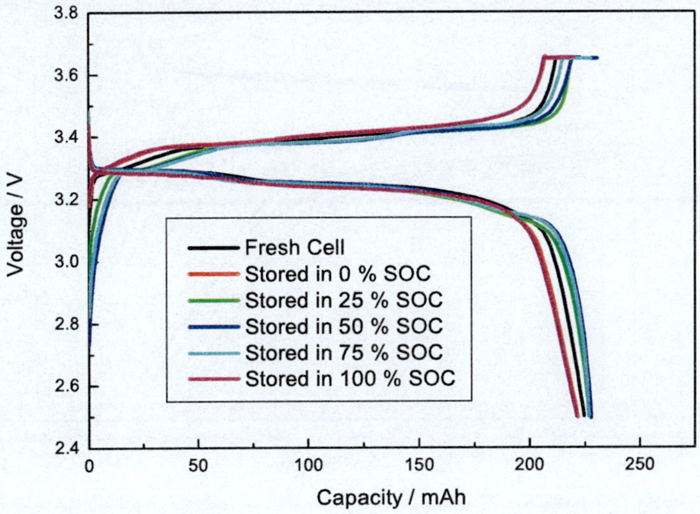

Fig. 19.2 0.2C (40 mA) charge–discharge curve of 2H-graphite/LiFePO$_4$ battery stored at 55 °C for 135 days

The law of capacity change of 2H-graphite/LiFePO$_4$ battery during storage has not been reported in the literature. In order to further study the change of charge–discharge characteristics, the 0.2C (40 mA) charge–discharge curve of 2H-graphite/LiFePO$_4$ battery stored for 135 days is plotted in Fig. 19.2. As can be seen from the figure, due to the coexistence of two phases of LiFePO$_4$/FePO$_4$ in the positive electrode during charge and discharge, the charge and discharge platform of the battery is very flat, which is also one of the advantages of LiFePO$_4$ battery.

At the same time, it can be found that after 25–75%SOC storage, not only the capacity of the battery increases, but also its charge and discharge performance is improved, which shows that the charging platform voltage decreases and the discharge platform voltage increases, which indicates that the internal polarization of the battery decreases after storage. After measuring its AC internal resistance, it was found that it was stored at 25, 50 and 75%SOC. The AC internal resistance of the battery decreased by 1.5%, 2.5%, and 1.5%, respectively, which was consistent with the decrease of polarization of the battery. At the same time, an interesting phenomenon can be found by careful observation. After 25–75%SOC storage, the discharge curve of the battery appears a new discharge platform around 3.15 V, which leads to the increase of battery capacity. It is well known that during the charge–discharge process of LiFePO$_4$, only a single valence change of Fe^{2+}/Fe^{3+} occurs in Fe ions, so there is usually only one discharge platform. The appearance of this platform near 3.15 V indicates that some structural changes have taken place in the electrode active material after storage. This phenomenon has not been reported in the literature, and this phenomenon will be further discussed later. At the same time, it can also be found from Fig. 19.2 that the battery capacity attenuates while

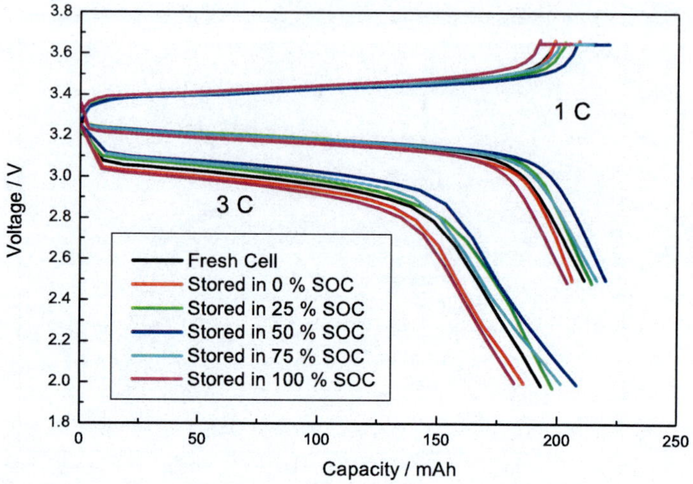

Fig. 19.3 1C (200 mA) charge–discharge curve and 3C (600 mA) discharge curve of graphite/$LiFePO_4$ battery stored at 55 °C for 135 days

the charge–discharge platform voltage difference (lag) increases after storage at 0% and 100%SOC, indicating that the internal polarization of the battery increases. The AC internal resistance was measured and found that the internal resistance increased by 2.5% and 2.0%, respectively.

Figure 19.3 shows 1C (200 mA) charge–discharge curve and 3C (600 mA) discharge curve of graphite/$LiFePO_4$ battery after 135 days 55 °C storage. It can be found that the capacity of the battery decreases gradually with the increase of the charge–discharge rate, but the capacity changes are similar to the 0.2C rate after storage. At the same time, it can be found that the 3C discharge platform voltage of the battery increased significantly after 25–75%SOC storage, which indicated that the power performance of the battery was also improved to a certain extent, while the decrease of 3C discharge platform voltage of the battery stored at 0% and 100%SOC also reflected the decline of its power performance.

19.1.2 Changes in the Safety of Graphite/$LiFePO_4$ Batteries After Storage

After storage, $LiFePO_4$ materials are favored by people because of their excellent safety. Research found that large polyanion group in its structure has strong binding force on *O*. So the material will not be overcharged, even at a voltage of 12 V, Materials do not decompose and release oxygen like lithium cobalt oxide and ternary materials. As mentioned in Li's doctoral paper [1], stages *C* and *D* do not occur when Graphite/$LiCo_2$ batteries are overcharged, which prevents the battery from

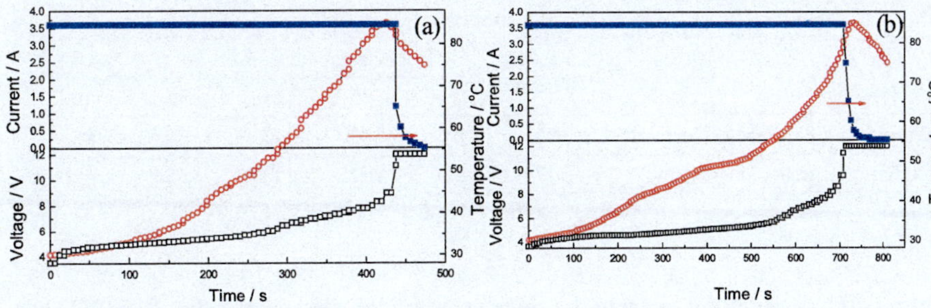

Fig. 19.4 3C-12 V overcharge curve of graphite/$LiFePO_4$ battery before (**a**) and after (**b**) at 55 °C stored 135 days

being thermally out of control due to severe oxidation of the electrolyte and negative electrodes. However, at present, the reports on the safety performance of lithium iron phosphate are only for the newly prepared $LiFePO_4$ battery, and there are no reports about the effect of storage on the safety performance of lithium iron phosphate battery. In this section, the safety performance of lithium iron phosphate battery before and after storage was studied by overcharge test and ARC experiment.

Figure 19.4 shows the 2H-graphite/$LiFePO_4$ symmetrical 3C-12 V overcharge curve before storage and after 55 °C storage 135 days in 50%SOC. It can be seen from the figure that the surface temperature increases with the increase of voltage.

When the voltage rises to 12 V, the surface temperature of the battery reaches 84 °C.

Since then, with the extension of time, the surface temperature of the battery decreases continuously, because $FePO_4$ formed after $LiFePO_4$ Li removal, conductivity of $FePO_4$ is 1/3 times that of $LiFePO_4$. Therefore, in the process of overcharging, the internal resistance of the battery increases gradually, and the temperature of the battery increases gradually through the Joule heat generated by the current. When the voltage reaches 12 V, the battery is charged at a constant voltage, the current decreases rapidly, and the temperature of the battery also decreases. During the whole overcharging process, the battery did not explode or catch fire. It can be seen that lithium iron phosphate has good overcharge safety performance.

Compared with the overcharge voltage curve of the battery before and after storage, it can be found that the overcharge resistance time of $LiFePO_4$ battery stored at 55 °C in 50%SOC for 135 days is nearly 350 s longer than that of the battery before storage, which indicates that its overcharge resistance has been improved to a certain extent after storage. The test results of other batteries are listed in Table 19.1; it can be found that the overcharging experiments of batteries stored at 25 and 75% show a similar phenomenon. This is contrary to the change of overcharge performance of $LiCoO_2$ and $Li(Ni_{1/3}Co_{1/3}Mn_{1/3})O_2$ batteries after storage, which has not been reported in other literature. However, the overcharge time of the battery did not increase after 0% and 100%SOC storage, but in the end, there was no thermal runaway. This shows that even after high temperature storage for a long time, $LiFePO_4$

Table 19.1 Results of 3C-12 V overcharge test of batteries stored at 55 °C for 135 days

	Fresh battery	SOC electrical states (%)				
		0	25	50	75	100
Maximum temperature of surface (°C)	85	84	83	85	88	85
Overcharge time (s)	474	455	625	810	684	420
Results	No fire	No fire	No fire	No fire	No fire	No fire

battery still has excellent overcharge resistance, which also proves that $LiFePO_4$ has great advantages in safety.

Figure 19.5 shows the variation curve of the self-heating rate of graphite/$LiFePO_4$ battery with temperature in the ARC experiment before storage and after 50% 55 °C and 135 days 100%SOC storage. Different from lithium cobalt oxide and ternary material batteries, the change of self-heating rate of graphite/$LiFePO_4$ battery with temperature is not obvious. When the temperature rises to 160 °C, the self-heating rate of the battery begins to increase, which indicates that the self-heating reaction begins to occur in the battery, and the self-heating rate of the battery increases slowly as the temperature continues to rise. When it reaches the maximum, the self-heating rate of the $LiFePO_4$ battery decreases gradually as the temperature continues to rise, and the self-heating rate of the $LiFePO_4$ battery remains at a relatively low level (<0.6 °C/min) in the whole temperature range. The increase in the self-heating rate of the battery may be due to the short circuit of the positive and negative electrodes in contact with the battery caused by the melting of the diaphragm after the temperature reaches the diaphragm melting temperature (higher than 130 °C). The instantaneous increase of current produces heat to increase the battery temperature, and the increase in temperature may also lead to the decomposition and exothermic heat of the electrolyte, but because of the structural stability of $LiFePO_4$ itself, oxygen will not be decomposed and precipitated at high temperature, so it will not promote the violent oxidation reaction between the electrolyte and the electrode. As a result, when the internal short-circuit discharge of the battery is completed and the electrolyte is decomposed, the self-heating rate of the battery decreases gradually, and the final safety problems such as thermal runaway will not occur in the battery. It can be found from Fig. 19.5 that the self-heating rate of the battery stored in 50%SOC is lower than that of the no-stored battery in the whole temperature range, while the self-heating rate of the battery stored in 100%SOC is higher than that of the no-stored battery. This shows that the thermal stability of graphite/$LiFePO_4$ battery is improved after stored 135 days in 55 °C 50%SOC, while the thermal stability of graphite/$LiFePO_4$ battery is decreased after 100%SOC storage.

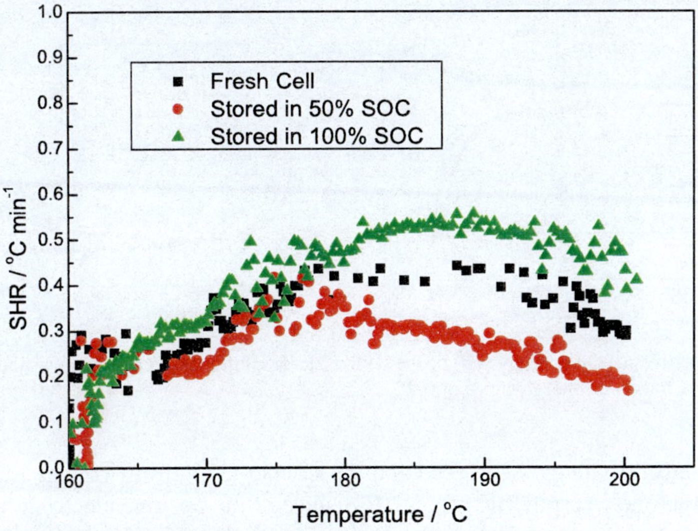

Fig. 19.5 Variation curve of self-heating rate of $LiFePO_4$ battery with temperature before storage and after 55 °C storage 135 days

19.2 Change of Electrode Characteristics of 2H-Graphite/$LiFePO_4$ Battery After Storage [1]

19.2.1 Capacity Change of Positive and Negative Electrodes of 2H-Graphite/$LiFePO_4$ Battery After Storage

In order to explore the reasons for the performance change of 2H-graphite/ $LiFePO_4$ battery after storage, The battery stored in several charged states for 135 days was 0.2C discharged to 2.5 V (0%SOC) at room temperature and then dissected in a glove box filled with argon, the positive and negative plates were removed, and the positive and negative plates were cut off the small pieces of 1 cm^2, respectively, washed with isopropanol several times to remove the residual electrolyte of the electrode, then dried naturally and weighed. The 2025 button battery was assembled with the positive and negative electrodes before and after storage as the research electrode and the lithium sheet as the counter electrode, and the charge and discharge of the battery was tested.

The 0.2C charge–discharge curve of button battery before and after positive and negative electrode storage is shown in Fig. 19.6. As can be seen from Fig. 19.6a, the specific capacity of $LiFePO_4$ before storage is 147.2 $mAhg^{-1}$, which is lower than its theoretical capacity 170 $mAhg^{-1}$. The platform voltage is 3.5 V and 3.4 V, respectively, and the platform is stable. Similar to the change rule of the capacity of the whole battery after storage, after 25–75%SOC storage, the positive capacity increased, and its specific capacity reached 148.6, 149.1, and 148.5 $mAhg^{-1}$, respectively, while the

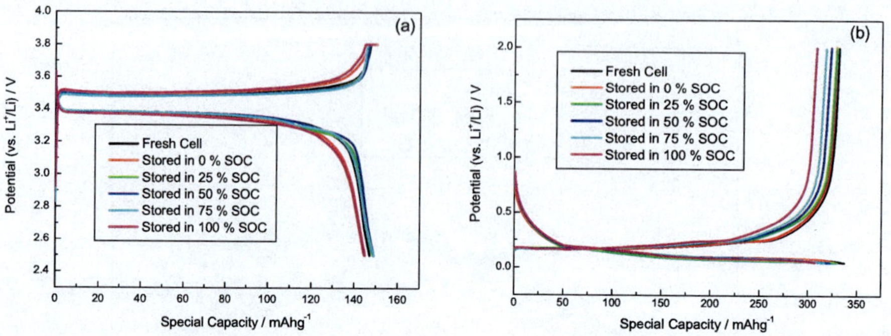

Fig. 19.6 Positive (**a**) and negative (**b**) of the button 2H-graphite/$LiFePO_4$ battery stored at 55 °C for 135 days. Charge–discharge curve of 0.2C

charging platform voltage decreased and the discharge platform voltage increased, indicating that the charge–discharge performance of the positive electrode was also improved. At the same time, it can also be found that after 25–75%SOC storage, the discharge curve of the positive electrode appears a discharge platform around 3.25 V, which is consistent with the change of the discharge curve of the whole battery after storage. After storage at 0 and 100%SOC, the positive capacity attenuates, and the charge–discharge platform voltage difference (lag) increases, which is consistent with the performance of the whole battery.

It can be seen from Fig. 19.6b that the specific capacity of graphite electrode before storage is 331 $mAhg^{-1}$, and after storage, its capacity decreases with the increase of charge state during storage, which is similar to the change of negative electrode capacity of graphite/$LiCoO_2$ and graphite/ternary material batteries after storage.

In order to clearly distinguish the effect of the change of the capacity of the positive and negative electrodes on the capacity of the whole battery after storage, Table 19.2 shows the 0.2C capacity change rate of graphite/$LiFePO_4$ battery and its positive $LiFePO_4$ and negative graphite after storage 135 days at 55 °C. It can be found from the table that the capacity change rate of the positive electrode and the whole battery after storage is very similar, which shows that the capacity increases after 25–75%SOC storage, in which the capacity increase is the largest after 50%SOC storage, which is 1.37%, while the capacity decreases after 0 and 100%SOC storage. The change rate of capacity after negative electrode storage decreases with the increase of charge state during storage. It can be seen that the $LiFePO_4$ positive plays a decisive role in the capacity change of graphite/$LiFePO_4$ battery after storage.

Table 19.2 0.2C capacity change rate of graphite/$LiFePO_4$ battery and positive and negative electrodes stored 135 days at 255 °C for

		0%SOC	25%SOC	50%SOC	75%SOC	100%SOC
Capacity change rate (%)	Full battery	98.73	101.08	101.37	100.92	98.50
	$LiFePO_4$	98.65	100.91	101.30	100.86	98.31
	Graphite	99.09	98.32	97.36	95.83	93.00

19.2.2 Change of Dynamic Performance of Positive and Negative Electrodes of 2H-Graphite/$LiFePO_4$ Battery After Storage

In order to further study the dynamic performance of positive and negative electrode of 2H graphite/$LiFePO_4$ battery after storage, AC impedance (EIS) and cyclic voltammetry (CV) were used to test the electrode. The test process and conditions are the same as those in Chap. 18.

Figure 19.7 shows the AC impedance Nyquist curve of the positive and negative electrodes before and after storage. It can be seen from the figure that the impedance of positive and negative electrode of graphite/$LiFePO_4$ battery is quite different, and the impedance of positive electrode is one order of magnitude larger than that of negative electrode, which is similar to that of graphite/$LiCoO_2$ and graphite/ternary material battery, indicating that the positive electrode has a great influence on the impedance of lithium-ion battery. For the $LiFePO_4$ electrode, the impedance spectrum before and after storage is composed of a semicircular arc located in the middle and high-frequency region and a straight line in the low-frequency region. At the same time, it can be found that the semicircular arc in the high-frequency region is composed of two semicircular arcs, and there is an no-obvious semicircular arc in the range higher than 103 Hz, which corresponds to the impedance of the passivation film on the electrode surface. The larger semicircular arc with lower frequency corresponds to the charge transfer impedance between the electrolyte and the electrode, and the straight line in the low-frequency region corresponds to the Warburg impedance of Li^+ diffusion. Therefore, it can be seen that the electrode reaction of $LiFePO_4$ includes three steps: the diffusion of lithium atom in the positive active material, the diffusion of lithium ion in the electrode surface film, and the charge exchange between electrolyte and electrode. However, from the impedance spectrum, it can be seen that, in contrast, the charge transfer impedance and Warburg impedance of the electrode are significantly larger than the film impedance, indicating that the whole electrode process is mainly controlled by charge transfer and Li^+ diffusion.

It can be found from the figure that the impedance of the positive electrode changes obviously after storage, and the impedance of the electrode decreases after 25–75%SOC storage, especially after 50%SOC storage, while the impedance of the electrode increases after 0% and 100%SOC storage. This rule is similar to the change of the capacity of the battery and electrode after storage, and the increase of the capacity of the electrode corresponds to the decrease of impedance. In order to

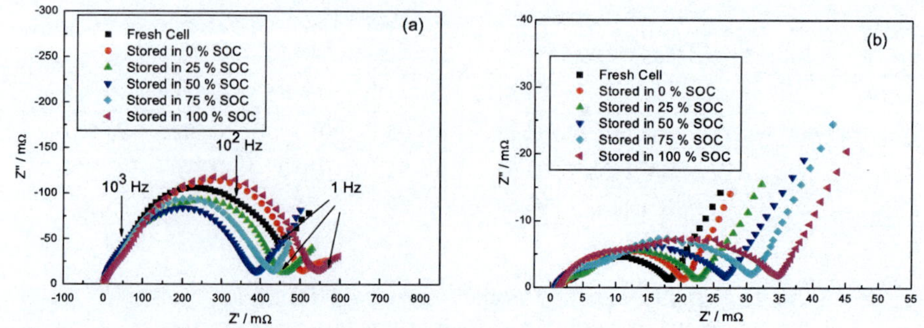

Fig. 19.7 Nyquist curve of AC impedance of 2H-graphite/$LiFePO_4$ battery after stored at 55 °C for 135 days. **a** positive and **b** negative

analyze the impedance change of the positive electrode more clearly, the impedance spectrum is fitted according to the equivalent circuit shown in Fig. 19.8a, where R_e represents ohmic resistance, R_f and R_{CT}, represent surface film resistance and charge transfer resistance respectively, and surface film capacitance, double layer capacitance and concentration impedance are represented by constant phase angle elements CPE_1, CPE_2, and CPE_3, respectively. CPE can characterize the non-ideal behavior of the composite electrode (the porosity of the material, the roughness of the electrode surface), so it is more suitable to simulate the lithium-ion battery electrode [6]. The admittance response expression of CPE is as follows:

$$Y = Y_0\omega^n \cos\left(\frac{n\pi}{2}\right) + jY_0\omega^n \sin\left(\frac{n\pi}{2}\right) \quad (19.1)$$

where ω is the angular frequency; j is the imaginary unit $\sqrt{-1}$. When $n = 0$, CPE is equivalent to a resistor; $n = 1$ is equivalent to a capacitor, and $n = 0.5$ is equivalent to Warburg impedance. In this study, when $0.5 < n < 1.0$, Y_0 is regarded as a false capacitor. The equivalent circuit can well fit the EIS experimental data obtained before and after electrode storage, and the fitting errors of each equivalent circuit parameters are less than 5%. The fitting results are listed in Table 19.3, from which it can be seen

Fig. 19.8 Equivalent circuit model of AC impedance of positive (**a**) and negative (**b**) electrodes of battery before and after storage

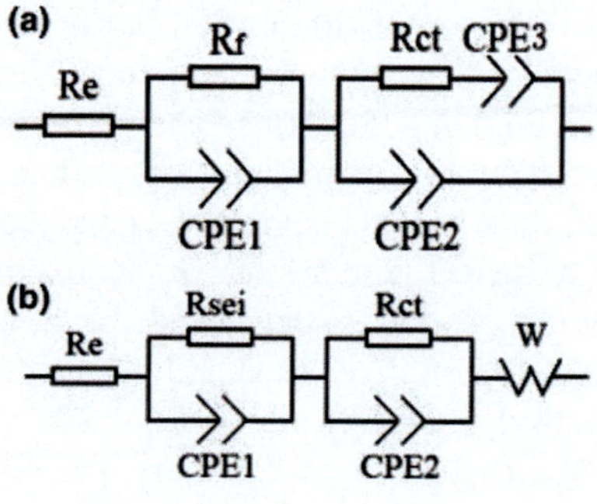

that the positive ohmic impedance R_e changes little before and after storage, while the surface film impedance Rf increases, which indicates that the positive $LiFePO_4$ has a relatively weak side reaction with the electrolyte during storage, which leads to the increase of impedance due to the change of positive surface film formation. At the same time, it was found that the charge transfer impedance R_{CT} of the positive electrode changed obviously after storage, and the R_{CT} of the electrode decreased after 25–75%SOC storage, especially after 50%SOC storage, while the R_{CT} of the electrode increased after 0% and 100%SOC storage, indicating that it was the change of R_{CT} of the electrode that led to the change of its overall impedance. Compared with the positive electrode, the impedance of the negative graphite electrode of the battery changes little after storage, which varies with the charge state during storage. This is similar to graphite/lithium cobalt and graphite/ternary batteries. It is fitted according to the equivalent circuit shown in Fig. 19.8b, and the fitting results are also shown in Table 19.3. It can be seen that, compared with the positive electrode, both the surface film impedance and the charge transfer impedance are 1–2 orders of magnitude lower. After storage, the SEI film impedance and charge transfer resistance of the negative electrode increase with the increase of the charge state during storage, and the change of R_{SEI} is more obvious, which indicates that the higher the amount of lithium embedded during storage, the more intense the reaction between the negative electrode and the electrolyte, thus the deposition of the reaction products increases the surface film thickening impedance and leads to the increase of the charge transfer impedance of the electrode.

Figure 19.9a and b is cyclic voltammetry curves of positive and negative electrodes of graphite/$LiFePO_4$ battery before and after storage at 55 °C for 135 days. It can be seen from the figure that the cyclic voltammetry curve of the positive electrode changes obviously before and after storage. After 25–75%SOC storage, the redox peak current of the electrode increases and the redox peak becomes sharp, indicating that the reaction kinetic performance of the electrode is improved, which is consistent with the decrease of R_{CT} after storage. At the same time, it can be found that the shape of the reduction peak changes from a single reduction peak before storage to two peaks, and a new peak appears around 3.3 V. Combined with Fig. 19.6a, it can be found that this peak corresponds to a new platform for the positive discharge curve after storage. This shows that after 25–75%SOC storage, the discharge reaction of

Table 19.3 Fitting results of impedance spectra of positive and negative electrodes of graphite/$LiFePO_4$ battery stored at 355 °C for 135 days

		Fresh	0%SOC	25%SOC	50%SOC	75%SOC	100%SOC
Positive	R_e	2.34	2.12	2.06	2.21	2.11	1.98
	R_f	152	164	165	168	169	167
	R_{CT}	455	511	402	362	395	532
Negtive	R_e	0.94	1.21	1.24	1.15	1.34	1.51
	R_{SEI}	3.34	4.56	7.32	10.22	11.54	15.01
	R_{CT}	15.32	19.87	22.12	24.52	28.55	31.45

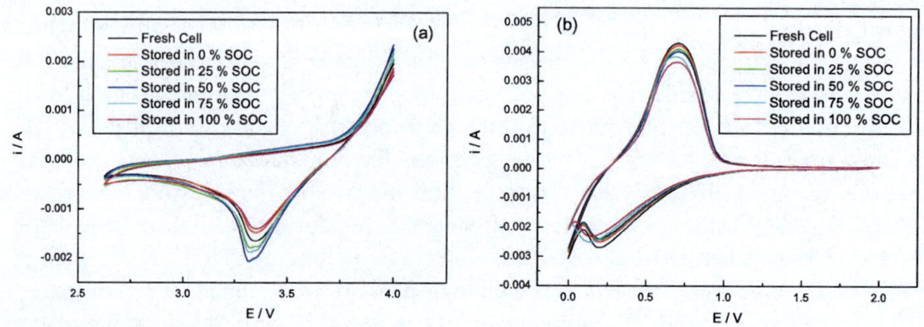

Fig. 19.9 Cyclic voltammetry curves of positive (**a**) and negative (**b**) electrodes of graphite/$LiFePO_4$ battery before and after storage at 55 °C for 135 days

$LiFePO_4$ electrode changes, and a new reduction reaction appears, which increases the capacity of the electrode and improves the kinetic performance. However, after 0% and 100%SOC storage, the redox peak current of the electrode decreased, indicating that 0% and 100%SOC storage had a certain effect on the reaction kinetics of the electrode. Combined with the EIS curve, it can be found that this is due to the change of the surface layer of the electrode during storage, which leads to the increase of R_f and R_{CT}, which makes the charge exchange between Li^+ and positive active materials difficult. It can also be found from the figure that compared with lithium cobalt dioxide battery, lithium iron phosphate electrode only showed a decrease in redox peak current after storage at 100%SOC, and its peak position did not change, indicating that the degree of reversibility of the electrode reaction did not occur. Change again shows the excellent storage performance of $LiFePO_4$.

From Fig. 19.9b, it can be found that the cyclic voltammetry curve of negative graphite after lithium-ion battery storage changes little compared with the positive electrode, and the redox peak current decreases slightly with the increase of storage charge state. After storage at 75% and 100%SOC, the redox peak current of the negative electrode decreases greatly, and the reduction peak position shifts to the negative potential. This shows that 75% and 100%SOC storage have some effects on the reaction kinetics and reaction reversibility of graphite negative electrode, but the degree of effect is much less than that of positive electrode. The results correspond to the impedance results, indicating that the dynamic performance of the positive electrode changes greatly after storage, which plays a decisive role in the performance of the battery.

19.2.3 Change of Surface Morphology of Positive and Negative Electrodes of Graphite/$LiFePO_4$ Battery After Storage

In order to further analyze the reasons for the change of battery performance after storage, the batteries stored in different charge states at 55 °C for 135 days were dissected, positive and negative plates and diaphragms were removed, and the surface morphology was observed by field emission scanning electron microscope (FE-SEM). It can be found that the change of the surface morphology of the negative electrode after storage is similar to that described in the previous two chapters, showing that the surface film is thickened; at the same time, there is no obvious change in the surface morphology of the positive and negative side diaphragm after storage. However, it is found that the surface morphology of the positive active materials is obviously different after being stored in different charged states, so in order to avoid redundancy, only the SEM photos of the battery positive active materials before and after storage at 55 °C for 135 days are listed, as shown in Fig. 19.10.

It can be found from Fig. 19.10 that before storage the surface of the positive active material particles is smooth and continuous, and there are no obvious deposits and microcracks; after 0%SOC and 100%SOC storage, there are fine speckle products on the particle surface, but the surface is still continuous and there are no microcracks. After 25–75%SOC storage, a small amount of speckled products also appeared on the particle surface, but unlike 0%SOC and 100%SOC storage, cracks appeared on the material particle surface, and the cracks were the most obvious after 50%SOC storage. Combined with the above, it can be seen that the improvement of the performance of $LiFePO_4$ electrode after 25–75%SOC storage has a lot to do with these microcracks on the surface of positive active materials. We know that during the charge and discharge of the battery, an electrochemical reaction occurs between the electrode and the electrolyte through electron and ion exchange on the solid/liquid interface between the electrode active material and the electrolyte. The larger the solid/liquid interface area is, the higher the reaction activity of the electrode is, and the better the performance of the battery is. After 25–75%SOC storage, the particle surface of the positive active material will crack and many fresh surfaces will appear, which will increase the contact area between the electrode and the electrolyte, thus increase the electrode reaction area and improve the electrode reaction activity, which is shown in the positive electrode characteristics, that is, the impedance decreases and the reaction kinetic performance is improved. It can be found from Fig. 19.10 that the surface cracks of positive active material particles are the most after 50%SOC storage, so the performance of battery and positive is improved most obviously after 50%SOC storage. Formation mechanism of these microcracks on the surface of positive active materials after 25–75%SOC storage will be analyzed more deeply in the next Sect. 19.3.

At the same time, in order to further analyze the composition of the surface products of the positive active material after battery storage, the positive electrode was analyzed by EDS. The analysis results of the positive electrode before and

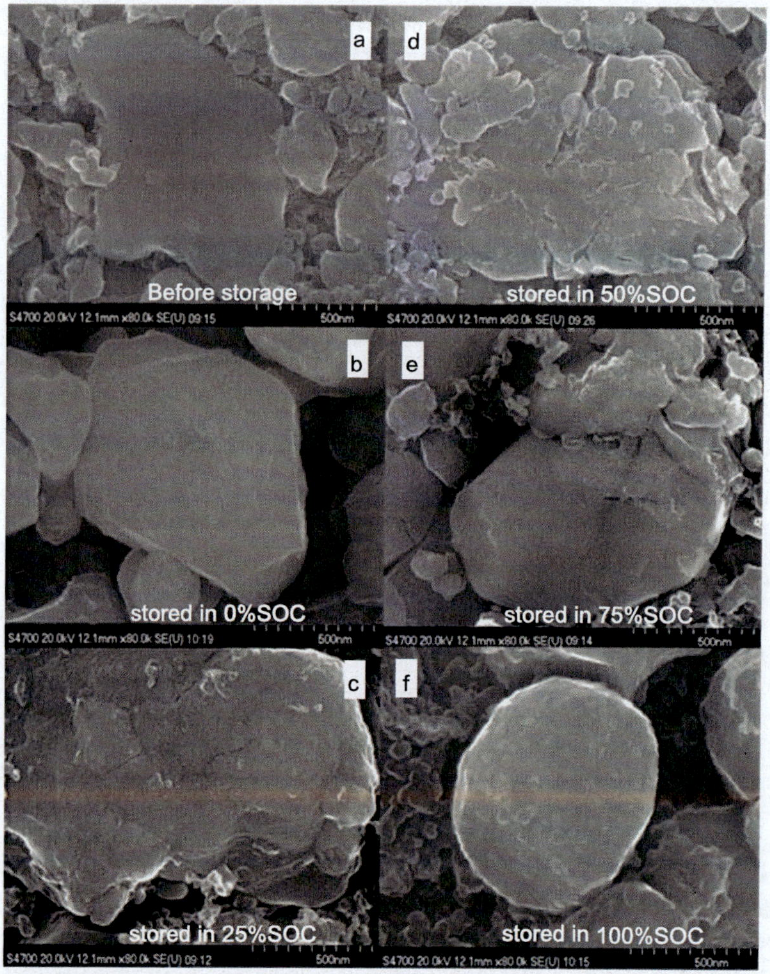

Fig. 19.10 SEM photos of positive active materials of graphite/$LiFePO_4$ battery stored at 55 °C for 135 days. **a** Before storage; after **b** 0%SOC storage; after **c** 25%SOC storage; **d** 50%SOC storage. e After 75%SOC storage; after **f** 100%SOC storage

after storing at 55 °C for 135 days are listed in Table 19.4. As can be seen from Fig. 19.10 and Table 20.4, similar to the changes of surface composition of $LiCoO_2$ and $Li(Ni_{1/3}Co_{1/3}Mn_{1/3})O_2$ before and after storage, there are no other elements on the surface of positive active materials before storage; *F* is detected on the surface of positive active materials after storage. This kind of material should also be the decomposition product of $LiPF_6$ during battery storage, and the product LiF formed after HF erodes the positive active material. It can be found that the increase of R_f of $LiFePO_4$ electrode after storage is caused by the *F*-containing substance formed on the surface of the electrode. However, different from the above two materials, the *F*

Table 19.4 Surface composition of positive active materials obtained from EDS before and after 135 days storage

	Composition (at %)				
	Fe	P	O	C	F
Before storage	15.32	19.23	56.87	14.58	0
0%SOC after storage 135 days	14.60	19.40	54.05	17.08	0.87
25%SOC after storage 135 days	12.45	14.57	56.42	15.44	1.12
50%SOC after storage 135 days	14.12	14.85	55.22	14.39	1.42
75%SOC after storage 135 days	12.25	19.32	54.15	19.26	1.02
100%SOC after storage 135 days	19.22	12.85	55.35	17.4	1.18

content on the surface of $LiFePO_4$ is lower after storage, and the stored charge state seems to have no obvious effect on the *F* content. On the one hand, it is affected by the reaction kinetics of the electrode. Combined with the EIS curve, it can be found that this is due to the change of the surface layer of the electrode during storage, which leads to the increase of R_f and R_{CT}, which makes the charge exchange between Li^+ and positive active materials difficult. It can also be found from the figure that compared with lithium cobalt dioxide battery, lithium iron phosphate electrode only showed a decrease in redox peak current after storage at 100%SOC, and its peak position did not change. This shows that the degree of reversibility of the electrode reaction has not changed. LiFePO4 has higher stability against HF corrosion.

On the other hand, because the electrode potential of $LiFePO_4$ is lower ($LiFePO_4$: 3.65 V versus Li^+/Li; $LiCoO_2$: 4.3 V versus Li^+/Li at storage 100%SOC), at high charge its oxidizability is obviously lower than that of the other two materials, so the accelerated effect on the decomposition of $LiPF_6$ is not obvious. This shows that, especially at high charge state, the stability of $LiFePO_4$ in electrolyte is higher than that of $LiCoO_2$ and $Li(Ni_{1/3}Co_{1/3}Mn_{1/3})O_2$, so the storage performance of graphite/$LiFePO_4$ is better.

19.3 Fine Structure Changes of Positive Active Materials During Battery Charge–Discharge Before and After Storage

In order to further analyze the reasons for the performance improvement of graphite/$LiFePO_4$ battery stored in 25–75%SOC, the structural changes of positive active materials of graphite/$LiFePO_4$ battery during charge and discharge were systematically studied by means of quasi-dynamic X-ray diffraction before and after 135 days storage of 55 °C in 50%SOC.

The study believes that the Li-atoms de-intercalation behavior of $LiFePO_4$ is actually a pseudo-structural phase transition process in which $LiFePO_4$ de-Li, gradually transforms $LiFePO_4$ into $FePO_4$ (see Chap. 12). It can also be said that it is a two-phase reaction process that forms the interface between $FePO_4$ and $LiFePO_4$. Newman [3], Yamada [4] and Dodd [5] have systematically studied the phase transition process in the charge–discharge process of $LiFePO_4$. During charging, the lithium ion migrates from the FeO_6 layers and enters the negative electrode through the electrolyte, resulting in the oxidation reaction of $Fe^{2+} \rightarrow Fe^{3+}$. In order to maintain the charge balance, the electron arrives at the negative electrode from the external circuit. The reduction reaction occurs during discharge, which is opposite to the above process. That is:

$$\text{Charge: } LiFePO_4 \rightarrow Li^+ + FePO_4 + e^-$$

$$\text{Discharge: } FePO_4 + Li^+ + e^- \rightarrow LiFePO_4$$

19.3.1 Structural Changes of $LiFePO_4$ During Charge and Discharge Before Storage

Figure 19.11A shows the XRD patterns of several stages of the charge and discharge process of $LiFePO_4$ positive active material before battery storage and is indexed by $LiFePO_4$ and $FePO_4$ with orthogonal structure, represented by the letters T and H, respectively. It can be found that in the charging process, with the increase of charging depth, the content of $LiFePO_4$ phase gradually decreases, while the content of $FePO_4$ increases gradually; when the charge reaches 80%, there is still a trace of $LiFePO_4$, and until the charge reaches 100%, the peak of $LiFePO_4$ phase in the XRD spectrum completely disappears. During the discharge process, 20% of the discharge appears $LiFePO_4$, and its content increases with the increase of the discharge depth, up to 80% of the discharge, it is mainly $LiFePO_4$ phase, but there is still a certain content of $FePO_4$. It can be seen that there is a phase transition of $LiFePO_4 \rightarrow FePO_4$ in the positive active material during the charge–discharge process. At the same time, it is found that after the activation of the battery, even the discharge positive active material is not a single phase of $LiFePO_4$, but contains a small amount of $FePO_4$ phase.

In order to show the change of the two phases during the charge and discharge process more clearly, the positions of the characteristic peaks of the two phases in Fig. 19.11a are locally magnified, and the XRD spectra of the same charged state in the process of charge and discharge are compared, as shown in Fig. 19.12. It can be found that in the process of charge and discharge, even in the same charged state, there is an obvious difference in the relative intensity of the two phases in the XRD pattern. In the same charged state, the intensity of $FePO_4$ phase in the charged positive active material is higher than that in the discharge state, while the intensity of the

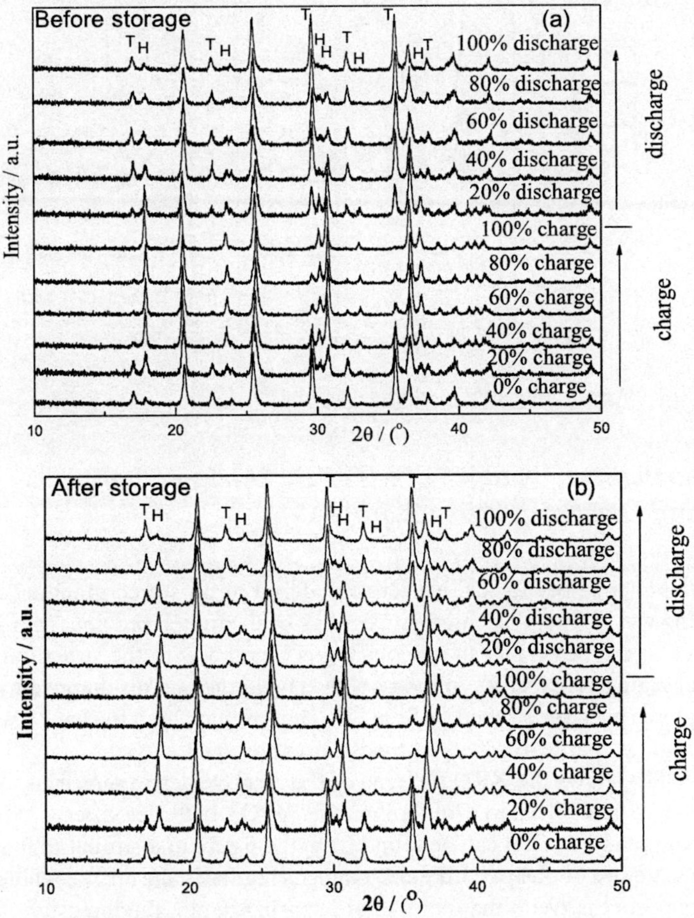

Fig. 19.11 Charge and discharge process of graphite/$LiFePO_4$ battery before (**a**) and after (**b**) storage in 55 °C 50%SOC 135 days

$LiFePO_4$ phase in the discharged positive active material is higher than that in the charged state. This shows that in the process of charging, the disappearance rate of $LiFePO_4$ phase and the increase rate of $FePO_4$ phase are faster, while in the process of discharge, the disappearance rate of $FePO_4$ phase and the increase rate of $LiFePO_4$ phase are faster. This shows that there is a certain hysteresis or asymmetry [2] in the phase change of $LiFePO_4$ positive active materials during charge–discharge.

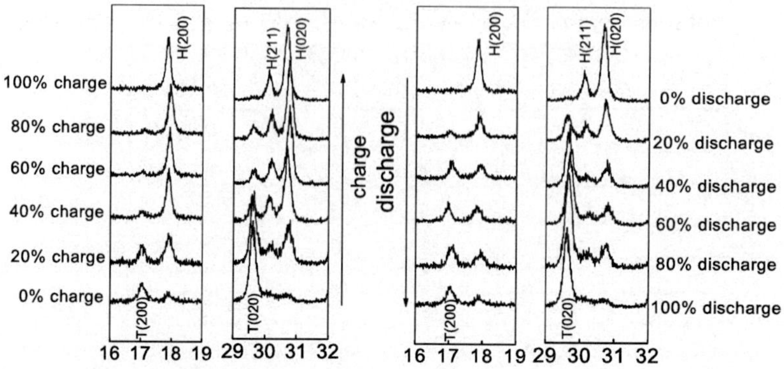

Fig. 19.12 Partial magnification of Fig. 19.11

19.3.2 *Structural Changes of $LiFePO_4$ During Charge and Discharge After Storage*

The structural change of the positive active material of no-stored graphite/$LiFePO_4$ battery during the charge–discharge process has been studied previously. It is found that the discharge process has an obvious lag compared with the structural change during the charging process; Asymmetry between charge and discharge during high rate cycling,have been researched by Shin [2]. Then, What about the law of structural change during the charge–discharge process after storage?

Figure 19.11b shows the XRD patterns of the positive active materials at typical stages of charge and discharge of graphite/$LiFePO_4$ batteries after 135 days of 50%SOC storage in 55 °C. It can be found from the figure that similar to that before storage, the structure of positive $LiFePO_4$ changes regularly during the whole charge and discharge process. With the increase of charging depth, the intensity of $FePO_4$ peak increases, the intensity of $LiFePO_4$ peak decreases, and the discharge process is opposite. From the surface, it is difficult to see the difference in the law of structural change before and after storage. However, after careful comparison, it can be found that the intensity ratios of $FePO_4$ and $LiFePO_4$ peaks in the diffraction patterns with the same charge–discharge state before and after storage are different, which seems to indicate that hysteresis of phase change of $LiFePO_4$ positive active materials during charge–discharge after storage has changed.

In order to further analyze the lag difference of the phase transition of positive active materials before and after storage, the relative contents of $FePO_4$ and $LiFePO_4$ in positive active materials at each stage of charge and discharge were quantitatively calculated by Zerwin's non-standard sample method [6, 7]. By formula

$$\begin{cases} \left[\left(1 - \frac{I_{LiFePO_4B}}{I_{LiFePO_4A}}\right)128.28x_{LiFePO_4A}\right] + \left[\left(1 - \frac{I_{FePO_4B}}{I_{FePO_4A}}\right)128.25x_{FePO_4A}\right] = 0 \\ x_{LiFePO_4A} + x_{FePO_4A} = 1 \end{cases} \quad (19.2)$$

The relative integral intensities of the two phases in the XRD diffraction spectra of each charge–discharge state are obtained by fitting the Jade6.5 software and substituted into the Eq. (19.2). By solving them, the relative contents of $FePO_4$ and $LiFePO_4$ in the positive active materials in each charge–discharge state can be obtained. The results are shown in Fig. 19.13.

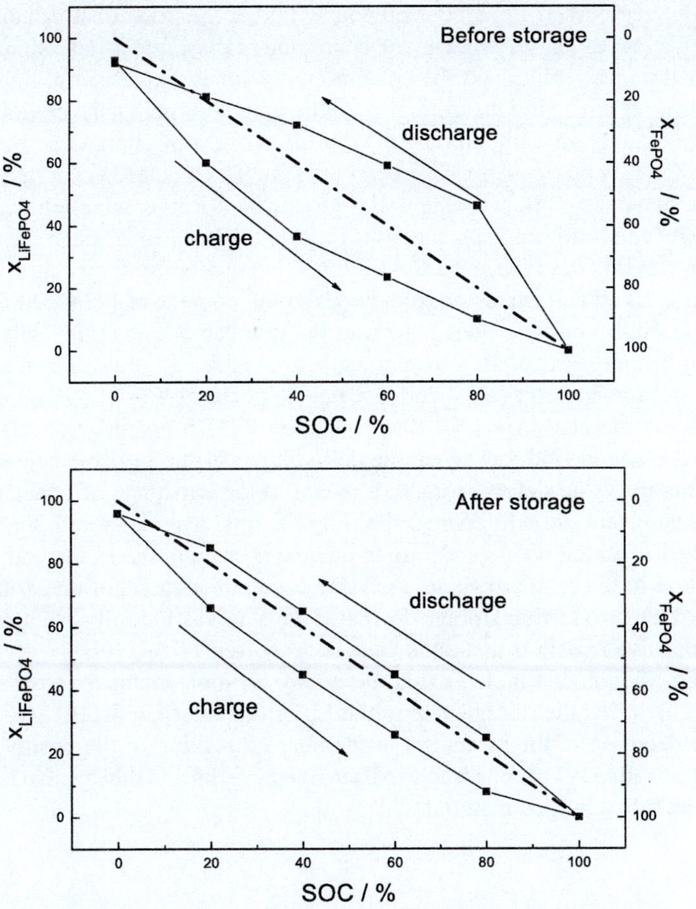

Fig. 19.13 During the charge–discharge process of $LiFePO_4$ electrode before and after 135 days 55 °C 50%SOC storage. Relative content change of $LiFePO_4$ and $FePO_4$ phases

It can be seen from the figure that even in the discharge state before storage, the positive active material is not entirely $LiFePO_4$ phase, in which there is a small amount of $FePO_4$ phase (calculated to be about 8%), which is caused by the low ion conduction capacity of $LiFePO_4$ material itself and the formation of inactive $FePO_4$ shell during the first charge. However, it can be found that after 135 days of 55 °C storage in 50%SOC, the content of $FePO_4$ phase in the positive active material decreased to about 4% when it was in the discharge state, indicating that the inactive $FePO_4$ shell was reduced after storage. Compare the curve of the relative content of $LiFePO_4$ and $FePO_4$ phase with the depth of charge and discharge before and after storage (the dotted line in the figure is the ideal curve of the relative content of $LiFePO_4$ and $FePO_4$ phase during the charge–discharge process (the equation is $y = x$)). It can be found that before storage, the increase rate of $FePO_4$ phase content at the beginning of charge is obviously higher than the ideal value, and the disappearance rate of $FePO_4$ phase at the beginning of discharge is also significantly higher than the ideal value, and the content change of two phases during charge–discharge process have obvious difference, indicating that there is an obvious lag in the structural change of cathode materials in the process of charge and discharge. The reason for this phenomenon has been explained earlier. However, the change curve of the two phases after storage is close to the ideal curve, which indicates that the lag is reduced after storage. At the same time, in the XRD diagram of the positive electrode in the full state, because the $LiFePO_4$ is covered by the peak of $FePO_4$, the content of $LiFePO_4$ cannot be calculated, so the content of $FePO_4$ in the fully filled state is 100%, but this does not mean that the cathode is completely $FePO_4$ phase at this time. By calculating the unit cell parameters of $FePO_4$, it can be found that before storage, the unit cell volume of $FePO_4$ phase is $V_{cell} = 277.64$, storage, and its cell volume is $V_{cell} = 274.12$ and $V_{cell} = 273.75$ given by standard PDF card. It can be understood that when the cell volume of the positive pole is larger than the standard value before storage, it is due to the existence of undetached Li in the positive electrode, while when the fully charged state is stored, the content of undetached Li in the positive electrode decreases, which makes the cell volume decrease, close to the standard value. This shows that the content of unexfoliated Li in the positive electrode after storage decreases when it is in the full state; that is, the content of inactive $LiFePO_4$ mentioned above decreases.

The above phenomena indicate that there may be some changes in the positive active material during the storage of graphite/$LiFePO_4$ battery with 50%SOC, which leads to the decrease of the hysteresis of the phase transition of the positive active material, the inactive $FePO_4$ phase in the discharge material and the inactive $LiFePO_4$ content in the fully charged material.

19.4 Analysis of Battery Performance Degradation Mechanism

19.4.1 The Relationship Between the Change of Battery Performance and the Fine Structure of Battery Active Material

In order to reveal the law of the decline of storage performance, the research results of performance and structure of 19.1–19.3 are sorted out as shown in Table 19.5.

Summing up Table 19.5, it is not difficult to find:

(1) After storage at high temperature (55 °C), the charge state of the battery also has a great influence on the storage performance of graphite/$LiFePO_4$ battery, which is similar to that of lithium cobalt and ternary material battery mentioned above.
(2) However, the difference is that the capacity of graphite/$LiFePO_4$ battery increases at first and then decreases during 25–75%SOC storage, and this phenomenon is the most obvious during 50%SOC storage; during storage, the highest point of battery capacity appears at 135 days, at which time, the power performance and safety of the battery have been improved to some extent. During storage at 0% and 100%SOC, the capacity of the battery decreases with the extension of storage time, but its capacity decay rate is also much lower than that of the other two types of batteries.
(3) After storage, the capacity change rate of the positive electrode and the whole battery is very close. At the same time, the impedance of the positive electrode of the battery decreases and the dynamic performance improves after 25–75%SOC storage, while after 0% and 100%SOC storage, the impedance of the positive electrode of the battery increases and the dynamic performance decreases, which is consistent with the performance change of the battery, indicating that the $LiFePO_4$ positive electrode plays a decisive role in the performance change of graphite/$LiFePO_4$ battery after storage.
(4) The surface morphology of $LiFePO_4$ electrode after storage was observed, and the composition was determined. It was found that, similar to lithium cobalt oxide and ternary materials, a small amount of fluorine-containing precipitates also appeared on the surface of $LiFePO_4$ particles after storage, which was the main reason for the decline of battery performance after storage at 0% and 100%SOC. At the same time, it was found that microcracks appeared in the particles of positive active materials after 25–50%SOC storage. It is found that this phenomenon has a lot to do with the improvement of battery performance after storage
(5) There is an obvious hysteresis effect in the phase transition of $LiFePO_4$ positive active material during charge and discharge before storage, but this hysteresis effect is greatly reduced after storage.

Table 19.5 Summary of the results of the study in 19.1–19.3

Item		Content summary
1	55 °C storage capacity of 0.2C	The changes of 55 °C storage capacity with storage days are as follows: 1. The SOC of 0% and 100% is a monotonous slow attenuation 2. When SOC is 25, 50, 75%, the capacity increases at first and then decreases with the increase of storage days. The maximum value appears at 135 days, and 50%SOC is the most obvious
2	Charge/discharge characteristics of 0.2C after 135 days storage in 55 °C	1. The charging and discharging platform before storage is very flat 2. When the SOC is 25–75%, the battery capacity increases, the discharge characteristics improve, the charging platform voltage decreases, and the discharge voltage platform increases 3. After the storage of 25, 50, and 75% of the charge for 135 days, the AC internal resistance decreased by 1.5%, 2.5%, and 1.5%, respectively 4. After 135 days of 25–75%SOC storage, the discharge curve of the battery showed a new discharge platform at 3.15 V
3	Charge/discharge characteristics of 1C 3C after 135 days storage in 55 °C	The battery capacity decreased gradually with the increase of multiple ratio, and the performance of 25–75%SOC storage improved to a certain extent. After 0% and 100%SOC storage, the battery capacity decreased gradually with the increase of multiple ratio. At 3C, the discharge platform increased at first and then decreased with the increase of SOC
4	Safety performance	According to the overcharge curve of 55 °C storage for 135 days, the surface temperature of the battery increases with the increase of voltage, the temperature reaches 85 °C when it reaches 12 V, the charging time is prolonged, and the surface temperature decreases, so it has good safety performance
5	Stability	From the curve of the relationship between the self-heating rate and temperature of the battery stored in 55 °C for 135 days, it is known that from 160 °C, the self-heating rate of the battery increases with the increase of temperature, and after reaching the maximum value, the self-heating rate decreases gradually with the increase of temperature. The results show that the battery has good overcharge resistance
6	Comparison of 0.2C capacity change rate between battery and negative electrode after storage 135 days at 55 °C	Both the whole battery and $LiFePO_4$ increased at first and then decreased with the increase of SOC, and the maximum value was at 50%SOC, while the capacity of negative electrode decreased with the increase of SOC
7	Cyclic voltammetric characteristics	The kinetic properties of 25–75%SOC after storage are improved. The kinetic properties of 25–75%SOC after storage are improved, and the peak shape is reduced from a single peak before storage to two peaks, which corresponds to the emergence of a new platform in the positive charge–discharge curve after storage

(continued)

Table 19.5 (continued)

Item		Content summary
8	Impedance of positive and negative electrodes	After storage, the impedance of positive electrode changes obviously, the impedance of 25–75%SOC decreases, the impedance of 50%SOC is the most obvious, and the impedance of 0% and 100%SOC increases; the impedance of negative electrode changes little and increases with the increase of SOC during storage; the impedance of positive electrode is one order of larger than impedance of negative electrode
9	The surface morphology of the positive and negative electrode and the surface morphology of the diaphragm on one side of the positive electrode	The surface film of the negative electrode is thickened; the surface morphology of the positive side diaphragm has no obvious change after storage; but the surface of the positive electrode changes obviously with SOC, and there is a small product on the surface after 0% and 100%SOC storage, but the surface is still continuous and free of microcracks. After 25–75%SOC storage, there are also small products and surface cracks, especially in 50%SOC
10	Phase transition and asymmetry of phase transition before and after 135 days storage in 55 °C	There is an pseudo-structure phase transition of $LiFePO_4 \leftrightarrow FePO_4$ in the charge–discharge process, but there is an obvious hysteresis effect in the phase transition of $LiFePO_4$ positive active materials before storage, that is, the disappearance speed of $LiFePO_4$ phase is faster than that of $FePO_4$ phase in the process of charging, while the increasing speed of $LiFePO_4$ phase is faster than vanishing speed of $FePO_4$ phase in the process of discharge Through the quantitative analysis of $LiFePO_4$ and $FePO_4$ phases in the charge and discharge process of the two kinds of batteries before and after storage, it is found that the hysteresis effect is greatly reduced after 135 days of 55 °C storage, which shows that the deviation from the ideal value of the relative content curve of charge and discharge phases before and after storage is greatly reduced

(6) During the charged storage of $LiFePO_4$, the two phases of $LiFePO_4$ and $FePO_4$ coexist in the material particles, and there is a volume difference between the two phases. The internal stress on the interface makes the grains of the two phases split along the interface direction, which leads to cracks. The crack will produce a fresh interface and make the inactive core which originally exists in the active material particles have lithium removal activity again, so the dynamic performance of the positive electrode is improved and the capacity is increased; at the same time, the particle size will be reduced after cracking, which leads to the lag of $LiFePO_4/FePO_4$ phase transition of the positive material during charge and discharge after storage.

19.4.2 Discussion on the Mechanism of Storage Performance Change of Battery

As in Chap. 18, the self-discharge and the chemical/electrochemical interaction between the positive electrode surface, the negative electrode surface, the two surfaces of the diaphragm and the electrolyte also occur during battery storage. Therefore, the mechanism discussed in Chap. 18 also applies here, which can solve the storage performance changes of 0% and 100%SOC. However, the storage performance of 25–75% of the charge cannot be explained, which is discussed below.

As mentioned earlier, when the $LiFePO_4$ electrode is not activated, it is $LiFePO_4$ single phase. After the first charge, the single-phase $LiFePO_4$ is transformed into two-phase $LiFePO_4$/$FePO_4$, and then the residual $FePO_4$ phase appears in the electrode after discharge. This phase cannot be reversibly de-intercalated lithium in the subsequent charge–discharge cycle, thus reducing the capacity of the battery. When the battery is stored in the charged state, the two phases of $LiFePO_4$/$FePO_4$ coexist in the positive electrode. It is found from the manual that $LiFePO_4$ belongs to the orthorhombic, Pnma (No. 62 space group of lithium iron phosphate, while $FePO_4$ belongs to the orthorhombic Pnma (No. 62) space group of ferromanganese. The lattice parameters are listed in Table 19.6. It can be seen that the lattice parameters of $LiFePO_4$ and $FePO_4$ are quite different. The *a* and *b* values of $LiFePO_4$ are larger than those of $FePO_4$, and its *c* value is less than $FePO_4$. After calculating the cell volume of the two phases, it is found that there is a large volume difference between the two cells, and $LiFePO_4$ is 6.46% larger than $FePO_4$. From the lithium de-intercalation model of $LiFePO_4$ charge and discharge process proposed by Andersson [7], it can be known that In the charged $LiFePO_4$ particles, $LiFePO_4$ and $FePO_4$ form continuous phases with layered or mosaic distribution, respectively, so that the volume difference between the two cells will further accumulate, resulting in a larger volume difference between the two phases, when a sharp interface will be formed between the two phases, and there is a strong internal stress on the interface. In the process of storage, the interface stress will split the two-phase grains along the interface direction, resulting in cracks, as shown in Fig. 19.10.

The crack will produce a fresh interface, so that the inactive core that originally exists in the active material particles will be exposed and come into contact with the electrolyte, thus making it have lithium removal activity again, so that the positive electrode will have a new discharge platform after storage. To increase the capacity

Table 19.6 Lattice parameters and cell volumes of $LiFePO_4$ and $FePO_4$

	Lattice parameters (Å)				
	a	*b*	*c*	*V*	$(V_T - V_H)/V_T$ (%)
$LiFePO_4$	6.018	10.34	4.703	292.65	0.00
$FePO_4$	5.824	9.821	4.786	273.75	−6.46

of the battery, the appearance of the fresh interface will increase the electrochemical reaction activity of the electrode, reduce the RCT, of the positive electrode as shown in Table 19.3, and improve the kinetic performance of the electrode, thus reducing the polarization of the battery during charge and discharge (especially at high current). At the same time, the cracking of the particles will reduce the particle size and shorten the diffusion path of lithium ions in the material, so that the thickness of the $FePO_4$ (charging) and $LiFePO_4$ (discharging) phases preferentially formed in the outer layer of the particles is reduced during the charge and discharge process, so that the blocking effect on the X-ray diffraction of the inner $LiFePO_4$ (charging) and $FePO_4$ (discharge) phases is reduced in the X-ray diffraction experiment, so that the barrier effect on the X-ray diffraction of the inner $LiFePO_4$ (charging) and $FePO_4$ (discharge) phases is reduced. After storage, it is observed in the XRD that the hysteresis or asymmetry of the phase change in the charge-discharge process of the positive active material has been reduced pattern.

At the same time, it can also be found that when the battery is 50%SOC, the ratio of $LiFePO_4$ to $FePO_4$ in the positive active material is about 1 $FePO_4$, the interface between the two phases in the positive electrode is the largest, the particles of the positive active material have the most microcracks after storage (as shown in Fig. 19.10), and the fresh surface area inside the particles is the largest, so the battery performance is improved the most after storage. When the battery is 25% and 75%SOC, the ratio of two-phase content in the positive active material is about 3:1 and 1:3, and the interface area between the two phases in the two-electrode active material is similar, but smaller than that in the 50%SOC positive active material, so the improvement of battery performance is small after storage.

However, when the battery is stored at 0% or 100%SOC, the positive active material is basically single phase, so the particles will not crack. At this time, the interaction between the electrolyte and the positive active material leads to the deposition of F-containing material on the cathode surface, thus changing the surface characteristics of the positive electrode becomes the main factor affecting the performance of the electrode. Therefore, after storage at 0% or 100%SOC, the positive electrode impedance increases and the dynamic performance decreases.

This results in a decline in the performance of the battery.

Comparing the storage performance of three different positive active materials, graphite/$LiCoO_2$ and graphite/$Li(Ni_{1/3}Co_{1/3}Mn1_{/3})O_2$ graphite/$LiFePO_4$, it can be found that the capacity decay of graphite/$LiFePO_4$ battery is the slowest in the process of storage, the performance decline is the smallest after storage, and the storage performance is the best, and the storage performance can be improved when the charge is 25–75%. Especially when the charge is 50%, the storage is the most favorable.

References

1. Li, Jia. 2010. Study on storage performance of lithium-ion battery. Doctoral thesis of Shanghai Institute of Microsystems and Information Technology, 6.
2. Shin, Ho Chul, Kyung Yoon Chung, Won Sik Min, Dong Jin Byun, Ho Jang, and Byung Won Cho. 2008. Asymmetry between charge and discharge during high rate cycling in $LiFePO_4$—in situ X-ray diffraction study. *Electrochemistry Communications* 10: 536–540.
3. Newman, J., and V. Srinivasan. 2004. Discharge model for the lithium iron-phosphate electrode. *Journal of Electrochemical Society* 151 (10): A1517–A1529.
4. Yamada, A., H. Koizumi, N. Sonoyama, and R. Kanno. 2005. Phase change in Li_xFePO_4. *Electrochemical and Solid-State Letters* 8 (8): A409–A413.
5. Dodd, J.L., R. Yazami, and B. Fultz. 2006. Phase diagram of Li_xFePO_4. *Electrochemical and Solid-State Letters* 9 (3): A151–A155.
6. Cheng, G.F., C.Z. Yang, and Y.H. Huang. 2010. X-ray analysis of nanomaterials, Chap. 5. Chemical Industry Press, 6; Yang, C.Z., D.C. Xie, K.Z. Chen, and F.M. Zhong. 1989. Phase diffraction, Chap. 4. Metallurgical Industry Press.
7. Zevin, L.S. 1977, 1979. *Journal of Applied Crystallography* 10: 147; 12: 582.
8. Andersson, A.S., and J.O. Thomas. 2001. The source of first-cycle capacity loss in $LiFePO_4$. *Journal of Power Sources* 97–98: 498–502.

Chapter 20
Effect and Action Mechanism of β-$Ni(OH)_2$ Additive in MH/Ni Battery

Chapter 13 has mentioned that the purpose of studying the degradation mechanism of battery cycle performance is to seek and improve the performance of MH-Ni battery and improve cycle performance. This chapter studies and discusses this problem.

The design of MH/Ni battery generally follows the concept of positive limit capacity, and the positive electrode generally accounts for 35% of the battery quality and 36% of the volume, so improving the performance of the positive electrode is one of the keys to the manufacture of high-performance MH/Ni batteries. Because additives can significantly improve the performance of nickel electrodes, it has always been an important research direction of MH/Ni batteries [1, 2]. Among them, the calcium additive can increase the oxygen evolution potential of the nickel electrode and significantly improve the charging efficiency of the nickel electrode at 20–70 °C, thus improving the utilization rate of nickel hydroxide [3–7].

In this chapter, the effects of positive material β-$Ni(OH)_2$ additive on the charge–discharge capacity and cycle performance of MH/Ni battery were introduced, and the effects of additives on the microstructure of positive β-$Ni(OH)_2$ and negative AB_5 hydrogen storage alloys were studied by using the least square method [8, 9] to separate the X-ray diffraction broadening effect of microcrystal–micro-strain double and microcrystal–micro-strain–stacking fault triple. The effect of additives on the microstructure of positive β-$Ni(OH)_2$ and negative AB_5 hydrogen storage alloys was studied, and the properties and microstructure parameters were linked to explore the action mechanism of β-$Ni(OH)_2$ additive.

C. Yang et al., *Materials and Working Mechanisms of Secondary Batteries*,
https://doi.org/10.1007/978-981-19-5955-4_20

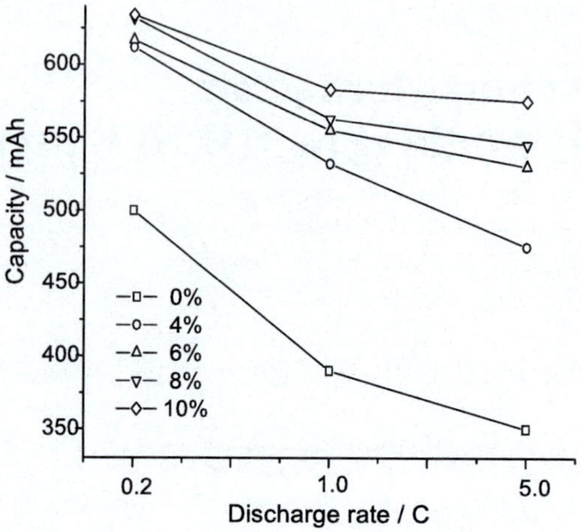

Fig. 20.1 Relationship between CoO content in positive electrode and rate discharge capacity

20.1 Effect of Adding CoO [4]

20.1.1 Effect of CoO Addition on Battery Performance

Figures 20.1 and 20.2 show the effects of different weight ratios of CoO additives on battery charge–discharge capacity at different rates and 1 C charge–2 C discharge capacity in the battery cycle, respectively. It can be seen that CoO additives have obvious effects on battery charge–discharge capacity and cycle performance and increase significantly with the increase of CoO content.

20.1.2 Effect of CoO Addition on Microstructure Parameters of Battery Positive and Negative Active Materials

In order to explore and analyze the reasons for the difference of battery performance with different CoO content, the microstructure of positive and negative electrode materials is studied. Figure 20.3 shows the XRD pattern of β-Ni(OH)$_2$ after the battery cycle 400 period. It can be seen that the XRD pattern of β-Ni(OH)$_2$ powder without CoO is obviously different. The diffraction peaks of 101 and 100 are not completely separated, especially the asymmetry of the diffraction peaks of 101 in the low angle direction, which indicates that the crystal structure of β-Ni(OH)$_2$ without CoO is obviously distorted. However, β-Ni(OH)$_2$ with CoO can keep its original crystal structure during the cycle. The analysis results of the relevant data are shown in Figs. 20.4 and 20.5 as a curve.

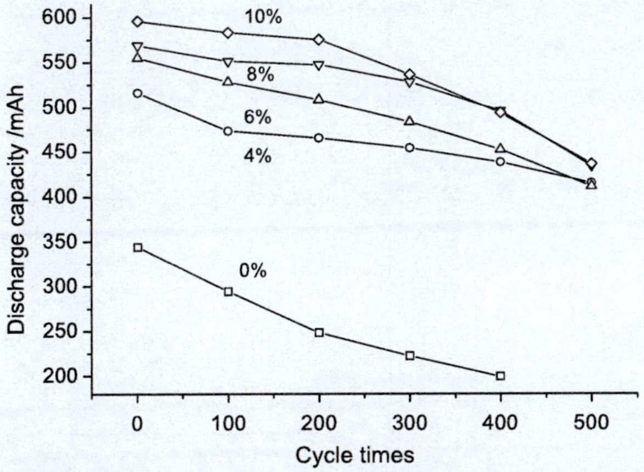

Fig. 20.2 Relationship between CoO content in positive electrode and 1 C charge 2 C discharge capacity in the battery cycle

Fig. 20.3 XRD pattern of β-$Ni(OH)_2$ powder in AAA battery with different CoO content in cathode after 400 cycles

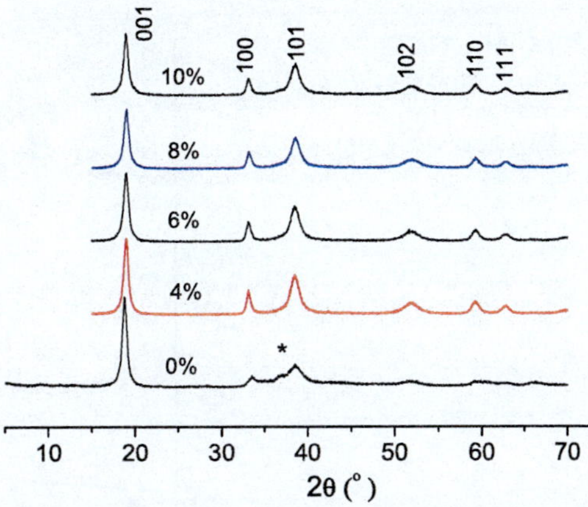

Figure 20.6 shows the XRD pattern of AB_5 hydrogen storage alloy in AAA battery with different CoO content after cycle 400 periods. They're almost the same. Except for the diffraction peaks of AB_5 alloy, a new diffraction peak appears near $2\theta =$ 27–29°. The results of Jade program search/matching analysis show that the new phases are written as $A(OH)_3$ and B, where A represents La, B represents Ni, and the relevant data analysis results are shown in Figs. 20.7 and 20.8.

It can be seen from the above that effect of CoO content on average grain size, micro-strain and stacking fault probability f_D, f_T, $f_D + f_T$ in the positive active materials has no significant effect. The average grain size and micro-strain of negative electrode materials increase with the increase of CoO. This indicates that the grain fining of the negative active material is not inhibited, but the precipitation of $A(OH)_3$ and B phase is greatly inhibited.

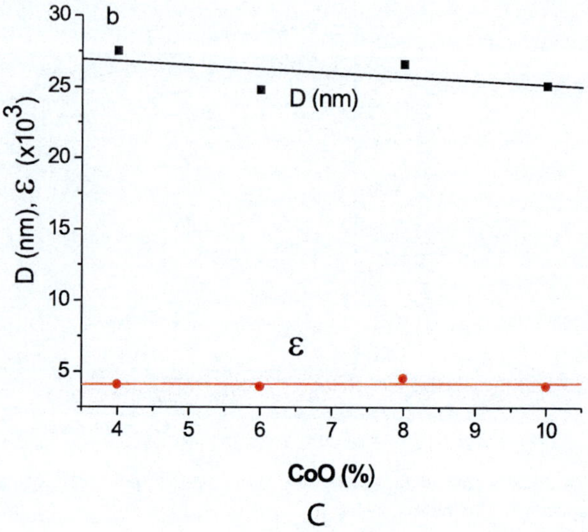

Fig. 20.4 Effect of CoO content in the positive on the microstructure parameters of β-$Ni(OH)_2$ after cycle 400 periods

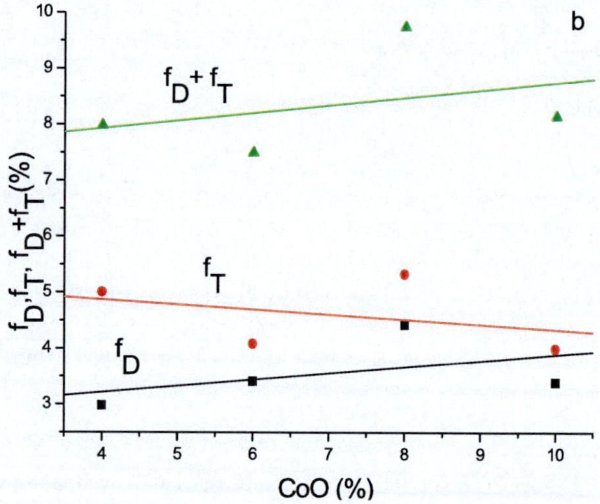

Fig. 20.5 Effect of CoO content in positive electrode on stacking fault in β-$Ni(OH)_2$ after cycle 400 periods

Considering that the cycle performance of AAA battery with 8 and 10% CoO is basically the same after more than 300 periods, 8% has a higher probability of stacking fault, as well as cost factors, so the following study uses β-$Ni(OH)_2$ with 8% additive as positive material to further study the effects of other additives.

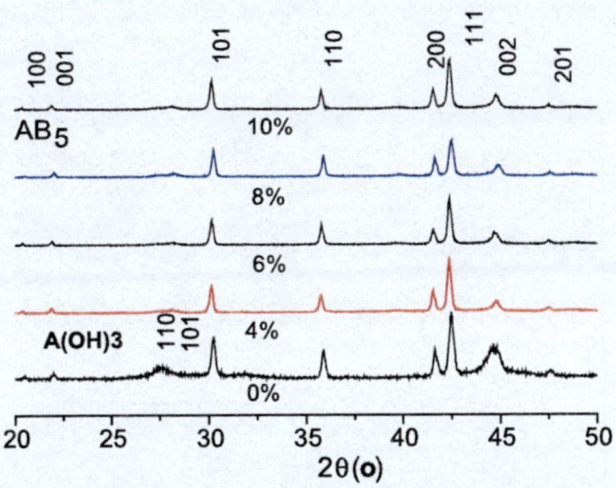

Fig. 20.6 XRD patterns of hydrogen storage alloys in AAA batteries with different CoO contents after cycle 400 periods

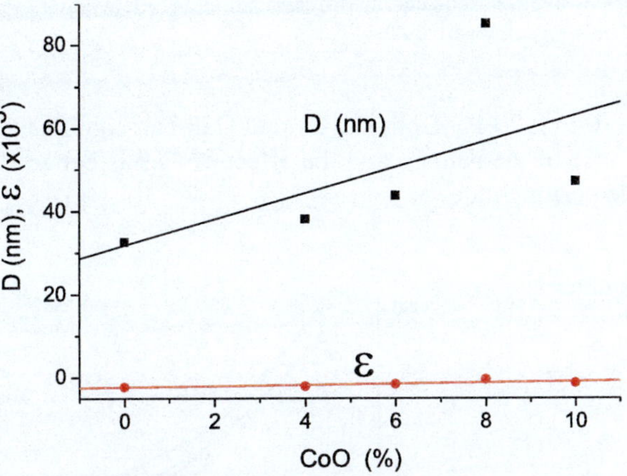

Fig. 20.7 Effect of CoO content in positive electrode on microstructure parameters D and ε of hydrogen storage alloy AB_5 after cycle 400 periods

20.2 Effects of Different Types of Additives [4]

20.2.1 Effects of Different Types of Additives on Battery Performance

Different kinds of additives are mechanically evenly mixed with the same β-$Ni(OH)_2$ according to 3% by weight, and then they are used as positive materials and negative AB_5 hydrogen storage alloys to make AAA batteries. Its rate performance is shown in Fig. 20.9. It can be seen that at low rates, Lu_2O_3 and CaF increase the initial capacity, while Lu_2O_3 increases the capacity significantly under all charging rates. The cycling

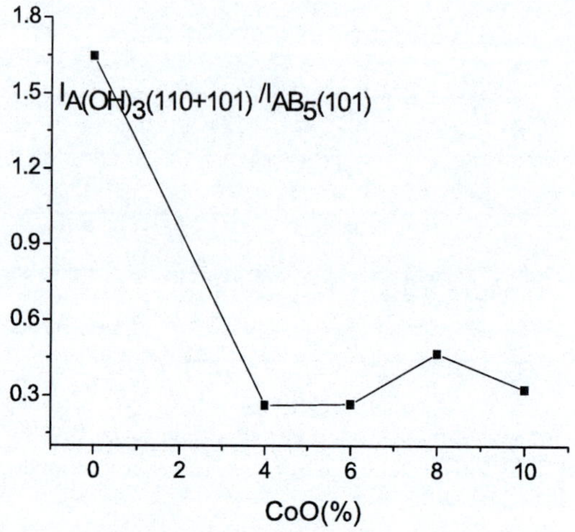

Fig. 20.8 Effect of positive electrode adding different amounts of CoO after cycle 400 periods of on $I_{A(OH)3(110+101)}/I_{AB5(101)}$ of negative AB_5

energy is shown in Fig. 20.10. Both Lu_2O_3 and CaF can significantly improve the cycling performance of the battery, and the effect of CaF is better. However, Y_2O_3 deteriorates cycle performance.

Fig. 20.9 Effect of different additives on the rate performance of battery

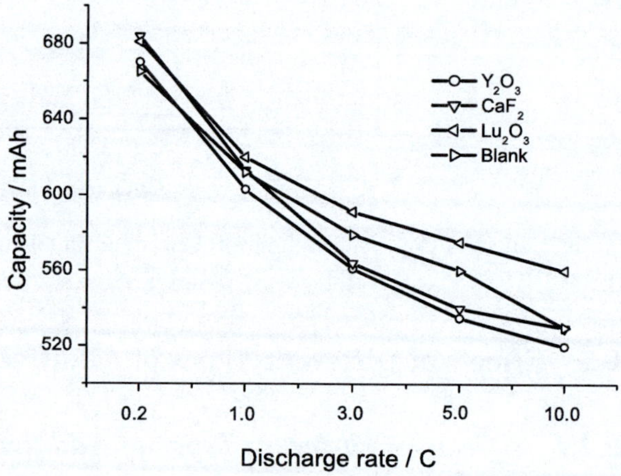

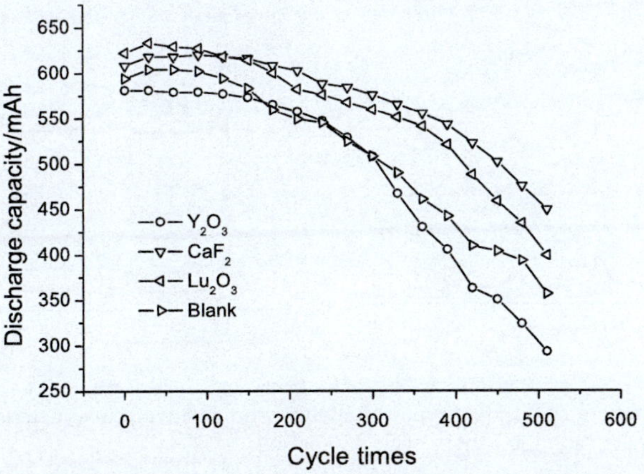

Fig. 20.10 Effect of different additives on cycle performance of battery

20.2.2 *Effects of Different Additives on Structure and Microstructure of Positive Active Materials*

In order to further explore the effect of additives on the microstructure parameters of positive active materials, the structural changes of positive active materials after activation and cycle 400 period were studied by X-ray diffraction. The XRD patterns after activation and cycle 400 periods of the battery are shown in Fig. 20.11a, b, respectively. The effects of different additives on the microstructure parameters of positive active materials are listed in Table 20.1. It is found that there is no regularity in the activated data, but after 400 cycles, the average grain size of adding Lu_2O_3 and CaF_2 is only 1.36 times of that of unadded, while that of adding Y_2O_3 is 3.69 times that of unadded, indicating that the activity of cathode materials added with Y_2O_3 is very poor, while that of adding Lu_2O_3 and CaF_2 decreases little; and the micro-strain of the former two is also smaller than that of the latter. In addition, the stacking fault probability ($f_D, f_T,$ and $f_D + f_T$) of adding CaF_2 is also smaller than that of unadded, which indicates that CaF_2 can inhibit the microstructure changes in the process of activation and circulation.

20.2.3 *Effects of Different Additives on the Structure and Microstructure of Negative Active Materials*

The XRD pattern of negative AB_5 alloy is shown in Fig. 20.12, and the results of parameter analysis of its structure and microstructure are listed in Table 20.2. It can be seen that compared with the unadded samples, the change rate of lattice parameters a and c of the negative active material AB_5 with Lu_2O_3 and CaF_2 are much larger than

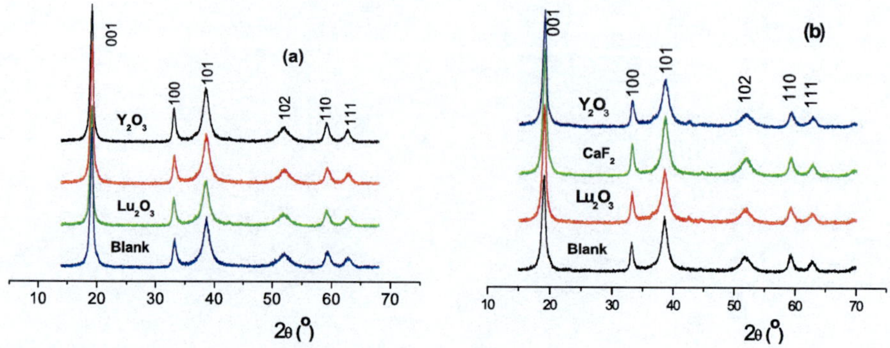

Fig. 20.11 XRD pattern of different types of additives on positive active material β-$Ni(OH)_2$. **a** After AAA battery activation, **b** battery cycle 400 periods

Table 20.1 Effects of different positive additives on microstructure parameters of positive active materials

		No additive	Lu_2O_3	Y_2O_3	CaF_2
After activation	a (Å)	2.9816	3.1194	3.1058	2.9685
	c (Å)	4.3533	4.6061	4.5846	4.3313
	$\overline{D}$ (nm)	47.5	81.8	46.2	48.8
	ε ($\times 10^{-3}$)	6.13	6.63	4.84	5.66
	f_D (%)	5.77	4.66	4.90	6.07
	f_T (%)	3.91	4.58	3.32	1.42
	$f_D + f_T$ (%)	9.68	9.24	8.22	7.49
After cycle	a (Å)	2.9732	3.1001	3.0706	3.1214
	c (Å)	4.3623	4.5779	4.5488	4.6335
	$\overline{D}$ (nm)	24.6	33.4	90.9	33.6
	ε($\times 10^{-3}$)	3.44	5.03	7.84	4.62
	f_D (%)	3.54	6.52	5.67	3.87
	f_T (%)	4.25	0.29	1.99	2.95
	$f_D + f_T$ (%)	7.69	6.81	7.66	6.82

those with Y_2O_3; but after cycle 400 periods, it is much smaller, and the micro-strain is also smaller. However, the most obvious is the change of the intensity ratio of $I_{A(OH)3(110+101)}/I_{AB5(101)}$. The values without and without Lu_2O_3, Y_2O_3, and CaF_2 are 0.541, 1.007, and 0.249, respectively (see the last line of Table 20.2). It can be seen that the addition of Lu_2O_3 and CaF_2 can obviously inhibit the precipitation of new phase, especially CaF_2. These are the reasons why the addition of Lu_2O_3 and CaF_2 can significantly improve the first discharge capacity and cycle performance of the battery. The addition of Y_2O_3 promotes the precipitation of new phase.

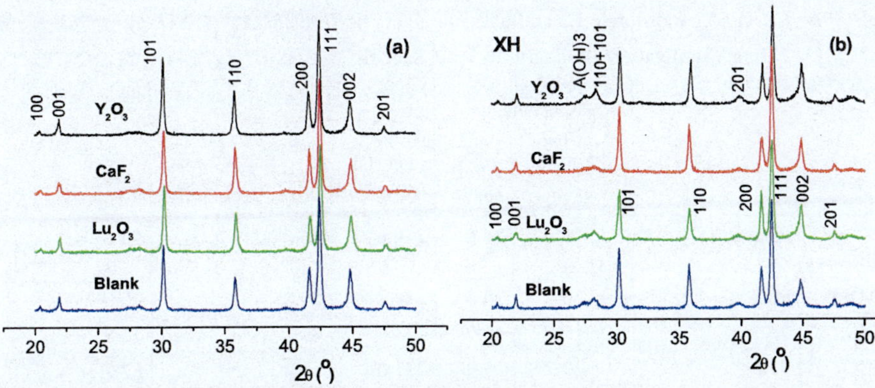

Fig. 20.12 XRD pattern of negative electrode active material AB_5 alloy with different types of additives. **a** After battery activation, **b** battery cycle 400 times

Table 20.2 Effect of cathode additives on microstructure parameters of hydrogen storage alloy AB_5

		No additives	Lu_2O_3	Y_2O_3	CaF_2
After activation	a (Å)	4.9869	5.0326	4.9786	5.0179
	c (Å)	4.0230	4.0589	4.0098	4.0441
	$\overline{D}$ (nm)	190.7	68.1	49.8	31.7
	$\varepsilon(\times 10^{-3})$	0.79	0.19	− 0.66	− 1.95
After cycle 400 period	a (Å)	4.9421	4.9765	5.0677	4.9873
	c (Å)	3.9829	4.0095	4.0857	4.0183
	$\overline{D}$ (nm)	99.8	35.2	36.0	110.8
	$\varepsilon\ (\times 10^{-3})$	1.02	− 1.04	1.62	0.53
	$\frac{I_{A(OH)_3 110+101}}{I_{AB_5 101}}$	0.631	0.541	1.007	0.249

20.3 Effects of Different Ca Additives [4, 10]

In order to further prove the good effect of CaF_2 additive, different Ca additives: $Ca(OH)_2$, CaF_2, and $CaCO_3$ are selected and added at the weight ratio of 1% of β-$Ni(OH)_2$, and the battery is made according to the method used in the 20.2 section. The performance after charge and discharge is shown in Table 20.3. It can seen that the addition of CaF_2 and $Ca(OH)_2$ has higher specific capacity and higher β-$Ni(OH)_2$ utilization. The data analysis results of the positive materials after cycle 200 periods of 1 C cycle are shown in Table 20.4. According to the analysis of the above data, we can see that:

(1) The compounds containing Ca do not change the activation to transform the chunky grains into near-equiaxed grains, but hinder the process and degree of

Table 20.3 Effect of 1% different Ca additives on battery performance

	$Ni(OH)_2$ utilization rate (%)	Battery internal resistance (mΩ)		Specific capacity of cathode material (mAh/g)	C_{600C}/C_{200C} (%)	C_{190}/C_1 (%)
		No cycle	After cycle			
No additives	93.5	15.0	26.9	219.8	46.0	90.5
$Ca(OH)_2$	96.2	15.3	34.0	224.2	50.8	89.0
CaF_2	96.1	15.0	46.9	224.0	49.8	75.8
$CaCO_3$	94.8	20.7	25.9	221.0	51.4	87.2

Table 20.4 Microstructure of positive materials with 1% different calcium additives

D_{hke} (nm)	001	100	110	111	$\overline{D}$	$\varepsilon(\times 10^{-3})$	f_D	f_T	$f_D + f_T$
No additives	19.5	16.3	9.3	11.0	36.6	4.92	1.84	4.00	5.84
$Ca(OH)_2$	24.5	22.5	10.5	12.3	89.6	5.12	3.11	2.49	5.60
CaF_2	19.5	19.6	8.8	10.4	61.1	5.20	3.39	2.47	5.58
$CaCO_3$	19.5	19.3	10.2	11.0	41.9	4.68	3.32	2.44	5.56

grain fining and increase the average grain size. The addition of $Ca(OH)_2$ is the most obvious.

(2) Ca-containing compounds promote the transformation from growth stacking fault to deformed stacking fault, but the total stacking fault probability does not change much.

(3) The compounds containing Ca also change the existence situation of micro-strain caused by the reversible de-intercalation of H atoms. Comparing change of performance and microstructure, the influence of $Ca(OH)_2$ and CaF_2 is the most significant, while the influence of $CaCO_3$ is smaller, which is related to the activity of these three compounds.

In order to further prove the effect of the above addition of Ca-containing compounds, we added $Ca(OH)_2$, CaF_2, and $CaCO_3$ at a weight ratio of 3%, respectively, and found that CaF_2 also greatly improved the cycle performance (see Fig. 20.13).

On the contrary, $CaCO_3$ can improve the discharge capacity at higher rates, but significantly deteriorate the cycle performance. In order to explore the reason, the microstructure X-ray studies of the cathode and negative materials in the initial and final states of the cycle are carried out. The results are listed in Table 20.5. The XRD spectra of the final state cycles of the negative materials are shown in Fig. 20.14.

A comprehensive analysis of the above data clearly shows that:

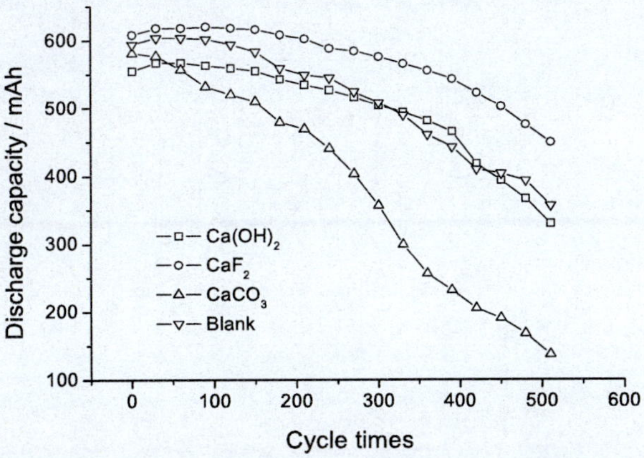

Fig. 20.13 Effect of different calcium additives on cycle performance of AAA battery (1 C charge 2 C discharge)

(1) The cycle further fines the microcrystals, regardless of whether the positive electrode is added or not, but for the negative electrode, the unadded ones are still fined, and the added ones make the grains grow.
(2) Additives promote the transformation of deformation stacking faults into growth stacking faults, but the total stacking fault probability does not change much.
(3) From the rightmost column of data in Fig. 20.13 and Table 20.5, it can be seen that the addition of CaF_2 and $Ca(OH)_2$ can affect of adding $Ca(OH)_2$ on battery performance greatly inhibit the formation of A $(OH)_3$ and B phase, especially the addition of CaF_2 is the most obvious, but $CaCO_3$ has no effect.

The above results are consistent with the results of the first half of this section (adding 1% Ca compound) and further prove that CaF_2 additive can improve the cycle performance of the battery.

20.4 Physical Mechanism of the Action of β-$Ni(OH)_2$ Additive [4]

20.4.1 Behavior of Electrode Active Material During Battery Cycle

For MH/Ni batteries with β-$Ni(OH)_2$ as positive electrode and hydrogen storage alloy AB_5 as negative electrode, it is generally thought that the following electrochemical reactions take place during charge and discharge:

Table 20.5 Effect of 3% calcium additives on microstructure of cathode and anode materials after only activation and cycle 400 periods

		Cathode material					Anode material		
		$\overline{D}$ (nm)	$\varepsilon(\times 10^{-3})$	f_D (%)	f_T (%)	$f_D + f_T$ (%)	$\overline{D}$ (nm)	$\varepsilon(\times 10^{-3})$	$I_{A(OH)_3}/I_{AB_5}$
No additive	Only activation	47.5	6.13	5.77	3.91	9.68	190.7	0.79	
	Cycle	24.6	3.44	3.45	4.24	7.69	99.8	1.02	0.631
CaF_2	Only activation	60.4	5.84	5.78	2.13	7.91	31.7	−1.95	
	Cycle	33.6	4.62	3.87	2.95	6.82	110.8	0.53	0.249
$Ca(OH)_2$	Only activation	60.4	5.84	5.78	2.13	7.91	63.0	0.22	
	Cycle	25.4	4.19	4.96	2.07	7.03	76.0	0.11	0.340
$CaCO_3$	Only activation	56.7	5.87	5.96	1.99	7.95	42.9	-0.94	
	Cycle 400	28.4	5.67	7.94	-1.76	6.18	93.9	0.68	0.636

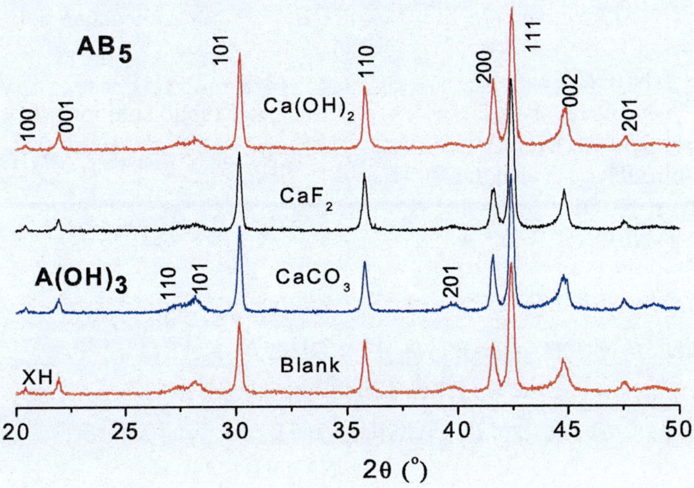

Fig. 20.14 XRD pattern of hydrogen storage alloy in AAA battery with 3% calcium additive added to the positive electrode after 500 cycles

Positive pole	Negative pole
Charge β-$Ni(OH)_2 + OH^-$ $\rightarrow \beta$-$NiOOH + H_2O$	$AB_5 + H_2O + e^- \rightarrow AB_5H_x + OH^-$
Discharge β-$NiOOH + H_2O$ $\rightarrow \beta$-$Ni(OH)_2 + OH$	$AB_5H_x + OH^- \rightarrow AB_5 + H_2O + e^-$
Total reaction	β-$Ni(OH)_2 + AB_5 \leftrightarrow \beta$-$NiOOH + AB_5H_x$

The above electrochemical reaction also includes the structural phase transition of positive and negative electrode materials during charge and discharge and seems to be "completely" reversible, so it is inferred that the service life of MH/Ni is infinite, which is not consistent with the practical facts. According to the results of Chap. 10, the following reactions have been written:

At Positive pole | At negative pole

Charge $\beta\text{-Ni(OH)}_2 \rightarrow \beta\text{-Ni(OH}_{1-x})_2 + 2 \times H^+ + e^-$

$AB_5 + 2 \times H^+ + e^- \rightarrow AB_5 - 2 \times H$ (solid solution) $0 < X < 0.50$

Over Charge $2\beta\text{-Ni(OH)}_2 \rightarrow \beta\text{-Ni(OH)}_2 + \gamma\text{-NiOOH} + H^+ + e^-$

$AB_5 + H^+ + e^- \rightarrow AB_5H$ (Hydride)

Discharge $\beta\text{-Ni(OH}_{1-x})_2 + 2 \times H^+ + e^- \rightarrow \beta\text{-Ni(OH)}_2$

$AB_5 - 2 \times H \rightarrow AB_5 + 2 \times H^+ + e^-$

Total reaction $\beta\text{-Ni(OH)}_2 + AB_5 \leftrightarrow \beta\text{-Ni(OH}_{1-x})_2 + AB_5 - 2 \times H$ (Solid solution)

Total reaction of over charge: $\beta - \text{Ni(OH)}_2 + AB_5 \rightarrow \beta\text{-Ni(OH)}_2 + \gamma\text{-NiOOH} + AB_5H_x$

Why the service life of battery is always limited and closely related to raw materials and charge–discharge cycle test (use) conditions. This is an issue that people have been paying close attention to all the time. The key of this problem is not the structural phase transition of positive and negative materials, but the change beyond the structural phase transition of positive and negative materials, that is, the irreversible change of microstructure.

Through the study, it is found that:

(1) After the activation of the battery, the microstructure of the positive material changes greatly, the short and fat columnar microcrystal turns into polyhedron or near-equiaxed microcrystal, the grain is obviously refined, and the material changes from no micro-strain state to micro-strain state. The type and number of stacking faults have also changed [10, 11].

(2) When we studied the relationship between cycle performance and microstructure of MH/Ni battery, we found the following points [10]:

   (i) The lattice parameters, average grain size, micro-strain and total stacking fault probability of cyclic positive materials at 20 °C are all decreases with the increase of the number of cycles, indicating that the formation of Ni vacancies leads to the gradual decrease of the activity of β-Ni(OH)$_2$ and the monotonous attenuation of cycle performance.

   (ii) The lattice parameters of the negative material decrease slightly with the increase of the number of cycles, indicating that A (or/and B) vacancy may be produced. The average grain size and micro-strain increase with the increase of cycle periods, especially the new phases A(OH)$_3$ and B increase with the increase of cycle periods, which is the result of corrosion of AB$_5$ alloy and the surface cracking and shedding of negative plate.

   (iii) Due to the transfer of Ni ions in positive electrode, A and B ions in the negative electrode, and corrosion products A(OH)$_3$ and B phase transfer to electrolyte, not only the electrolyte is consumed, but also the properties of the electrolyte are changed.

The joint action of the above three aspects is the real reason for the decline of cycle performance. The change process of the positive electrode, negative electrode, and electrolyte mentioned above is related to the temperature of the cycle life test and the charge–discharge rate during the cycle, and the life decay trend is also different.

20.4.2 On the Action Mechanism of β-$Ni(OH)_2$ Additive

After clarifying the relationship between the microstructure changes of the positive and negative materials and the decline of their cycle performance, how to improve the charge–discharge capacity and cycle performance of the battery by restraining the microstructure changes of the positive and negative materials is one of the purposes of this study. Through the study of adding different types of additives Lu_2O_3, Y_2O_3, CaF_2 and different Ca compound additives CaF_2, $Ca(OH)_2$, $CaCO_3$ to β-$Ni(OH)_2$, it is found that: (1) It can increase the charge–discharge capacity and improve the cycle performance, such as Lu_2O_3; (2) It cannot increase the charge–discharge capacity, but can obviously improve the cycle performance, such as CaF_2; (3) It can neither increase the charge–discharge capacity nor improve the cycle performance, such as $CaCO_3$.

So what is the action mechanism of these three kinds of additives? This is another task of this study. Through the above experimental results, the following facts can be summarized:

(1) The grain size, micro-strain, and stacking fault probability of β-$Ni(OH)_2$ with the addition of Lu_2O_3 and CaF_2 are smaller than those without addition, while the change rate of β-$Ni(OH)_2$ increases obviously with the addition of Y_2O_3 and $CaCO_3$, which indicates that CaF_2 and Lu_2O_3 inhibit the change of microstructure in the process of activation and cycle process, while Y_2O_3 plays a certain promoting role;
(2) The most obvious effects on negative active materials are CaF_2, Lu_2O_3, and $Ca(OH)_2$, they inhibited obviously the formation of new phase $A(OH)_3$ and B phase; CaF_2 was the most obvious, $CaCO_3$ did not have this effect, and Y_2O_3 promoted the formation of new phase.
(3) Due to the above two effects, the additives of CaF_2 and Lu_2O_3 can obviously improve the cycle performance of the battery.

As to why additives can play such a role, a preliminary discussion is described below.

(1) **The effect of additives on β-$Ni(OH)_2$ as positive active material**

Positive additives CoO, Lu_2O_3, and CaF_2 are added by mechanical mixing, molecules or ions of additives is impossible to enter the lattice of β-$Ni(OH)_2$, and it is only possible that β-$Ni(OH)_2$ particles are partially covered by additives, resulting in a micro-potential difference between additives and β-$Ni(OH)_2$ particles. It not only does not affect the leave of H atom from β-$Ni(OH)_2$ lattice, but also compensates the strain of β-$Ni(OH)_2$ lattice caused by H atom leaving β-$Ni(OH)_2$ lattice, which inhibits the refinement of β-$Ni(OH)_2$ microcrystals and slows down the decrease of total stacking fault probability in the process of cycle.

Although positive additives Y_2O_3, $Ca(OH)_2$ and effect of adding $Ca(OH)_2$ on battery performance $CaCO_3$ can partially cover β-$Ni(OH)_2$ particles in additives, their covered performance is poor, and the micro-potential difference between them is too small, to compensate for the strain of β-$Ni(OH)_2$ lattice caused by H atom leaving β-$Ni(OH)_2$ lattice. Therefore, it cannot play the role of improvement.

(2) **The effect of additives on negative electrode active material AB_5 alloy**

Positive additives CoO, $Lu_2O_{3,}$ and CaF_2 may be partially dissociated during the cycling process, Under the action of electric field, positive ions Co^+, Ca^+, Lu^+ migrate to the surface of negative electrode and passivates the surface of negative electrode. It does not affect of H atom embedded into AB_5 alloy lattice, but hinders the alkalization of AB_5 alloy, slows down the formation of $A(OH)_3$ and B phases, and then improves the cycle performance of MH-Ni battery.

Additive $CaCO_3$ can only be decomposed into $CaO + CO_2$ at high temperature and cannot dissociate active Ca^+ ions in electrolyte. Therefore, it cannot hinder the alkalization of AB_5 alloy and can also not slow down the formation of $A(OH)_2$ and B phase. On the other hand, the partial dissociation of Y_2O_3 into Y^+ can also reach the negative electrode surface, which may be due to its low chemical potential, which not only cannot hinder the alkalization of AB_5 alloy, but also promote the alkalization of AB_5 alloy and the formation of $A(OH)_3$ and B phase. Therefore, $CaCO_3$ and Y_2O_3 additives cannot improve the cycle performance of MH-Ni battery.

20.5 Conclusion

Based on the comprehensive analysis of the experimental results and discussion in Sects. 20.1–20.4 of this chapter, we get the conclusion of additive effect.

(1) For the AAA MH/Ni battery made by adding different contents of CoO to the positive material, the CoO content increased from 0 to 10%, the battery 0.2 C capacity increased 134 mAh, the 5 C discharge capacity increased by 225 mAh, the self-discharge rate decreased by 55%, and the cycle life increased by 18%. The CoO content of 8% is the best value.

(2) For the positive material β-$Ni(OH)_2$, the change rate of grain size, micro-strain, and stacking fault of Lu_2O_3 and CaF_2 are added is smaller than that without addition, while the change rate of the addition of Y_2O_3 and $CaCO_3$ is much larger, which indicates that CaF_2 and Lu_2O_3 play a passivating role in the microstructure change of positive material β-$Ni(OH)_2$ in the process of activation and cycling. Y_2O_3 does not passivate, but plays a catalytic role.

(3) On the side of negative materials, CaF_2, Lu_2O_3, and $Ca(OH)_2$ obviously inhibit effect of adding $Ca(OH)_2$ on battery performance the formation of new phase $A(OH)_3$ and B, especially CaF_2 has the best effect, $CaCO_3$ has no effect, and Y_2O_3 promotes the formation of new phase.
(4) Due to the effects of (2) and (3) above, especially the third point, the electrolyte is consumed and changed electrolyte performance. Therefore, CaF_2 and Lu_2O_3 additives significantly improve the cycle performance of the battery, because CaF_2 is cheap, particularly effective, and easy to popularize.

References

1. Sakai, T., I. Uehara, and H. Ishikawa. 1999. R On metal hydride materiale and Ni-MH batteries in Japan. *Journal of Alloys and Compounds* 293–295: 762–769.
2. Pralong, V., Y. Chabre, A. Delaye-Vidal, et al. 2002. Study of the contribution of cobalt additive to the behavior of the nickel oxy-hydroxide electrode by potential-dynamic techniques. *Solid State Lonics* 147: 73–84.
3. Provazi, K., M.J. Giz, L.H. Dall Antonia, et al. 2001. The effect of Cd Co, and Zn as additives on nickel hydroxide opto-electrochemical behavior. *Journal of Power Sources* 102: 224–232.
4. Lou, Y.W., C.Z. Yang, and B.J. Xia. 2009. Effect of β-$Ni(OH)_2$ additive on electrode material in MH-Ni battery. *Power supply technology* 33 (6): 449–453.
5. Oshitani, M., M. Watada, K. Shodai, et al. 2001. Effect of lanthanide oxide additives on the high-temperature characteristics. *Journal of the Electrochemical Society* 148 (1): A67–A73.
6. Li, F., Y.W. Lou, B.J. Xia, et al. 2005. Application of Ca additives in high capacity Ni-MH battery. *Power Supply Technology* 29 (5): 312–314.
7. Xia, B.J., Z.Q. Lin, L.P. Ma, et al. 2003. Effect of cathode Additives on High temperature charging behavior of MH/Ni Battery. *Battery* 33 (2): 68–70.
8. Qin, P., Y.W. Lou, and C.Z. Yang. 2006. A new method and calculation program for separating the multiple linewidth effect of X-ray diffraction. *Acta Physica Sinica* 55 (3): 314–325.
9. Yang, C.Z., and J. Zhang. 2008. Some progress in the study of microstructure of nanomaterials by X-ray diffraction. *Progress in Physic* 28 (3): 280–313.
10. Lou, Y.W., C.Z. Yang, D.N. He, and B.J. Xia. 2008. Study on cycle performance and electrode material microstructure of Ni-MH battery. *Journal of Chemistry* 66 (10): 1173–1180.
11. Lou, Yu-Wan. 2007. *XRD study on the microstructure of active material for positive and negative electrodes of metal hydride-nickel battery.* Doctoral thesis of Shanghai Institute of Microsystems and Information Technology

Chapter 21
The Method and Action Mechanism of Improving the Performance of Secondary Battery

21.1 Methods to Improve the Performance of Secondary Battery

Chapter 20 has introduced the role of adding CoO, Y_2O_3, CaF_2 and $CaCO_3$ to the positive active material β-$Ni(OH)_2$ of MH/Ni battery [40–42]. The results show that change rate of the grain size, micro-strain and stacking fault probability of β-$Ni(OH)_2$ adding Lu_2O_3 and CaF_2 are smaller than those without addition, while the change rates of β-$Ni(OH)_2$ adding Y_2O_3 and $CaCO_3$ are much higher, which indicates that CoO, CaF_2, and Lu_2O_3 play a passivating role in the microstructure change of positive material β-$Ni(OH)_2$ in the process of activation and cycling. Y_2O_3 does not passivate, but plays a catalytic role. On the side of negative materials, CaF_2, Lu_2O_3, and $Ca(OH)_2$ obviously inhibit the formation of new phase $A(OH)_3$ and B, especially CaF_2 has the best effect, $CaCO_3$ has no effect, and Y_2O_3 promotes the formation of new phase.

Chapter 7 also introduces the main methods to improve the electrical conductivity of $LiFePO_4$ materials, including: (1) particle nanocrystallization; (2) surface coating of conductive layer, such as nanocarbon layer; (3) bulk doping of $LiFePO_4$; (4) formation of Fe_2P, Fe_3P, and $Fe_{15}P_3C_2$ phases with good electronic conductivity on the surface of $LiFePO_4$ during synthesis; (5) improvement of surface morphology of $LiFePO_4$.

Chapter 14 also introduces that $Li(Ni,Co,Mn)O_2$ coated with Al_2O_3 can greatly improve the properties of graphite/$Li(Ni,Co,Mn)O_2$ battery. It can be seen that the methods to improve and raise the electrochemical performance of positive and negative active materials, cycle and storage performance of batteries are as follows:

As far as positive active materials are concerned, there are positive additives, coating [43], doping, nano-crystallization [44] and surface-modified [45] et al. of materials.

C. Yang et al., *Materials and Working Mechanisms of Secondary Batteries*,
https://doi.org/10.1007/978-981-19-5955-4_21

As far as the negative materials of lithium-ion batteries are concerned, most of the commercial negative materials of lithium-ion batteries use a variety of lithium-intercalated carbon materials, such as artificial graphite or mesophase carbon microspheres (MCMB). However, there are still some shortcomings when used as negative materials for lithium-ion batteries, such as low first charge and discharge efficiency, and the potential of carbon material is very close to that of lithium metal. When the battery is overcharged, the metal lithium may precipitate on the surface of the carbon electrode to form lithium dendrite and cause short circuit; there is an obvious voltage lag phenomenon when it interacts with the electrolyte; complicated preparation method and so on.

Compared with carbon negative electrode materials, alloy negative materials have higher specific capacity, but alloy electrodes have large volume effect in the process of repeated intercalation and de-intercalation of lithium, so the cycle performance is poor. Compared with the above negative materials for lithium-ion batteries, spinel lithium titanate ($Li_4Ti_5O_{12}$) shows its unique advantages as negative materials for lithium-ion batteries [1–5]. There are mainly three aspects as follows:

(1) As a zero-stress intercalation material, the crystal structure can be highly stable in the process of intercalation-detachment of lithium atom, which makes it have excellent cycle stability and stable discharge voltage platform.
(2) It has a high operating voltage (~1.5 V), which avoids the decomposition of electrolyte to form solid electrode interphase film (SEI).
(3) The source of raw materials is abundant, and the production cost is easy to be controlled.

Compared with commercial carbon negative materials, it usually has better safety performance; compared with alloy negative materials, it has high cycle stability, material easy preparation, and low cost. Therefore, spinel lithium titanate ($Li_4Ti_5O_{12}$), as an alternative material for commercial lithium-ion battery negative materials in the future, has been widely concerned by researchers [6].

In addition, hard carbon is also used, and its cycle performance is much better than that of 2H-graphite, as introduced in Sect. 21.2.

Electrolyte [7, 8]. As the electrolyte of lithium-ion battery, lithium salt should have the following requirements: (1) good chemical stability and electrochemical stability. There is no chemical reaction with electrode active material or current collector; (2) high solubility of lithium salt in organic solvent to ensure sufficient electrical conductivity; (3) good thermal stability.

According to the different anions of lithium salts used in lithium-ion batteries, they can be divided into two categories: inorganic anionic lithium salts and organic anionic lithium salts. Among inorganic lithium salts, $LiClO_4$ has the characteristics of high electrical conductivity, good thermal stability, low price and easy preparation and purification, but its strong oxidation brings safety risks; $LiAsF_6$ has good thermal stability and is not easy to hydrolyze, but it is expensive and toxic; the electrical conductivity of LiBF is lower, and its electrochemical stability is lower than that of $LiAsF_6$ and $LiPF_6$, but its performance at high and low temperature is better than that of $LiPF_6$; $LiPF_6$ is the lithium salt with the best comprehensive performance at present, with high electrical conductivity, good electrochemical stability and no corrosion of aluminum collector, but its price is high, and its heat resistance and hydrolysis resistance are not ideal.

In the single-solvent DMC electrolyte system, the oxidation potentials of several lithium salts change according to the following rules: $LiPF_6 > LiBF_4 > LiAsF_6 > LiClO_4$; in ec/dmc electrolyte system, the conductivity decreases in the order of $LiAsF_6 \approx LiPF_6 > LiClO_4 > LiBFO_4$; in order to further improve the low-temperature performance of lithium-ion battery, Zhangjiagang Guotai Huarong 5160 A electrolysis was used as low-temperature co-solvent (PA), whose core was to reduce the content of EC in the electrolyte system get better results.

However, various methods have been mentioned above to raise and improve the performance of secondary batteries. Because the positive active materials play a leading role in the performance degradation, there are many researches on positive active materials.

21.2 Application of Hard Carbon as Negative Electrode Active Material [9]

In recent years, some people try to use hard carbon (see Chap. 8) as the negative active material of lithium-ion battery. Figure 21.1 shows a comparison between the cycle performance of the battery using Japanese "Carbontron P" as negative active material and the cycle performance of graphite as negative electrode with or without conductive auxiliaries. It can be seen from Fig. 21.1a, b that: (1) even when the termination voltage is low, the cycle performance of hard carbon can be much better than that of graphite; (2) the cycle performance of hard carbon without conductive auxiliaries is much better than that of graphite with conductive auxiliaries. The reason is that the stability of the crystal structure of "hard carbon" is better than that of graphite (Fig. 21.2a) and the stability of the electrode structure (see Fig. 21.2b).

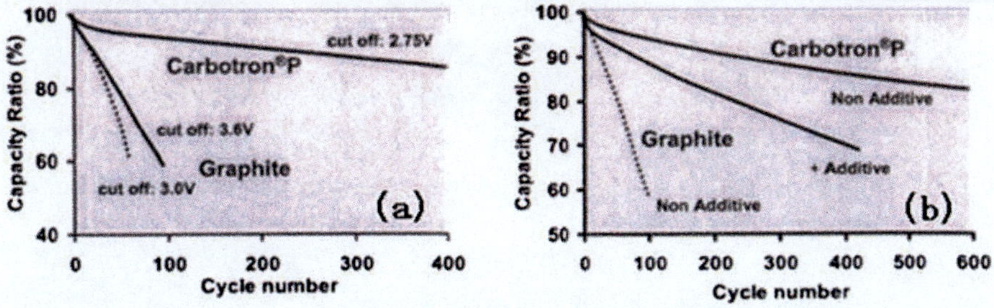

Fig. 21.1 Comparison of cycle performance of lithium-ion batteries using hard carbon and graphite as negative active materials

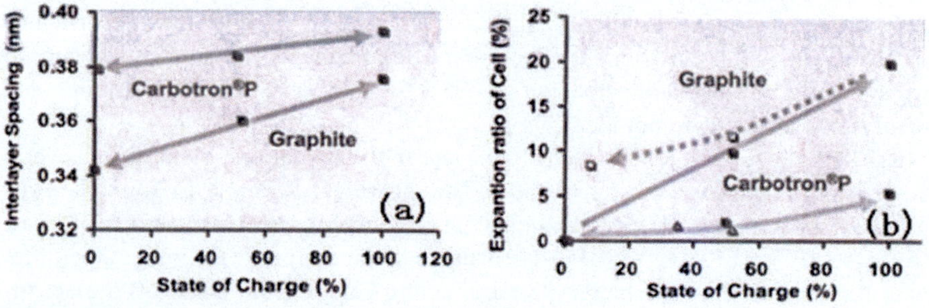

Fig. 21.2 Comparison of structural stability of hard carbon and graphite anode crystal (**a**) and electrode structure stability (**b**)

Figure 21.3 shows a schematic diagram of the conductive path between "hard carbon" and graphite particles. It can be seen that the binding mode of "Carbontron P" particles is almost unchanged before charging, after charging and after discharge, while graphite particles can expand during charging and shrink during discharge, so the former has stable conductivity, while the latter may be cut off to the electrical path. Some performance parameters of hard carbon of "Carbontron P" are listed in Table 21.1.

It should be noted that according to the measurement and analysis of Sect. 8.7 of Chap. 8, it is known that hard carbon is a mixture of amorphous carbon and distorted face-centered cubic carbon, where there are performance parameters such as two-phase ratio, particle size, specific surface area, layer spacing, and so on. Their role is improving the cycle performance of the battery. In order to get real application, the preparation of hard carbon, the preparation of electrode, and the preparation of battery need to be further studied.

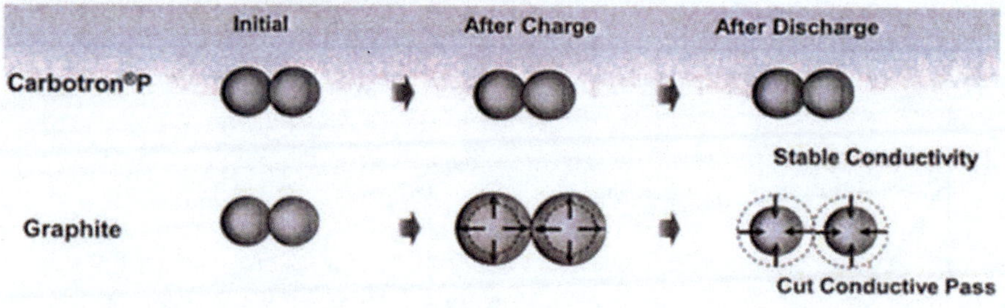

Fig. 21.3 Schematic diagram of the conductive path between "Carbontron P" and graphite particles

Table 21.1 Comparison of some performance parameters between Carbontron P and hexagonal graphite

Performance	Unit	Carbontron P	Hexagonal graphite
Specific surface area	m^2/g	6	
Layer spacing d_{002}	nm	0.38*	
Crystal size L_{002}	nm	1.1*	0.354–0.344
Average particle size	μm	9	A few to a dozen
Discharge capacity	Ah/kg	430	
Irreversible capacity	Ah/kg	70	

* It seems unscientific and reasonable to determine d_{002} and layer spacing according to the position of the scattering peak of the first amorphous carbon, as well as the FWHM of the amorphous peak. The same is true for calculating the size of microcrystals. Because the peak position and peak area of the amorphous dispersion pattern after Fourier transformation represent the distance and the number of atoms in the first shell of the local structure, respectively

21.3 The Effect of the Additive of Positive $Li(Ni_{1/3}Co_{1/3}Mn_{1/3})O_2$ [10]

21.3.1 *$Al(OH)_3$ Additive*

In order to prove this, $Al(OH)_3$ was first added as an additive in slurry process of the positive active material, and then the battery was made and the performance was tested. Type 18,650 battery (rated capacity 1200 mAh) was prepared by using $Li(Ni_{1/3}Co_{1/3}Mn_{1/3})O_2$ and graphite as positive and negative active materials, respectively. Coating amount of reference positive active material [11], The addition amount of $Al(OH)_3$ is 4 wt%.

The performance of positive electrode of lithium-ion battery with graphite as negative electrode changes greater during storage, which plays a decisive role in the performance of the whole battery after storage.

The main tactics to improve the storage performance of lithium-ion battery are: (1) reduce the self-discharge effect during storage; (2) reduce the side reaction between positive active material with electrolyte; (3) reduce the corrosion of positive active material by HF; (4) Inhibit the formation of fluorine-containing material on its surface.

The researchers coated the positive active material to improve its surface stability; the main purpose is to improve the cycle performance of the battery. Therefore, in order to evaluate the effect of this method, the cycle performance of the battery was tested at first. Graphite/$Li(Ni_{1/3}Co_{1/3}Mn_{1/3})O_2$ battery has excellent cycle performance (between 4.2 and 2.75 V at room temperature, 1 C cycle is 500 times, capacity loss is only 10%, so the battery is tested in 1 C cycle between 4.3 and 2.75 V at 55 °C. Figure 21.4 shows the cycle performance comparison of graphite/$Li(Ni_{1/3}Co_{1/3}Mn_{1/3})O_2$ batteries of the positive electrode coated without

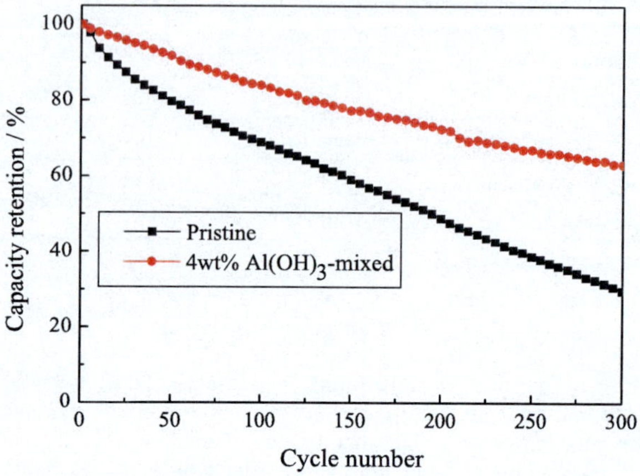

Fig. 21.4 1 C cycle performance of graphite/Li($Ni_{1/3}Co_{1/3}Mn_{1/3}$)O_2 battery between 4.3 and 2.75 V at 55 °C

and with 4 wt% $Al(OH)_3$. It can be seen that after increasing the temperature and the charging upper limit voltage, the capacity attenuation of the battery is accelerated during the 1 C cycle, and the capacity of the battery decreases by 70% after 300 cycles; at the same time, it can be seen that the cycle performance of the battery is significantly improved after adding $Al(OH)_3$ to the positive electrode, and the capacity attenuation rate is reduced to 37% after 300 cycles. This shows that the effect of this method is similar to that of surface coating in improving the cycle performance of positive active materials.

Compared with the cycle performance, whether this method can improve the storage performance of the battery is more concerned in this section. At 80 °C, the battery was stored with 100% SOC, and the 0.2 °C charge–discharge curve of the battery adding with and without $Al(OH)_3$ before and after 20 days storage was shown in Fig. 21.5. It can be seen that after storage, the capacity of the battery without $Al(OH)_3$ decreases greatly, and the platform voltage difference between charge and discharge increases; while the capacity attenuation of the battery with $Al(OH)_3$ decreases obviously, and the platform voltage difference between charge and discharge has little change compared with that before storage. This shows that this method is beneficial to the maintenance of battery capacity and charge–discharge performance during storage.

However, comparing the charge–discharge curve of the battery before storage (see Fig. 21.5), it can be seen that the initial charge–discharge platform voltage difference of the battery increased and the capacity decreased after the addition of $Al(OH)_3$. This is mainly due to the fact that $Al(OH)_3$ is a poor conductor of electrons and lithium ions, which affects the integrity and continuity of the positive conductive network, thus increasing the electrode impedance and increasing the polarization of the battery during charge and discharge. At the same time, the addition of inactive substances reduces the content of active substances in the electrode and reduces the capacity of the battery.

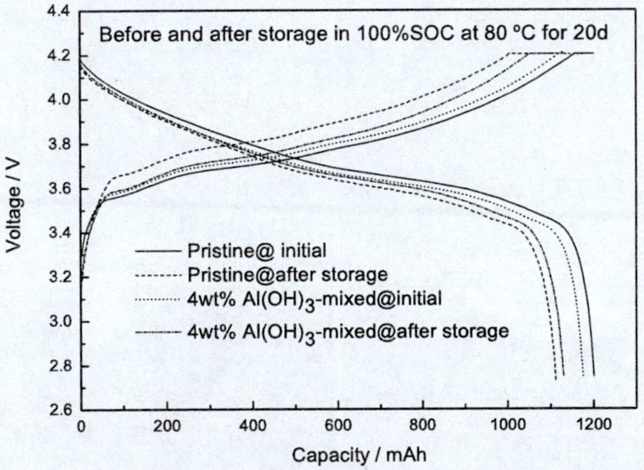

Fig. 21.5 Charge–discharge curve of graphite/$Li(Ni_{1/3}Co_{1/3}Mn_{1/3})O_2$ battery before and after storage 20 days

In order to reduce the effect of additives on the initial performance of the battery and improve the storage performance of the battery, the amount of $Al(OH)_3$ added is considered below, and the mechanism of improving the storage performance of the battery is further analyzed.

21.3.2 Effect of $Al(OH)_3$ Adding Amount on Battery Performance

1, 2, 3, and 5 wt% $Al(OH)_3$ were added separately to the positive slurry to make the electrode sheet, and then, the lithium plate was used as the opposite electrode to make 2025 button battery. The storage condition of the battery is stored in 100%SOC at 80 °C for 10 days. The distribution of additives in the electrode has an important influence on its performance. As can be seen from the SEM photos, because the shape of the conductive agent and small particle size active material particles in the electrode is similar to that of the additive particles, it is difficult to see the existence of additives from the surface morphology. The surface scan of the electrode is carried out by EDS, and the distribution of additives in the electrode can be determined by the distribution of Al elements. It can be seen that when the amount of additive is small, the distribution of additives in the electrode is relatively uniform, but with the increase of the amount of additives, the distribution of additives appears aggregation. Figure 21.6 shows the SEM photo and the corresponding Al element scan image of the positive plate with 2 and 5 wt% $Al(OH)_3$ added. It can be seen that when 2 wt% $Al(OH)_3$ is added, it is uniformly distributed in the electrode, while when the addition amount is 5 wt%, the distribution of $Al(OH)_3$ in the electrode is no longer uniform and aggregation occurs, which will adversely affect the performance of the electrode.

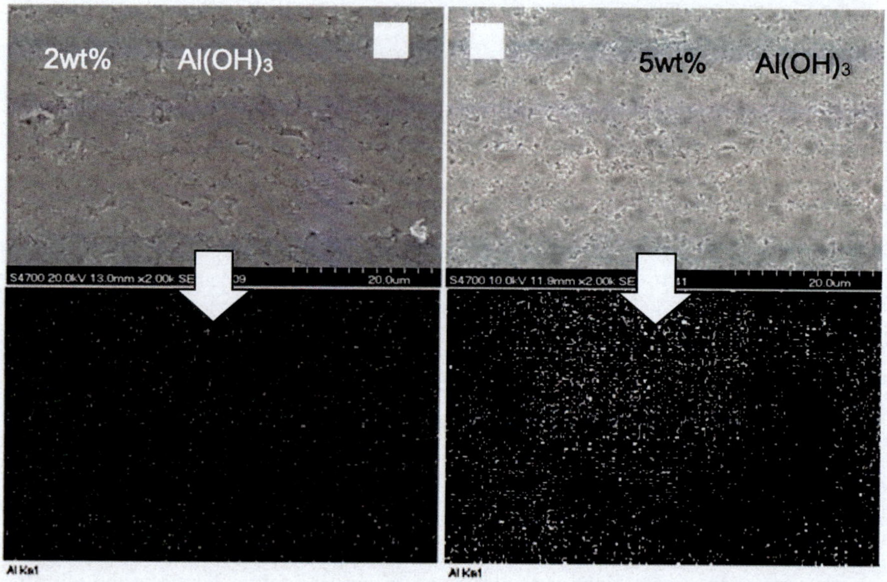

Fig. 21.6 SEM photos with $Al(OH)_3$ positive plates and their Al plane scan images

Figure 21.7 shows the charge–discharge curve of $Li/Li(Ni_{1/3}Co_{1/3}Mn_{1/3})O_2$ battery added different amounts of $Al(OH)_3$ to the positive electrode before and after storage. It can be seen that, When $Al(OH)_3$ is added to the positive electrode, the voltage difference between the charging and discharging platform of the battery increases, indicating that the polarization of the battery increases, and this trend is more obvious with the increase of the amount of $Al(OH)_3$. At the same time, with the increase of the amount of $Al(OH)_3$, the capacity of the battery decreases slightly. As mentioned earlier, this is mainly due to the increase of the impedance of the positive electrode and the decrease of the content of active material in the positive electrode caused by electrochemical and electronic inert $Al(OH)_3$; at the same time, when the addition amount is increased, $Al(OH)_3$ is unevenly distributed in the electrode, which may further worsen the local conductivity of the positive electrode, thus destroying the charge and discharge performance of the battery and reducing the capacity of the battery. However, when the amount of $Al(OH)_3$ added is not more than 2 wt%, this effect is not obvious.

Comparing Fig. 21.7a, b, we can see that the capacity decay rate of the battery without $Al(OH)_3$ after storage is 8.6%, and the difference between the platform voltage of charge and discharge is increased; for the battery with 1, 2, 3, and 5 wt% $Al(OH)_3$, the capacity decay rate after storage is 6.8, 3.1, 2.9, and 3%, respectively. It can be seen that after the addition of $Al(OH)_3$, the capacity decay rate of the battery decreases after storage, and the decreasing range decreases with the increase of its addition. But when the addition amount is more than 2 wt%, the capacity decay rate of the battery after storage is basically the same. The comparison of the charge

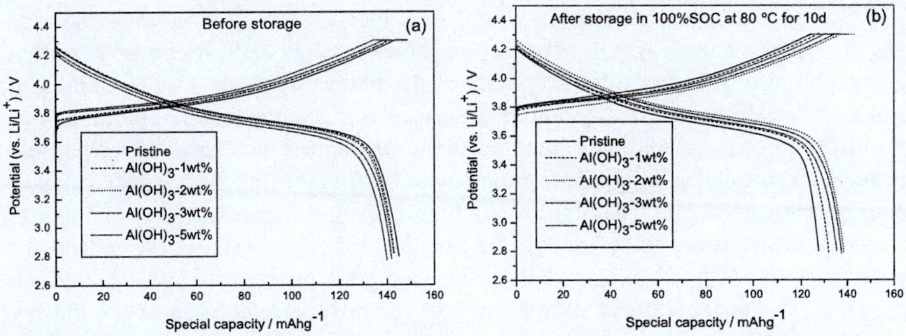

Fig. 21.7 0.2 C charge–discharge curve of $Li/Li(Ni_{1/3}Co_{1/3}Mn_{1/3})O_2$ battery containing $Al(OH)_3$ before and after storage

and discharge performance of each battery after storage shows that the battery with 2 wt% $Al(OH)_3$ has the best charge and discharge performance after storage.

The increase of positive electrode impedance of graphite/$Li(Ni_{1/3}Co_{1/3}Mn_{1/3})O_2$ battery after storage is the main reason for the degradation of battery performance. The batteries with and without $Al(OH)_3$ were dissected before and after storage, and the positive electrode was tested by EIS. The results are shown in Fig. 21.8.

It can be seen that before storage, the impedance of $Li(Ni_{1/3}Co_{1/3}Mn_{1/3})O_2$ electrode increased due to the addition of $Al(OH)_3$. And the increase amplitude increases with the increase of the amount of $Al(OH)_3$. At the same time, after storage, the positive electrode impedance of the battery without $Al(OH)_3$ increased obviously; however, the positive electrode impedance increased slightly after adding $Al(OH)_3$. After storage, the positive impedance of $Al(OH)_3$ with 2 wt% is the lowest.

To sum up, the storage performance of the battery can be improved by adding $Al(OH)_3$ to the cathode, but too much $Al(OH)_3$ will decrease the initial performance of the battery, and the optimum addition should be about 2 wt%.

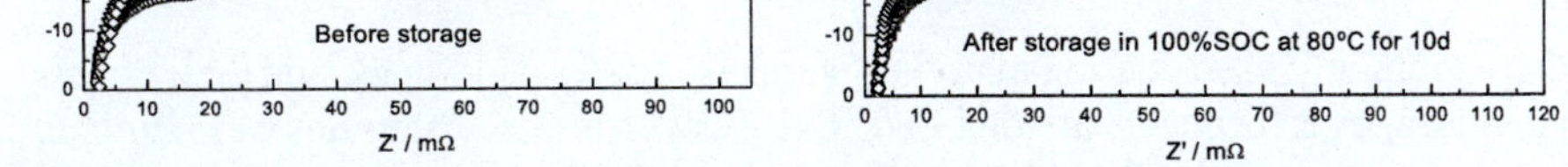

Fig. 21.8 Nyquist curves of $Li(Ni_{1/3}Co_{1/3}Mn_{1/3})O_2$ electrodes containing different amounts of $Al(OH)_3$ before and after storage

The SEM picture of $Li(Ni_{1/3}Co_{1/3}Mn_{1/3})O_2$ electrode before and after storage with or without 2 wt% $Al(OH)_3$ battery is shown in Fig. 21.9. It can be seen that the morphology of the positive electrode of the battery without $Al(OH)_3$ changed significantly before and after storage. The surface of the active material particles was smooth before storage, and the white flocculent substance was formed on the particle surface after storage, and higher brightness which indicated that the conductivity was poor and the impedance was high. It is a high impedance compound containing F, and its deposition on the surface of the positive active material is the reason for the increase of the positive impedance after storage. However, it can be seen from Fig. 21.9c, d that the surface morphology of the positive electrode of the battery with 2 wt% $Al(OH)_3$ does not change obviously before and after storage. The white particle in the picture is $Al(OH)_3$. During battery storage, $Al(OH)_3$ self-sacrifices and reacts with HF in the electrolyte to reduce its content, thereby reducing its erosion of the active material.

In order to further prove this point, the area shown in Fig. 21.9b, d is scanned by EDS for the F element. It can be seen that after storage, the distribution of F element on the surface of the positive electrode without $Al(OH)_3$ corresponds well with the particle shape of the active material, indicating that F is distributed on the surface of the particles and has a higher content, while the distribution of F element on the surface of the electrode with 2 wt% $Al(OH)_3$ is in good agreement with the shape of $Al(OH)_3$ particles after storage. The results show that F element is mainly distributed on the surface of $Al(OH)_3$ particles, while the F content on the surface of positive active materials is less. This shows that during storage, $Al(OH)_3$ in the positive electrode reacts with HF in the electrolyte, thus reducing the surface HF content of the active material, thus reducing the erosion of the positive active material, reducing the formation of high impedance material containing F on the surface of the positive active material, and reducing the impedance of the positive electrode. When the addition of $Al(OH)_3$ is more than 2 wt%, the improvement effect on the storage performance of the battery is no longer increased. This may be due to the fact that 2 wt% $Al(OH)_3$ is enough to consume HF in the electrolyte, and the existence of excess $Al(OH)_3$ is not only unnecessary, but also makes the initial performance of the battery worse. Therefore, according to the previous experimental results, after adding 2 wt% $Al(OH)_3$, the initial performance and storage performance of the battery are the best.

21.3.3 Effect of Different Additives on Battery Performance

Other commonly used coating materials were also tried as additives, and their effects on the storage performance of graphite/$Li(Ni_{1/3}Co_{1/3}Mn_{1/3})O_2$ batteries were studied, including Al_2O_3, ZrO_2, Zn_O, MgO, and so on. The addition amount of other additives is calculated according to the equivalent ratio of 2 wt% $Al(OH)_3$ to HF, as follows:

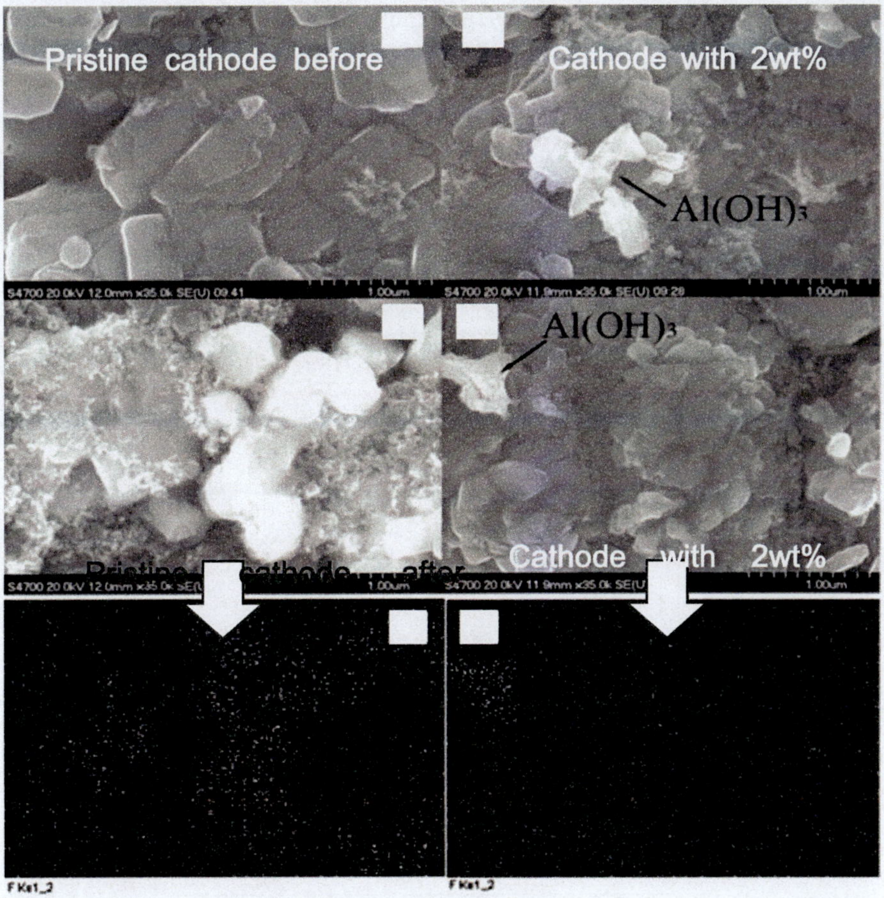

Fig. 21.9 SEM photos of $Li(Ni_{1/3}Co_{1/3}Mn_{1/3})O_2$ electrodes with or without $Al(OH)_3$ batteries before and after storage and its F element surface scan image

$$m = \frac{2\,\text{wt\%} \cdot 3 \cdot M_B}{M_{Al(OH)_3} \cdot n} \tag{21.1}$$

$M_{Al(OH)_3}$ is the molecular weight of $Al(OH)_3$, M_B is the molecular weight of additive B, and n means that a B molecule can react with at most n HF molecules. The addition amounts of Al_2O_3, ZrO_2, Zn_O, and MgO were 1.3, 1.77, 3.15, and 1.54 wt%, respectively. The electrodes were made by adding the above additives to the positive electrode, and the 18,650 cylindrical battery (rated capacity 1200 mAh) was assembled with graphite as the negative electrode. The storage condition is that it is stored in 100% SOC at 80 °C for 20 days.

Figure 21.10 shows the 0.2 C charge–discharge curve of graphite/$Li(Ni_{1/3}Co_{1/3}Mn_{1/3})O_2$ battery with different additives in the positive before and after storage. It can be seen that after adding additives, the initial charge and discharge performance of the battery does not change much. The platform voltage difference of the battery with additives increases slightly, which is mainly due to the electron of additives and lithium-ion insulation; at the same time, due to the difference between added quantity and conductivity, the effects of different additives on the initial charge and discharge performance of the battery are different.

Compared with Fig. 21.10a, b, it can be seen that the capacity of the battery without additives attenuates obviously after storage (the attenuation rate is 9.8%), and the charge–discharge platform voltage difference increases obviously, while after adding additives, the change of the charge–discharge curve of the battery before and after storage is not obvious. The capacity decay rates of batteries with $Al(OH)_3$, Al_2O_3, ZrO_2, ZnO, and MgO after storage were 4.1, 4.9, 5.8, 5.0, and 5.1%, respectively. It

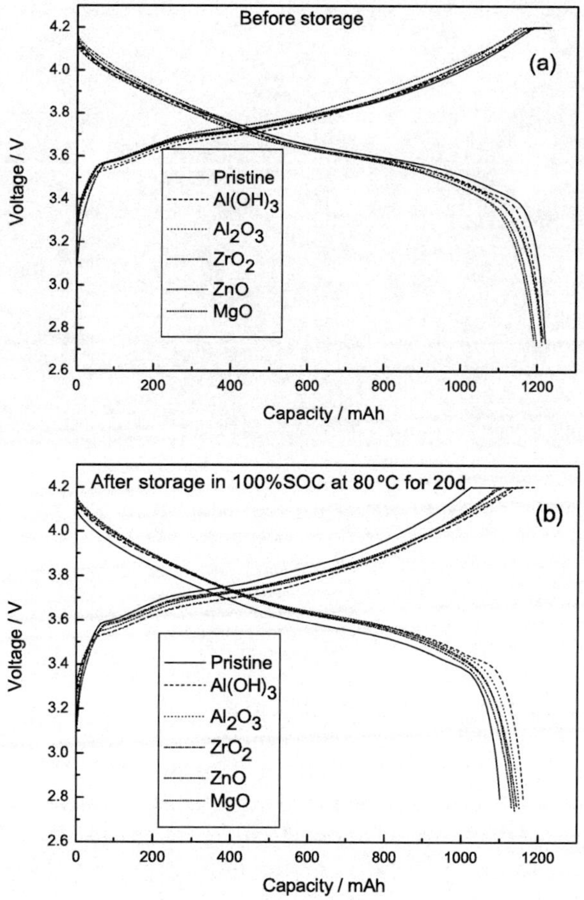

Fig. 21.10 Charge–discharge curve of graphite/$Li(Ni_{1/3}Co_{1/3}Mn_{1/3})O_2$ battery with different additives in positive electrode before and after storage

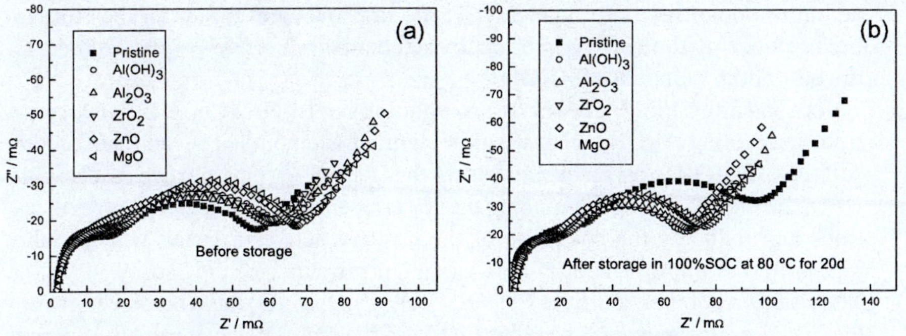

Fig. 21.11 Nyquist curves of $Li(Ni_{1/3}Co_{1/3}Mn_{1/3})O_2$ electrodes containing different additives before and after storage

can be seen that after adding additives to the positive electrode, after the battery is stored. The capacity decay rate and the charge–discharge platform voltage difference of the battery decreased obviously, It can be seen that after adding additives to the positive electrode, the capacity decay rate of the battery decreases obviously, and the charge–discharge platform voltage difference decreases obviously, which shows that the addition of these additives can also restrain the decline of the capacity and charge–discharge performance of the battery during storage.

The batteries with different additives before and after storage were dissected and their positive electrodes were tested by EIS. The results are shown in Fig. 21.11. It can be seen that before storage, due to the addition of additives, the impedance of $Li(Ni_{1/3}Co_{1/3}Mn_{1/3})O_2$ electrode increased, but the increase was not obvious. The combination of Fig. 21.10 shows that this is the main reason why the initial charge and discharge performance of the battery becomes worse after adding additives. At the same time, after storage, the positive electrode impedance of the battery without additives increased obviously. However, with the addition of additives, the increase of the positive impedance of the battery decreased obviously.

21.3.4 Summary of Research on Additives

Summing up the research in this section, the following main conclusions are drawn.

(1) $Al(OH)_3$ as an additive can inhibit the decline of capacity and charge–discharge performance during storage; however, it has a certain adverse effect on the initial charge–discharge performance of the battery.
(2) The addition quantity of $Al(OH)_3$ has a great influence on the initial performance and storage performance of the battery. With the increase of the amount of $Al(OH)_3$, the impedance of $Li(Ni_{1/3}Co_{1/3}Mn_{1/3})O_2$ electrode increases, the initial charge–discharge performance of the battery decreases, and the improvement effect on the storage performance of the battery increases, but when the

addition amount is more than 2 wt%, the improvement effect on the storage performance of the battery is basically unchanged. 2 wt% was determined as the best adding amount of $Al(OH)_3$.

(3) Before and after storage, the surface morphology of $Li(Ni_{1/3}Co_{1/3}Mn_{1/3})O_2$ electrode containing $Al(OH)_3$ changes little, while F element is concentrated on the surface of $Al(OH)_3$ particles, indicating that $Al(OH)_3$ self-sacrifice reacts with HF in the electrolyte, thus reducing the HF concentration near the positive electrode and reducing the corrosion of the positive active material, which is the reason for the improvement of the storage performance of the battery.

(4) The addition of other additives (Al_2O_3, ZrO_2, ZnO, MgO) has a certain effect on the initial performance of graphite/$Li(Ni_{1/3}Co_{1/3}Mn_{1/3})O_2$ battery, but it is not obvious. Similar to $Al(OH)_3$, the addition of these additives can also inhibit the decline of the capacity and charge–discharge performance of the battery during storage, improve the cycle performance of the battery after storage, reduce the increase of DC internal resistance of the battery after storage, and restrain the increase of positive impedance. The results show that this method can improve the storage performance of graphite/$Li(Ni_{1/3}Co_{1/3}Mn_{1/3})O_2$ battery.

21.4 Overview of Coating of Positive Active Material for Li-Ion Battery

The performance of positive electrode for lithium-ion battery with $LiCoO_2$, $Li(Ni_{1/3}Co_{1/3}Mn_{1/3})O_2$ and $LiFePO_4$ as positive electrode and 2H-graphite as negative electrode changes greatly during storage, which plays a decisive role in the performance of the whole battery after storage. The main reason for the performance decay of the positive electrode during storage is that the electrolyte decomposes on the surface of the positive active material, and the corrosion of HF, etc., acidic substance in the electrolyte leads to the positive active material, and formatting LiF, etc., high impedance substance on the surface of the positive active material, which leads to the attenuation of the positive capacity and the increase of impedance after storage, resulting in the performance attenuation of the whole battery. Based on this mechanism, people think that the main way to improve the storage performance of lithium-ion battery is to reduce the side reaction between positive active materials and electrolyte, reduce the corrosion of positive active materials by HF, and inhibit the formation of fluorine-containing materials on the surface of battery electrode. The issue of carbon inclusion in $LiFePO_4$ has been mentioned in Chap. 7.

At present, in the research of raise the surface stability of positive active materials, the surface coating is mainly used to reduce the corrosion of electrolyte to positive active materials. The coating materials are mainly metal oxides (including Al_2O_3 [12–14], ZrO_2 [15–17], ZnO [18, 19], Y_2O_3 [20], etc.) and inorganic salts (including AlF_3 [21–23], $Al(OH)_3$ [24, 25], phosphide [26], and phosphate [27–30], etc.). At present, the consistent coating modification mechanism is as follows: First, the coating isolates the direct contact between the active material and the electrolyte, thus preventing the corrosion of the decomposition products of the electrolyte to the active material; second, the coating material self-sacrifices and reacts with the acidic substances (mainly HF) in the electrolyte, so as to reduce the erosion of the active material. The former requires the coating to be as dense as possible and completely cover the particle surface of the material. However, recent studies have shown that even a very dense coating cannot completely prevent the spontaneous reaction between the positive active material and the electrolyte; on the contrary, the loose and incomplete coating can also improve the performance of the material, so more researchers believe that the effect of the second mechanism mentioned above is more obvious.

At present, there are many methods of surface coating, but there are many disadvantages, such as complex process, difficult to control, high cost and damage to active materials. Based on the second mechanism mentioned above, if it is not necessary to coat the material with complex process, but the above oxide (or hydroxide) is directly added to the cathode slurry as an additive, it can also inhibit the erosion of positive active materials, which is proved by the introduction of 21.3.

The following will introduce the methods and mechanisms of surface coating to raise and improve the performance of lithium-ion batteries.

21.4.1 Coating of Positive Active Material for Li-Ion Battery

Coating is considered to be one of the most simple and effective ways to improve the performance of cathode materials, which has aroused widespread concern in the industry. The coating materials are mainly oxides, fluoride, phosphate, and so on. It is briefly described as follows.

(1) **Oxide-coated**

Oxide has the advantages of simple synthesis and good stability, so it is widely used in the coating of cathode materials, such as Al_2O_3 [12–14], TiO_2 [31–33], ZrO_2 [15–17], Y_2O_3 [20], ZnO [18, 19], and so on. Because of their different cations, they have different surface layer doping characteristics and two-phase interface characteristics. At the same time, the dispersion uniformity of different oxides on the surface of positive active materials is different.

Al_2O_3 has a series of advantages, such as rich Al resources, good uniformity, simple method, good stability, and so on, so it is widely used. At present, the main methods of coating Al_2O_3 on the cathode are precipitation method, nanoassembly method and sol–gel method. Liu et al. adopt precipitation method to coat the surface of $LiCoO_2$ with an amorphous layer of Al_2O_3. After 20 cycles of 4.5–3 V, the capacity retention rate increases from 60 to 97.6%. It is considered that the existence of the amorphous layer of Al_2O_3 prevents the dissolution of Co in the $LiCoO_2$ material. Wang et al. also synthesized Al_2O_3-coated $LiCoO_2$, by precipitation method, but SEM showed that the coating was not uniform. In order to form a more uniform coating layer, Liu et al. used nanoparticle assembly method to

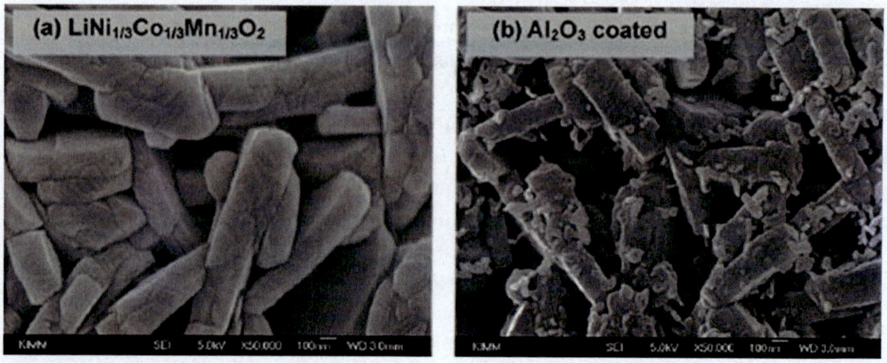

Fig. 21.12 SEM diagram of $Li(Ni_{1/3}Co_{1/3}Mn_{1/3})O_2$ before and after Al_2O_3 coating

coat $Li(Ni_{0.42}Co_{0.16}Mn_{0.42})O_2$ with Al_2O_3, ZnO, Bi_2O_3 and $AlPO_4$, respectively. A comparative study found that $Li(Ni_{0.42}Co_{0.16}Mn_{0.42})O_2$ coated with Al_2O_3 has the best cycle stability and initial capacity, while $Li(Ni_{0.42}Co_{0.16}Mn_{0.42})O_2$ coated with Bi_2O_3 has the best rate performance. After 5–3.5 V and 50 cycles, the capacity retention showed the following rule: uncoated $Li(Ni_{0.42}Co_{0.16}Mn_{0.42})O_2 < AlPO_4$ coated < ZnO coated < Al_2O_3 coated < Bi_2O_3 coated. When the nanoparticle assembly method is coated in a small amount, the morphology of the coating layer can be controlled uniformly. However, the dispersion of nanoparticles is difficult to achieve in the process of large-batch coating, and the preparation and re-dispersion of nanomaterials are more difficult. Therefore, it is not suitable for large-scale coating at present. The sol–gel method has excellent film-forming properties. Kim et al. synthesized Al_2O_3-coated $Li(Ni_{1/3}Co_{1/3}Mn1/3)O_2$ materials by alkoxide hydrolytic gel method. The capacity retention of 4.5–2.8 V and 1 C cycles increased from 82% before coating to 93% after coating. However, it can also be seen from the SEM photo that there is still obvious agglomeration of Al_2O_3 on the surface, as shown in Fig. 21.12. It can be seen that the sol–gel process is still very necessary to optimize.

Wu et al. [32] TiO_2 was coated on the surface of $Li(Ni_{1/3}Co_{1/3}Mn_{1/3})O_2$ by Tetrabutyl titanate hydrolysis to raise the cycle stability of $Li(Ni_{1/3}Co_{1/3}Mn_{1/3})O_2$. The results show that 1 wt% TiO_2-coated $Li(Ni_{1/3}Co_{1/3}Mn_{1/3})O_2$ has the best cycle performance and rate retention ability. Zheng et al. [33] used the method of adding tetrabutyl titanate solution to $Li[Li_{0.2}Mn_{0.54}Ni_{0.13}Co_{0.13}]O_2$ suspension to coat TiO_2, to improve the cycle performance and thermal stability of the material. Although both of them were coated by sol–gel method, the SEM photo showed that there was non-uniformity of TiO_2 on the surface of both materials, as shown in Figs. 21.13 and 21.14. However, by comparison, it can be seen that the uniformity of TiO_2 coating on the prepared $Li(Ni_xCo_yMn_{1-x-y})O_2$ material is difference with different feeding sequence. It is found that the uniformity of TiO_2 coating prepared by Wu et al. [32] by forming sol first and then adding $Li(Ni_xCo_yMn_{1-x-y})O_2$ is obviously better than that prepared by Zheng et al. [33].

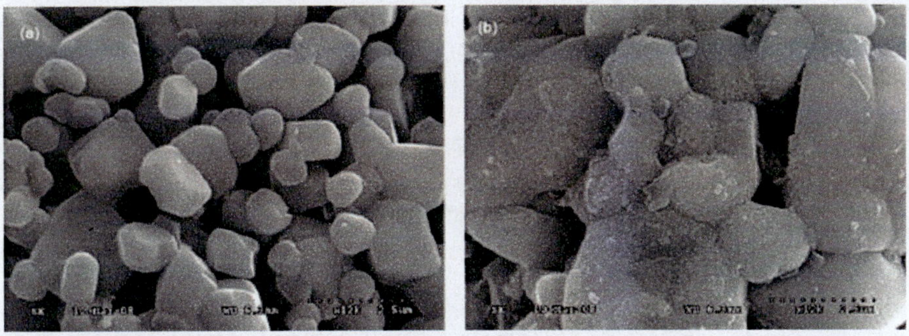

Fig. 21.13 SEM photograph of $Li(Ni_{1/3}Co_{1/3}Mn_{1/3})O_2$ before and after TiO_2 coating [32]

Fig. 21.14 SEM diagram of $Li(Li_{0.2}Mn_{0.54}Ni_{0.13}Co_{0.13})O_2$ before and after TiO_2 coating. **a** uncoated, **b** 1.0 mol% coated, **c** 3.0 mol% coated, **d** 5.0 mol% coated [33]

ZrO_2 is mainly coated in two ways: One is to coat the precursor, and the other is to coat the surface of the synthesized cathode material. Huang et al. [16] realized the enrichment of the surface layer of Zr compounds by coating ZrO_2 in the precursor $Ni_{1/3}Co_{1/3}Mn_{1/3}$ $(OH)_2$ stage, then mixing lithium and sintering. The XRD results show that there is a part of Li_2ZrO_3 phase on the surface. The results of 4.5–3 V and 0.5 C cycles show that the capacity retention of $Li(Ni_{1/3}Co_{1/3}Mn_{1/3})O_2$ enriched on the surface of ZrO_2 is much higher than that before coating after 100 cycles. Hu et al.

[17] mixed $Li(Ni_{1/3}Co_{1/3}Mn_{1/3})O_2$ with zirconium isopropoxide solution, formed sol at 80 °C, and sintered $Li(Ni_{1/3}Co_{1/3}Mn_{1/3})O_2$ coated with ZrO_2. TEM results show that a ZrO_2 coating layer of 10 nm is formed on the surface of $Li(Ni_{1/3}Co_{1/3}Mn_{1/3})O_2$. The capacity retention rate of 50 cycles of 4.5–3 V and 1 C cycles increased from 62% before coating to 90% after coating. The DSC test showed that the heat release decreased from 320.5 J/g before coating to 287.3 J/g after coating.

Wu et al. [20] used precipitation method to coat the surface of $Li(Ni_{1/3}Co_{1/3}Mn_{1/3})O_2$ with Y_2O_3. The results show that the coating layer is not uniform. Although the coating improves the cycle performance of $Li(Ni_{1/3}Co_{1/3}Mn_{1/3})O_2$, the effect is not very significant. Chang et al. [18] coated ZnO on the surface of $LiCoO_2$ by chemical vapor deposition (CVD) using diethyl zinc as raw material. SEM shows that ZnO is mainly distributed in the form of nanoparticles on its surface, but the distribution is uneven and not dense, as shown in Fig. 21.15. The results of 4.5–3 V and 1 C cycles show that its capacity retention rate has been improved to a certain extent.

(2) **Fluoride coating** [21–23]

Fluoride, like oxide, is not easy to react with electrolyte. AlF_3 has strong resistance to HF, and F partially replaces O in the surface layer of AlF_3 coating and cathode materials during heat treatment, which can inhibit the precipitation of active O.

Sun et al. [21] controlled the Al:F = 3 ratio to control the deposition rate of AlF_3 on the surface. The capacity retention of $Li(Ni_{1/3}Co_{1/3}Mn_{1/3})O_2$ coated with AlF_3, 4.5–3 V and 0.5 C cycle increased from 75% before coating to 93% after coating. Park et al. [22] coating AlF_3 on the surface of $Li(Ni_{1/3}Co_{1/3}Mn_{1/3})O_2$ improves the cycle performance in the range of 4.5–3 V. DSC results show that the coated $Li(Ni_{1/3}Co_{1/3}Mn_{1/3})O_2$ has a higher initial temperature of exothermic peak. Woo et al. [22] used F:Al = 6 proportion to control deposition, and deposited AlF_3, on the surface of $Li(Ni_{0.8}Co_{0.1}Mn_{0.1})O_2$ significantly raise the cycle performance. After 1000 cycles under the charge-discharge condition of 4.2–3 V and 1 C, the capacity retention rate reached 93%, and the initial temperature of the exothermic peak increased from 217 to 248 C after coating.

(3) **Phosphate coating** [27–30]

Due to the existence of polyanion $PO_4{}^{3-}$, phosphate has a stable structure and is a good coating material. Kim et al. [26] $AlPO_4$, $FePO_4$, $CePO_4$, and $SrHPO_4$ were coated on the surface of $LiCoO_2$ by nanoparticle assembly method. The results showed that these coatings improved the cycling performance of $LiCoO_2$. Due to the difference of film formation, the coating effect showed the rule of $AlPO_4$ coating > $CePO_4$ coating > $SrHPO_4$ coating > $FePO_4$ coating > uncoated. This is consistent with the coating uniformity of the four materials on the surface of $LiCoO_2$. Eom et al. [27] $Mg_3(PO_4)_{2-}$ and $Zn_3(PO_4)_2$-coated $LiCoO_2$ materials were prepared by nanoparticle assembly method, and their storage properties at 90 °C were studied. The results show that the storage structure stability of $Mg_3(PO_4)_{2-}$ and $Zn_3(PO_4)_2$-coated $LiCoO_2$ at 90 °C is obviously better than that of Mg- and Zn-doped $LiCoO_2$ materials.

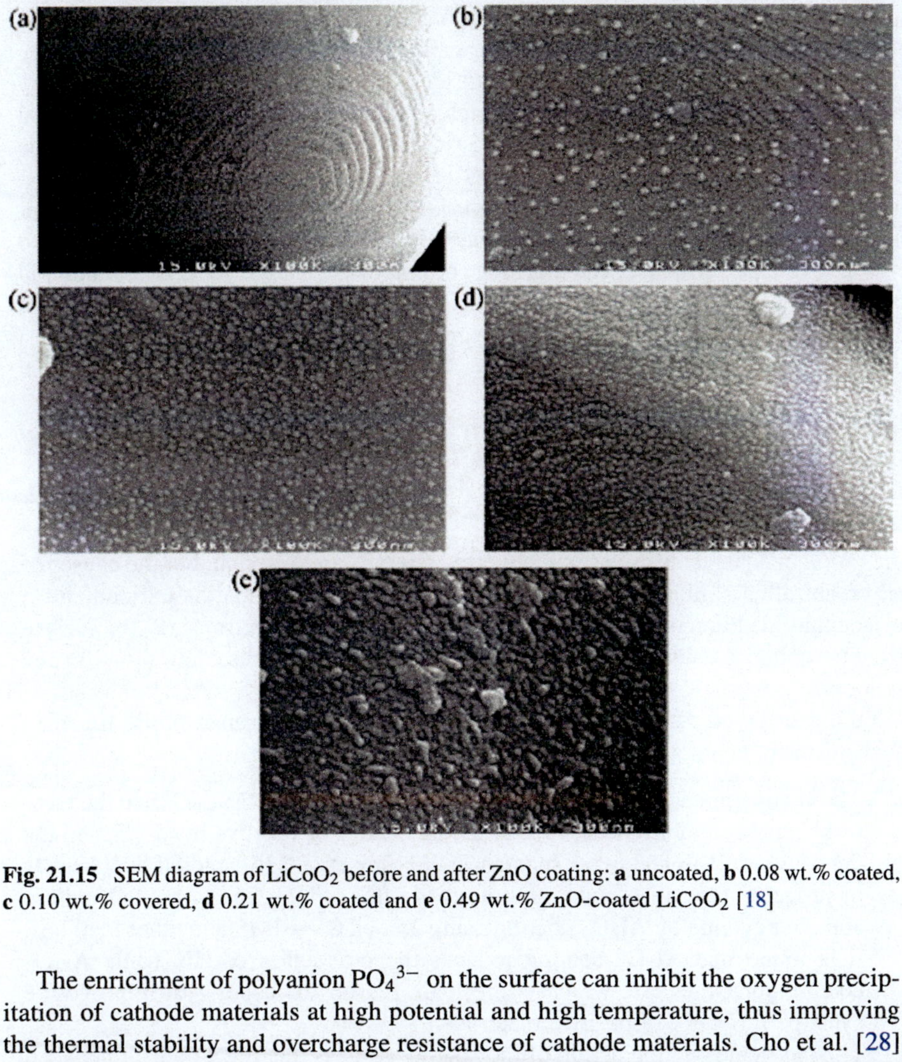

Fig. 21.15 SEM diagram of $LiCoO_2$ before and after ZnO coating: **a** uncoated, **b** 0.08 wt.% coated, **c** 0.10 wt.% covered, **d** 0.21 wt.% coated and **e** 0.49 wt.% ZnO-coated $LiCoO_2$ [18]

The enrichment of polyanion PO_4^{3-} on the surface can inhibit the oxygen precipitation of cathode materials at high potential and high temperature, thus improving the thermal stability and overcharge resistance of cathode materials. Cho et al. [28] after $Li(Ni_{0.8}Co_{0.1}Mn_{0.1})O_2$ was coated with $AlPO_4$, the exothermic test of DSC and the change of material structure of overcharge test were less than that of uncoated $LiCoO_2$. Cho et al. [28, 29] after $LiCoO_2$ was coated with $AlPO_4$, the battery assembled with graphite negative electrode did not show thermal out of control in 1 C and 10 V anti-overcharge test, and the highest surface temperature was only 60 °C, while thermal out of control occurred when it was not coated.

(4) **Coating of other materials** [34–36]

Lithium salts such as Li_3PO_4, $LiAlO_2$, and Li_2ZrO_3 have high lithium-ion conduction capacity. Song et al. [30] used Li_3PO_4 to coat $Li(Ni_{0.4}Co_{0.3}Mn_{0.3})O_2$ to raise the cycle and rate performance. It was found that the properties of the coating materials prepared at different pH values. Their properties vary with the uniformity of the coating layer. Among them, the Li_3PO_4_coated $Li(Ni_{0.4}Co_{0.3}Mn_{0.3})O_2$ prepared at pH = 2 has the best cycle and rate performance.

Composite oxides and compound fluoride can also be used as cathode coating materials. Sun et al. [34] improved the high temperature cycle stability and thermal stability of $Li(Ni_{1/3}Co_{1/3}Mn_{1/3})O_2$ at 55 °C by surface coating $(NH_4)_3AlF_6$, the exothermic peak increased from 289 °C before coating to 301 °C after coating, and the exothermic heat decreased from 523 J/g before coating to 429 J/g after coating. Chen et al. [35] coated the surface of $LiCoO_2$ with a layer of $Y_3Al_5O_{12}$ to raise its cycle performance, calculated according to the charge capacity attenuation to 80%, cycle under the condition of 4.4–2.75 V, 0.2 C charge and discharge, and the number of $LiCoO_2$ cycles after coating can reach 164 times.

21.4.2 Study on the Mechanism of Surface Coating Modification of Cathode Materials

The failure mechanisms of various positive materials are different, but the causes of failure are often similar. It can be attributed to two aspects: (1) the enrichment of surface fluoride hinders the conduction of lithium ions and electrons; (2) the surface structure collapse caused by the surface reaction including dissolution or oxygen evolution.

Taking the layered LiMe (Me = Ni, Co, Mn)O_2 materials as an example, the main mechanisms of coating modification are as follows:

(1) The coating materials reduce the production of HF and consume HF in the electrolyte and to inhibit the surface structure change of positive materials and the deposition of LiF and other substances. Myung et al. [36] studied the changes of surface composition and morphology of $Li(Li_{0.05}Ni_{0.4}Co_{0.15}Mn_{0.4})O_2$ before and after cycling by Al_2O_3 coating using XPS, TOF–SIMS and other methods. It is found that Al_2O_3 coating reduces the production of HF, while Al_2O_3 coating consumes HF, which is the main reason for coating to improve the cycle performance of positive materials.
(2) The physical isolation action of the coating reduces the direct contact between the electrolyte and the positive active material, thus protecting the positive material. Sun et al. [37] raise the cycle performance and thermal stability of $Li(Ni_{1/3}Co_{1/3}Mn_{1/3})O_2$ at high potential by AlF_3 coating, and think that AlF_3, as a physical barrier layer, hinders the direct contact between electrolyte and positive material is the key to raise performance.
(3) Due to the Lewis acid characteristics of the coating layer material, the surface of the positive active material is corroded and the fresh surface is constantly exposed, which reduces the conduction impedance of lithium ion. Bai et al. [38] soaked the Al_2O_3 particles in the electrolyte and found that the battery assembled with the soaked electrolyte had better cycle stability. Through SEM observation, it was found that the surface of the positive $LiCoO_2$ was smoother after the cycle

of the battery assembled with the soaked electrolyte. From this, it was inferred that the key to improving the performance of the positive material was the acid characteristics of the coating layer, which was caused by the continuous corrosion of the positive material surface.

To sum up, there is no consistent view on the modification mechanism of positive coating at present. The existing viewpoints have some limitations, and the study of the mechanism of positive coating modification is the key to further guide the related work, so the research on it is still of great significance.

21.5 Comparison of Storage Performance of 2H-Graphite/Li(Ni,Co,Mn)O_2 Battery Before and After Al_2O_3 Coating [11]

The comparison of cycle performance of 2H-graphite/Li(Ni,Co,Mn)O_2 battery before and after Al_2O_3 coating has been introduced in Chap. 14. The comparative study of storage performance and overcharge performance is introduced below.

Figure 21.16 shows the 0.2 C charge–discharge curves of coated and uncoated batteries before and after storage. It can be seen that the capacity attenuation of the coated battery is less than that of the uncoated battery, and the storage capacity retention of the battery before and after coating is 94.7% electric quantity of the constant voltage part is different before and after and 97.6%, respectively. It can also be seen from the figure that the charge storage, and the higher the value, the worse the performance of the charging rate. The uncoated battery increased from 122 mAh before storage to 128 mAh, while the coated battery increased from 109 mAh before storage to 112 mAh.

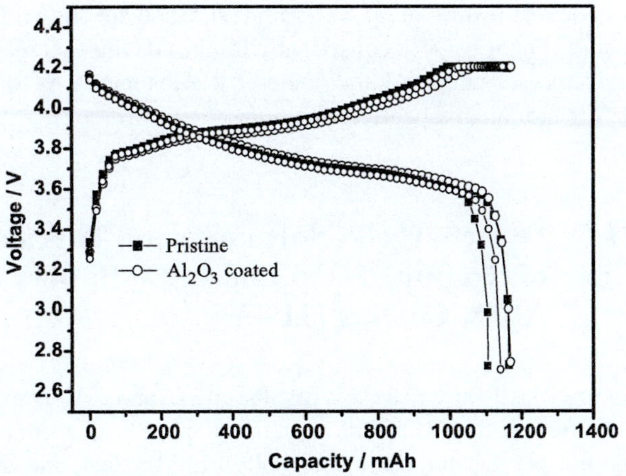

Fig. 21.16 Charge–discharge curves of coated and uncoated Al_2O_3 batteries before and after storage

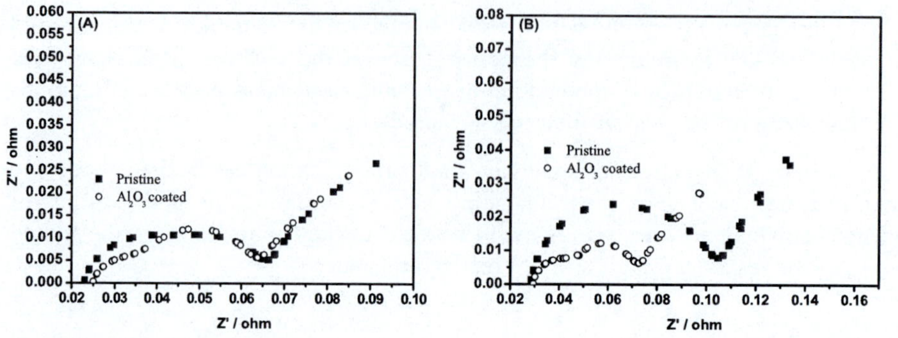

Fig. 21.17 AC impedance spectra of coated and uncoated Al_2O_3 batteries before and after storage

Table 21.2 Impedance of coated and uncoated Al_2O_3 batteries before and after storage

Unit/mΩ		Before storage	After storage
Unpackaged batteries	R_e	24	28
	R_f	20.3	21.6
	R_{CT}	24.3	79.2
Packaged batteries	R_e	24.9	28.1
	R_f	17.6	21.3
	R_{CT}	25.4	32.3

Figure 21.17 shows the AC impedance Nyquist spectra of coated and uncoated batteries before and after storage. The results of fitting according to the equivalent circuit are shown in Table 21.2. It can be seen that the change of R_{ct} is obvious. After storage, the R_{ct} of uncoated battery increases to 3.27 times of that before storage, while that of coated battery only increases to 1.27 times.

The test results of DC internal resistance are shown in Fig. 21.18. It can be seen that the change law is consistent with the AC internal resistance, and the DC internal resistance of uncoated and coated batteries increases to 1.88 and 1.13 times of that before storage, respectively.

21.6 Overcharge Performance and Thermal Stability of Graphite/Li(Ni,Co,Mn)O_2 Battery Before and After Al_2O_3 Coating [11 –39]

Figure 21.19 shows the test results of 3 C and 5 V overcharge of the battery before and after coating. It can be seen that the battery before and after coated with Al_2O_3 did not get thermal out of control during the test, and the temperature increased at first and then decreased. At the same time, it can be seen that the highest surface

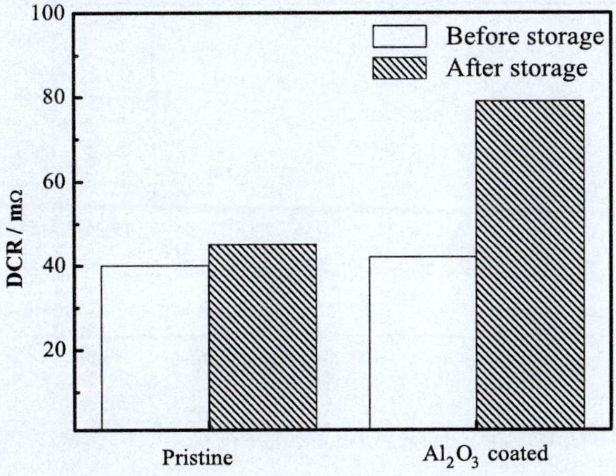

Fig. 21.18 Changes of DC internal resistance of coated and uncoated Al_2O_3 batteries before and after storage

temperature of the coated battery is 46 °C, which is significantly lower than that of the uncoated battery at 63 °C, indicating that the Al_2O_3 coating of NCM in the positive electrode suppresses release thermal of the battery during overcharge.

Figure 21.20 shows the results of 3 C and 10 V overcharge tests of two kinds of batteries, both of battery occurred thermal out of control in 3 C and 10 V overcharge tests. But it can also be seen from the figure that the temperature rising rate of the coated battery is significantly lower than that of the uncoated battery, which also shows that the coated battery improves the anti-overcharge performance of the battery.

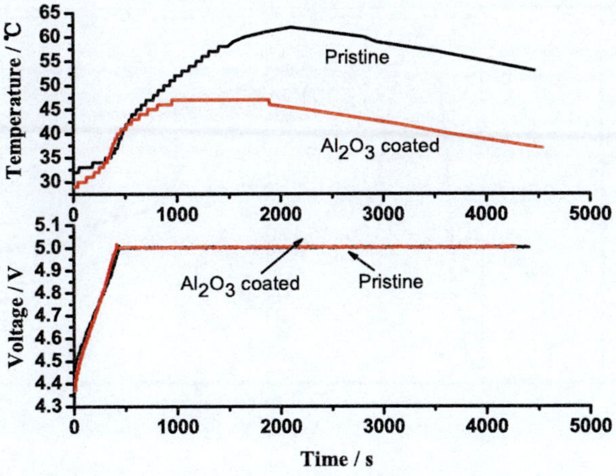

Fig. 21.19 Battery 3 C, 5 V overcharge test curve

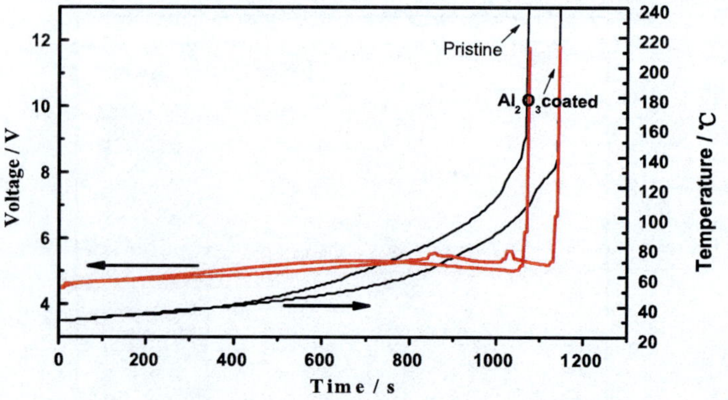

Fig. 21.20 Battery 3 C, 10 V overcharge test curve

Figure 21.21 shows the potentiostatic polarization curves of two kinds of batteries before and after $Li(Ni_{0.4}Co_{0.2}Mn_{0.4})O_2$ (NCM) coating at 25 °C. Considering that the polarization of negative lithium is small, it is used as reference electrode and auxiliary electrode. It can be seen that 3.5–4.3 V versus Li/Li^+ is A zone, corresponding to the oxidation of Ni^{2+} to Ni^{4+} in NCM, 4.3–5.0 V is B zone; corresponding to Co^{3+} to Co^{4+} oxidation; and C zone is electrolyte oxidation zone. It can be seen that there is little difference between A and B zone, while the oxidation current of uncoated NCM in C zone is higher.

In order to further analyze its related mechanism, the above two kinds of batteries were charged to 4.5 V, the positive electrode was taken out in a glove box filled with argon, washed in DMC for three times, and then dried under the protection of Ar gas. The powder was taken to test the TG curve, and the results were as shown in

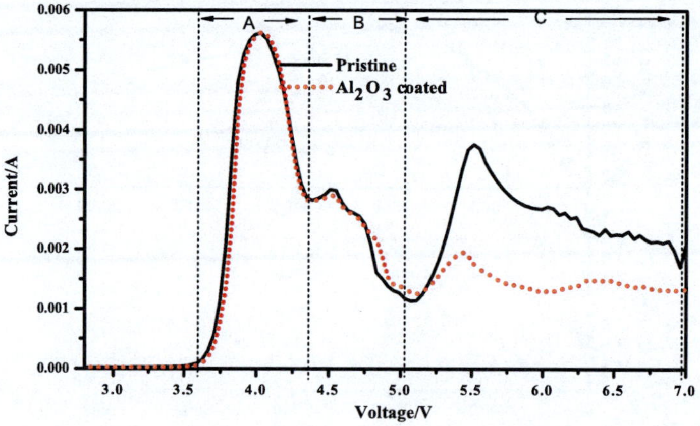

Fig. 21.21 Potential polarization curve of $Li(Ni_{0.4}Co_{0.2}Mn_{0.4})O_2$ battery before and after Al_2O_3 coating under 25 °C

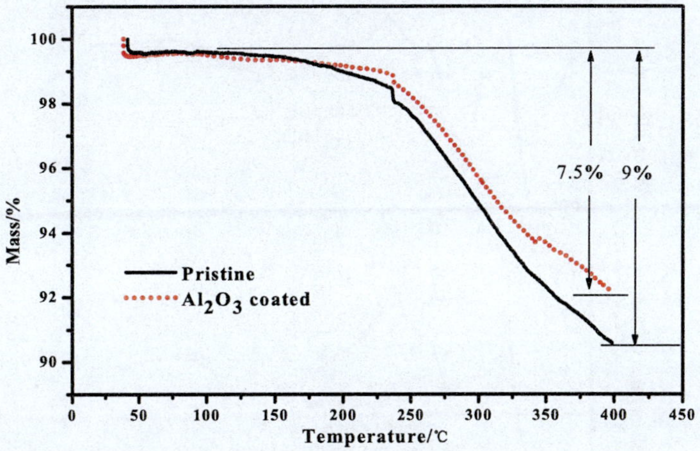

Fig. 21.22 Thermo-gravimetric curve of $Li(Ni_{0.4}Co_{0.2}Mn_{0.4})O_2$ before and after Al_2O_3 coating [45]

21.22. Because the SP in the cathode is stable at 400 C, but the content of PVDF in the cathode is the same before and after coating. Therefore, it can be inferred that the oxygen evolution of NCM in the range of 400 C is the main reason for the mass loss of the electrode. It can be seen that the coating also significantly reduces the oxygen evolution of NCM without the participation of electrolyte. Therefore, the viewpoint that the physical isolation action of the coating layer reduces the amount of oxygen evolution is not comprehensive enough, but should be further analyzed from the point of view of the stability of NCM itself (Fig. 21.22).

Then the local environmental changes of bulk phase and surface oxygen in NCM before and after coating were analyzed. The fluorescence yield (FY) and total electron yield (TEY) of in X-ray absorption near-limit structure (XANES) were used to detect the NCM before and after coating; the results shown in Fig. 21.23. The K-edge absorption spectrum of O is an effective method to characterize the unoccupied O2p state. Figure 21.23a shows the K-edge absorption spectrum of O in FY mode. The spectral lines of the two cases almost coincide, indicating that the coating does not change the electronic structure of oxygen in the NCM bulk phase, which is consistent with the above XRD results. Figure 21.23b shows the K-edge absorption spectrum of O in TEY mode. The peak from 525 to 534 eV is spectrum of O2p and the 3d orbital hybrid spectrum of transition metal in the NCM surface layer, while the wider peak above 534 eV is the spectrums of O2p and 4sp orbital hybrid of transition metal in the surface layer NCM. It can be seen that the hybridization of O2p with 3D orbitals of transition metals in the surface layer of NCM after coating decreases, while hybridizes with 4sp orbitals of transition metals increases. According to Yoon et al., O hybridized with 4sp orbitals has higher binding energy (above 534 eV, while the peak of hybridization with 3D orbitals is below 534 eV). Therefore, the O on the

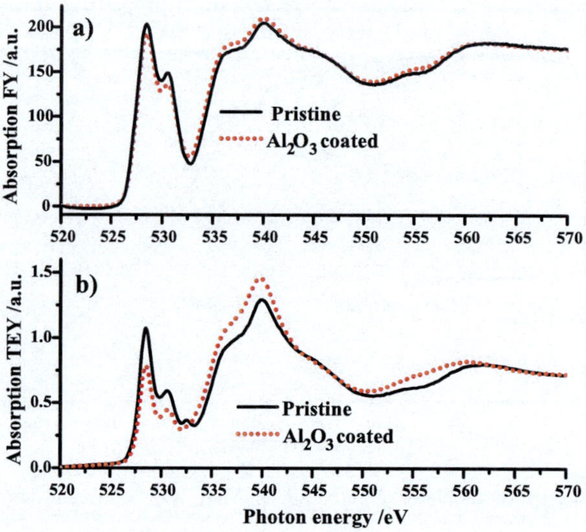

Fig. 21.23 K-edge X-ray absorption near-edge structure spectrum of O [39]. (**a**) fluorescence yield (FY) mode; (**b**) total electron yield (TEY) mode

surface of the coated NCM material is more stable, so it has higher stability at high temperature (TG curve) and high potential (Fig. 21.21 area C) [39].

21.7 Research Conclusion of Graphite/$Li(Ni_{0.4}Co_{0.2}Mn_{0.4})O_2$ Battery Coated by Al_2O_3

The comparative study results of graphite/Al_2O_3-coated $Li(Ni_{0.4}Co_{0.2}Mn_{0.4})O_2$ battery are summarized as follows:

(1) There is little difference in specific surface area, vibrating density and diffusion coefficient of $Li(Ni_{0.4}Co_{0.2}Mn_{0.4})O_2$ before and after coating. The coating of Al_2O_3 on the surface reduces the basicity of $Li(Ni_{0.4}Co_{0.2}Mn_{0.4})O_2$, which is beneficial to the coating of electrode active materials.
(2) There is a 10.9% difference in discharge capacity attenuation at 1 C after 4.5–2.8 V, 1 C, and cycle 100 period of 2025 battery before and after coating, but the difference of 0.2 C discharge capacity is only 3.1%. It shows that the key to the cyclic attenuation of the battery lies in the increase of impedance. SEM observed that the surface of the positive active material of the uncoated battery was rougher than that of the coated one after cycling, indicating that there were more attachments on the surface. The EDS results show that the fluoride content on the surface of the uncoated positive is higher. The XPS results show that more

NiO is formed on the uncoated surface, which is one of the main reasons for the increase of impedance in the cycle process.

(3) After 400 cycles of 4.2–2.75 V and 2 C current of 18,650 battery before and after coating, the capacity of uncoated battery decreased by 17.8%, but the capacity attenuation of positive and negative electrodes was 7.1 and 3.4%, respectively, which was much less than 17.8% of that of batteries. It shows that the attenuation of battery capacity is due to the mismatch between positive and negative charge and discharge capacities. The ICP detection of active lithium shows that the loss of active lithium is the key to its capacity attenuation during the cycle. The ICP test of the negative electrode material showed that the deposition amount of Ni, Co, and Mn which promoted the formation of SEI film and the loss of active lithium in the coated anode was significantly lower than that in the uncoated battery, which was the reason for improving the cycle performance of the battery.

(4) The measurement results of the positive and negative electrode impedance of the 18,650 battery during the cycle show that the key to the decline of the rate performance lies in the positive electrode, and the coating reduces the impedance of both positive and negative electrodes, thus improving the rate retention ability of the battery.

(5) The results of the 18,650 battery stored at 60 °C for 60 days showed that the positive coating improved the storage performance of the battery, the capacity retention rate increased from 94.7 to 97.6%, the increase amplitude of R_{CT} decreased from 3.27 times to 1.27 times; at the same time, the deposition of Ni, Co, and Mn in the negative electrode was reduced.

References

1. Ohzuku, T., A. Ueda, and N. Yamamoto. 1995. Zero-strain insertion materials of $Li(Li_{1/3}Ti_{5/3})O_4$ for rechargeable lithium cells. *Journal of the Electrochemical Society* 142: 1431–1435.
2. Wang, G.X., D.H. Brandhurst, S.X. Dou, et al. 1999. Spine $Li_2(Li_{1/3}Ti_{5/3})O_4$ as an anode material for lithium ion batteries. *Journal of Power Sources* 83: 156–161.
3. Scharner, S., Weppner, W., et al. 1999. Evidence of two-phase formation upon lithium insertion into the $Li_{1.33}Ti_{1.67}O_4$ spinel. *Journal of the Electrochemical Society* 146 (3): 857–861.
4. Kanamura, K., T. Umegaki, H. Naito, et al. 2001. Structural and electrochemical characteristics of $Li_{4/3}T_{i5/3}O_4$ as an anode material for rechargeable lithium batteries. *Journal of Applied Electrochemistry* 31: 73–78.
5. Liu, Wei. 2014. Study on preparation and properties of lithium titanat. *Doctoral paper of Shanghai Institute of Microsystems and Information Technology.*
6. Qian, Wang. 2015. *Development of lithium titanate lithium ion battery and study on its gas expansion mechanism.* Doctoral Paper of Shanghai Institute of Microsystems and Information Technology.
7. Xie, Xiao-Hua. 2008. Study on the performance of organic liquid electrolyte for lithium-ion battery at low temperature. *Doctoral paper of Shanghai Institute of Microsystems and Information Technology.*

8. Xie, Xiao-Hua. 2014. Study on low temperature performance of lithium iron phosphate lithium ion power battery. *Postdoctoral research report of Shanghai Institute of Microsystems and Information Technology*.
9. RUREHA. 2010. *Carbontron P*
10. Jia, Li. 2010. Study on storage performance of lithium-ion battery. *Doctoral paper of Shanghai Institute of Microsystems and Information Technology*.
11. Liu, Hao-Ran. 2012. Study on coating modification and application of cathode material LiNixCoyMn1-x-yO_2 for Li-ion Battery. *Doctoral paper of Shanghai Institute of Microsystems and Information Technology*.
12. Oh, Seungsuk, J.K. Lee, D. Byun, W.I. Cho, and B.W. Cho. 2004. Effect of Al_2O_3 coating on electrochemical performance of $LiCoO_2$ as cathode materials for secondary lithium batteries. *Journal of Power Sources*, 132 (1–2): 249–255.
13. Van Landschoot, N., E.M. Kelder, P.J. Kooyman, et al. 2004. Electrochemical performance of Al_2O_3-coated Fe doped $LiCoVO_4$. *Journal of Power Sources* 138 (1–2): 262–270.
14. Kim, Y., H.S. Kim, and W.M. Steve. 2006. Synthesis and electrochemical characteristics of Al_2O_3-coated $Li(Ni_{1/3}Co_{1/3}Mn_{1/3})O_2$ cathode materials for lithium ion batteries. *Electrochimica Acta* 52: 1316–1322.
15. Liu, G.Q., H.T. Kuo, R.S. Liu, et al. 2010. Study of electrochemical properties of coating ZrO_2 on $LiCoO_2$. *Journal of Alloys and Compounds* 496 (1–2): 512–516.
16. Huang, Y.Y., J.T. Chen, J.F. Ni, H.H. Zhou, and X.X. Zhang. 2009. A modified ZrO_2-coating process to improve electrochemical performance of $Li(Ni_{1/3}Co_{1/3}Mn_{1/3})O_2$. *Journal of Power Sources* 188: 538–545.
17. Hu, S.K., G.H. Cheng, M.Y. Cheng, et al. 2009. Cycle life improvement of ZrO_2-coated spherical $Li(Ni_{1/3}Co_{1/3}Mn_{1/3})O_2$ cathode material for lithium ion batteries. *Journal of Power Sources* 188: 564–569.
18. Chang, W., J.W. Choi, J.C. Im, et al. 2010. Effects of ZnO coating on electrochemical performance and thermal stability of $LiCoO_2$ as cathode material for lithium-ion batteries. *Journal of Power Sources* 195 (1): 320–326.
19. Chang, W., J.W Choi, J. Choo. Im, et al. 2010. Effects of ZnO coating on electrochemical performance and thermal stability of $LiCoO_2$ as cathode material for lithium-ion batteries. *Journal of Power Sources* 195 (1): 320–326.
20. Wu, F., M. Wang, Y.F. Su, et al. 2009. Surface modification of $Li(Co_{1/3}Ni_{1/3}Mn_{1/3})O_2$ with Y_2O_3 for lithium-ion battery. *Journal of Power Sources* 189 (1): 743–747.
21. Sun, Y.K., S.W. Cho, S.W. Lee, et al. 2007. AlF_3-coating to improve high voltage cycling performance of $Li(Ni_{1/3}Co_{1/3}Mn_{1/3})O_2$ cathode materials for lithium secondary batteries. *Journal of The Electrochemical Society* 154 (3): A168–A172.
22. Park, B.C., H.B. Kim, S.T. Myung, et al. 2008. Improvement of structural and electrochemical properties of AlF_3-coated $Li(Ni_{1/3}Co_{1/3}Mn_{1/3})O_2$ cathode materials on high voltage region. *Journal of Power Sources* 178: 826–831.
23. Woo, S.U., C.S. Yoon, K. Amine, et al. 2007. Significant improvement of electrochemical performance of AlF_3-coated $Li(Ni_{0.8}Co_{0.1}Mn_{0.1})O_2$ cathode materials. *Journal of the Electrochemical Society* 154 (11): A1005–A1009.
24. Kang, Y.J., J.H. Kim, S.W. Lee, et al. 2005. The effect of $Al(OH)_3$ coating on the $Li(Li_{0.2}Ni_{0.2}Mn_{0.6})O_2$ cathode material for lithium secondary battery. *Electrochimica Acta* 50 (24): 4784–4791.
25. Jang, S.B., S.H. Kang, K. Amine, et al. 2005. Synthesis and improved electrochemical performance of $Al(OH)_3$-coated $Li(Ni_{1/3}Mn_{1/3}Co_{1/3})O_2$ cathode materials at elevated temperature. *Electrochimica Acta* 50 (20): 4168–4173.
26. Kim, J., M.J. Noh, J. Cho, et al. 2005. Controlled nanoparticle metal phosphates (metal = Al, Fe, Ce, and Sr) coatings on $LiCoO_2$ cathode materials. *Journal of The Electrochemical Society* 152 (6): A1142–A1148.
27. Eom, J., J. Cho. 2008. $M_3(PO_4)_2$-nanoparticle-coated $LiCoO_2$ vs $Li(Co_{0.96}M_{0.04})O_2$ (M = Mg and Zn) on electrochemical and storage characteristics. *Journal of the Electrochemical Society* 155 (3): A201–A205.

28. Cho, J., H. Kim, and B. Park. 2004. Comparison of overcharge behavior of $AlPO_4$-coated $LiCoO_2$ and $LiNi_{0.8}Co_{0.1}Mn_{0.1}O_2$ cathode materials in Li-ion cells. *Journal of the Electrochemical Society* 151 (10): A1707–A1711.
29. Cho, J. 2004. Dependence of $AlPO_4$ coating thickness on overcharge behaviour of $LiCoO_2$ cathode material at 1 and 2 C rates. *Journal of Power Sources* 126: 186–189.
30. Song, H.G., J.Y. Kim, K.T. Kim, et al. 2011. Enhanced electrochemical properties of $Li(Ni_{0.4}Co_{0.3}Mn_{0.3})O_2$ cathode by surface modification using Li_3PO_4-based materials. *Journal of Power Sources* 196: 6847–6855.
31. Fey, G.T., C.Z. Lu, T.P. Kumar, et al. 2005. TiO_2 coating for long-cycling $LiCoO_2$: A comparison of coating procedures. *Surface and Coatings Technology* 199 (1): 22–31.
32. Wu, F., M. Wang, Y.F. Su, S. Chen, and B. Xu. 2009. Effect of TiO_2-coating on the electrochemical performances of $Li(Co_{1/3}Ni_{1/3}Mn_{1/3})O_2$. *Journal of Power Sources* 191: 628–632.
33. Zheng, J.M., J. Li, Z.R. Zhang, et al. 2008. The effects of TiO_2 coating on the electrochemical performance of $Li(Li_{0.2}Mn_{0.54}Ni_{0.13}Co_{0.13})O_2$ cathode material for lithium-ion battery. *Solid State Ionics* 179: 1794–1799.
34. Sun, Y.K., S.T. Myung, C.S. Yoon, et al. 2009. Improvement of high voltage cycling performances of $Li(Ni_{1/3}Co_{1/3}Mn_{1/3})O_2$ at 55 degrees C by a $(NH_4)_3AlF_6$ coating. *Electrochemical and Solid-State Letters* 12 (8): A163–A166.
35. Chen, J.M., Y.D. Cho, C.L. Hsiao, et al. 2009. Electrochemical studies on $LiCoO_2$ surface coated with $Y_3Al_5O_{12}$ for lithium-ion cells. *Journal of Power Sources* 189: 279–287.
36. Myung, S.T., K. Izumi, S. Komaba, et al. 2005. Role of alumina coating on Li-Ni-Co-Mn-O particles as positive electrode material for lithium-ion batteries. *Chemistry of Materials* 17 (14): 3695–3704.
37. Sun, Y.K., S.W. Cho, S.W. Lee, et al. 2007. AlF_3-coating to improve high voltage cycling performance of $Li[Ni_{1/3}Co_{1/3}Mn_{1/3}]O_2$ cathode materials for lithium secondary batteries. *Journal of The Electrochemical Society* 154 (3): A168–A172.
38. Bai, Y., Y. Yin, N. Liu, et al. 2007. New concept of surface modification to $LiCoO_2$. *Journal of Power Sources* 174: 328–334.
39. Liu, H.H., J. Zhang, Y.W. Lou, et al. 2012. XANES study on the stability of coated $Li(N_{i0.4}Co_{0.2}Mn_{0.4})O_2$ materials. *Acta Chemica Sinica* 70 (9): 1055–1058.
40. Li, F., Y.W. Lou, and B.J. Xia. 2005. Application of Ca additives in high capacity Ni-MH battery. *Power Supply Technology* 20 (5): 312–314.
41. Xia, Bao-Jia, Qing-Ze Lin, Li-Ping Ma, et al. 2003. Effect of cathode additives on high temperature charging behavior of MH/Ni Battery. *Battery* 33 (2): 68–70.
42. Lou, Y.W., C.Z. Yang, and B.J. Xia. 2009. Effect of β-$Ni(OH)_2$ additive on electrode material in MH-Ni Battery. *Power Supply Technology* 33 (6): 449–453.
43. Li, D.C., Y. Kato, K. Kobayakawa, H. Noguchi, and Y. Sato. 2006. Preparation and electrochemical characteristics of $LiNi_{1/3}Mn_{1/3}Co_{1/3}O_2$ coated with metal oxides coating. *Journal of Power Sources* 160 (2): 1342–1348.
44. Liu, J.Y., N. Liu, D.T. Liu, Y. Bai, L.H. Shi, Z.X. Wang, L.Q. Chen, V. Hennige, and A. Schuch. 2007. Improving the performances of $LiCoO_2$ cathode materials by soaking nano-alumina in commercial electrolyte. *Journal of the Electrochemical Society*, 154 (1): A55–A63.
45. Kannan, A.M., L. Rabenberg, and A. Manthiram. 2003. High capacity surface-modified $LiCoO_2$ cathodes for lithium-ion batteries. *Electrochemical and Solid-State Letters* 6 (1): A16–A18, A30.

The Recommendation of Three Full Professorial Experts

Recommendation of Professor Li Guo-Xin, at Shanghai Space Power Research Institute

"I have been engaged in the research of chemical power supply for a long time in Shanghai Space Power Research Institute, and many books on Ni-MH and Li-ion batteries have been published at home and abroad. My colleagues and I have also edited and published an《**Introduction to New Chemical Power supply Technology**》(Shanghai Science and Technology Press, 2007). However, there is no book devoted to the characterization of battery materials and the mechanism of electrode process, especially the monograph "**Material characterization and Mechanism of electrode process for green Secondary Battery**", which uses X-ray diffraction to study battery materials and electrode processes. Therefore, I would like to recommend this book to be published in time. The reasons are as follows:

(1) This book is rich, comprehensive, and systematic, including the preparation of electrode active materials for Ni-MH battery and Li-ion battery and X-ray diffraction characterization. The physical mechanism of the electrode process of this kind of battery such as charge and discharge, cycle, and storage are also described and discussed in detail. It is a novel monograph on chemical power supply.

(2) In the study of the electrode process of charge and discharge, a new ion de-intercalation theory and conductive mechanism are proposed on the basis of experiment and conductive phase transition theory, which has a higher theoretical level.

(3) Many innovative achievements have been made in the research on the cycle process and storage process of the related battery, which reveals the physical mechanism of the electrode process and the attenuation mechanism of the cycle performance and storage performance, which brings the understanding of the electrode process to a new level with a higher academic level.

C. Yang et al., *Materials and Working Mechanisms of Secondary Batteries*,
https://doi.org/10.1007/978-981-19-5955-4

(4) On the basis of explaining the attenuation mechanism of cycle performance and storage performance, the author describes many experimental exploration and research results of improving battery performance, which are closely related to the production and application of chemical power supply. It has higher engineering practical value.

In a word, 《**Material characterization and Mechanism of electrode process for green Secondary Battery**》 is a rich, novel, and systematic work, which has high theoretical academic level and practical application value. Therefore, I highly recommend 《**Material characterization and Mechanism of electrode process for green Secondary Battery**》 is published as soon as possible, which can be used as a reference for scientific and technological workers and graduate students".

(2014.7).

Recommendation of Professor Jiang Chuan-Hai, School of Materials Science and Engineering, Shanghai Jiao Tong University

"I have been engaged in the teaching and research of metal materials and X-ray diffraction for a long time, and I am well aware of the importance between X-ray diffraction and material characterization and research. After retiring, Professor Yang Chuan-Zheng applied to the Shanghai Institute of Microsystems and Information Technology to combine X-ray diffraction with the characterization of battery active materials and the study of battery electrode process and achieved fruitful innovative results. More than 40 papers have been published with collaborators. 《**Material characterization and Mechanism of electrode process for green Secondary Battery**》 focuses on the research results in this field, so I unreservedly recommend the publication of this book. The reasons are as follows:

(1) This book is the result of close cooperation between experts in material physics, X-ray diffraction, and electrochemistry chemical power supply and is an innovative achievement of interdisciplinary research, which is worthy of support and recommendation.
(2) The authors and collaborators have proposed many new X-ray analysis methods for the characterization of battery materials, which have been applied to the testing of related battery materials and the study of electrode processes, and many innovative achievements have been obtained, such as the solution of multiple microstructure parameters, the experimental determination of graphite disorder degree, and the solution of Li/Ni mixed occupying parameters in Li $(Ni_xCo_yMn_{1-x-y})\ O_2$ materials.
(3) In the study of the electrode process of charge and discharge, the ion de-intercalation theory and conductive mechanism are expounded from the

new point of view of X-ray stress analysis, which has both theoretical and experimental basis.

(4) The application of X-ray analysis method to the study of battery cycle and storage process not only reveals the physical mechanism of electrode process, but also has a deeper understanding of the attenuation mechanism of cycle performance and storage performance.

The method of improving battery performance has made a series of innovative achievements, which is a good book that combines theory with practice".

(2014.6).

The Recommendation of Professor Wang Ran, Shanghai Aowei Technology Development Co., Ltd.

"The authors applied X-ray diffraction experimental analysis technology to the characterization of active materials related to Ni-MH battery, graphite/$LiMeO_2$, and graphite/$LiFePO_4$ battery and the study of electrode processes such as charge–discharge, cycle, and storage and achieved a series of important innovations. I actively and unreservedly recommend the publication of the book 《**Material characterization and Mechanism of electrode process for green Secondary battery**》 for the following reasons:

(1) X-ray diffraction experimental analysis technology is applied to the characterization of battery materials and the study of electrode process mechanism, and a series of new X-ray diffraction methods for characterizing battery active materials are proposed, and the application of these new methods to the study of electrode processes such as charge and discharge, cycle, and storage of batteries is worthy of support and praise.

(2) The de-intercalation theory of conductive ions is put forward in the study of charge–discharge electrode process, which makes us have a better understanding of the charge–discharge process mechanism and conductive mechanism.

(3) When introducing the research results of the cycle and storage process of this kind of battery, on the basis of the experimental results, it describes not only the physical mechanism of the electrode process, but also the attenuation mechanism of cycle performance and storage performance. On this basis, the methods to improve the performance of this kind of battery are introduced. The book not only has great innovation in theory and academic, but also has important guidance and reference significance for battery production and practical application.

In a word, 《**Material characterization and Mechanism of electrode process for green Secondary Battery**》 is a rich, novel, systematic, and comprehensive academic monograph, and it is a new book with high theoretical academic value and practical significance in chemical power supply (battery) industry. It is hoped that it will be published as soon as possible for reference by colleagues".

(2014.7).

Subject Index

C. Yang et al., *Materials and Working Mechanisms of Secondary Batteries*,
https://doi.org/10.1007/978-981-19-5955-4